AIRCRAFT PERFORMANCE AND DESIGN

AIRCRAFT PERFORMANCE AND DESIGN

John D. Anderson, Jr.
University of Maryland

Boston   Burr Ridge, IL   Dubuque, IA   Madison, WI
New York   San Francisco   St. Louis
Bangkok   Bogotá   Caracas   Lisbon   London   Madrid   Mexico City
Milan   New Delhi   Seoul   Singapore   Sydney   Taipei   Toronto

WCB/McGraw-Hill

*A Division of The **McGraw·Hill** Companies*

AIRCRAFT PERFORMANCE AND DESIGN

Copyright © 1999 by The McGraw-Hill Companies, Inc. All rights reserved. Printed in the United States of America. Except as permitted under the United States Copyright Act of 1976, no part of this publication may be reproduced or distributed in any form or by any means, or stored in a data base or retrieval system, without the prior written permission of the publisher.

This book is printed on acid-free paper.

6 7 8 9 10 11 12 DOC/DOC 0 9 8 7 6 5

ISBN 0-07-001971-1

Vice president/Editor-in-chief: *Kevin T. Kane*
Publisher: *Thomas Casson*
Senior sponsoring editor: *Jonathan W. Plant*
Marketing manager: *John T. Wannemacher*
Senior project manager: *Denise Santor-Mitzit*
Senior production supervisor: *Heather D. Burbridge*
Freelance design coordinator: *JoAnne Schopler*
Supplement coordinator: *Marc Mattson*
Compositor: *Techsetters, Inc.*
Typeface: *Times Roman*
Printer: *R. R. Donnelley & Sons Company*

Library of Congress Cataloging-in-Publication Data

Anderson, John David.
    Aircraft performance and design / John D. Anderson, Jr.
      p.   cm.
    Includes bibliographical references and index.
    ISBN 0-07-001971-1
    1. Airplanes–Performance. 2. Airplanes–Design and construction.
    I. Title.
    TL671.4.A528 1999
    629.134′1–dc21

                                     98-48481
                                          CIP

http://www.mhhe.com

To Sarah-Allen, Katherine, and Elizabeth
for all their love and understanding
John D. Anderson, Jr.

ABOUT THE AUTHOR

John D. Anderson, Jr., was born in Lancaster, Pennsylvania, on October 1, 1937. He attended the University of Florida, graduating in 1959 with high honors and a Bachelor of Aeronautical Engineering Degree. From 1959 to 1962, he was a lieutenant and task scientist at the Aerospace Research Laboratory at Wright-Patterson Air Force Base. From 1962 to 1966, he attended the Ohio State University under the National Science Foundation and NASA Fellowships, graduating with a Ph.D. in aeronautical and astronautical engineering. In 1966, he joined the U.S. Naval Ordnance Laboratory as Chief of the Hypersonic Group. In 1973, he became Chairman of the Department of Aerospace Engineering at the University of Maryland, and since 1980 has been professor of Aerospace Engineering at Maryland. In 1982, he was designated a Distinguished Scholar/Teacher by the University. During 1986–1987, while on sabbatical from the university, Dr. Anderson occupied the Charles Lindbergh chair at the National Air and Space Museum of the Smithsonian Institution. He continues with the Museum in a part-time appointment as special assistant for aerodynamics. In addition to his appointment in aerospace engineering, in 1993 he was elected to the faculty of the Committee on the History and Philosophy of Science at Maryland, and is an affiliate faculty member in the Department of History.

Dr. Anderson has published seven books: *Gasdynamic Lasers: An Introduction*, Academic Press (1976), *A History of Aerodynamics and Its Impact on Flying Machines*, Cambridge University Press (1997), and with McGraw-Hill, *Introduction to Flight*, 3d edition (1989), *Modern Compressible Flow*, 2d Edition (1990), *Fundamentals of Aerodynamics*, 2d edition (1991), *Hypersonic and High Temperature Gas Dynamics* (1989), and *Computational Fluid Dynamics: The Basics with Applications* (1995). He is the author of over 120 papers on radiative gasdynamics, re-entry aerothermodynamics, gas dynamic and chemical lasers, computational fluid dynamics, applied aerodynamics, hypersonic flow, and the history of aerodynamics. Dr. Anderson is in *Who's Who in America*, and is a Fellow of the American Institute of Aeronautics and Astronautics (AIAA). He is also a Fellow of the Washington Academy of Sciences, and a member of Tau Beta Pi, Sigma Tau, Phi Kappa Phi, Phi Eta Sigma, The American Society for Engineering Education (ASEE), The Society for the History of Technology, and the History of Science Society. He has received the Lee Atwood Award for excellence in Aerospace Engineering Education from the AIAA and the ASEE, and the Pendray Award for Aerospace Literature from the AIAA.

PREFACE

There are a number of books on airplane performance, and a number of books on airplane design. Question: Where does the present book fit into the scheme of things? Answer: Overlapping and integrating both subjects. On one hand, this book gives a presentation of airplane performance at the college level. It covers both static and accelerated performance topics. On the other hand, this book also gives a presentation of airplane design, with an emphasis on the *philosophy* and *methodology* of design. Some emphasis is also placed on historical material and design case studies in order to illustrate this philosophy and methodology.

This book is not a handbook for airplane design. It is intended to be used in courses in airplane performance as a main text, and in courses in airplane design as an introduction to the philosophy of design, and hence in conjunction with an existing detailed text on airplane design. To paraphrase a popular television commercial, this book is not intended to make a course in handbook engineering design—rather it is intended to make a course in handbook engineering design **better**. This author hopes that such intentions are indeed achieved in the present book.

The major features of this book are as follows.

1. This book is unique in that it is the first to provide an integrated introductory treatment of both aircraft performance and aircraft design—two subjects that are so closely connected that they can be viewed as technological Siamese twins.

2. This book is intentionally written in a conversational style, much like the author's previous texts, in order to enhance the readers' understanding and enjoyment.

3. The book is divided into three parts. Part I contains introductory material that is important for an understanding of aircraft performance and design. Chapter 1 deals with the history of aircraft design. It is important for students and practitioners of aircraft design to understand this history because the design of a new airplane is usually evolutionary; a new airplane is frequently an evolutionary extension of one or more previous designs. Even the most revolutionary of new airplane designs contain some of the genes of almost all previous aircraft. Hence, Chapter 1 is an essential part of this book. Other historical notes appear elsewhere in the book. Chapters 2 and 3 are overviews of aerodynamics and propulsion, respectively. These chapters focus on only those aspects of aerodynamics and propulsion that are necessary for an understanding and application of both aircraft performance (Part II) and aircraft design (Part III). However, they serve a secondary function; they provide a self-contained overview of theoretical and applied aspects of aerodynamics and propulsion that help the reader obtain a broader perspective of these subjects. So Chapters 2 and 3, in addition to being essential to the material in Parts II and III, have intrinsic educational value in and of themselves, no matter what may be the reader's background.

4. Part II deals with static and accelerated aircraft performance. The basic equations of motion are derived in Chapter 4. These equations are then specialized for the study of static performance (no acceleration) in Chapter 5, and are used in Chapter 6 in their more general form for performance problems involving acceleration. The material is presented in two parallel tracks: (1) graphical solutions, and (2) closed-form analytical solutions. The value of each approach is emphasized.

5. Parts I and II provide the material for a complete one-semester course on aircraft performance at the junior-senior level.

6. Parts I and II are sprinkled with "design cameos"—boxed discussions of how the material being discussed at that stage is relevant to aircraft design. These design cameos are a unique feature of the present book. They are part of the overall integrated discussion of performance and design that is a hallmark of this book. In addition, many worked examples are included in Parts I and II as a means to support and enhance the reader's understanding of and comfort level with the material. Homework problems are included at the end of most of the chapters, as appropriate to the nature of each chapter.

7. Part III is all about aircraft design, but with a different style and purpose than existing airplane design texts. Chapter 7 lays out an intellectual, almost philosophical road map for the process of aircraft design. Then the methodology is applied to the design of a propeller-driven airplane in Chapter 8, and jet-propelled airplanes in Chapter 9. In addition, Chapters 8 and 9 are enhanced by important case histories of the design of several historic airplanes—another dose of history, but with a powerful purpose, namely to drive home the philosophy and methodology of aircraft design. Part III is not a design handbook; rather, it provides an intellectual perspective on design—a perspective that all airplane designers, past and present, exhibit, whether knowingly or subconsciously. Part III is intended for the first part of a senior design course. The complete book—Parts I, II, and III—is intended to provide a unique "pre-design" experience for the reader. I wanted to create a book that would work synergistically with existing main-line design texts. As mentioned earlier, this book is not intended to constitute a complete course in aircraft design; rather, its purpose is to make such a course a *better* and *more rewarding* experience for the student.

8. Although "history" is not in the title of this book, another unique aspect is the extensive discussions of the history of airplane design in Chapter 1 and the extensively researched historical case studies presented in chapters 8 and 9. In this vein, the present book carries over some of the tradition and historical flavor of the author's previous books, in particular some of the historical research contained in the author's recent book, *The History of Aerodynamics, and Its Impact on Flying Machines* (see Reference 8).

9. There are carefully selected homework problems at the end of most of the chapters—not an overpowering number, but enough to properly reinforce the material in the chapter. There is a Solutions Manual for the use of instructors. Permission is granted to copy and distribute these solutions to students at the discretion of the instructor. In addition, the answers to selected problems are given at the end of the book.

10. Computer software for aircraft performance and design calculations is intentionally not provided with this book. This may be seen as bucking current trends with engineering textbooks. But I want this book to provide a comfortable intellectual experience for the reader, unencumbered by the need to learn how to use someone else's software. The reader's experience with software for these subjects will most likely come soon enough in the classroom. However, much of the material in this book is ideally suited to the creation of simple computer programs, and the reader should enjoy the creative experience of writing such programs as he or she wishes.

I wish to acknowledge the author Enzo Angelucci and his wonderful book *Airplanes From the Dawn of Flight to the Present Day*, published in English by McGraw-Hill in 1973. The airplane drawings that appear in Chapter 1 of the present book are taken from his book.

I wish to thank the many colleagues who have provided stimulating discussions during the time that this book was being prepared, as well as the reviewers of the manuscript. I also thank Sue Cunningham, who has provided some expert word processing for the manuscript. And most of all I thank Sarah-Allen Anderson for being such a supportive and understanding wife during the long time it has taken me to finish this project.

So here it is—this integrated treatment of aircraft performance and design. Try it on for size. I hope that it fits comfortably and serves you well. If it does, then all my labors will not have been in vain.

John D. Anderson, Jr.
September 1998

CONTENTS

AIRCRAFT PERFORMANCE
AND DESIGN

1

PRELIMINARY CONSIDERATIONS

Part 1 consists of three chapters which set the stage for our subsequent discussion of airplane performance and design. Chapter 1 is a short history of the evolution of the airplane and its design; its purpose is to set the proper philosophical perspective for the material in this book. Chapters 2 and 3 cover aspects of applied aerodynamics and propulsion, respectively, insofar as they *directly relate* to the performance and design considerations to be discussed in the remainder of this book.

The Evolution of the Airplane and Its Performance: A Short History

Instead of a palette of colors, the aeronautical engineer has his own artist's palette of options. How he mixes these engineering options on his technological palette and applies them to his canvas (design) determines the performance of his airplane. When the synthesis is best it yields synergism, a result that is dramatically greater than the sum of its parts. This is hailed as "innovation." Failing this, there will result a mediocre airplane that may be good enough, or perhaps an airplane of lovely external appearance, but otherwise an iron peacock that everyone wants to forget.

> Richard Smith, Aeronautical Historian
> From *Milestones of Aviation*,
> National Air and Space Museum, 1989

1.1 INTRODUCTION

The next time you are outside on a clear day, look up. With some likelihood, you will see evidence of an airplane—possibly a small, private aircraft hanging low in the sky, slowly making its way to some nearby destination (such as the Cessna 172 shown in Fig. 1.1), or maybe a distinct white contrail high in the sky produced by a fast jet transport on its way from one end of the continent to the other (such as the Boeing 777 shown in Fig. 1.2). These airplanes—these *flying machines*—we take for granted today. The airplane is a part of everyday life, whether we simply see one, fly in one, or receive someone or something (package, letter, etc.) that was delivered by one. The invention and development of the airplane are arguably one of the three most important technical developments of the twentieth century—the other two being the electronics revolution and the unleashing of the power of the atom. The airplane has

Figure 1.1   Cessna 172. (Courtesy of Cessna Aircraft.)

Figure 1.2   Boeing 777. (Courtesy of Boeing.)

transformed life in the twentieth century, and this transformation continues as you read these words.

However, the airplane did not just "happen." When you see an aircraft in the sky, you are observing the resulting action of the natural laws of nature that govern flight. The human understanding of these laws of flight did not come easily—it has evolved over the past 2,500 years, starting with ancient Greek science. It was not until a cold day in December 1903 that these laws were finally harnessed by human beings to a degree sufficient to allow a heavier-than-air, powered, human-carrying machine to execute a successful sustained flight through the air. On December 17 of that year, Orville and Wilbur Wright, with pride and great satisfaction, reaped the fruits of their labors and became the first to fly the first successful flying machine. In Fig. 1.3, the *Wright Flyer* is shown at the instant of liftoff from the sands of Kill Devil Hill, near Kitty Hawk, North Carolina, at 10:35 on that morning, on its way to the first successful flight—you are looking at the most famous photograph in the annals of the history of aeronautics. At that moment, the Wright brothers knew they had accomplished something important—a feat aspired to by many before them, but heretofore never achieved. But they had no way of knowing the tremendous extent to which their invention of the first successful airplane was to dominate the course of the twentieth century—technically, socially, and politically.

The *airplane* is the subject of this book—its performance and its design. The purpose of this book is to pass on to you an appreciation of the laws of flight, and the embodiment of these laws in a form that allows the understanding and prediction of how the airplane will actually perform in the air (airplane performance) and how to approach the creation of the airplane in the first place in order to achieve a desired performance or mission (the creative process of airplane design). By 1903, the Wright brothers had achieved a rudimentary understanding of the principles of airplane performance, and they had certainly demonstrated a high degree of creativity in their inventive process leading to the design of the *Wright Flyer*. (See the book by Jakab, Ref. 1, for a definitive analysis of the Wrights's process of invention.) Today, our analyses of airplane performance have advanced much further, and the modern process of airplane design demands even greater creativity. The processes of airplane *performance* and airplane *design* are intimately coupled—one does not happen without the other. Therefore, the purpose of this book is to present the elements of both performance and design in an integrated treatment, and to do so in such fashion as to give you both a *technical* and a *philosophical* understanding of the process.

Figure 1.3   The *Wright Flyer*, at the moment of liftoff on its first flight, December 17, 1903.

Hopefully, this book will give you a better idea of how the aeronautical engineer mixes "engineering options on his technological palette and applies them to his canvas," as nicely stated by Richard Smith in the quotation at the beginning of this chapter.

1.2 FOUR HISTORICAL PERIODS OF AIRPLANE DESIGN CHARACTERISTICS

Before we proceed to the technical aspects of airplane performance and design, it is useful to briefly survey the historical evolution of these aspects, in order to have a better appreciation of modern technology. In this section, the technical evolution of the airplane is divided into four eras: (1) pre-Wright attempts, (2) strut-and-wire biplanes, (3) mature propeller-driven airplanes, and (4) jet-propelled airplanes. We have room for only short discussions of these eras; for a more detailed presentation, see Ref. 2.

If you like aeronautical history, this chapter is for you. However, if you do not particularly want to read about history or do not see the value in doing so, this chapter is *especially* for you. Whether you like it or not, good airplane design requires a knowledge of previous designs, that is, a knowledge of history. Even the *Wright Flyer* in 1903 was as much evolutionary as it was revolutionary, because the Wright brothers drew from a prior century of aeronautical work by others. Throughout the twentieth century, most new airplane designs were evolutionary, depending greatly on previous airplanes. Indeed, even the most recent airplane designs, such as the Boeing 777 commercial transport and the F-22 supersonic military fighter, contain the "genes" of 200 years of flying machine design. If you are interested in learning about airplane design, you need to know about these genes. So no matter what your innate interest in reading history may be, this chapter is an essential part of your education in airplane design. Please read it and benefit from it, in this spirit.

1.2.1 Pre-Wright Era

Before the Wright brothers's first flight, there were *no* successful airplane designs, hence *no* successful demonstrations of airplane performance. However, there were plenty of attempts. Perhaps the best way of gaining an appreciation of these attempts is to go through the following fanciful thought experiment. Imagine that you were born on a desolate island somewhere in the middle of the ocean, somehow completely devoid of any contact with the modern world—no television, radio, newspapers, magazines, etc. And imagine that for some reason you were possessed with the idea of flying through the air. What would you do? Would you immediately conceive of the idea of the modern airplane with a fixed wing, fuselage, and tail, propelled by some separate prime mover such as a reciprocating or jet engine? Certainly not! Most likely you would look at the skies, watch the birds, and then try to emulate the birds. To this end, you would fashion some kind of wings out of wood or feathers, strap these wings to your arms, climb to the roof of your hut, and jump off, flapping wildly. However, after only a few of these attempts (maybe only after one such trial), you would most certainly conclude that there had to be a better way. Indeed, history is full of such accounts of people attempting to fly by means of wings strapped to their arms and/or legs—the aeronautical historians call such people *tower jumpers*. They were all singularly unsuccessful. So perhaps you on your desolate island might take the next evolutionary step; namely, you might design some mechanical mechanism that you could push or pull with your hands and arms, or pump with your legs, and this mechanical mechanism would have wings that would flap up and down. Such mechanisms are called *ornithopters*. Indeed, no less a great mind than Leonardo da Vinci designed numerous such ornithopters in the period from 1486 to 1490; one of da Vinci's own drawings from his voluminous notebooks is reproduced in Fig. 1.4. However, one look at this machine shows you that it has no aerodynamically redeeming value! To this day, no human-powered ornithopter has ever successfully flown. So, after a few trials with your own mechanical device on your desolate island, you would most likely give up your quest for flight altogether. Indeed, this is what happened to most would-be aviators before the beginning of the nineteenth century.

So we pose the question: Where and from whom did the idea of the modern configuration airplane come? The modern configuration, that which we take for granted today, is a flying machine with fixed wings, a fuselage, and a tail, with a separate mechanism for propulsion. This concept was first pioneered by Sir George Cayley (Fig. 1.5) in England in 1799. In that year, Cayley inscribed on a silver disk two sketches that were seminal to the development of the airplane. Shown at the left in Fig. 1.6 is the sketch on one side of the silver disk; it illustrates for the first time in history a flying machine with a fixed wing, a fuselage, and a tail. Cayley is responsible for conceiving and advancing the basic idea that the mechanisms for lift

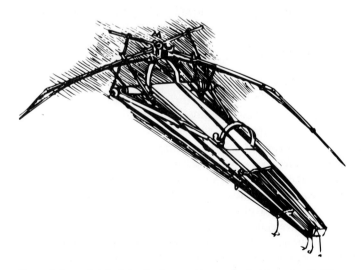

Figure 1.4   Original sketch of an ornithopter by da Vinci, circa 1492.

Figure 1.5   Sir George Cayley (1773–1857).

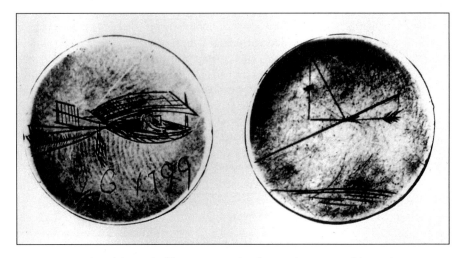

Figure 1.6   Silver disk inscribed by George Cayley showing the concept of the modern configuration airplane, 1799.

and thrust should be separated, with fixed wings moving at an angle of attack through the air to generate lift and a separate propulsive device to generate thrust. He recognized that the function of thrust was to overcome aerodynamic drag. In his own words, he stated that the basic aspect of a flying machine is "to make a surface support a given weight by the application of power to the resistance of air." On the flip side of the disk, shown at the right of Fig. 1.6, Cayley drew the first lift-drag vector diagram in the history of aeronautical engineering. Here we see the edge view of a flat-plate wing at an angle of attack to the relative wind (the relative wind is shown as a horizontal arrow, pointing toward the left). The resultant aerodynamic force is shown as the line inclined upward and backward, perpendicular to the plate. This resultant force is then resolved into components perpendicular and parallel to the relative wind, that is, the lift and drag, respectively. This silver disk, no larger than a U.S. quarter, is now in the collection of the British Science Museum in London. In this fashion, the concept of the modern configuration airplane was born. (More extensive discussions by the author of George Cayley's contributions to aeronautics can be found in Refs. 3 and 4; definitive studies of his life and contributions are contained in Refs. 5 and 6.)

To key on Cayley's seminal ideas, the nineteenth century was full of abortive attempts to actually build and fly fixed-wing, powered, human-carrying flying machines. Cayley himself built several full-size aircraft over the span of his long life (he died in 1857 at the age of 83), but was unsuccessful in achieving sustained flight. Some of the most important would-be inventors of the airplane—such as William Samuel Henson and John Stringfellow in England, Felix Du Temple in France, and Alexander Mozhaiski in Russia—are discussed in chapter 1 of Ref. 3; hence no further elaboration will be given here. They were all unsuccessful in achieving sustained flight. In regard to the nature of airplane performance and design, we note that these enthusiastic but unsuccessful inventors were obsessed with horsepower (or thrust). They were mainly concerned with equipping their aircraft with engines powerful enough to accelerate the machine to a velocity high enough that the aerodynamic lift of the wings would become large enough to raise the machine off the ground and into the air. Unfortunately, they all suffered from the same circular argument—the more powerful the engine, the heavier it weighs; the heavier the machine is, the faster it must move to produce enough lift to get off the ground; the faster the machine must move, the more powerful (and hence heavier) the engine must be—which is where we entered this circular argument. A way out of this quandary is to develop engines with more power *without* an increase in engine weight, or more precisely, to design engines with larger horsepower-to-weight ratios. Later, we will find this ratio, or more importantly the thrust-to-weight ratio T/W for the entire aircraft, to be a critical parameter in airplane performance and design. In the nineteenth century, inventors of flying machines functioned mainly on the basis of intuition, with little quantitative analysis to guide them. They knew that, to accelerate the aircraft, thrust had to be greater than the drag; that is, $T - D$ had to be a positive number. And the larger the thrust and the smaller the drag, the better things were. In essence, most of the nineteenth-century flying machine inventors were obsessed with *brute force*—given enough thrust (or horsepower) from the engine, the airplane could be wrestled into the air. The aviation historians call such people "chauffeurs." They were so busy trying to get the flying machine off the ground that they paid little attention to how the machine would be controlled once it got into the air—their idea was that somehow the machine could be "chauffeured" in the air much as a carriage driven on the ground. This philosophy led to failure in all such cases.

Perhaps the epitome of the chauffeurs was Sir Hiram Maxim, a U.S. expatriate from Texas living in England. To the general world, Maxim is known as the inventor of the first fully automatic machine gun. Developed by Maxim in England around 1884, the guns were manufactured by Vickers in England and were used by every major army around the globe. The wealth thus derived allowed Maxim to explore the design of a flying machine. From results obtained from his own wind tunnel tests, Maxim designed the huge airplane shown in Fig. 1.7. Built in 1893, the machine was powered by two 180-horsepower (180-hp), lightweight (for their day) steam engines of Maxim's design, driving two propellers. The total weight of the flying machine, including its three-person crew, was about 8,000 pounds (lb). On July 31, 1894, on the grounds of the rented Baldwyns Park in Kent, the Maxim airplane actually took off, although in a very limited sense. The airplane had a four-wheel undercarriage of steel wheels which ran along a straight, specially laid, railway track of 1,800 feet (ft) in length. Above the track was a wooden guardrail which engaged the undercarriage after about a 2-ft rise of the machine; Maxim was careful to not damage the aircraft, and hence he limited its height after takeoff to about 2 ft. On that day in July, the Maxim flying machine rolled down the track for 600 ft and lifted off. Almost immediately it engaged the guardrail, and Maxim quickly shut the steam off to the two engines. The Maxim flying machine came to a stop—but not without demonstrating that the engines were powerful enough to accelerate the machine to a high enough velocity that sufficient lift could be generated to raise the aircraft off the ground. With this demonstration, Maxim quit his aeronautical investigations until 1910—well after the stunning, successful demonstrations of almost "effortless" flight by the Wright brothers. As with all chauffeurs, Maxim's work was of little value to the state of the art of airplane design, and his aeronautical activities were soon forgotten by most people. This is in spite of the fact that Maxim supported this work out of his own pocket, spending over £30,000. In the words of the famous British aviation historian

Figure 1.7   Hiram Maxim and his flying machine, 1894.

Charles H. Gibbs-Smith (Ref. 7), commenting on Maxim's efforts: "It had all been time and money wasted. Maxim's contribution to aviation was virtually nil, and he influenced nobody." This indeed was the fate of all would-be inventors of the airplane during the nineteenth century who followed the chauffeur's philosophy. We mention Maxim here only because he is a classic example of this philosophy.

The antithesis of the chauffeur's philosophy was the "airman's" approach. This latter philosophy simply held that, in order to design a successful flying machine, it was necessary to first get up in the air and experience flight with a vehicle unencumbered by a power plant; that is, you should learn to fly *before* putting an engine on the aircraft. The person who introduced and pioneered the airman's philosophy was Otto Lilienthal, a German mechanical engineer, who designed and flew the first successful gliders in history. Lilienthal first carried out a long series of carefully organized aerodynamic experiments, covering a period of about 20 years, from which he clearly demonstrated the aerodynamic superiority of cambered (curved) airfoils in comparison to flat, straight surfaces. His experiments were extensive and meticulously carried out. They were published in 1890 in a book entitled *Der Vogelflug als Grundlage der Fliegekunst* (*Bird Flight as the Basis of Aviation*); this book was far and away the most important and definitive contribution to the budding science of aerodynamics to appear in the nineteenth century. It greatly influenced aeronautical design for the next 15 years, and was the bible for the early work of the Wright brothers. Among other contributions, Lilienthal presented drag polars in his book—the first drag polars to be published in the history of aeronautical engineering. (We will define and discuss drag polars in Chapter 2—they reflect all the aerodynamic information necessary for the performance analysis of an airplane.) Lilienthal's aerodynamic research led to a quantum jump in aerodynamics at the end of the nineteenth century. (See the extensive analysis of Lilienthal's aerodynamics contained in Ref. 8.)

During the period from 1891 through 1896, Lilienthal designed, built, and flew a number of gliders. With these successful glider flights, over 2,000 during the 5-year period, he personified the airman's philosophy. A photograph of Lilienthal on one of his gliders is shown in Fig. 1.8; he supported himself by grasping a bar with his arms, and the part of his body below his chest and shoulders simply dangled below the wings. He controlled his gliders by swinging his body—he was indeed the inventor of the hang glider. With these glider flights, Lilienthal advanced the cause of aeronautics by leaps and bounds. Many of his flights were public demonstrations; his fame spread far and wide. In the United States, stories and photographs of his flights were carried in popular magazines—the Wright brothers read about Lilienthal in *McClure's* magazine, a popular periodical of that day. Lilienthal was a professional mechanical engineer with a university degree, hence he had some credibility—and there he was, gliding through the air on machines of his own design. With this, the general public moved a little closer to acceptance that the quest for powered, heavier-than-air flight was respectable and serious.

On Sunday, August 9, 1896, Otto Lilienthal was once again flying from Gollenberg Hill in the Rhinow mountains, about 100 kilometers (km) northwest of Berlin. His first flight went well. However, during his second glide he encountered an unexpected sharp gust of air; his glider pitched up and stalled. Lilienthal violently swung

Figure 1.8   Otto Lilienthal flying one of his monoplane gliders, 1894.

his body to regain control of the glider, but to no avail. He crashed and died the next day in a Berlin clinic from a broken back.

At the time of his death, Lilienthal was working on a power plant for one of his glider designs. There is some feeling that, had he lived, Lilienthal may have preempted the Wright brothers and been the first to fly a powered machine. However, upon further investigation (Ref. 8) we find that his engine was intended to power a flapping motion of the outer wing panels—shades of the ornithopter concept. This author feels that had Lilienthal continued to pursue this course of action, he would have most certainly failed.

The last, and perhaps the most dramatic, failure of the pre-Wright era was the attempt by Samuel P. Langley to build a flying machine for the U.S. government. Intensely interested in the physics and technology of powered flight, Langley began a series of aerodynamic experiments in 1887, using a whirling arm apparatus. At the time, he was the director of the Allegheny Observatory in Pittsburgh. Within a year he seized the opportunity to become the third Secretary of the Smithsonian Institution in Washington, District of Columbia. Once on the Smithsonian's mall, Langley continued with his aeronautical experiments, including the building and flying of a number of elastic-powered models. The results of his whirling arm experiments were published in his book *Experiments in Aerodynamics* in 1890—a classic treatise that is well worth reading today. In 1896, the same year that Lilienthal was killed, Langley was successful in flying several small-scale, unmanned, powered aircraft, which he called *aerodromes*. These 14-ft-wingspan, steam-powered aerodromes were launched from the top of a small houseboat on the Potomac River, and they flew for about a minute, covering close to 1-mile (mi) over the river. These were the first steam-powered, heavier-than-air machines to successfully fly—an historic event in the history of aeronautics that is not always appreciated today.

However, this was to be the zenith of Langley's success. Spurred by the exigency of the Spanish-American War, Langley was given a $50,000 grant from the War Department to construct and fly a full-scale, person-carrying aerodrome. He hired an assistant, Charles Manly, who was a fresh, young graduate of the Sibley School of Mechanical Engineering at Cornell University. Together, they set out to build the required flying machine. The advent of the gasoline-powered internal-combustion engine in Europe convinced them that the aerodrome should be powered by a gasoline-fueled reciprocating engine turning a propeller. Langley had calculated that, for his new aerodrome, he needed an engine that would produce at least 12 hp and weigh no more than 120 lb. This horsepower-to-weight ratio was well above that available in any engine of the time; indeed, Balzer Company in New York, under subcontract from Langley to design and build such an engine, went bankrupt trying. Manly then personally took over the engine design in the basement of the original "castle" building of the Smithsonian, and by 1901 he had assembled a radically designed five-cylinder radial engine, shown in Fig. 1.9. This engine weighed only 124 lb and produced a phenomenal 52.4 hp. It was to be the best airplane power plant designed until the beginning of World War I. The full-scale aerodrome, equipped with Manly's engine,

Figure 1.9 The first radial engine for an aircraft, developed by Charles Manly, 1901.

was ready in 1903. The first attempted flight—on October 7, 1903—with Manly at the controls resulted in the aerodrome's falling into the river moments after its launch by a catapult mounted on top of a new houseboat on the Potomac River. Undaunted, Langley rationalized that the aerodrome was fouled in the catapult mechanism at the instant of launch. The aerodrome, somewhat damaged, was fished out of the river (Manly was fortunately unhurt) and returned to the Smithsonian for repairs. On December 8, 1903, they were ready to try again; the scenario was the same—the same pilot, the same aerodrome, the same houseboat. A photograph of the Langley aerodrome, taken just a moment after launch, is shown in Fig. 1.10. Here we see the Langley aerodrome going through a 90° angle of attack; the rear wings of the tandem wing design have collapsed totally. Again, Manly was retrieved from the river unhurt, but this was the rather unglorious end of Langley's aeronautical work.

Langley's aerodrome and the fate that befell it, as shown in Fig. 1.10, are an excellent study in the basic aspects of airplane design. The aircraft had a superb power plant. Its aerodynamic design, based mainly on 14 years of experimentation by Langley, was marginally good—at least it was sufficient for Langley's purposes.

Figure 1.10   The Langley aerodrome an instant after launch, December 8, 1903.

However, a recent in-house study by Dr. Howard Wolko, a mechanical and aerospace engineer now retired from the National Air and Space Museum, showed that the Langley aerodrome was *structurally* unsound—a result certainly in keeping with the aerodrome's failure, shown in Fig. 1.10. This illustrates a basic tenet of any system design, such as an airplane or a stereo system, namely, that the system is no better than its weakest link. In Langley's case, in spite of excellent propulsion and adequate aerodynamics, it was the poor structural design that resulted in failure of the whole system.

In spite of this failure, Langley deserves a lot of credit for his aeronautical work in the pre-Wright era. You experience some of the legacy left by Langley's name everytime you walk into the Langley Theater at the National Air and Space Museum in Washington, or visit the NASA Langley Research Center, built right beside Langley Air Force Base in Hampton, Virginia.

The story of Langley's aeronautical work is covered in much greater detail in Refs. 8 through 10, among others. In particular, Ref. 8 contains a detailed discussion of Langley's aerodynamics, and Ref. 10 has an extremely interesting and compelling presentation of the human dynamics associated with Langley's overall quest to build a flying machine.

1.2.2 Era of Strut-and-Wire Biplanes

The 1903 *Wright Flyer* ushered in the era of successful strut-and-wire biplanes—an era that covers the general period from 1903 to 1930. Unlike Langley's full-scale aerodrome, there were no fatal "weak links" in the design of the *Wright Flyer*. There is no doubt in this author's mind that Orville and Wilbur Wright were the first *true* aeronautical engineers in history. With the 1903 *Wright Flyer*, they had gotten it all right—the propulsion, aerodynamic, structural, and control aspects were carefully calculated and accounted for during its design. The Wright brothers were the first to fully understand the airplane as a whole and complete system, in which the individual components had to work in a complementary fashion so that the integrated system would perform as desired. A three-view drawing of the 1903 *Wright Flyer* is shown in Fig. 1.11.

Volumes have been written about the Wright brothers and their flying machines— their story is one of the greatest success stories in the history of technology. In our brief review of the evolution of the airplane in this chapter, it is perhaps better to defer to these volumes of literature than to attempt to relate the Wright brothers's story—we simply do not have space to do it justice. You are referred particularly to the authors's discussions of the Wright brothers in chapter 1 of Ref. 3, and in Ref. 9. The study by Jakab (Ref. 1) is an excellent portrait of the inventive processes of the Wright brothers as they created the first successful airplane. Similarly, Tom Crouch in his book *The Bishop's Boys* (Ref. 11) has painted an excellent humanistic portrait of the Wright brothers and their family as people caught up in this whirlwind of inventiveness—Crouch's book is the most definitive biography of the Wrights to date. For an extensive discussion of the Wright brothers's aerodynamics, see Ref. 8.

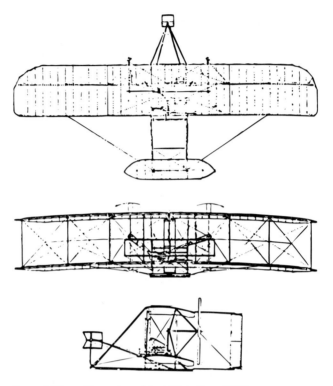

Figure 1.11   Three-view of the *Wright Flyer*, 1903.

Instead, let us dwell for a moment on the *Wright Flyer* itself as an airplane design. In Figs. 1.3 and 1.11, you see all the elements of a successful flying machine. Propulsion was achieved by a four-cylinder in-line engine designed and built by Orville Wright with the help of their newly hired mechanic in the bicycle shop, Charlie Taylor. It produced close to 12 hp and weighed 140 lb—barely on the margin of what the Wrights had calculated as the minimum necessary to get the flyer into the air. This engine drove two propellers via a bicyclelike chain loop. The propellers themselves were a masterpiece of aerodynamic design. Wilbur Wright was the first person in history to recognize the fundamental principle that a propeller is nothing more than a twisted wing oriented in a direction such that the aerodynamic force produced by the propeller was predominately in the thrust direction. Wilbur conceived the first viable propeller theory in the history of aeronautical engineering; vestiges of Wilbur's analyses carry though today in the standard "blade element" propeller theory. Moreover, the Wrights had built a wind tunnel, and during the fall and winter of 1901 to 1902, they carried out tests on hundreds of different airfoil and wing shapes. Wilbur incorporated these experimental data in his propeller analyses; the result was a propeller with an efficiency that was close to 70% (propeller efficiency is the power output from the propeller compared to the power input to the propeller from the engine shaft). This represented a dramatic improvement of propeller performance over contemporary practice. For example, Langley reported a propeller efficiency of only 52% for his aerodromes. Today, a modern, variable-pitch propeller can achieve efficiencies as high as 85% to 90%. However, in 1903, the Wrights's propeller efficiency of 70% was simply phenomenal. It was one of the lesser-known but most compelling reasons for the success of the *Wright Flyer*. With their marginal engine linked to their highly efficient propellers, the Wrights had the propulsion aspect of airplane design well in hand.

The aerodynamic features of the *Wright Flyer* were predominately a result of their wind tunnel tests of numerous wing and airfoil shapes. The Wrights were well aware that the major measure of aerodynamic efficiency is the *lift-to-drag ratio L/D*. They knew that the lift of an aircraft must equal its weight in order to sustain the machine in the air, and that almost any configuration could produce enough lift if the angle of attack were sufficiently large. But the secret of "good aerodynamics" is to produce this lift with as small a drag as possible, that is, to design an aircraft with as large an L/D value as possible. To accomplish this, the Wrights did three things:

1.   They chose an airfoil shape that, based on the collective data from their wind tunnel tests, would give a high L/D. The airfoil used on the *Wright Flyer* was a thin, cambered shape, with a camber ratio (ratio of maximum camber to chord length) of 1/20, with the maximum camber near the quarter-chord location. (In contrast, Lilienthal favored airfoils that were circular arcs, i.e., with maximum camber at midchord.) It is interesting that the precise airfoil shape used for the *Wright Flyer* was never tested by the Wright brothers in their wind tunnel. By 1903, they had so much confidence in their understanding of airfoil and wing properties that, in spite of their characteristic conservative philosophy, they felt it unnecessary to test that specific shape.

2.   They chose an aspect ratio of 6 for the wings. They had experimented with gliders at Kitty Hawk in the summers of 1900 and 1901, and they were quite disappointed in their aerodynamic performance. The wing aspect ratio of these early gliders was 3. However, their wind tunnel tests clearly indicated that higher-aspect-ratio wings produced higher values of L/D. (This was not a new discovery; the advantage of high-aspect-ratio wings had been first theorized by Francis Wenham in 1866. Langley's whirling arm data, published in 1890, proved conclusively that better performance was obtained with higher-aspect-ratio wings. It is a bit of a mystery why the Wrights, who were very well read and had access to these results, did not pick up on this important aerodynamic feature right from the start.) In any event, based on their own wind tunnel results, the Wrights immediately adopted an aspect ratio of 6 for their 1902 glider, and the following year for the 1903 flyer. At the time, the Wrights had no way of knowing about the existence of induced drag; this aerodynamic phenomenon was not understood until the work of Ludwig Prandtl in Germany 15 years later. The Wrights did not know that, by increasing the aspect ratio from 3 to 6, they reduced the induced drag by a factor of 2. They only knew from their empirical results that the L/D ratio of the 6-aspect-ratio wing was much improved over their previous wing designs.

3.   The Wrights were very conscious of the importance of parasite drag, which in their day was called *head resistance*. They used empirical formulas obtained from Octave Chanute to estimate the head resistance for their machines. (Octave Chanute was a well-known civil and railroad engineer who had become very interested in aeronautics. He published in 1893 an important survey of past aeronautical work from around the world in a book entitled *Progress in Flying Machines*. It has become a classic—you can still buy reprinted copies today. From 1900, Octave Chanute was a close friend and confidant of the Wright brothers, giving them much encouragement during their intensive inventive work in 1900 to 1903.) The Wrights's choice of lying prone while flying their machines, rather than sitting up, or even dangling underneath as Lilienthal had done, was simply a matter of decreasing head resistance. In early 1903, they even tested a series of wooden struts in an airstream in order to find the cross-sectional shape that gave minimum drag. Unfortunately, they did not appreciate the inordinately high drag also produced by the supporting wires between the two wings. Of course, today we recognize that the struts and wires necessary to structurally strengthen biplane wing configurations are a major source of drag. In any event, the Wrights were very conscious of head resistance, and they attempted to keep it as low as they possibly could, given the state of their knowledge at that time.

The Wrights never quoted a value of L/D for their 1903 *Wright Flyer*. Modern wind tunnel tests of models of the *Wright Flyer* carried out in 1982 and 1983 as reported in the paper by Culick and Jex (Ref. 12) indicate a maximum L/D of 6. This value is totally consistent with values of $(L/D)_{max}$ measured by Gustave Eiffel in 1910 in his large wind tunnel in Paris for models of a variety of aircraft of that time (see Ref. 8). Also, Loftin (Ref. 13) estimated a value for $(L/D)_{max}$ of 6.4 for the Fokker E-III, an early World War I aircraft. *Hence, in 1903 the Wrights had achieved a value of $(L/D)_{max}$ with their flyer that was as high as that for aircraft designed 10 years later.* Clearly, the Wrights had the aerodynamic aspect of airplane design well in hand.

The control features of the *Wright Flyer* are also one of the basic reasons for its success. The Wright brothers were the first to recognize the importance of flight control around all three axes of the aircraft. Pitch control, obtained by a deflection of all or part of the horizontal tail (or the forward canard such as on the *Wright Flyer*), and yaw control, obtained by deflection of the vertical rudder, were features recognized by investigators before the Wrights; for example, Langley's aerodrome had pitch and yaw controls. However, no one except the Wrights appreciated the value of roll control. Their novel idea of differentially warping the wing tips to control the rolling motion of the airplane, and to jointly control roll and yaw for coordinated turns, was one of their most important contributions to aeronautical engineering. Indeed, when Wilbur Wright finally carried out the first public demonstrations of their flying machines in LeMans, France, in August 1908, the two technical features of the Wright machines most appreciated and immediately copied by European aviators were their roll control and their efficient propeller design. Clearly, the Wrights had the flight control aspect of airplane design well in hand.

Finally, the structural features of the *Wright Flyer* were patterned partly after the work of Octave Chanute and partly after their own experience in designing bicycles. Chanute, inspired by the gliding flights of Lilienthal, carried out tests of gliders of his own design beginning in 1896. The most important technical feature of Chanute's gliders was the sturdy and lightweight Pratt-truss method of rigging a biplane structure. (Chanute was a civil engineer, noted in part for designing and building bridges; it is natural that he would have exercised his expertise in the structural design of flying machines.) The Wright brothers adopted the Pratt-truss system for the *Wright Flyer* directly from Chanute's work. Other construction details of the *Wright Flyer* took advantage of the Wrights's experience in designing and building sturdy but lightweight bicycles. When it was finished, engine included, the empty weight of the *Wright Flyer* was 605 lb. With a 150-lb person on board, the empty weight–gross weight ratio was 0.8. By comparison, the empty weight of the Fokker E-III designed 10 years later was 878 lb, and the empty weight–gross weight ratio was 0.65, not greatly different from that of the *Wright Flyer*. Considering that 10 years of progress in aircraft structural design had been made between the 1903 flyer and the Fokker E-III, the structural design of the 1903 *Wright Flyer* certainly seems technically advanced for its time. And the fact that the flyer was structurally sound was certainly well demonstrated on December 17, 1903. Clearly, the Wrights had the structural aspect of airplane design well in hand.

In summary, the Wright brothers had gotten it right. All the components of their system worked properly and harmonically—propulsion, aerodynamics, control, and structures. There were no fatal weak links. The reason for this was the natural inventiveness and engineering abilities of Orville and Wilbur Wright. The design of the *Wright Flyer* is a classic first study in good aeronautical engineering. There can be no doubt that the Wright brothers were the first true aeronautical engineers.

The *Wright Flyer* ushered in the era of strut-and-wire biplanes, and it basically set the pattern for subsequent airplane design during this era. The famous World War I fighter airplanes—such as the French Nieuport 17 and the SPAD XIII, the German Fokker D. VII, and the British Sopwith Camel—were in many respects "souped-up" Wright flyers. The principal technical improvements contained in these later biplanes are described below.

First, the wing-warping method of roll control used by the Wrights was quickly supplanted by ailerons in most other aircraft. (The idea of flaplike surfaces at the trailing edges of airplane wings can be traced to two Englishmen: M. P. W. Boulton, who patented a concept for lateral control by ailerons in 1868; and Richard Harte, who also filed for a similar patent in 1870. In both cases, neither man truly understood the function of these devices, and since there existed no successful flying machines at that time which could demonstrate the use of ailerons, the idea and the patents were quickly forgotten.) Ailerons in the form of triangular "winglets" that projected beyond the usual wingtips were used in 1908 by Glenn Curtiss on his *June Bug* airplane; flying the *June Bug*, Curtiss won the Scientific American Prize on July 4, 1908, for the first public flight of 1000 meters (m) or longer. By 1909, Curtiss had designed an improved airplane, the *Gold Bug*, with rectangular ailerons located midway between the upper and lower wings, as seen in Fig. 1.12. Finally, in 1909 the Frenchman Henri Farman designed a biplane named the *Henri Farman III*, which included a flaplike aileron at the trailing edge of all four wingtips; this was the true ancestor of the conventional modern-day aileron. Farman's design was soon adopted by most designers, and wing warping quickly became passé. Only the Wright brothers clung to their old concept; a Wright airplane did not incorporate ailerons until 1915, six years after Farman's development.

Second, the open framework of the fuselage, such as seen in the *Wright Flyer* and the Curtiss *Gold Bug*, was in later designs enclosed by fabric. The first such airplane to have this feature was a Nieuport monoplane built in 1910, shown in Fig. 1.13. This was an attempt at "streamlining" the airplane, although at that time the concept of streamlining was only an intuitive process rather than the result of real technical knowledge and understanding about drag reduction.

Third, the demands for improved airplane performance during World War I gave a rebirth to the idea of "brute force" in airplane design. In relation to the thrust minus

Figure 1.12   Glenn Curtiss flying his *Gold Bug*. Note the midwing ailerons.

Figure 1.13   Nieuport monoplane, 1910.

drag expression $T - D$, designers of World War I fighter airplanes, in their quest for faster speeds and higher rates of climb, increased the thrust rather than decreased the drag. The focus was on more powerful engines. The SPAD XIII (Fig. 1.14), one of the best and most famous aircraft from World War I, had a Hispano-Suiza engine that produced 220 hp—the most powerful engine used on a fighter aircraft at that time. Because of this raw power, the SPAD XIII had a maximum velocity of 134 miles per hour (mi/h), which made it one of the fastest airplanes during the war. Examining Fig. 1.14, we see the typical strut-and-wire biplane; the struts and wires produced large amounts of drag, although this was not fully understood by most airplane designers at that time. In fact, in the March 1924 issue of the *Journal of the Royal Aeronautical Society*, the noted British aeronautical engineer Sir Leonard Bairstow was prompted to say, "Our war experience showed that, whilst we went forward as regard to horsepower, we went backwards with regard to aerodynamic efficiency." The demands of wartime did not allow the airplane designer the luxury of obtaining "aerodynamic efficiency." Aircraft design during World War I was an intuitive "seat-of-the-pants" process. Some designs were almost literally marked off in chalk on the concrete floor of a factory, and the completed machines rolled out the front door two weeks later. Clearly, there was plenty of room for improvement.

Strut-and-wire biplanes were the mainstay of aeronautics for the decade following World War I. To be sure, there were gradual design improvements, mainly involving some streamlining to reduce drag and gradual increases in engine power to increase thrust. However, airplane designers, then as now, tended to be conservative, and hence advancement in the design of new aircraft was a gradual, evolutionary process; each new aircraft design tended to be closely related to the previous design. For this reason, airplane designers clung to the strut-and-wire biplane, some into the early 1930s—well beyond the effective lifetime of this configuration. The conservative philosophy of airplane designers in the 1920s and 1930s is summarized by Miller and Sawers (Ref. 14) as follows: "Designers had acquired the attitude of the practical man, who knew how airplanes should be designed, because that was how they had designed them for the previous 20 years."

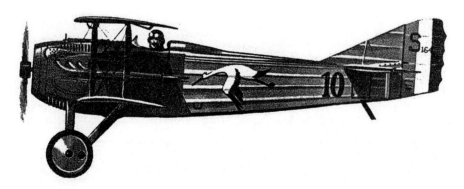

Figure 1.14   SPAD XIII, 1917.

However, some designers had vision; during the 1920s some knew what had to be done to greatly improve airplane performance. For example, the concept of streamlining to reduce drag was a major topic of discussion. The famous French airplane designer Louis Brequet, in a talk given to the Royal Aeronautical Society on April 6, 1922, showed his appreciation of the value of streamlining the airplane when he said

> The conclusion is that one must bring to the minimum the value of D/L. It can be obtained by choosing the best possible profile for the wings, the best designs for the body, empennage, etc. Moreover, the undercarriage should be made to disappear inside the body on the wings when the aeroplane is in flight.

Here we have Brequet calling for retractable landing gear, something not seen on any contemporary aircraft of that day.

There were exceptions to the tried-and-proven way of evolutionary airplane design during the 1920s. Air races, with prizes for speed, were popular. The international Schneider Cup races were perhaps the most important and seminal of them all. On December 5, 1912, the French industrialist Jacques Schneider announced a competition to promote the development of seaplanes. He offered an impressive trophy to the first nation that could win the race three times out of a series of five successive yearly events. Starting in Monaco in 1913, the Schneider Cup races continued on an almost annual basis (interrupted by World War I). Winning the Schneider Cup race became a mattter of national prestige for some countries; as a result, every effort was made to increase speed. Specialized high-power engines were designed and built, and extreme (for that time) measures were taken to reduce drag. For example, the 1925 winner of the Schneider race was Lieutenant Jimmy Doolittle, flying an Army Curtiss R3C-2 biplane, as shown in Fig. 1.15. The high degree of streamlining in this aircraft is clearly evident; powered by a 619-hp Curtiss V-1400 engine, the R3C-2 achieved a speed of 232.57 mi/h over the course of the race. The Schneider Trophy was finally permanently acquired in 1931 by Britain, winning the last three races

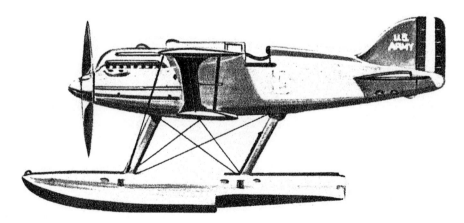

Figure 1.15   Curtiss R3C-2, flown by Jimmy Doolittle, winner of the 1925 Schneider Cup race.

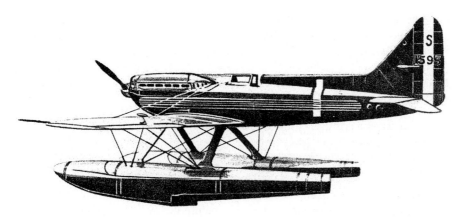

Figure 1.16   Supermarine S.6B, the airplane that won the Schneider Cup permanently for Great Britain, 1931.

with Supermarine Monoplane aircraft which were precursers to the famous British Spitfire of World War II. The final Schneider Cup winner was the Supermarine S.6B (Fig. 1.16) with a speed of 340.08 mi/h; both the S.6B and the Schneider Trophy are now displayed in the British Science Museum in South Kensington, London. The winning speed of 340 mi/h was phenomenal for that time; it is even more phenomenal considering that the aircraft was equipped with large pontoons. Jacques Schneider had initiated the competition to promote the development of seaplanes; all entries had to take off from and land on water.

Specially designed racing airplanes, such as those for the Schneider Cup races, were indeed exceptions to the tried-and-proven way of evolutionary airplane design. They are perfect examples of single-purpose, point-design aircraft. However, they provided the incentive for innovative thinking and new aeronautical research and development. Ultimately they substantially contributed to the demise of the era of strut-and-wire biplanes and to the beginning of the era of the *mature*, propeller-driven airplane.

1.2.3   Era of the Mature, Propeller-Driven Airplane

The period from 1930 to 1950 can be classified as the era of the mature, propeller-driven airplane. During this time, airplane design matured, new technical features were incorporated, and the speed, altitude, efficiency, and safety of aircaft increased markedly. In particular, the 1930s are considered by many aviation historians as the "golden age of aviation" (indeed, there is currently a gallery at the National Air and Space Museum with this title). Similarly, the 1930s might be considered as a golden age for aeronautical engineering—a period when many improved design features, some gestating since the early 1920s, finally became accepted and incorporated on "standard" aircraft of the age.

The maturity of the propeller-driven airplane is due to nine major technical advancements, all of which came to fruition during the 1930s. These technical advancements are discussed below.

First, the cantilevered-wing monoplane gradually replaced the strut-and-wire biplane. The main reason for the dominance of the biplane in early airplane design was structural strength. The struts and wires had a purpose; two wings of relatively short span, trussed together as a stiff box, were structurally sounder than if the same total wing area were spread out over a single wing with larger span. Moreover, the moment of inertia about the roll axis was smaller for the shorter-span biplanes, leading to more rapid rolling maneuverability. For these reasons, pilots and airplane designers were reluctant to give up the biplane; for example, it was not until 1934 that the British Air Ministry ordered monoplane fighters for the first time. This is not to say that monoplanes did not exist before the 1930s; quite the contrary, a number of early monoplane designs were carried out before World War I. When Louis Bleriot became the first person to fly across the English Channel on July 25, 1909, it was in a monoplane of his own design (although there are some reasons to believe that the airplane was designed in part by Raymond Saulnier). Because of the publicity following Bleriot's channel crossing, the monoplane experienced a surge of popularity. Bleriot himself sold hundreds of his Bleriot XI monoplanes, and it dominated the aviation scene until 1913. Its popularity was somewhat muted, however, by an inordinate number of crashes precipitated by structural failure of the wings, and ultimately helped to reinforce distrust in the monoplane configuration.

However, the monoplane began its gradual climb to superiority when in 1915 Hugo Junkers, at that time the Professor of Mechanics at the Technische Hochschule in Aachen, Germany, designed and built the first all-steel cantilever monoplane in history. This initiated a long series of German advancements in cantilever-wing monoplanes by both Junkers and Anthony Fokker through the 1920s. In the United States, the first widely accepted monoplane was the Ford Trimotor (Fig. 1.17) introduced in 1926; this aircraft helped to establish the civil air transport business in the United States. (However, the public's faith in the Ford Trimotor was shaken when the

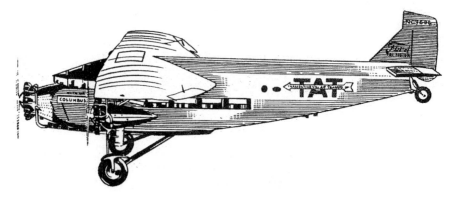

Figure 1.17   Ford Trimotor, 1926.

famous Notre Dame football coach Knute Rockne was killed on March 31, 1931, in a crash of a trimotor.) However, the monoplane configuration really came into its own with the Boeing Monomail of 1930, shown in Fig. 1.18. This airplane embodied two other important technical developments; it had all-metal, stressed skin construction, and its landing gear was retractable. In addition, it was one of the first to use wing fillets in an effort to smooth the airflow at the wing-fuselage juncture. The airplane you are looking at in Fig. 1.18 is certainly a proper beginning to the era of the mature, propeller-driven airplane.

A major technical development during this era was the National Advisory Committee for Aeronautics (NACA) cowling for radial piston engines. Such engines have their pistons arranged in a circular fashion about the crankshaft, and the cylinders themselves are cooled by airflow over the outer finned surfaces. Until 1927, these cylinders were usually directly exposed to the main airstream of the airplane, causing inordinately high drag. Engineers recognized this problem, but early efforts to enclose the engines inside an aerodynamically streamlined shroud (a cowling) interfered with the cooling airflow, and the engines overheated. One of the earliest aeronautical engineers to deal with this problem was Colonel Virginius E. Clark (for whom the famous CLARK-Y airfoil is named). Clark designed a primitive cowling in 1922 for the Dayton-Wright XPS-1 airplane; it was marginal at best, and besides Clark had no proper aerodynamic explanation as to why a cowling worked. The first notable progress was made by H. L. Townend at the National Physical Laboratory in England. In 1927, Townend designed a ring of relatively short length which wrapped around the outside of the cylinders. This resulted in a noticeable decrease in drag, and at least it did not interfere with engine cooling. Engine designers who were concerned with the adverse effect of a full cowling on engine cooling were more ready to accept a ring. The Boeing Monomail was equipped with a Townend ring, which is clearly seen in Fig. 1.18.

However, the major breakthrough in engine cowlings was due to the National Advisory Committee for Aeronautics in the United States. Beginning in 1927, at the insistence of a group of U.S. aircraft manufacturers, the NACA Langley Memorial Laboratory at Hampton, Virginia, undertook a systematic series of wind tunnel tests with the objective of understanding the aerodynamics of engine cowlings and designing an effective shape for such cowlings. Under the direction of Fred E. Weick at Langley Laboratory, this work quickly resulted in success. Drag reduction larger than that with a Townend ring was obtained by the NACA cowling. In 1928, Weick

Figure 1.18   Boeing Monomail, 1930, with a Townend ring.

published a report comparing the drag on a fuselage-engine combination with and without a cowling. Compared with the uncowled fuselage, a full cowling reduced the drag by a stunning 60%! Moreover, by proper aerodynamic design of the cowling, the airflow between the engine and the inside of the cowling resulted in *enhanced* cooling of the engine. Hence, the NACA cowling was achieving the best of both worlds. One of the first airplanes to use the NACA cowling was the Lockheed Vega, shown in Fig. 1.19. Early versions of the Vega without a cowling had a top speed of 135 mi/h; after the NACA cowling was added to later versions, the top speed increased to 155 mi/h. The Lockheed Vega went on to become one of the most successful airplanes of the 1930s. The Vega 5, equipped with the NACA cowling and a more powerful engine, had a top speed of 185 mi/h. It was used extensively in passenger and corporate service. In addition, Amelia Earhart and Wiley Post became two of the most famous aviators of the 1930s—both flying Lockheed Vegas. Not only is the Vega a classic example of the new era of mature propeller-driven airplanes, but also its aesthetic beauty supported the popular adage "If an airplane looks beautiful, it will also fly beautifully."

Before the 1930s, a weak link in all propeller-driven aircraft was the propeller itself. As mentioned earlier, Wilbur Wright was the first to recognize that a propeller

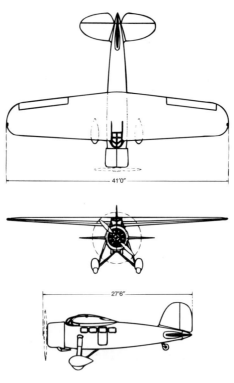

Figure 1.19   Lockheed Vega, 1929, with an NACA cowling.

is essentially a twisted wing oriented in such a fashion that the principal aerodynamic force is in the thrust direction. For a propeller of fixed orientation, the twist of the propeller is designed so that each airfoil section is at its optimum angle of attack to the relative airflow, usually that angle of attack that corresponds to the maximum lift-to-drag ratio of the airfoil. The relative airflow seen by each airfoil section is the vector sum of the forward motion of the airplane and the rotational motion of the propeller. Clearly, when the forward velocity of the airplane is changed, the angle of attack of each airfoil section changes relative to the local flow direction. Hence, a fixed-pitch propeller is operating at maximum efficiency only at its design speed; for all other speeds of the airplane, the propeller efficiency decreases. This is a tremendous disadvantage of a fixed-pitch propeller. Indeed, the Boeing Monomail shown in Fig. 1.18 had a fixed-pitch propeller, which greatly compromised its performance at off-design conditions. Because of the reduced thrust from the fixed-pitch propeller, the Monomail reached a top speed of only 158 mi/h, partially negating the advantage of the reduced drag obtained with its Townend ring and retracted landing gear. Its propeller problem was so severe that the Monomail never entered serial production.

The solution to this problem was to vary the pitch of the propeller during the flight so as to operate at near-optimum conditions over the flight range of the airplane—a mechanical task easier said than done. The aerodynamic advantage of varying the propeller pitch during flight was appreciated as long ago as World War I, and Dr. H. S. Hele-Shaw and T. E. Beacham patented such a device in England in 1924. However, the first practical and reliable mechanical device for varying propeller pitch was designed by Frank Caldwell of Hamilton Standard in the United States. The first production order for Caldwell's design was placed by Boeing in 1933 for use on the Boeing 247 transport (Fig. 1.20). The 247 was originally designed in 1932 with fixed-pitch propellers. However, when it started flying in early 1933, Boeing found that the airplane had inadquate takeoff performance from some of the airports high in the Rocky Mountains. By equipping the 247 with variable-pitch propellers, this problem was solved. Moreover, the new propellers increased its rate of climb by 22% and its cruising velocity by over 5%. To emphasize the impact of this development on airplane design, Miller and Sawyer (Ref. 14) stated, "After this demonstration of its advantages and its successful service on the 247, no American designer could build a high-performance airplane without a variable-pitch propeller." Later in the 1930s, the variable-pitch propeller, which was controlled by the pilot, developed

Figure 1.20   Boeing 247, 1933.

into the constant-speed propeller, where the pitch was automatically controlled so as to maintain constant revolutions per minute (rpm) over the flight range of the airplane. Because the power output of the reciprocating engine varies with rotational speed, by having a propeller in which the pitch is continuously and automatically varied to maintain constant engine speed, the net power output of the engine-propeller combination can be maintained at an optimum value.

Another important advance in the area of propulsion was the development of high-octane aviation fuel, although it was eclipsed by the more visibly obvious break-throughs in the 1930s such as the NACA cowling, retractable landing gear, and the variable-pitch propeller. Engine "pinging," an audible local detonation in the engine cylinder caused by premature ignition, had been observed as long ago as 1911. An additive to the gasoline, tetraethyl lead, was found by C. F. Kettering of General Motors Delco to reduce this engine knocking. In turn, General Motors and Standard Oil formed a new company, Ethyl Gasoline Corporation, to produce "ethyl" gasoline with a lead additive. Later, the hydrocarbon compound of octane was also found to be effective in preventing engine knocking. In 1930, the Army Air Corps adopted 87-octane gasoline as its standard fuel; in 1935, this standard was increased to 100 octane. The introduction of 100-octane fuel allowed much higher compression ratios inside the cylinder, and hence more power for the engine. For example, the introduction of 100-octane fuel, as well as other technological inprovements, allowed Curtiss-Wright Aeronautical Corporation to increase the power of its R-1820 Cyclone engine from 500 to 1,200 hp in the 1930s—by no means a trivial advancement.

In subsequent chapters, we will come to appreciate that, when a new airplane is designed, the choice of wing area is usually dictated by speed at takeoff or landing (or alternatively by the desired takeoff or landing distances along a runway). The wing area must be large enough to provide sufficient lift at takeoff or landing; this criterion dictates the ratio of airplane weight to wing area, that is, the wing loading W/S—one of the most important parameters in airplane performance and design. After the airplane has taken off and accelerated to a much higher cruising speed, the higher-velocity airflow over the wing creates a larger pressure difference between the upper and lower wing surfaces, and hence the lift required to sustain the weight of the airplane can be created with a smaller wing area. From this point of view, the extra wing area required for takeoff and landing is extra baggage at cruising conditions, resulting in higher structural weight and increased skin friction drag. The design of airplanes in the era of strut-and-wire biplanes constantly suffered from this compromise. However, a partial solution surfaced in the late 1920s and 1930s, namely, the development of high-lift devices—flaps, slats, slots, etc. Figure 1.21 illustrates some of the standard high-lift devices employed on aircraft since the 1920s, along with a scale of lift coefficient indicating the relative increase in lift provided by each device. By employing such high-lift devices, sufficient lift can be obtained at takeoff and landing with wings of smaller area, hence allowing airplane designers the advantage of high wing loadings at cruise. High-lift devices were one of the important technical developments during the era of the mature propeller-driven airplane. Let us examine the history of that development in greater detail.

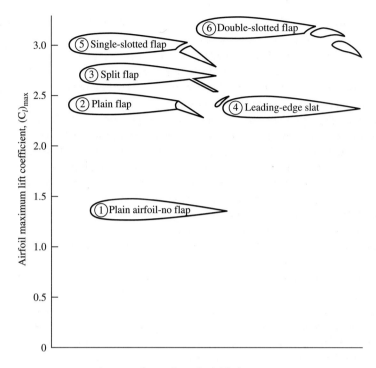

Figure 1.21 Schematic of some basic high-lift devices.

The basic plain flap (labeled 2 in Fig. 1.21) evolved directly from the trailing-edge ailerons first used by Henri Farman in the autumn of 1908 in France. However, designers of the relatively slow World War I biplanes were not inclined to bother with flaps. Plain flaps were first used on the S.E.-4 biplane built by the Royal Aircraft Factory in 1914; they became standard on airplanes built by Fairey from 1916 onward. Even the pilots of these aircraft rarely bothered to use flaps.

The single-slotted flap (labeled 5 in Fig. 1.21) was developed around 1920 independently by three different people in three different places. One person was G. V. Lachmann, a young German pilot who ran smoke tunnel tests in 1917 on the single-slotted flap and then filed for a patent on the concept. The patent was rejected on the basis that the slot would destroy the lift on the wing, rather than enhance it. (The reviewers of the patent application did not realize that the high-energy jet of air through the slot produced by the higher pressure on the bottom surface and the lower pressure on the top surface helped to prevent the boundary layer from separating on the top surface, hence increasing lift.) The second person to develop the slotted flap was Sir Frederick Handley Page in England, who claimed that it increased lift by 60%. When Lachmann in Germany read about Page's work, he convinced Ludwig Prandtl at Gottingen University to run wind tunnel tests on the slotted flap. Prandtl, skeptical at first, ran the tests and found a 63% increase in lift. Lachmann got his patent, and he pooled rights with Page in 1921. (Much later, in 1929, Lachmann went to work for Handley Page Company.) The third person to develop the slotted flat was O. Mader, an engineer working for Junkers in Germany. Mader first tested the concept in a wind tunnel, and then during the period of 1919 to 1921 he made flight tests on airplanes equipped with a single-slotted flap. This work was carried out independently of either Lachmann or Page. To no surprise, however, when Mader applied for a German patent in 1921, it was held to infringe upon Lachmann's patent. The viability of the single-slotted flap was finally established beyond a doubt by a 2-year wind tunnel testing program carried out at the National Physical Laboratory (NPL) in England starting in 1920. The NPL data showed that flaps enhanced the lift most effectively on *thick* airfoils, which helped to explain why they were rarely used on World War I biplanes with their exceptionally thin airfoil profiles. In spite of the favorable NPL data, and the tests by Lachmann, Page, and Mader, single-slotted flaps were slow to be implemented. During the 1920s, the only aircraft to be equipped with such flaps were those designed by Page and Lachmann.

Flap development in the United States was spurred by the invention of the split flap (labeled 3 in Fig. 1.21) by Orville Wright in 1920. Working with J. M. H. Jacobs at the Army Air Corps's technical laboratory at McCook Field in Dayton, Ohio, Orville showed that the split flap increased both lift *and* drag. The increase in drag is actually beneficial during landing; the associated decrease in lift-to-drag ratio L/D results in a steeper glide slope during landing, hence reducing the overall landing distance. Whether or not it had anything to do with nationalistic pride, the first type of flap to be used on an airplane designed in the United States was the split flap, and this was not until 1932 when Jack Northrop used it on his Northrop Gamma and Lockheed Orion designs (Northrop was a designer working for Lockheed in the early 1930s). In 1933 Douglas designed the pioneering DC-1 with split flaps; the use of such flaps carried through to the venerable Douglas DC-3 (Fig. 1.22) in the mid-1930s. By the late 1930s, split flaps were being used on most civil and military aircraft.

The next major advancement in flap development came in 1924, also in the United States. Harlan D. Fowler, an engineer with the Army Air Corps, working independently as a private venture with his own money, developed the Fowler flap, sketched in Fig. 1.23. The Fowler flap combined two advantageous effects. The deflection of the flap increased the effective camber of the wing, hence increasing

Figure 1.22   Douglas DC-3, 1935.

Flap undeployed   Fowler flap deployed

Figure 1.23   Schematic of a Fowler flap.

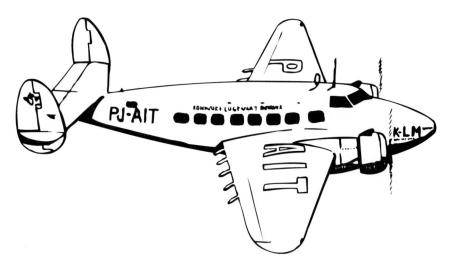

Figure 1.24   Lockheed L-14 Super Electra, 1937. First production airplane to incorporate Fowler flaps.

lift; this feature it shared with the other types of flaps. However, in addition, the Fowler flap, when deployed, was mechanically extended beyond the trailing edge of the wing, as shown in Fig. 1.23. This resulted in an increase in wing area, hence increasing the lift. In an interesting parallel with Lachmann's experience with the slotted flap, Fowler was unable to generate much interest in his invention, until 1932 when NACA ran a series of wind tunnel tests which proved the value of the Fowler flap. In 1933, Fowler traveled to Baltimore, Maryland, and convinced Glenn L. Martin of the effectiveness of the flap. Fowler was hired by Martin Company to help design flaps for Martin's series of new aircraft. The Fowler flap was first used on the Martin 146 bomber design in 1935. The airplane was not ordered by the Army for production; ironically, the first production airplane to use the Fowler flap was the Lockheed L-14 Super Electra commerical twin-engine transport in 1937 (Fig. 1.24).

Since the late 1930s, the only major advancement in flap design has been the combination of the Fowler flap with the slotted flap, and with the use of more than one slot. The double-slotted flap was first developed in Italy by G. Pegna of Piaggio Aircraft Company, and it was first used on the Italian M-32 bomber in 1937. In the late 1940s, the double-slotted Fowler flap was widely used on airliners such as the Douglas DC-6 and DC-7. A double-slotted Fowler flap is shown in Fig. 1.21, labeled number 6. Finally, the triple-slotted Fowler flap was developed by Boeing for use on the 727 jet transport in the 1960s.

Examine the Douglas DC-3 and the Lockheed L-14 shown in Figs. 1.22 and 1.24, respectively. These airplanes epitomize the mature, propeller-driven aircraft of the 1930s. Here you see cantilever wing monoplanes powered by radial engines enclosed in NACA cowlings, and equipped with variable-pitch propellers. They are all-metal airplanes with retractable landing gear, and they use flaps for high lift during takeoff and landing. It is for these reasons that the 1930s can be called the golden age of aeronautical engineering.

Three other technical developments of the late 1930s are worth mentioning. One is the advent of the pressurized airplane. Along with the decrease in atmospheric pressure with increasing altitude, there is the concurrent decrease in the volume of oxygen necessary for human breathing. Hence, the useful cruising altitude for airplanes was limited to about 18,000 ft or lower. Above this altitude for any reasonable length of time, a human being would soon lose consciousness due to lack of oxygen. The initial solution to the problem of sustained high-altitude flight was the pressure suit and the auxiliary oxygen supply breathed through an oxygen mask. The first pilot to use a pressure suit was Wiley Post. Looking like a deep-sea diver, Post set an altitude record of 55,000 ft in his Lockheed Vega in December 1934. However, this was not a practical solution for the average passenger on board an airliner. The answer was to pressurize the entire passenger cabin of the airplane, so as to provide a shirtsleeve environment for the flight crew and passengers. The first airplane to incorporate this feature was a specially modified and structurally strengthened Lockheed 10E Electra for the Army Air Corps in 1937. Designated the XC-35 (it looked much like the Lockheed L-14 in Fig. 1.24), this airplane had a service ceiling of 32,000 ft. It was the forerunner of all the modern pressurized airliners of today.

Along with pressurization for the occupants, high-altitude aircraft needed "pressurization" for the engine. Engine power is nearly proportional to the atmospheric density; without assistance, engine power simply dropped too low at high altitudes, and this was the major *mechanical* obstacle to high-altitude flight. However, assistance came in the form of the supercharger, a mechanical pump that simply compressed the incoming air before it went into the engine manifold. Supercharger development was a high priority during the 1930s and 1940s; it was a major development program within NACA. All high-performance military aircraft during World War II were equipped with superchargers as a matter of necessity.

Finally, we mention an interesting development in aerodynamic design which took place during the era of the mature, propeller-driven airplane, but which was to have an unexpected impact well beyond that era. It has to do with the boundary layer on a surface in an airstream—that thin region adjacent to the surface where the mechanism of air friction is dominant. Ever since Ludwig Prandtl in Germany introduced the concept of the boundary layer in 1904, it has been recognized that two types of flow were possible—laminar flow and turbulent flow—in the boundary layer. Moreover, it was known that the friction drag is higher for a turbulent boundary layer than for a laminar boundary layer. Since mother nature always moves toward the state of maximum disorder, and turbulent flow is much more disorderly than

Figure 1.25   North American P-51 Mustang. First production airplane to use a laminar-flow wing.

laminar flow, about 99% of the boundary layer along the wings and fuselage of typical airplanes in flight is turbulent, creating high skin-friction drag. However, in the late 1930s, by means of proper design of the airfoil shape, NACA developed a series of *laminar-flow airfoils* that encouraged large regions of laminar flow and reduced airfoil drag by almost 50%. Such a laminar-flow wing was quickly adopted in 1940 for the design of the new North American P-51 Mustang (Fig. 1.25). However, in practice, these wings did not generate the expected large laminar flow. The NACA wind tunnel experiments were conducted under controlled conditions using models with highly polished surfaces. In contrast, P-51 wings were manufactured with standard surface finishes that were rougher than the almost jewellike wind tunnel models. Moreover, these wings were further scored and scratched in service. Roughened surfaces encourage turbulent flow; even insect smears on the wing can cause the flow to change from laminar to turbulent. Hence, in practice, the laminar-flow wing never created the large regions of laminar flow required to produce the desired low level of skin-friction drag. However, totally unexpectedly, the laminar-flow airfoil shape turned out to be a very good *high-speed* airfoil. It had a much higher critical Mach number than a conventional airfoil did, and hence it delayed the onset of compressibility problems encountered by many high-speed airplanes in the early 1940s. A technological development from the era of the mature, propeller-driven airplanes resulted instead in paving the way for the next era—the era of the jet-propelled airplane.

1.2.4   Era of the Jet-Propelled Airplane

Many types of aircraft gained fame during World War II. Typical of these were the Lockheed P-38 Lightning (Fig. 1.26) and the Republic P-47 Thunderbolt (Fig. 1.27) as well as the P-51 Mustang (Fig. 1.25). However, these aircraft exhibited no inherently new features compared to the mature, propeller-driven airplanes from the late 1930s; they were simply more refined and more powerful, with correspondingly improved performance. Indeed, virtually all the important airplanes that participated in World War II were *designed* well before the United States entered the war; this includes the aircraft in Figs. 1.25 to 1.27. In fact, just a few months before the U.S. entry

Figure 1.26   Lockheed P-38 Lightning.

Figure 1.27   Republic P-47 Thunderbolt.

into the war, development of new airplane types was frozen in order to concentrate on mass production of existing models. The situation was summed up by James H. Kindelberger, president of North American Aircraft, in a paper entitled "The Design of Military Aircraft" in the *Aeronautical Engineering Review*, December 1953: "As far as United States military aviation is concerned, the Second World War may be characterizied as a period of intensive design improvements and refinement rather than as a period of innovation." Clearly, the policy of the U.S. government relative to the aircraft industry in 1941 was to concentrate on the development of designs that already existed. On the other hand, the Germans and the British had a somewhat different perspective, out of which was born the jet-propelled airplane.

The invention of the jet engine caused a revolution in airplane design and performance. Besides the invention of the first practical airplane by the Wright brothers, no other technical development has had the impact on aeronautics as that of the jet engine. The conventional propeller-driven airplane had a major speed limitation; when the speed of the propeller tips approached or exceeded the speed of sound, shock waves would form at the tips, and propeller efficiency would plummet. Jet propulsion dispensed with the propeller and opened the way toward efficient transonic and supersonic flight.

The jet engine was invented independently by two people—Frank Whittle (now Sir Frank Whittle) in England and Dr. Hans von Ohain in Germany. In 1928, as a student at the Royal Air Force technical college at Cranwell, Frank Whittle wrote a senior thesis entitled "Future Developments in Aircraft Design" in which he expounded on the virtues of jet propulsion. It aroused little interest. Although Whittle patented his design for a gas-turbine aircraft engine in 1930, it was not until 5 years later that he formed, with the help of friends, a small company to work on jet engine development. Named Power Jets Ltd., this company was able to successfully bench-test a jet engine on April 12, 1937—the first jet engine in the world to successfully operate in a practical fashion. However, it was not the first to fly. Quite independently, and completely without the knowledge of Whittle's work, Dr. Hans von Ohain in Germany developed a similar gas-turbine engine. Working under the private support of the famous airplane designer Ernst Heinkel, von Ohain started his work in 1936. On August 27, 1939, a specially designed Heinkel airplane, the He 178, powered by von Ohain's jet engine, successfully flew—it was the first gas turbine-powered, jet-propelled airplane in history to fly. It was strictly an experimental airplane, but von Ohain's engine of 838 lb of thrust pushed the He 178 to a maximum speed of 360 mi/h. (Five days later, Germany invaded Poland, and World War II began.) It was not until almost 2 years later that a British jet flew. On May 15, 1941, the specially designed Gloster E.28/39 airplane took off from Cranwell, powered by a Whittle jet engine. It was the first to fly with a Whittle engine. With these first flights in Germany and Britain, the jet age had begun.

Although the He 178 had flown just before World War II, and the Gloster E.28/39 early in the war, it was not until 1944 that a jet aircraft was deployed in any numbers. That was the German Messerschmitt Me 262 (Fig. 1.28). The Me 262 first flew on July 18, 1942, but Hitler's attempt to convert the fighter to a vengeance bomber, problems with the turbojet engines, a deteriorating fuel situation, and Allied air attacks on the aircraft assembly factories combined to delay its appearance in combat until September 1944. Powered by two Junkers Jumo 004 turbojets with a thrust of 1984 lb each, the Me 262 had a maximum speed of 540 mi/h. In total, 1433 Me 262s were

Figure 1.28   Messerschmitt Me 262A, the first operational jet airplane, 1944.

produced before the end of the war. It was the first practical operational jet-propelled aircraft.

The era of jet-propelled aircraft is characterized by a number of design features unique to airplanes intended to fly near, at, or beyond the speed of sound. One of the most pivotal of these design features was the advent of the *swept wing*. For a subsonic airplane, sweeping the wing increases the airplane's critical Mach number, hence allowing it to fly closer to the speed of sound before encountering the large drag rise caused by the generation of shock waves somewhere on the surface of the wing. For a supersonic airplane, the wing sweep is designed such that the wing leading edge is inside the Mach cone from the nose of the fuselage; if this is the case, the component of airflow velocity perpendicular to the leading edge is subsonic (called a *subsonic leading edge*), and the resulting wave drag is not as severe as it would be if the wing were to lie outside the Mach cone. In the latter case, called the *supersonic leading edge*, the component of flow velocity perpendicular to the leading edge is supersonic, with an attendant strong shock wave generated at the leading edge. In either case, high subsonic or supersonic, an airplane with a swept wing will be able to fly faster than one with a straight wing, everything else being equal.

The concept of the swept wing for high-speed aircraft was first introduced in a public forum in 1935. At the fifth Volta Conference, convened on September 30, 1935, in Rome, Italy, the German aerodynamicist Adolf Busemann gave a paper in which he discussed the technical reasons why swept wings would have less drag at high speeds than conventional straight wings. Although several Americans were present, such as Eastman Jacobs from NACA and Theodore von Karman from Cal Tech, Busemann's idea went virtually unnoticed; it was not carried back to the United States with any sense of importance. Not so in Germany. One year after Busemann's presentation at the Volta Conference, the swept-wing concept was classified by the German Luftwaffe as a military secret. The Germans went on to produce a large bulk of swept-wing research, including extensive wind tunnel testing. They even designed a few prototype swept-wing jet aircraft. Many of these data were confiscated by the United States after the end of World War II, and made available to U.S. aircraft companies and government laboratories. Meanwhile, quite independently of this German research, Robert T. Jones, an NACA aerodynamicist, had worked out the elements of swept-wing theory toward the end of the war. Although not reinforced by definitive wind tunnel tests in the United States at that time, Jones's work served as a second source of information concerning the viability of swept wings.

In 1945, aeronautical engineers at North American Aircraft began the design of the XP-86 jet fighter; it had a straight wing. However, the XP-86 design was quickly changed to a swept-wing configuration when the German data, as well as some of the German engineers, became available after the war. The prototype XP-86 flew on October 1, 1947, and the first production P-86A flew with a 35° swept wing on May 18, 1948. Later designated the F-86, the swept-wing fighter had a top speed of 679 mi/h, essentially Mach 0.9—a stunning speed for that day. Shown in Fig. 1.29, the North American F-86 Sabre was the world's first successful operational swept-wing aircraft. (In this author's opinion, the F-86 is asthetically one of the most beautiful airplanes ever designed.)

Figure 1.29   North American F-86 Sabre, 1949.

Figure 1.30   Lockheed F-104 Starfighter, 1954. The first airplane designed for sustained flight above Mach 2.

By the time the F-86 was in operation, the sound barrier had already been broken. On October 14, 1947, Captain Charles (Chuck) Yeager became the first human being to fly faster than the speed of sound in the Bell X-1 rocket-powered airplane. Eight years later, in February 1954, the first fighter airplane capable of sustained flight at Mach 2, the Lockheed F-104 Starfighter, made its first appearance. The F-104 (Fig. 1.30) exhibited the best qualities of good supersonic aerodynamics—a sharp, pointed nose, slender fuselage, and extremely thin and sharp wings. The airfoil section on the F-104 is less then 4% thick (maximum thickness compared to the chord length). The wing leading edge is so sharp that protective measures must be taken by maintenance people working around the aircraft. The purpose of these features is to reduce the strength of shock waves at the nose and leading edges, hence reducing supersonic wave drag. The F-104 also had a straight wing with a very low aspect ratio rather than a swept wing. This exhibits an alternative to supersonic airplane designers; the wave drag on straight wings of *low aspect ratio* is comparable to that on swept wings with high aspect ratios. Of course, this low-aspect-ratio wing gives poor aerodynamic performance at subsonic speeds, but the F-104 was point-designed for maximum performance at Mach 2. (This is just another example of the many compromises embodied in airplane design.) With the F-104, supersonic flight became an almost everyday affair, not just the domain of research aircraft.

The delta wing concept was another innovation to come out of Germany during the 1930s and 1940s. In 1930, Dr. Alexander Lippisch designed a glider with a delta configuration; the leading edges were swept back by 20°. Of course, the idea had nothing to do with high-speed flight at that time; the delta configuration had some stability and control advantages associated with its favorable center-of-gravity location. However, when Busemann introduced his swept-wing ideas in 1935, Lippisch and his colleagues knew they had a potential high-speed wing in their delta configuration. Lippisch continued his resarch on delta wings during the war, using small models in German supersonic wind tunnels. By the end of the war, he was starting to design a delta wing ramjet-powered fighter. Along with the German swept-wing data, this delta wing technology was transferred to the United States after the war; it served as the basis for an extended wind tunnel test program on delta wings at NACA Langley Memorial Laboratory. All this long-term aerodynamic research in Germany on swept and delta wings ironically did not help Germany in the outcome of the war. However, as stated by Miller and Sawers (Ref. 14), it was a "gift to the victors."

The first practial delta wing aircraft was the Convair F-102 (Fig. 1.31). The design of this aircraft is an interesting story in its own right—a story of the interplay between design and research, and between industry and NACA. The F-102 was designed as a supersonic airplane. However, much to the embarrassment and frustration of the Convair engineers, the prototype F-102 being tested at Edwards Air Force Base during October 1953 and then again in January 1954 exhibited poor performance and was unable to go supersonic. At the same time, Richard Whitcomb at NACA Langley was conducting wind tunnel tests on his "area rule" concept, which called for the cross-sectional area of the fuselage to be reduced in the vicinity of the wing. By so doing, the transonic drag was substantially reduced. The Convair engineers quickly adopted this concept on a new prototype of the F-102, and it went supersonic on its second flight. The area-rule design feature is clearly seen in Fig. 1.31; at the top is the original design of the F-102, and at the bottom is the area-rule design. The "Coke bottle shape" of the fuselage caused by area rule is clearly evident. Convair went on to produce 975 F-102s; the practical delta wing airplane was finally a reality.

The area rule was one of the most important technical developments during the era of jet-propelled airplanes. Today, almost all transonic and supersonic aircraft incorporate some degree of area rule. For his work on the area rule, Whitcomb received the Collier Trophy, the highest award given in the field of aeronautics.

One of the most tragic stories in the annals of airplane design occurred in the early 1950s. Keying on England's early lead in jet propulsion, de Havilland Aircraft Company designed and flew the first commercial jet transport—the de Havilland Comet (Fig. 1.32). Powered by four de Havilland Ghost jet engines, the Comet carried 36 passengers for 2,000 mi at a speed of 460 mi/h, cruising at relatively high altitudes near or above 30,000 ft. The passenger cabin was pressurized; indeed, the Comet was the first pressurized airplane to fly for extended periods at such high altitudes. Inasmuch as good airplane design is an evolutionary process based on preceding aircraft, the de Havilland designers had little precedent on which to base the structural design of the pressurized fuselage. The Comet entered commercial service with BOAC (a forerunner of British Airways today) in 1952. However, in 1954, three Comets disintegrated in flight, and the airplane was quickly withdrawn from service. The problem was later found to be structural failure of the fuselage while pressurized. De Havilland used countersunk rivets in the construction of the

(a)

(b)

Figure 1.31 (a) Convair F-102, without area rule. (b) Convair F-102A, with area rule.

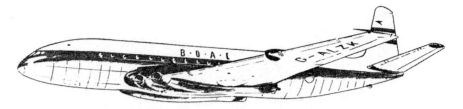

Figure 1.32   The de Havilland Comet 1, 1952, the first commerical jet transport. It was withdrawn from service in 1954 after three catastrophic in-flight disintegrations.

Figure 1.33   Boeing 707, 1958. The first successful commercial jet airliner.

Comet; reaming the holes for the rivets produced sharp edges. After a number of pressurization cycles, cracks in the fuselage began to propagate from these sharp edges, leading eventually to catastrophic failure. At the time, de Havilland had a massive lead over all other aircraft companies in the design of commercial jet aircraft. Moreover, while it was in service, the Comet was very popular with the flying public, and it was a money earner for BOAC. Had these failures not occurred, de Havilland and England might have become the world's supplier of commercial jet aircraft rather than Boeing and the United States. But it was not to be.

In 1952, the same year as the ill-fated de Havilland Comet went into service, the directors of Boeing Company made a bold and risky decision to privately finance and build a commercial jet prototype. Designated the model 367-80, or simply called the *Dash 80* by the Boeing people, the prototype first flew on July 15, 1954. It was a bold design which carried over to the commerical field Boeing's experience in building swept-wing jet bombers for the Air Force (the B-47 and later the B-52). Later renamed the Boeing 707, the first production series of aircraft were bought by Pan American Airlines and went into service in 1958. The Boeing 707 (Fig. 1.33), with its swept wings and podded engines mounted on pylons below the wings, set the standard design pattern for all future large commercial jets to present day. The design of the 707 was evolutionary because it stemmed from the earlier experience at Boeing with jet bombers. But it was almost revolutionary in the commercial field, because no airliner (not even the Comet) looked like it before. Boeing's risky gamble paid off, and it transformed a predominately military airplane company into the world's leader in the design and manufacture of commercial jet transports.

Boeing made another bold move on April 15, 1966, when the decision was made to "go for the big one." Boeing had lost the Air Force's C-5 competition to Lockheed; the C-5 at the time was the largest transport airplane in the world. Taking their losing design a few steps further, Boeing engineers conceived of the 747—the first wide-body commerical jet transport. Bill Allen, president of Boeing at that time, and Juan Trippe, president of Pan American Airlines, shared the belief that the large, wide-body airplane offered economic advantages for the future airline passenger market, and they both jointly made the decision to pursue the project. This was an even bolder decision than that concerning the 707. In the words of the authoritative aeronautical historian James Hansen (Ref. 15),

> In the opinion of many experts, the 747 was the greatest gamble in the history of the aircraft business. At risk were the lives of both companies, as well as the solvency of several private lending institutions.
>   Financed with private money, if the 747 had failed, half the banks west of the Mississippi would have been badly shaken. Another important meaning of big aircraft is thus clear; big dollars go along with them.

The gamble paid off. The Boeing 747 (Fig. 1.34) first flew in February 1969, and it entered service for the first time in January 1970 on Pan American's New York–London route. At the time of this writing, some 25 years later, 747s are still being produced by Boeing.

The 747 set the design standard for all subsequent wide-body transports. It has done much more. It opened the opportunity for huge numbers of people to fly quickly and relatively cheaply across oceans, and to travel to all parts of the globe. The 747 has had a tremendous sociological impact. It has brought people of various nations closer to one another. It has fostered the image of the "global village." It has had a direct impact on society, business, and diplomacy in the last third of the twentieth century. It is a wonderful example of the extent to which airplane design can favorably mold and influence society in general.

Examine Figs. 1.33 and 1.34; here we see examples of subsonic and transonic commercial airplane designs that are a major part of the era of the jet-propelled

Figure 1.34   Boeing 747, 1970.

aircraft. But what about commercial transportation at *supersonic* speeds? In the 1960s this question was addressed in Russia, the United States, England, and France. The Tupolev Design Bureau in Russia rushed a supersonic transport design into production and service. The Tu-144 supersonic transport first flew on December 31, 1968. More than a dozen of these aircraft were built, but none entered extended service, presumably due to unspecified problems. One Tu-144 was destroyed in a dramatic accident at the 1973 Paris Air Show. In the United States, the government orchestrated a design competition for a supersonic transport; the Boeing 2707 was the winner in December 1966. The design turned into a nightmare for Boeing. For 2 years, a variable-sweep wing supersonic transport (SST) configuration was pursued, and then the design was junked. Starting at the beginning again in 1969, the design was caught up in an upward spiral of increased weight and development costs. When the predictions for final development costs hit about $5 billion, Congress stepped in and refused to appropriate any more funds. In May 1971, the SST development program in the United States was terminated. Only in England and France was the SST concept carried to fruition.

The first, and so far only, supersonic commercial transport to see long-term, regular service is the Anglo-French Concorde (Fig. 1.35). In 1960 both the British and French independently initiated design studies for a supersonic transport. It quickly became apparent that the technical complexities and financial costs were beyond the abilities of either country to shoulder alone. Hence, on November 29, 1962, England and France signed a formal treaty aimed at the design and construction of a supersonic transport. (By the way, this reality is becoming more and more a part of modern airplane design; when certain projects exceed the capability of a given company or even a given country, the practical solution is sometimes found in national or international consortia. It might be worthwhile for future airplane designers in the United States to learn to speak French, German, or Japanese.) The product of this treaty was the Aerospatiale–British Aerospace Corporation's Concorde. Designed to cruise at Mach 2.2 carrying 125 passengers, the Concorde first flew on March 2, 1969. It first exceeded Mach 1 on October 1, 1969, and Mach 2 on November 4, 1970. Originally, orders for 74 Concordes were anticipated. However, when the airlines were expected to place orders in 1973, the world was deep in the energy crises. The skyrocketing costs of aviation jet fuel wiped out any hope of an economic return from flying the Concorde, and no orders were placed. Only the national airlines of France and Britain, Air France and British Airways, went ahead, each signing up for

Figure 1.35   The Concorde supersonic transport, 1972.

seven aircraft after considerable pressure from their respective governments. After a long development program, the Concorde went into service on January 21, 1976. In the final analysis, the Concorde was a technical, if not financial, success. It has been in regular service since 1976. It represents an almost revolutionary (rather than evolutionary) airplane design in that no such aircraft existed before it. However, the Concorde designers were not operating in a vacuum. Examining Fig. 1.35, we see a supersonic configuration which incorporates good supersonic aerodynamics— a sharp-nosed slender fuselage and a cranked delta wing with a thin airfoil. The Concorde designers had at least 15 years of military airplane design experience with such features to draw upon. Today, we know that any future second-generation SST will have to be economical in service and environmentally acceptable. The design of such a vehicle is one of the great challenges in aeronautics. Perhaps some of the readers of this book will someday play a part in meeting this challenge.

Today, we are still in the era of the jet-propelled airplane, and we will be there for the indefinite future. The evolution of this era can be seen at a glance just by flipping through Figs. 1.28 to 1.35. Here we see subsonic jet planes, some with straight wings and others with swept wings, all with high aspect ratios. We also see supersonic jet planes, some with straight wings and others with delta wings, all with low aspect ratios. In their time, the designs of all these airplanes were driven by the quest for speed and altitude, mitigated in some cases by the realities of economic and environmental constraints. In the future, we will continue the quest for speed and altitude, while at the same time these constraints (and possibly others) will become even more imposing. In the process, the challenges to be faced by future airplane designers will only become more interesting.

1.3   UNCONVENTIONAL DESIGNS (INNOVATIVE CONCEPTS)

We end this chapter with a mention, albeit brief, of the design of certain aircraft that do not "fit the mold" of previous, conventional airplanes, that is, unconventional airplane designs. Section 1.2 focused on airplanes that set the standard for airplane design— airplanes that came to be accepted as representative of the *conventional airplane*. However, this is not to downgrade the importance of unconventional thinking for the design of new aircraft that look different and/or fly differently. A case can be made that George Cayley's concept of what today we call the modern configuration airplane (Fig. 1.6) was, in its time, quite "unconventional" when viewed against the panorama of flapping-wing ornithopter concepts that preceded it. For this reason, we might also entitle this section "Innovative Concepts," because most unconventional designs are derived from innovative thinking.

Airplanes that take off and land *vertically* are unconventional airplanes. Any such airplane is classified as a *vertical-takeoff-and-landing* (VTOL) airplane. (We are considering fixed-wing VTOL airplanes here, not helicopters, which are a completely different consideration.) One of the best examples of a successful VTOL airplane, one that has been in continuous service since the 1970s, is the Harrier jet propelled fighter aircraft, shown in Fig. 1.36. The Harrier is a British design; first conceived by Hawker Aircraft, a prototype called the P-1227 Kestrel first flew in 1960. Later, the production version, called the Harrier, was built in numbers for the Royal Air Force and the Royal Navy. A version of the Harrier, the AV-8, was adapted and manufactured by McDonnell–Douglas in the United States in the early 1980s, and it is in service with the U.S. Marine Corps. There are many approaches to providing the vertical thrust for a VTOL craft. In the case of the Harrier, the jet exhaust from the single Rolls-Royce Pegasus jet engine passes through four nozzles, two located on each side of the engine. Vanes in these nozzles deflect the exhaust in the downward, vertical direction for vertical takeoff and landing, and in the horizontal, backward direction for conventional, forward flight.

Another unconventional airplane concept is the *flying wing*. From a purely aerodynamic viewpoint, a fuselage is mainly a drag-producing element of the airplane; its lift-to-drag ratio is much smaller than that of a wing. Hence, if the whole airplane were simply one big wing, the maximum aerodynamic efficiency could be achieved. The idea for such flying wings is not new. For example, the famous airplane designer Jack Northrop began working with flying-wing designs in the early 1930s. During and just after World War II, Northrop built several flying-wing bombers. A photograph of one, the YB-49 jet bomber, is shown in Fig. 1.37. However, the longitudinal stability and control normally provided by the horizontal tail and elevator at the end of a fuselage of a conventional airplane must instead be provided by flaps and unusual curvature of the camber line near the trailing edge of the flying wing. This caused stability and control problems for flying-wing aircraft—problems severe enough that no practical flying wings were produced until recently. In the modern aeronautical engineering of today, airplanes *can* be designed to be unstable, and the airplane is flown with the aid of a computer that is constantly deflecting the control surfaces to keep the airplane on its intended flight path—the fly-by-wire concept. Such new

Figure 1.36   The Hawker Siddeley Harrier, 1969, the first production
vertical-takeoff-and-landing airplane.

Figure 1.37 Northrop YB-49 flying-wing bomber, 1948. (Courtesy of Northrop-Grumman Corporation.)

flight management systems now make practical the design and operation of flying wings. One spectacular modern example, the B-2, is discussed next.

A class of highly unconventional aircraft has come on the scene in recent years, namely, *stealth aircraft*. Here, the primary design objective is to have the smallest radar cross section possible, in order to make the airplane virtually invisible on any enemy's radar screen. Two modern stealth airplanes are shown in Figs. 1.38 and 1.39, the Northrop B-2 and Lockheed F-117, respectively. Look at these aircraft—you see configurations with sharp edges and flat angled surfaces, all designed to reflect radar waves *away* from the source rather than back toward it. Moreover, these airplanes are made of special radar-absorbing material. The design features you see in Figs. 1.38 and 1.39 are dictated mainly by radar reflection considerations, and not by aerodynamic considerations. Good subsonic aerodnamic design is embodied by rounded leading edges, smoothly curving surfaces, and slender, streamlined shapes. You do not see these features in the B-2 and F-117. Here is an extreme example of the compromises that always face airplane designers. The overriding design concern for these stealth aircraft is very low radar cross section; good aerodynamics had to take a back seat. Sometimes these airplanes are jokingly referred to as "airplanes designed by electrical engineers." This is not far from the truth. However, the fact that both the B-2 and the F-117 have acceptable aerodynamic performance implies that the aeronautical engineer faced up to and partially solved a very difficult problem—that of integrating the electrical engineering features with the aeronautical engineering features to produce an effective flying machine. Finally, in reference to the previous

Figure 1.38   Northrop B-2 Stealth bomber. (Courtesy of Northrop-Grumman Corporation.)

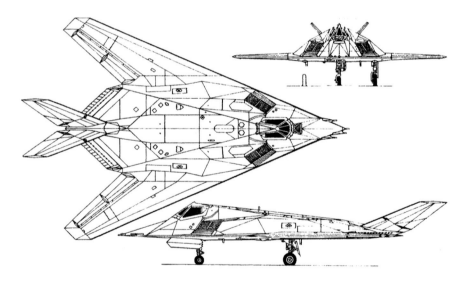

Figure 1.39   Lockheed F-117 Stealth fighter.

paragraph, note that the B-2 is indeed a flying wing, made possible by the advanced fly-by-wire technology of today.

There are many other unconventional concepts for airplanes, too numerous for us to treat in any detail. For example, since the 1930s, the concept of a combined automobile and airplane—the autoplane—has come and gone several times, without

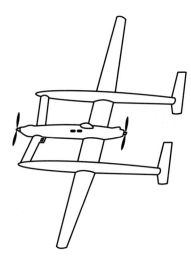

Figure 1.40   The *Voyager,* the first airplane to fly around the world without refueling, 1986.

any real success. Another idea, one that has been relatively successful, is the ultralight airplane—essentially an overgrown kite or parafoil, with a chair for the pilot and an engine equivalent to that of a lawn mower for power. These ultralights are currently one of the latest rages at the time of this writing. Another concept, not quite as unconventional, is the *uninhabited air vehicle* (UAV), an updated label for what used to be called a *remotely piloted vehicle* (RPV). For the most part, these UAVs are essentially overgrown model airplanes, although some recent UAV designs for high-altitude surveillance are large aircraft with very high-aspect-ratio wings, and wingspans on the order of 80 ft. And then there are airplanes that are so narrowly point-designed that they are good for only one thing, and this makes such airplanes somewhat unconventional. A case in point is the *Voyager* designed by Burt Rutan, and flown by Dick Rutan and Jeana Yeager in their record nonstop flight around the world, finishing on December 23, 1986. The *Voyager* is shown in Fig. 1.40; the airplane you see here is a somewhat unconventional configuration for a somewhat unconventional purpose.

1.4 SUMMARY AND THE FUTURE

With all the previous discussion in mind, return to Fig. 1.3, showing the *Wright Flyer* on its way toward historic destiny. That flight took place less than 100 years ago—a scant speck in the whole time line of recorded history. The exponential growth of aeronautical technology that has taken place since 1903 is evident just by leafing through the remaining figures in this chapter. In retrospect, the only adjective that can properly describe this progress is *mind-boggling*. Indeed, there are those who describe aeronautics as a *mature* technology today. This may be so, but just as a mature person is in the best position to decide his or her own future destiny, the mature aeronautical technology of today is in its best position ever to determine its destiny in the twenty-first century. I envy the readers of this book who will influence and guide this destiny.

A case in point is hypersonic flight. During the late 1980s and early 1990s, work on hypersonic airplanes was vigorously carried out in several countries, including the United States. The U.S. effort was focused on the concept of an *aerospace plane*, an aircraft that would take off from a normal runway as a normal airplane would, and then accelerate to near-orbital speeds within the atmosphere, using air-breathing propulsion (in this case, supersonic combustion ramjet engines). In the United States, this work was intended to produce an experimental hypersonic flight vehicle, the X-30 (Fig. 1.41). Although much technical progress was made during this design effort, the program floundered because of the projected enormous cost to bring it to the actual flight vehicle stage. However, in this author's opinion, this hiatus is just temporary. If the history of flight has told us anything, it has shown us that aeronautics has always been paced by the concept of faster and higher. Although this has to be somewhat mitigated today by the need for economically viable and environmentally safe airplanes, the overall march of progress in aeronautics will continue to be faster and higher. In some sense, practical, everyday hypersonic flight may be viewed as the final frontier of aeronautics. This author feels that most young readers of this book will see, in their lifetime in the twenty-first century, much pioneering progress toward this final frontier.

And hypersonic flight is not the only challenge for the future. As long as civilization as we know it today continues to exist in the world, we will always design and build new and improved airplanes for all the regimes of flight—low-speed, subsonic,

Figure 1.41   Artist's sketch of the X-30, a transatmospheric hypersonic vehicle.

transonic, supersonic, and hypersonic. From our viewpoint at the end of the twentieth century, we see unlimited progress and opportunities in the enhancement of airplane performance and design in the twenty-first century, and you will be in a position to be part of this action.

The stage is now set for the remainder of this book. The principles of airplane performance and design discussed in the following chapters will give you a better appreciation of past airplane designs, an understanding of present designs, and a window into future designs. If you are interested in obtaining such appreciation and understanding, and if you are anxious to jump through the window into the future, simply read on.

chapter
2

Aerodynamics of the Airplane: The Drag Polar

The results which we reach by practical flying experiments will depend most of all upon the shapes which we give to the wings used in experimenting. Therefore, there is probably no more important subject in the technics of flying than that which refers to wing formation.

Otto Lilienthal, Berlin, 1896

2.1 INTRODUCTION

Without aerodynamics, airplanes could not fly, birds could not get off the ground, and windmills would never work. Thus, in considering the performance and design of airplanes, it is no surprise that aerodynamics is a vital aspect. That is why this chapter is devoted exclusively to aerodynamics. Our purpose here is *not* to give a short course in aerodynamics; rather, only those aspects of aerodynamics necessary for our subsequent consideration of airplane performance and design are reviewed and discussed. Moreover, an understanding of what constitutes "good aerodynamics" is central to the design of good airplanes. In the following sections we discuss the lift and drag of various components of the airplane, as well as the overall lift and drag of the complete vehicle. We emphasize the philosophy that good aerodynamics is primarily derived from *low drag*; it is generally not hard to design a surface to give the requisite amount of lift, and the challenge is to obtain this lift with as small a drag as possible. A barn door at the angle of attack will produce a lot of lift, but it also produces a lot of drag—this is why we do not fly around on barn doors.

2.2   THE SOURCE OF AERODYNAMIC FORCE

Grab hold of this book with both hands, and lift it into the air. You are exerting a force on this book, and the force is being communicated to the book because your hands are in direct contact with the cover of the book. Similarly, the aerodynamic force exerted on a body immersed in an airflow is due to the two hands of nature which are in direct contact with the surface of the body; these two hands of nature are the pressure and shear stress distributions acting all over the exposed surface of the body. The pressure and shear stress distributions exerted on the surface of an airfoil due to the airflow over the body are sketched qualitatively in Fig. 2.1; pressure acts locally perpendicular to the surface, and shear stress acts locally parallel to the surface. The net aerodynamic force on the body is due to the pressure and shear stress distributions *integrated* over the total exposed surface area. Let us make this idea more quantitative. Let point A be any point on the surface of the body in Fig. 2.2. Let $\mathbf{n}$ and $\mathbf{k}$ be unit vectors normal and tangent, respectively, to the surface at point A, as shown in Fig. 2.2; also let dS be an infinitesimally small segment of surface area surrounding point A. If p and τ are the local pressure and shear stress at point A, then the resultant aerodynamic force $\mathbf{R}$ on the body can be written as

$$\mathbf{R} = -\iint\limits_{S} p\mathbf{n}\,dS + \iint\limits_{S} \tau\mathbf{k}\,dS \qquad \textbf{[2.1]}$$

Force due   Force due
to pressure   to friction

The two hands of nature that grab the body

where the integrals in Eq. (2.1) are *surface integrals*. It is always useful to keep in mind that, no matter how complex the flow may be, or what the shape of the body may be, the *only* two sources of aerodynamic force felt by the body are the integral of the perssure over the surface and the integral of the shear stress over the surface; that is, the first and second terms, respectively, in Eq. (2.1).

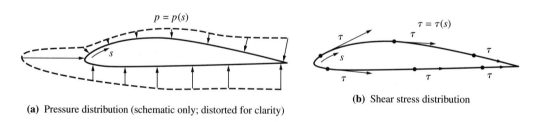

$p = p(s)$

(a) Pressure distribution (schematic only; distorted for clarity)

$\tau = \tau(s)$

(b) Shear stress distribution

Figure 2.1   **(a)** Schematic of the pressure distribution over an airfoil. *Note:* The relative magnitudes of the pressure, signified by the length of each arrow, are distorted in this sketch for the sake of clarity. In reality, for low speed subsonic flight, the minimum pressure is usually only a few percent below the freestream pressure. **(b)** Shear stress distribution.

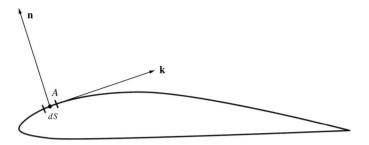

Figure 2.2   Sketch of the unit vectors.

2.3   AERODYNAMIC LIFT, DRAG, AND MOMENTS

Consider the body sketched in Fig. 2.3, oriented at an angle of attack α to the free-stream direction. The free-stream velocity is denoted by V_∞ and is frequently called the *relative wind*. The resultant aerodynamic force **R**, given by Eq. (2.1), is inclined rearward from the vertical, as shown in Fig. 2.3. (Note that, in general, **R** is *not* perpendicular to the chord line. There were several investigators during the nineteenth century who assumed erroneously that the resultant force was perpendicular to the chord; some definitive measurements by Otto Lilienthal published in 1890 were the first to prove that such an assumption was wrong.) By definition, the component of **R** perpendicular to the free-stream velocity is the *lift L*, and the component of **R** parallel to the free-stream direction is the *drag, D*.

For the body shown in Fig. 2.3, imagine that you place an axis perpendicular to the page at any arbitrary point on the body. Just for the sake of discussion, we choose the point one-quarter of the distance behind the leading edge, measured along the chord line, as shown in Fig. 2.4**a**. This point is called the *quarter-chord point*; there is nothing inherently magic about this choice—we could just as well choose any other point on the body. Now imagine that the axis perpendicular to the page through the point is rigidly attached to the body, and that you suspend the body in an airstream, holding the axis with your hand. Due to the pressure and shear stress distributed all over the surface of the body, there will be a tendency for the axis to *twist* in your hand; that is, there will be in general a *moment* about the axis. (See chapter 1 of Ref. 16 for the integral expressions due to pressure and shear stress which create this moment.) In this case, since the axis is located at the quarter-chord point, we call such a moment the *moment about the quarter chord, $M_{c/4}$*. If we had chosen instead to put the axis at the leading edge, as shown in Fig. 2.4**b**, then we would still feel a twisting action, but it would be a different magnitude from above. In this case, we would experience the *moment about the leading edge M_{LE}*; even though the surface pressure and shear stress distributions are the same

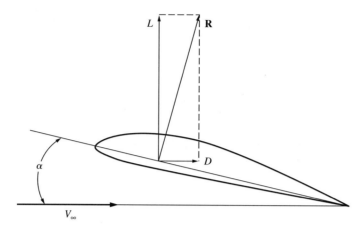

Figure 2.3   Lift, drag, and resultant aerodynamic force.

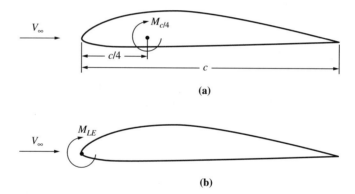

Figure 2.4   (**a**) Moment about the quarter-chord point. (**b**) Moment about the leading edge.

for parts (**a**) and (**b**) in Fig. 2.4, M_{LE} is different from $M_{c/4}$ simply because we have chosen a different point about which to take the moments. *Important:* By convention, a moment which tends to rotate the body so as to increase the angle of attack is considered *positive*. The moments shown in Fig. 2.4 are drawn in the positive sense; that is, they tend to pitch the nose upward. Depending on the shape of the body, the moments can be either positive or negative. (In reality, for the positively cambered airfoil shown in Fig. 2.4, the moments will be pitch-down moments; that is, $M_{c/4}$ and M_{LE} will be negative values and will act in the opposite direction from that shown in Fig. 2.4.)

Question: At what point on the body do the lift and drag act? For example, when we draw the lift and drag forces on a body such as in Fig. 2.3, through what point on the body should we draw these forces? To address this question, we note that the two hands of nature which grab the body—the pressure and shear stress distributions acting over the surface—are *distributed loads* which are impressed over the whole surface, such as sketched in Fig. 2.1. The *net effect* of these distributed loads is the production of the resultant concentrated force **R**, and hence the lift and drag, as shown in Fig. 2.3. In other words, the distributed loads create an aerodynamic force on the body, and we can *picture* this force as *equivalent* to a *single concentrated force vector* **R**, applied at a *point* on the body (as if nature were touching the body with only one finger at that point, instead of grabbing it all over the complete surface, as happens in reality).

This leads back to our original question: Through what point on the body should the single concentrated force **R** be drawn? One obvious answer would be to plot the distributed load on graph paper and find the *centroid* of this load, just as you would find the centroid of an area from integral calculus. The centroid of the distributed load on the body is the point through which the equivalent concentrated force acts. This point is called the *center of pressure*. The complete mechanical effect of the distributed aerodynamic load over the body can be exactly represented by the resultant force **R** (or equivalently the lift and drag) acting through the center of pressure. This is illustrated in Fig. 2.5. The actual distributed load is sketched in Fig. 2.5a. The mechanical effect of this distributed load is equal to the resultant lift and drag acting through the center of pressure, denoted by c.p., as shown in Fig. 2.5b. If we were to place an axis perpendicular to the page going through the center of pressure, there would be *no moment* about the axis. Hence, an alternate definition of the center of pressure is that point on the body about which the moment is zero.

However, we do not have to end here. Once we accept that the mechanical effect of the distributed load can be exactly represented by a concentrated force acting at the center of pressure, then we know from the principles of statics that the concentrated load can be shifted to any other part on the body, as long as we also specify the moments about that other point. For example, in Fig. 2.5c the lift and drag are shown acting through the quarter-chord point, with a moment acting about the quarter-chord point, namely, $M_{c/4}$. The mechanical effect of the distributed load in Fig. 2.5a can be exactly represented by a concentrated force acting at the quarter-chord point along with the specification of the moment about that point. Yet another choice might be to draw the lift and drag acting through the leading edge along with a specification of the moment about the leading edge, as sketched in Fig. 2.5d.

In summary, all four sketches shown in Fig. 2.5 are equivalent and proper representations of the same mechanical effect. Therefore, you should feel comfortable using any of them. For example, in airplane dynamics, the center of pressure is rarely used because it shifts when the angle of attack is changed. Instead, for airfoil aerodynamics, the concentrated force is frequently drawn at the quarter-chord point. Another choice frequently made is to apply the concentrated force at the aerodynamic center, a point on the body which we will define shortly.

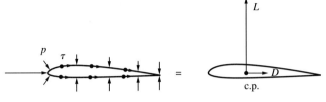

(a) Distributed load

(b) Concentrated force acting
through the center of pressure

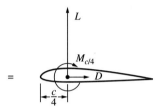

(c) Concentrated force acting through
the quarter-chord point, plus the
moment about the quarter-chord point

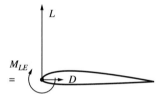

(d) Concentrated force acting through
the leading edge, plus the moment about
the leading edge

Figure 2.5 Three ways of representing the actual distributed load exerted by
pressure and shear stress on the surface of the airfoil by a
concentrated force at a point and the moment at that point.

Example 2.1

For a given set of free-stream conditions and angle of attack, the lift per unit span for a given airfoil is 200 pounds per foot (lb/ft). The location of the center of pressure is at $0.3c$, where c is the chord length; $c = 5$ ft. The force and moment system on the airfoil can be shown as sketched in Fig. 2.5**b**, namely, the aerodynamic force of 200 lb acting through the center of pressure, which is located at $0.3c$, with no moment about the center of pressure (by definition). What would the equivalent force and moment system be if the lift were placed at the quarter-chord point? At the leading edge? (Assume that the line of action of the drag D is close enough to the quarter-chord and leading-edge points that any moment about these points due to drag is negligible.)

Solution

Consider the quarter-chord point. The moment about the quarter-chord point is given by the lift acting through the center of pressure, with the moment arm $0.3c - 0.25c = 0.05c$. This moment is

$$M_{c/4} = -(0.05c)(200) = -10c = -10(5) = -50 \text{ ft·lb per unit span}$$

Note: Moments which cause a pitch-down motion (a decrease in angle of attack) are, by convention, negative. For the case above, the moment about the quarter-chord point causes a pitch-down action; hence it is a negative moment. The direction of this negative moment is the opposite direction from the arrow shown in Fig. 2.5c; that is, the moment calculated here is in the counterclockwise direction.

The equivalent force and moment system is shown in Fig. 2.5c, where the lift of 200 lb acts through the quarter-chord point, and a moment equal to -50 ft·lb exists about the quarter-chord point.

Consider the leading edge point. The moment about the leading edge point is given by the lift acting through the center of pressure, with the moment arm $0.3c$. This moment is

$$M_{LE} = -(0.3)(200) = -60c = -60(5) = -300 \text{ ft·lb per unit span}$$

The equivalent force and moment system is shown in Fig. 2.5**d**, where the lift of 200 lb acts through the leading edge point and a moment equal to -300 ft·lb exists about the leading edge point.

2.4 AERODYNAMIC COEFFICIENTS

The aerodynamic characteristics of a body are more fundamentally described by the force and moment *coefficients* than by the actual forces and moments themselves. Let us explain why.

Intuition, if nothing else, tells us that the aerodynamic force on a body depends on the velocity of the body through the air V_∞, the density of the ambient air ρ_∞, the size of the body, which we will denote by an appropriate reference area S, and the orientation of the body relative to the free-stream direction, for example, the angle of attack α. (Clearly, if we change the velocity, the aerodynamic force should change. Also, the force on a body moving at 100 feet per second (ft/s) through air is going to be smaller than the force on the same body moving at 100 ft/s through water, which is nearly a thousand times denser than air. Also, the aerodynamic force on a sphere of 1-inch (1-in) diameter is going to be smaller than that for a sphere of 1-ft diameter, everything else being equal. Finally, the force on a wing will clearly depend on how much the wing is inclined to the flow. All these are simply commonsense items.) Moreover, since friction accounts for part of the aerodynamic force, the force should depend on the ambient coefficient of viscosity μ_∞. Not quite so intuitive, but important nonetheless, is the compressibility of the medium through which the body moves. A measure of the compressibility of a fluid is the speed of sound in the fluid a_∞; the higher the compressibility, the lower the speed of sound. Hence, we can readily state the following relations for lift, drag, and moments of a body of *given shape:*

$$L = L(\rho_\infty, V_\infty, S, \alpha, \mu_\infty, a_\infty) \qquad \text{[2.2a]}$$

$$D = D(\rho_\infty, V_\infty, S, \alpha, \mu_\infty, a_\infty) \qquad \text{[2.2b]}$$

$$M = M(\rho_\infty, V_\infty, S, \alpha, \mu_\infty, a_\infty) \qquad \text{[2.2c]}$$

With the above in mind, let us go through the following thought experiment. Assume we want to find out *how* the lift on a given body varies with the parameters given in Eq. (2.2a). We could first run a series of wind tunnel tests in which the velocity is varied and everything else is kept the same. From this, we would obtain a stack of wind tunnel data from which we could extract how L varies with V_∞. Then we could run a second series of wind tunnel tests in which the density is changed and everything else is kept the same. From this we would obtain a second stack of wind tunnel data from which we could correlate L with ρ_∞. We could continue to run wind tunnel tests, varying in turn each one of the other parameters on the right-hand side of Eq. (2.2a), and obtain more stacks of wind tunnel data. When finished, we would have six separate stacks of wind tunnel data to correlate in order to find out how the lift varies for the given aerodynamic shape. This could be very time-consuming, and moreover, the large amount of wind tunnel time could be quite costly. However, there is a better way. Let us define the lift, drag, and moment *coefficients* for a given body, denoted by C_L, C_D, and C_M, respectively, as follows:

$$C_L = \frac{L}{q_\infty S} \qquad \text{[2.3]}$$

$$C_D = \frac{D}{q_\infty S} \qquad \text{[2.4]}$$

$$C_M = \frac{M}{q_\infty S c} \qquad \text{[2.5]}$$

where q_∞ is the *dynamic pressure*, defined as

$$q_\infty = \frac{1}{2}\rho V_\infty^2 \qquad \text{[2.6]}$$

and c is a characteristic length of a body (for an airfoil, the usual choice for c is the chord length). Let us define the following similarity parameters:

$$\text{Reynolds number (based on chord length): Re} = \frac{\rho_\infty V_\infty c}{\mu_\infty} \qquad \text{[2.7]}$$

$$\text{Mach number: } M_\infty = \frac{V_\infty}{a_\infty} \qquad \text{[2.8]}$$

The method of dimensional analysis—a very powerful and elegant approach used to identify governing nondimensional parameters in a physical problem—leads to the following result. For the give body *shape*, we have

$$C_L = f_1(\alpha, \text{ Re}, M_\infty) \qquad \textbf{[2.9a]}$$

$$C_D = f_2(\alpha, \text{ Re}, M_\infty) \qquad \textbf{[2.9b]}$$

$$C_M = f_3(\alpha, \text{ Re}, M_\infty) \qquad \textbf{[2.9c]}$$

[See chapter 5 of Ref. 3 and chaper 1 of Ref. 16 for a discussion of dimensional analysis and a derivation of Eqs. (2.9a) to (2.9c).] These results from dimensional analysis greatly simplify things for us. For example, let us once again go through our thought experiment in which we want to find out how lift on a given body varies. However, this time, in light of the results from Eq. (2.9a), we use the *lift coefficient*, not the lift itself, as the primary item. We could run a series of wind tunnel tests in which we obtain the lift coefficient as a function of α, keeping Re and M_∞ constant. In so doing, we would obtain a stack of wind tunnel data. Then we could run a second series of tests where Re is varied, keeping α and M_∞ constant. This would give us a second stack of wind tunnel data. Finally, we could run a third series of wind tunnel tests in which M_∞ is varied, keeping α and Re constant. This would give us a third stack of data. With only these three stacks of data, we could find out how C_L varies. This is a tremendous savings in time and money over our previous thought experiment, in which we generated six stacks of data to find out how L varies.

The above thought experiment is only one aspect of the value of C_L, C_D, and C_M. They have a more fundamental value, as follows. Take Eq. (2.9a), for example. This relationship shows that lift coefficient is a function of the angle of attack, Reynolds number, and Mach number. Imagine that we have a given body at a given angle of attack in a given flow, where ρ_∞, V_∞, μ_∞, and a_∞ are certain values. Let us call this the "green" flow. Now, consider another body of the same geometric shape (but not the same size) in another flow where ρ_∞, V_∞, μ_∞, and a_∞ are all different; let us call this flow the "red" flow.

Dimensional analysis, from Eq. (2.9a), tells us that even though the green flow and the red flow are two different flows, *if* the Reynolds number and the Mach number are the *same* for these two different flows, then the *lift coefficient* will be the same for the two geometrically similar bodies at the same angle of attack. If this is the case, then the two flows, the green flow and the red flow, are called *dynamically similiar flows*. This is powerful stuff! The essence of practical wind tunnel testing is built on the concept of dynamically similar flows. Say we want to obtain the lift, drag, and moment coefficients for the Boeing 747 flying at an altitude of 30,000 ft with a Mach number of 0.8. If we place a small-scale model of the Boeing 747 in a wind tunnel at the same angle of attack as the real airplane in flight, and if the flow conditions in the test section of the wind tunnel are such that the Reynolds number and Mach number are the same as for the real airplane in actual flight, then the lift, drag, and moment coefficients measured in the wind tunnel will be exactly the same values as those for the full-scale airplane in free flight. This principle has been a driving force in the design of wind tunnels. The ideal wind tunnel is one in which the proper Reynolds and Mach numbers corresponding to actual flight are simulated. This is frequently very difficult to achieve; hence most wind tunnel designs focus on the proper simulation of either one or the other—the simulation of either the high Reynolds numbers associated with flight or the proper Mach numbers. This is why most new airplane designs are tested in more than one wind tunnel.

A comment is in order regarding the reference area S in Eqs. (2.3) to (2.5). This is nothing other than just a *reference* area, suitably chosen for the definition of the force and moment coefficients. Beginning students in aerodynamics frequently want to think that S should be the total wetted area of the airplane. (Wetted area is the actual surface area of the material making up the skin of the airplane—it is the total surface area that is in actual contact with, i.e., wetted by, the fluid in which the body is immersed.) Indeed, the wetted surface area is the surface on which the pressure and shear stress distributions are acting; hence it is a meaningful geometric quantity when one is discussing aerodynamic force. However, the wetted surface area is not easily calculated, especially for complex body shapes. In contrast, it is much easier to calculate the *planform* area of a wing, that is, the projected area that we see when we look down on the wing. For this reason, for wings as well as entire airplanes, the *wing planform area* is usually used as S in the definitions of C_L, C_D, and C_M from Eqs. (2.3) to (2.5). Similarly, if we are considering the lift and drag of a cone, or some other slender, missile like body, then the reference area S in Eqs. (2.3) to (2.5) is frequently taken as the *base area* of the body. The point here is that S in Eqs. (2.3) to (2.5) is simply a reference area that can be arbitrarily specified. This is done primarily for convenience. Whether we take for S the planform area, base area, or any other area germane to a given body shape, it is still a measure of the relative size of different bodies which are geometrically similar. And what is important in the definition of C_L, C_D, and C_M is to divide out the effect of size via the definitions given by Eqs. (2.3) to (2.5). The moral to this story is as follows: *Whenever you take data for C_L, C_D, or C_M from the technical literature, make certain that you know what geometric reference area was used for S in the definitions and then use that same defined area when making calculations involving those coefficients.*

Example 2.2	The Boeing 777 (Fig. 1.2) has a wing planform area of 4605 square feet (ft²). (**a**) Assuming a takeoff weight of 506,000 lb and a takeoff velocity of 160 mi/h, calculate the lift coefficient at takeoff for standard sea-level conditions. (**b**) Compare the above result with the lift coefficient for cruise at Mach number 0.83 at 30,000 ft, assuming the same weight.

Solution

(**a**) For steady, level flight, the weight is equal to the lift. Hence, from Eq. (2.3),

$$C_L = \frac{L}{q_\infty S} = \frac{W}{q_\infty S}$$

The velocity must be expressed in consistent units. Since 60 mi/h = 88 ft/s (a convenient factor to remember),

$$V_\infty = 160 \left(\frac{88}{60} \right) = 234.7 \text{ ft/s}$$

From Appendix B, at standard sea level, $\rho_\infty = 0.002377$ slug per cubic foot (slug/ft^3), so

$$q_\infty = \frac{1}{2}\rho_\infty V_\infty^2 = \frac{1}{2}(0.002377)(234.7)^2 = 65.45 \text{ lb/ft}^2$$

Thus,

$$C_L = \frac{W}{q_\infty S} = \frac{506,000}{(65.45)(4,605)} = \boxed{1.68}$$

(b) At 30,000 ft, from Appendix B, $\rho_\infty = 8.907 \times 10^{-4}$ slug/ft^3 and $T_\infty = 411.86°R$. The speed of sound is

$$a_\infty = \sqrt{\gamma R T} = \sqrt{(1.4)(1,716)(411.86)} = 994.7 \text{ ft/s}$$

$$V_\infty = a_\infty M_\infty = (994.7)(0.83) = 825.6 \text{ ft/s}$$

$$q_\infty = \frac{1}{2}\rho_\infty V_\infty^2 = \frac{1}{2}(8.907 \times 10^{-4})(825.6)^2 = 303.56 \text{ lb/ft}^2$$

$$C_L = \frac{W}{q_\infty S} = \frac{506,000}{(303.56)(4,605)} = \boxed{0.362}$$

Note: The lift coefficient at the much higher cruise velocity is much smaller than that at takeoff, even though the density at 30,000 ft is smaller than that at sea level. It is sometimes convenient to think that the lift at high speeds is mainly obtained from the high dynamic pressure; hence only a small lift coefficient is required. In turn, at low speeds the dynamic pressure is lower, and in order to keep the lift equal to the weight in steady, level flight, the low dynamic pressure must be compensated by a high lift coefficient.

DESIGN CAMEO

As we will discuss in subsequent sections, the lift coefficient for a given aerodynamic shape is an intrinsic value of the shape itself, the inclination of the body to the free-stream direction (the angle of attack), the Mach number, and the Reynolds number. This intrinsic value has nothing to do with the weight of the body or its reference area. For example, it is common to calculate or measure the variation of lift coefficient for a given aerodynamic shape as a function of the angle of attack (for given Mach and Reynolds numbers). When we calculate the value of C_L which is necessary for flight of a given vehicle at a given weight, speed, and altitude, as in Example 2.2, then we hope that such a value of C_L lies within the intrinsic values associated with the vehicle shape; that is, we hope that the necessary value of C_L can be obtained at some reasonable angle of attack for the vehicle. This is not guaranteed; and if such a required C_L cannot be obtained, the design characteristics or design performance envelope for the flight vehicle must be modified. This is not a problem at high speeds, where the value of C_L is low and is readily obtainable. However, it can be a problem at the low speeds associated with takeoff or landing, where the required value of C_L is large. As we will see, this can have a major impact on airplane design, driving the designer to incorporate high-lift devices (flaps, slats, etc.) which "artificially" increase C_L beyond the intrinsic values

(*continued*)

for the basic vehicle shape. (Such high-lift devices are discussed in Chapter 5.) Also, the required high values of C_L at low speeds will influence the designer's choice of wing area for the airplane, because the required values of C_L can be reduced by increasing the wing area. However, a greater wing area may adversely affect other design characteristics of the airplane.

Also, in airplane design, the value of C_L does not always stand alone, a consideration by itself. In Example 2.2, we calculated that a value of $C_L = 0.362$ was required at the given high-speed condition. For the same condition, it is usually desirable for the drag coefficient C_D to be as small as possible. What is usually germane is the *lift-to-drag ratio* $L/D = C_L/C_D$. This ratio is a measure of the aerodynamic efficiency of a flight vehicle; the higher the value of L/D, the more pounds of lift are obtained per pound of drag. We will see that the design of an airplane for high-speed cruise is driven by a consideration of the highest possible L/D rather than a specific value of C_L by itself.

2.5   LIFT, DRAG, AND MOMENT COEFFICIENTS: HOW THEY VARY

Equations (2.9a) to (2.9c) indicate that C_L, C_D, and C_M for a given aerodynamic body shape vary with the angle of attack, Reynolds number, and Mach number. *Question:* What are these variations? There is no pat answer; first and foremost, the answer depends on the shape of the body itself. Whole volumes have been written about this question. In particular, the two books by Hoerner, one on drag (Ref. 17) and the other on lift (Ref. 18), have taken on the aura of Bibles in applied aerodynamics. They contain a wealth of information and data on aerodynamic coefficients for a wide variety of shapes. It is recommended that you own copies of these two books. Our purpose in this section is to address the question in a limited fashion, just to illustrate some typical variations, in order to give you a "feel" for the matter.

First, let us consider a conventional airfoil shape with positive camber (airfoil shape arched upward), such as the NACA 2412 airfoil shown at the top of Fig. 2.6**b**. This is a two-dimensional body, and it is customary in the aerodynamic literature to write the lift, drag, and moment coefficients for such two-dimensional shapes in lowercase letters, namely, c_l, c_d, and c_m, respectively. The variation of these coefficients with the angle of attack α and Reynolds number Re is shown in Fig. 2.6 for the NACA 2412 airfoil. These are actual experimental data obtained by NACA in the early 1940s in a specially designed wind tunnel for measuring airfoil properties (see the historical note in Section 2.11). Figure 2.6**a** is an answer to how the lift coefficient c_l and the moment coefficient taken about the quarter-chord point $c_{m_{c/4}}$ vary with α and Re.

First, consider the variation of the lift coefficient with angle of attack, as shown in Fig. 2.6**a**. Note that the curve of c_l versus α (the "lift curve") has the generic form as sketched in Fig. 2.7. Of greatest importance, the variation of c_l is essentially *linear* with α over most of the practical range of the angle of attack. The slope of this linear portion is called the *lift slope* and is designated by a_0. For thin airfoils, a theoretical value for the lift slope is 2π per radian, or 0.11 per degree. For most conventional

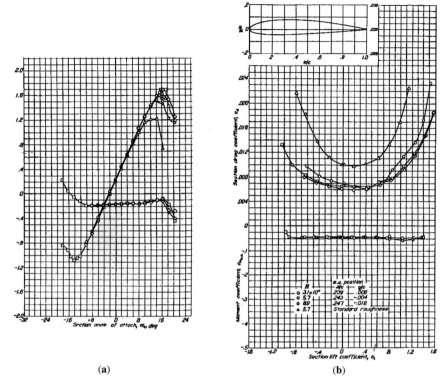

(a) (b)

Figure 2.6 Data for the NACA 2412 airfoil. (**a**) Lift coefficient and moment coefficient about the quarter-chord versus angle of attack. (**b**) Drag coefficient and moment coefficient about the aerodynamic center as a function of the lift coefficient. (From Abbott and von Doenhoff, Ref. 19.)

airfoils, the experimentally measured lift slopes are very close to the theoretical values. The experimental value for the lift slope for the NACA 2412 airfoil is easily measured from the data given in Fig. 2.6a; this author measures a value of $a_0 = 0.105$ (try it yourself). From the generic lift curve shown in Fig. 2.7, note that there is a finite value of c_l at zero angle of attack, and that the airfoil must be pitched down to some negative angle of attack for the lift to be zero. This angle of attack is denoted by $\alpha_{L=0}$ in Fig. 2.7; for the NACA 2412 airfoil, the data in Fig. 2.6a show that $\alpha_{L=0} = -2.2°$. All positively cambered airfoils have negative zero-lift angles of attack. In contrast, a symmetric airfoil has $\alpha_{L=0} = 0°$, and a negatively cambered airfoil (such as the NACA 2412 airfoil turned upside down) has a positive $\alpha_{L=0}$. (Negatively cambered airfoils are of little practical interest in aeronautics.)

At the other extreme, at high angles of attack, the lift coefficient becomes nonlinear, reaches a maximum value denoted by $(c_l)_{max}$ in Fig. 2.7, and then drops as α is further increased. The reason for this drop in c_l at high α is that flow separation

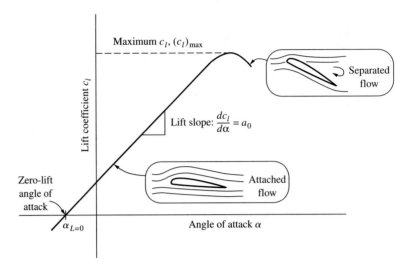

Figure 2.7   Sketch of a generic lift curve.

occurs over the top surface of the airfoil and the lift decreases (sometimes precipitously). In this condition, the airfoil is said to be *stalled*. In contrast, over the linear portion of the lift curve, the flow is attached over most of the airfoil surface. These two phenomena—attached and separated flow—are shown schematically for the appropriate portions of the lift curve in Fig. 2.7. (In the early part of the twentieth century, the great German aerodynamicist Ludwig Prandtl labeled attached and separated flows as "healthy" and "unhealthy" flows, respectively—an apt description.)

The variation of c_l with the Reynolds number is also shown in Fig. 2.6a. Note that the data are given for three different values of Re ranging from 3.1×10^6 to 8.9×10^6 (the code key for the different Reynolds numbers is given at the bottom of Fig. 2.6b). The data in Fig. 2.6a show virtually no effect of the Reynolds number on the linear portion of the lift curve; that is, $a_0 = dc_l/d\alpha$ is essentially insensitive to variations in Re. (This is true for the high Reynolds numbers associated with normal flight; however, at much lower Reynolds numbers, say, 100,000 encountered by model airplanes and many small uninhabited aerial vehicles, there is a substantial Re effect that reduces the lift slope below its high Reynolds number value.) On the other hand, the data in Fig. 2.6a show an important Reynolds number effect on $(c_l)_{max}$, with higher values of $(c_l)_{max}$ corresponding to higher Reynolds numbers. This should be no surprise. The Reynolds number is a similarity parameter in aerodynamics which governs the nature of *viscous flow*. The development of separated flow over the airfoil at high α is a viscous flow effect—the viscous boundary layer literally separates from the surface. Hence we would expect the value of $(c_l)_{max}$ to be sensitive to Re; such a sensitivity is clearly seen in Fig. 2.6a.

The variation of $c_{m_{c/4}}$ with α and Re is also shown in Fig. 2.6a. The angle-of-attack variation is sketched generically in Fig. 2.8. Note that the moment coefficient

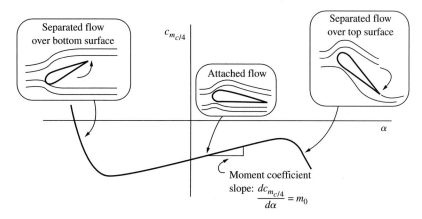

Figure 2.8   Sketch of a generic moment curve.

curve is essentially linear over most of the practical range of the angle of attack; that is, the slope of the moment coefficient curve, $m_0 = dc_{m_{c/4}}/d\alpha$ is essentially constant. This slope is positive for some airfoils (as shown here), but can be negative for other airfoils. The variation becomes nonlinear at high angle of attack, when the flow separates from the top surface of the airfoil, and at low, highly negative angles of attack, when the flow separates from the bottom surface of the airfoil. As was shown in the case of the lift curve, the linear portion of the moment curve is essentially independent of Re.

The variation of c_d with the lift coefficient is shown in Fig. 2.6**b**. Since the lift coefficient is a linear function of the angle of attack, you could just as well imagine that the abscissa in Fig. 2.6**b** could be α instead of c_l, and the shape of the drag curve would be the same. Hence, the generic variation of c_d with α is as shown in Fig. 2.9. For a cambered airfoil, such as the NACA 2412 airfoil, the minimum value $(c_d)_{\min}$ does not necessarily occur at zero angle of attack, but rather at some finite but small angle of attack. For the NACA 2412 airfoil considered in Fig. 2.6, the value of $(c_d)_{\min}$ for a Reynolds number of 8.9×10^6 is 0.006, and it occurs at an angle of attack of $-0.5°$ (i.e., the minimum value of c_d from Fig. 2.6**b** occurs at $c_l = 0.2$, which from Fig. 2.6**a** occurs at an angle of attack of $-0.5°$). The drag curve in Figs. 2.6**b** and 2.9 shows a very flat minimum—the drag coefficient is at or near its minimum value for a range of angle of attack varying from $-2°$ to $+2°$. For this angle-of-attack range, the drag is due to friction drag and pressure drag. In contrast, the rapid increase in c_d which occurs at higher values of α is due to the increasing region of separated flow over the airfoil, which creates a large pressure drag.

The variation of c_d with Reynolds number is also shown in Fig. 2.6**b**. Basic viscous flow theory and experiments show that the local skin-friction coefficient c_f on a surface, say, for a flat plate, varies as $c_f \propto 1/\sqrt{\text{Re}}$ for laminar flow, and approximately $c_f \propto 1/(\text{Re})^{0.2}$ for turbulent flow (see, e.g., Refs. 3 and 16). Hence, it is no surprise that $(c_d)_{\min}$ in Fig. 2.6**b** is sensitive to Reynolds number and is larger at the lower Reynolds numbers. Moreover, the Reynolds number influences the extent and characteristics of the separated flow region, and hence it is no surprise that c_d at the larger values of α is also sensitive to the Reynolds number.

Also shown in Fig. 2.6**b** is the variation of the moment coefficient about the aerodynamic center. By definition, the aerodynamic center is that point on the airfoil about which the moment is *independent of the angle of attack*. We discuss the concept of the aerodynamic center in greater detail in Section 2.6. However, note that, true to its definition, the experimentally measured value of $c_{m_{\text{a.c.}}}$ in Fig. 2.6**b** is essentially constant over the range-of-lift coefficient (hence constant over the range of angle of attack).

Returning to Eqs. (2.9a) to (2.9c), we note that the aerodynamic coefficients are a function of Mach number also. The data in Fig. 2.6 do not give us any information on the Mach number effect; indeed, these data were measured in a low-speed subsonic wind tunnel (the NACA Langley two-dimensional low-turbulence pressure tunnel) which had maximum velocities ranging from 300 mi/h when operated at one atmosphere (atm) and 160 mi/h when operated at 10 atm. Hence, the data in Fig. 2.6 are essentially incompressible flow data. *Question:* How do the aerodynamic coefficients vary when the free-stream Mach number M_∞ is increased to higher subsonic speeds and then into the supersonic regime? For a conventional airfoil, the generic variations of c_l and c_d with M_∞ are sketched in Figs. 2.10 and 2.11, respectively. Consider first the variation of c_l as shown in Fig. 2.10. At subsonic speeds, the "compressibility effects" associated with increasing M_∞ result in a progressive increase in c_l. The reason for this can be seen by recalling that the lift is mainly due to the pressure distribution on the surface. As M_∞ increases, the differences in pressure from one point to another on the surface become more pronounced. Hence, c_l increases as M_∞ increases. The Prandtl–Glauert rule, the first and simplest (and also the least accurate) of the several formulas for subsonic "compressibility corrections," predicts that c_l will rise inversely proportional to $\sqrt{1 - M_\infty^2}$ (see Refs. 3 and 16). Assuming an incompressible value of $c_l = 2\pi\alpha$ (the theoretical result for a flat plate in inviscid flow), the dashed line in the subsonic region of Fig. 2.10 shows the theoretical Prandtl–Glauert variation. In the supersonic region of Fig. 2.10, the dashed curve shows the theoretical supersonic variation for a thin airfoil, where $c_l = 4\alpha/\sqrt{M_\infty^2 - 1}$ (see Ref. 16). The solid curve illustrates a generic variation of c_l versus M_∞ for both the

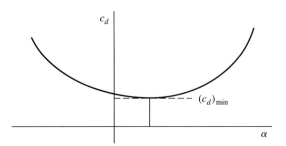

Figure 2.9   Sketch of a generic drag curve.

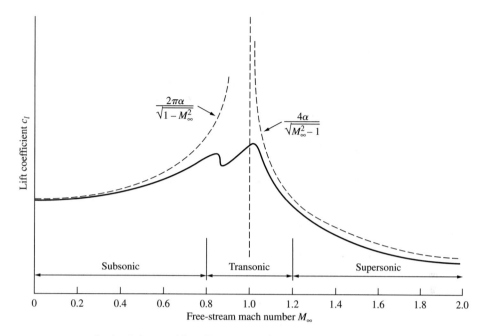

Figure 2.10   Sketch of a generic lift coefficient variation with Mach number.

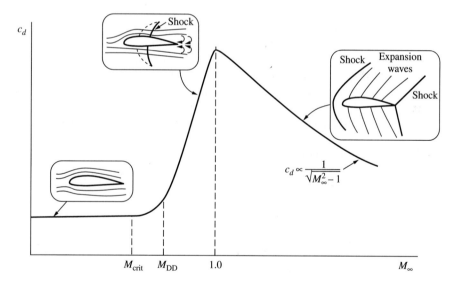

Figure 2.11   Sketch of a generic drag coefficient variation with Mach number.

subsonic and supersonic regions. The oscillatory variation of c_l near Mach 1 is typical of the transonic regime, and is due to the shock wave–boundary layer interaction that is prominant for transonic Mach numbers. Actual measurements made by NACA for the subsonic behavior of c_l versus Mach number for the NACA 2315 airfoil are shown in Fig. 2.12.

The generic variation of c_d with M_∞ is sketched in Fig. 2.11. Here, in contrast to c_l which increases with M_∞, c_d stays relatively constant with M_∞ up to, and slightly beyond, the critical Mach number—that free-stream Mach number at which sonic flow is first encountered at some location on the airfoil. The drag in the subsonic region is mainly due to friction, and the "compressibility effect" on friction in the subsonic regime is small. (In reality, the skin-friction drag coefficient *decreases* slightly as M_∞ increases, but we are ignoring this small effect.) The flow over the airfoil in this regime is smooth and attached, with no shock waves present, as sketched at the left in Fig. 2.11. As M_∞ increases above M_{crit}, a large pocket of locally supersonic flow forms above, and sometimes also below, the airfoil. These pockets of supersonic flow are terminated at the downstream end by shock waves. The presence of these shocks, by themselves, will affect the pressure distribution in such a fashion as to cause an increase in pressure drag (this drag increase is related to the loss of total pressure across the shock waves). However, the dominant effect is that the shock wave interacts with the boundary layer on the surface, causing the boundary layer to

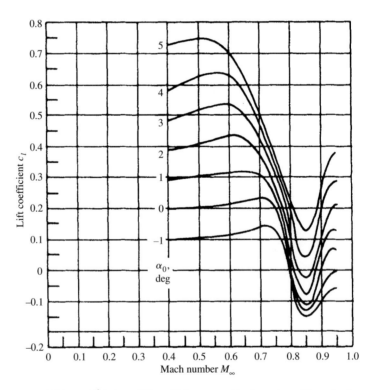

Figure 2.12   Variation of lift coefficient versus Mach number with angle of attack as a parameter for an NACA 2315 airfoil. (Wind tunnel measurements were taken at NACA Langley Memorial Laboratory.)

separate from the surface, and hence greatly increasing the pressure drag. This type of flow field is illustrated in the middle of Fig. 2.11; it is characteristic of transonic flows. As a result, the drag skyrockets in the transonic regime, as sketched in Fig. 2.11. This rapid divergence of drag occurs at a value of M_∞ slightly larger than M_{crit}; the free-stream Mach number at which this divergence occurs is called the *drag-divergence Mach number* $M_{drag\ div}$. Finally, in the supersonic regime, c_d gradually decreases, following approximately the variation $c_d \propto 1/\sqrt{M_\infty^2 - 1}$ (see Ref. 16). Actual measurements made by NACA for the subsonic and transonic behavior of c_d versus Mach number for the NACA 2315 airfoil are shown in Fig. 2.13. Note the drastic increase in c_d as Mach 1 is approached.

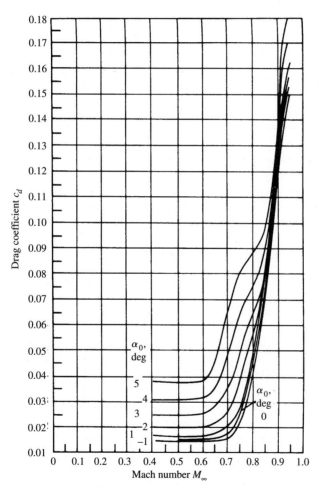

Figure 2.13   Variation of drag coefficient versus Mach number with angle of attack as a parameter for an NACA 2315 airfoil. (Wind tunnel measurements were made at NACA Langley Memorial Laboratory.)

Since the moment on the airfoil is due mainly to the surface pressure distribution, the variation of c_m with Mach number will qualitatively resemble the variation of c_l shown in Fig. 2.10; hence no more details will be given here.

Although a two-dimensional airfoil has been used in this section to illustrate the variation of the aerodynamic coefficients with α, Re, and M_∞, these results are qualitatively representative of the variations of C_L, C_D, and C_M for three-dimensional aerodynamic bodies. We will discuss the aerodynamics of three-dimensional shapes in subsequent sections. Also, the book by Abbott and Von Doenhoff (Ref. 19) is the definitive source of NACA airfoil data; it is recommended that you own a copy.

Example 2.3

For the NACA 2412 airfoil in Fig. 2.6, calculate the lift-to-drag ratios at $\alpha = 0°$, $6°$, and $12°$. Assume Re $= 8.9 \times 10^6$.

Solution

From Fig. 2.6**a**, at $\alpha = 0°$, $c_l = 0.25$. From Fig. 2.6**b**, at $c_l = 0.25$, we have $c_d = 0.006$. Hence,

$$\frac{c_l}{c_d} = \frac{0.25}{0.006} = \boxed{41.7} \qquad \text{at } \alpha = 0°$$

For $\alpha = 6°$, $c_l = 0.85$ and $c_d = 0.0076$. Hence,

$$\frac{c_l}{c_d} = \frac{0.85}{0.0076} = \boxed{111.8} \qquad \alpha = 6°$$

For $\alpha = 12°$, $c_l = 1.22$ and $c_d = 0.0112$. Hence,

$$\frac{c_l}{c_d} = \frac{1.22}{0.0112} = \boxed{109} \qquad \alpha = 12°$$

Note: (1) The values of L/D first increase with an increase in α, reach a maximum value at some angle of attack (in this case, somewhere between $6°$ and $12°$), and then decrease as α is further increased.

(2) The values of L/D for *airfoils* can exceed 100, as shown here. This is a large L/D ratio. However, for finite wings and complete airplane configurations, the maximum values of L/D are much smaller, typically in the range of 10 to 20, for reasons to be discussed later.

2.6   THE AERODYNAMIC CENTER

We have already defined the aerodynamic center as that point on a body about which the moments are independent of the angle of attack; that is, $C_{M_{\text{a.c.}}}$ is constant over the practical range of angle of attack. At first thought, such a concept intuitively seems strange. How can such a point exist, and how can it be found? We address these questions in this section.

Consider the front portion of an airfoil sketched in Fig. 2.14. We choose the lift and moment system on the airfoil to be specified by L and $M_{c/4}$ acting at the quarter-chord location, as shown in Fig. 2.14. (Recall from Section 2.3 and Fig. 2.5 that the resultant aerodynamic force can be visualized as acting through any point

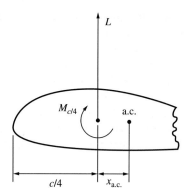

Figure 2.14 Schematic for finding the location of the aerodynamic center.

on the airfoil, as long as the corresponding moment about that point is also given.) The choice of the quarter-chord point for the application of lift in Fig. 2.14 is purely arbitrary—we could just as well choose any other point.

Does the aerodynamic center, as defined above, exist? For the time being, we assume its existence and denote its location on the airfoil by the fixed point labeled a.c. in Fig. 2.14. This point is located a distance $x_{\text{a.c.}}$ from the quarter-chord. Taking moments about the point a.c., we have

$$M_{\text{a.c.}} = Lx_{\text{a.c.}} + M_{c/4} \qquad \textbf{[2.10]}$$

Dividing Eq. (2.10) by $q_\infty Sc$, we have

$$\frac{M_{\text{a.c.}}}{q_\infty Sc} = \frac{L}{q_\infty S}\left(\frac{x_{\text{a.c.}}}{c}\right) + \frac{M_{c/4}}{q_\infty Sc}$$

or

$$c_{m_{\text{a.c.}}} = c_l \left(\frac{x_{\text{a.c.}}}{c}\right) + c_{m_{c/4}} \qquad \textbf{[2.11]}$$

Differentiating Eq. (2.11) with respect to angle of attack α gives

$$\frac{dc_{m_{\text{a.c.}}}}{d\alpha} = \frac{dc_l}{d\alpha}\left(\frac{x_{\text{a.c.}}}{c}\right) + \frac{dc_{m_{c/4}}}{d\alpha} \qquad \textbf{[2.12]}$$

Note that in Eq. (2.12) we are treating $x_{\text{a.c.}}$ as a *fixed* point on the airfoil, defined as that point about which moments are independent of the angle of attack. If such a fixed point does exist, it should be consistent with Eq. (2.12) where the derivative $dc_{m_{\text{a.c.}}}/d\alpha$ is set equal to zero (since $c_{m_{\text{a.c.}}}$ is constant with α, by definition of the aerodynamic center). In this case, Eq. (2.12) becomes

$$0 = \frac{dc_l}{d\alpha}\left(\frac{x_{\text{a.c.}}}{c}\right) + \frac{dc_{m_{c/4}}}{d\alpha} \qquad \textbf{[2.13]}$$

In Section 2.5, we saw how $dc_l/d\alpha$ and $dc_{m_{c/4}}/d\alpha$ are constant over the linear portions of the lift and moment curves; we denoted these constants by a_0 and m_0, respectively. Solving Eq. (2.13) for $x_{a.c.}/c$ yields

$$\frac{x_{a.c.}}{c} = -\frac{dc_{m_{c/4}}/d\alpha}{dc_l/d\alpha} = -\frac{m_0}{a_0} \qquad \textbf{[2.14]}$$

Hence, Eq. (2.14) proves that, for a body with linear lift and moment curves, where m_0 and a_0 are fixed values, the aerodynamic center does exist as a fixed point on the airfoil. Moreover, Eq. (2.14) allows the calculation of this point.

Example 2.4

For the NACA 2412 airfoil, calculate the location of the aerodynamic center.

Solution

From the airfoil data shown in Fig. 2.6a, we can find a_0 and m_0 as follows. First, examining the lift coefficient curve in Fig. 2.6a, we can read off the following data:

$$\text{At } \alpha = -8° \quad c_l = -0.6 \qquad \text{at } \alpha = 8° \quad c_l = 1.08$$

Hence,

$$a_0 = \frac{dc_l}{d\alpha} = \frac{1.08 - (-0.6)}{8° - (-8°)} = 0.105$$

Examining the moment coefficient curve in Fig. 2.6a, we can read off the following data:

$$\text{At } \alpha = -8° \quad c_{m_{c/4}} = -0.045 \qquad \text{at } \alpha = 10° \quad c_{m_{c/4}} = -0.035$$

Hence,

$$m_0 = \frac{dc_{mc/4}}{d\alpha} = \frac{-0.035 - (-0.045)}{10° - (-8°)} = 5.56 \times 10^{-4}$$

Thus, from Eq. (2.14),

$$\frac{x_{a.c.}}{c} = -\frac{5.56 \times 10^{-4}}{0.105} = \boxed{-0.0053}$$

Reflecting on Fig. 2.14, we see that the aerodynamic center is located 0.53% of the chord length ahead of the quarter-chord point. This is very close to the quarter-chord point itself. Moreover, this result agrees exactly with the measured value quoted on page 183 of Abbott and Von Doenhoff (Ref. 19).

The result of Example 2.4 is not uncommon. For most standard airfoil shapes, the aerodynamic center is quite close to the quarter-chord point. Indeed, the results of thin airfoil theory (see, e.g., Ref. 16) predict that, for a cambered airfoil, the quarter-chord point *is* the aerodynamic center.

2.7 NACA AIRFOIL NOMENCLATURE

Today, when new airplanes are designed, the shape of the airfoil section for the wings is usually custom-made. Most aircraft manufacturers have a stable of aerodynamic computer programs which allow them to customize the airfoil shape to specific design needs. In contrast, before the age of computers and computational aerodynamics, the aircraft industry relied primarily on series of airfoils empirically designed and tested by government agencies, such as the Royal Aircraft Establishment (RAE) in Britain and the National Advisory Committee for Aeronautics (NACA) in the United States. The work by these agencies in the period between 1920 and 1960 resulted in many families of "standard airfoils" from which the designer could choose. Many of these standard airfoils are used on airplanes still flying today, and the airfoils continue to provide a convenient selection for the designer who does not have the time or availability of the modern computer programs for custom-designing airfoil shapes. In particular, the many NACA families of airfoils have seen worldwide use. Because of the continued importance of the NACA standard airfoil designs, and the wide extent to which they have been used, it is worthwhile to discuss the appropriate NACA nomenclature for these airfoils. This is the purpose of this section.

Prior to 1930, an airfoil design was customized and personalized, with very little consistent rationale. There was no systematic approach or uniformity among the various designers and organizations in Europe or in the United States. This situation changed dramatically in the 1930s when NACA adopted a rational approach to the design of airfoils and carried out exhaustive and systematic wind tunnel measurements of the airfoil properties. The history of airfoil development is discussed in chapter 5 of Ref. 3; some additional historical comments are made in Section 2.11 of this book.

The NACA contributions started with the simple definition of airfoil geometric properties. These are sketched in Fig. 2.15. The major design feature of an airfoil is the *mean camber line*, which is the locus of points halfway between the upper and lower surfaces, as measured perpendicular to the mean camber line itself. The most

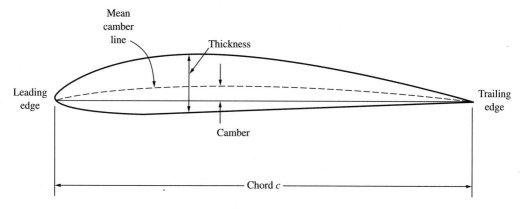

Figure 2.15   Airfoil nomenclature.

forward and rearward points of the mean camber line are the *leading* and *trailing edges*, respectively. The straight line connecting the leading and trailing edges is the *chord line* of the airfoil, and the precise distance from the leading to the trailing edge measured along the chord line is simply designated the *chord* of the airfoil, denoted by c. The *camber* is the maximum distance between the mean camber line and the chord line, measured perpendicular to the chord line. The camber, the shape of the mean camber line, and, to a lesser extent, the thickness distribution of the airfoil essentially control the lift and moment characteristics of the airfoil. The mean camber line is sketched in Fig. 2.16**a**. Then a thickness distribution is designed, which by itself leads to a symmetric shape, as sketched in Fig. 2.16**b**. The shape of the airfoil itself is essentially built up by designating first the shape of the mean camber line, which can be given as an analytic equation or simply as a tabulated set of coordinates. Then the ordinates of the top and bottom airfoil surfaces are obtained by superimposing the thickness distribution on the mean camber line, as shown in Fig. 2.16**c**; that is, the thickness distribution is laid perpendicular to the mean camber line. In this fashion, the final airfoil shape is obtained.

The first family of standard NACA airfoils was derived in the early 1930s. This was the four-digit series, of which the NACA 2412 airfoil shown in Fig. 2.6 was a member. The numbers in the designation mean the following: The first digit gives the maximum camber in percentage of chord. The second digit is the location of the maximum camber in tenths of chord, measured from the leading edge. The last two digits give the maximum thickness in percentage of chord. For example, the NACA 2412 airfoil has a maximum camber of 2% of the chord (or $0.02c$), located at $0.4c$ from the leading edge. The maximum thickness is 12% of the chord (or $0.12c$). In the four-digit series, a symmetric airfoil is designated by zeros in the first two digits; for example, the NACA 0012 airfoil is a symmetric airfoil with 12% thickness. The shapes of the NACA 2412 and 0012 airfoils are shown in Fig. 2.17**a** and **b**, respectively.

In the middle 1930s, the second family of NACA airfoils was developed, the five-digit series. This series was designed with the location of maximum camber closer to the leading edge than was the case for the four-digit series; it had been determined that the maximum lift coefficient increased as the maximum camber location was shifted forward. A typical NACA five-digit airfoil is the NACA 23012, shown in Fig. 2.17**c**. The numbers mean the following: The first digit, when multiplied by 3/2, gives the design lift coefficient in tenths (the design lift coefficient is defined and discussed in a subsequent paragraph). The design lift coefficient is an index of the

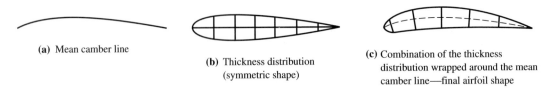

(a) Mean camber line

(b) Thickness distribution
(symmetric shape)

(c) Combination of the thickness
distribution wrapped around the mean
camber line—final airfoil shape

Figure 2.16   Buildup of an airfoil profile.

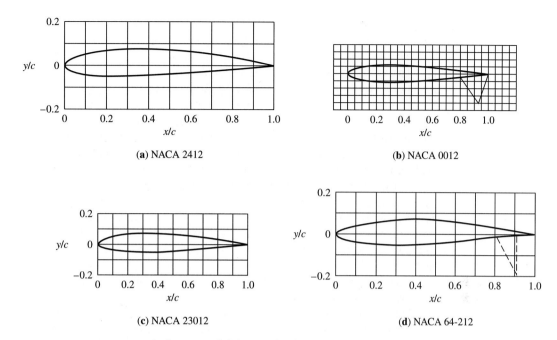

Figure 2.17   Various standard NACA airfoil shapes, all with 12% thickness.

amount of camber; the higher the camber, the higher the design lift coefficient. The second and third digits together are a number which, when multiplied by one-half, gives the location of maximum camber relative to the leading edge in percentage of chord. The last two digits give the maximum thickness in percentage of chord. For example, the NACA 23012 airfoil has a design lift coefficient of 0.3, the location of maximum camber at 15% of the chord (or 0.15c) from the leading edge, and a maximum thickness of 12% of the chord (or 0.12c).

During the late 1930s and early 1940s, NACA developed a series of airfoils designed to encourage laminar flow with the hope of reducing the skin-friction drag. The most successful of these laminar-flow airfoils was the 6-series sections. A typical NACA 6-series airfoil is the NACA 64-212 airfoil, shown in Fig. 2.17**d**. The numbers mean the following: The first digit is simply the series designation. The second digit is the location of the minimum pressure, in tenths of chord behind the leading edge, for the basic symmetric section at zero lift. (Recall that it is this symmetric thickness section which is wrapped around the mean camber line to generate the final airfoil shape. In the NACA numbering system for the 6-series airfoils, the second digit gives the location of the minimum pressure point on a symmetric airfoil with this thickness distribution and at zero angle of attack, rather than the minimum pressure point for the actual 6-series airfoil itself.) The third digit gives the design lift coefficient in tenths. The last two digits, as usual, give the maximum thickness in percentage of chord. For example, the NACA 64-212 airfoil is a member of the 6-series airfoils with a minimum pressure point (for the symmetric thickness distribution at zero angle of attack) at $0.4c$ from the leading edge. Its design lift coefficient is 0.2, and the maximum thickness is 12% of the chord (or $0.12c$).

The variation of c_d with c_l (or α) for the laminar-flow airfoils deserves some attention. The drag coefficient for the NACA 64-212 airfoil is shown in Fig. 2.18. Compare this with the drag curve shown in Fig. 2.6**b** for the NACA 2412. Note that, by comparison, the c_d curve for the NACA 64-212 airfoil has a dip located in the range of low angle of attack, resulting in a considerably lower minimum drag coefficient (0.004 for the NACA 64-212 compared to 0.006 for the NACA 2412). This dip in the drag curve is frequently called the *drag bucket*; all the 6-series airfoils exhibit a drag bucket. Clearly, in the NACA wind tunnel tests, the laminar-flow feature operated as planned and resulted in a 33% reduction in minimum drag coefficient. Because of this stunning improvement, many high-performance aircraft have utilized NACA 6-series airfoils. The first aircraft to use an NACA laminar-flow airfoil was the North American P-51 Mustang of World War II fame, shown in Fig. 1.25. [There is some controversy, due to lack of documentation, as to specifically *which* NACA laminar-flow section was used on the P-51. A later version, the P-51H, used a 6-series airfoil, but earlier versions apparently used an earlier, 4-series NACA laminar-flow airfoil. See the interesting paper by Lednicer and Gilchrist (Ref. 20) for more information; this paper describes a modern computational aerodynamic analysis of the P-51, and of course the authors needed the correct geometry of the airplane.] However, as described in Section 1.2.3, the laminar-flow airfoil, when manufactured in the factory and used in the field, was contaminated with surface roughness (in comparision to the NACA's jewellike wind tunnel models), and the expected benefit from laminar flow was not totally realized in practice. However, almost as a fluke, the NACA 6-series airfoils had relatively large critical Mach numbers compared to the earlier NACA airfoil families, and it is for this reason that the 6-series airfoils were used on many high-speed jet aircraft after World War II.

The numbering system for the NACA five-digit and 6-series airfoils involves in part the notion of the *design lift coefficient*. When applied in this context, the design lift coefficient for an airfoil is defined as follows: Imagine the airfoil replaced solely by its mean camber line, as sketched in Fig. 2.19. (This is the model used in the classic thin airfoil theory, such as described in Ref. 16.) There is only one angle of attack at which the local flow direction at the leading edge will be tangent to the camber line at that point; such a case is sketched in Fig. 2.19**a**. The theoretical lift coefficient for the camber line at this angle of attack is, by definition, the *ideal* or *design* lift coefficient. This definition was coined by Theodore Theodorsen, a well-known NACA theoretical aerodynamicist, in 1931. For any other angle of attack, the inviscid potential flow will have to curl around the leading edge, such as shown in Fig. 2.19*b*. Potential flow theory shows that when the flow passes over a sharp, convex corner, the velocity becomes infinite. There is only one angle of attack at which an infinite velocity is avoided at the leading edge, namely, that corresponding to the case shown in Fig. 2.19**a**. The lift coefficient for this case is, by definition, the design lift coefficient which is referenced in the NACA airfoil nomenclature.

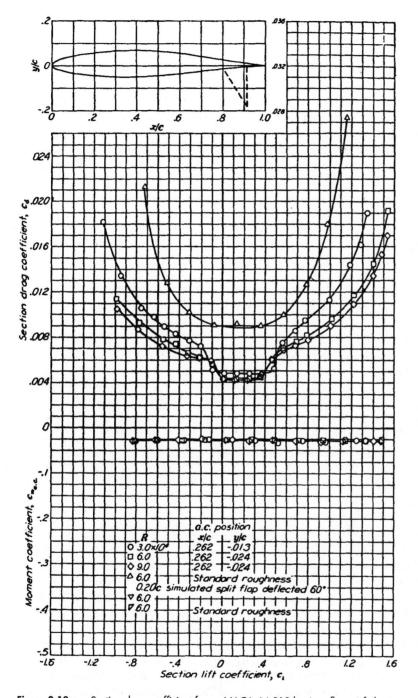

Figure 2.18 Section drag coefficient for an NACA 64-212 laminar-flow airfoil.

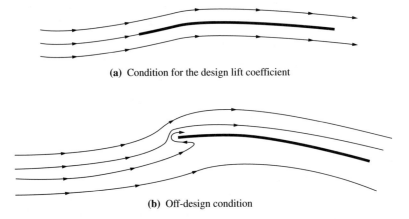

(a) Condition for the design lift coefficient

(b) Off-design condition

Figure 2.19   Sketch illustrating the definition of the theoretical design lift coefficient.

2.8   LIFT AND DRAG BUILDUP

The next time you see an airplane flying overhead, give a thought to the following concept. That airplane is more than just a flying machine, more than just an object, aesthetic as it may be. It is also a carefully designed synthesis of various aero-dynamic components—the wings, fuselage, horizontal and vertical tail, and other appendages—which are working harmoniously with one another to produce the lift necessary to sustain the airplane in the air while at the same time generating the smallest possible amount of drag, in a fashion so as to allow the airplane to carry out its mission, whatever that may be. The lift and drag exerted on the airplane are due to the pressure and shear stress distributions integrated over the *total* surface area of the aircraft—which certainly goes well beyond the concept of the lift and drag exerted on just the airfoil sections, as described in the previous sections. Therefore, in this section we expand our horizons, and we examine the lift and drag of various components of the airplane, both separately and collectively.

2.8.1   Lift for a Finite Wing

The airfoil properties discussed in Section 2.5 can be considered the properties of a wing with an infinite span; indeed, airfoil data are sometimes labeled as *infinite-wing data*. However, all real wings are finite in span (obviously). The planview (top view) of a finite wing is sketched in Fig. 2.20, where b is the wingspan and S is the planview area. An important geometric property of a finite wing is the *aspect ratio* AR, defined as $AR \equiv b^2/S$.

Question: Is the lift coefficient of the finite wing the same as that of the airfoil sections distributed along the span of the wing? For example, from Fig. 2.6a, the lift coefficient for the NACA 2412 airfoil at $4°$ angle of attack is 0.65. Consider a

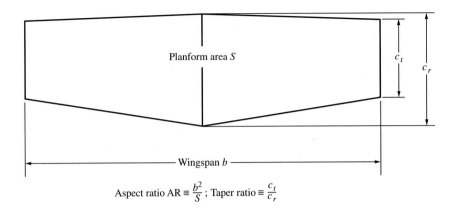

Aspect ratio $AR \equiv \dfrac{b^2}{S}$; Taper ratio $\equiv \dfrac{c_t}{c_r}$

Figure 2.20 Finite-wing geometry.

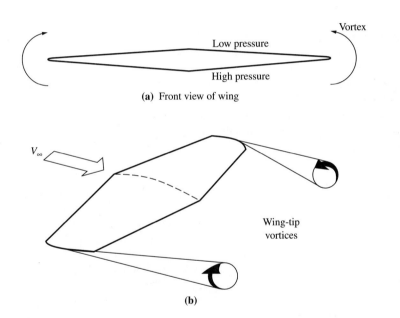

Figure 2.21 Wing-tip vortices.

finite wing made up of the NACA 2412 airfoil section, also at an angle of attack of 4°. Is the value of C_L for the wing also 0.65? The answer is *no*. The reason for the difference is that there are strong vortices produced at the wing tips of the finite wing, which trail downstream. These vortices are analogous to minitornadoes, and like a tornado, they reach out in the flow field and induce changes in the velocity and pressure fields around the wing. These wing-tip vortices are shown schematically in Fig. 2.21. Imagine that you are standing on top of the wing shown in Fig. 2.21.

You will feel a downward component of velocity over the span of the wing, induced by the vortices trailing downstream from both tips. This downward component of velocity is called *downwash*. Now imagine that you are a local airfoil section of the finite wing, as sketched in Fig. 2.22**a**. The local downwash at your location, denoted by w, will combine with the free-stream relative wind, denoted by V_∞, to produce a *local relative wind*, shown in Fig. 2.22**b**. This local relative wind is inclined below the free-stream direction through the induced angle of attack α_i. Hence, as shown in Fig. 2.22**a**, you are effectively feeling an angle of attack different from the actual geometric angle of attack α_g of the wing relative to the free stream; you are sensing a *smaller* angle of attack α_{eff}. So if the wing is at a geometric angle of attack of 5°, you are feeling an effective angle of attack which is smaller. Hence, the lift coefficient for the wing C_L is going to be smaller than the lift coefficient for the airfoil c_l. This explains the answer given to the question posed earlier.

We have just argued that C_L for the finite wing is smaller than c_l for the airfoil section used for the wing. The question now is: How much smaller? The answer depends on the geometric shape of the wing planform. For most airplanes in use today, the wing planform falls in one of four general categories: (1) high-aspect-ratio straight wing, (2) low-aspect-ratio straight wing, (3) swept wing, and (4) delta wing. Let us consider each of these planforms in turn.

High-Aspect-Ratio Straight Wing   The high-aspect-ratio straight wing is the choice for relatively low-speed subsonic airplanes, and historically it has been the type of wing planform receiving the greatest study. The classic theory for such wings was worked out by Prandtl during World War I, and it still carries through to today as the most straightforward engineering approach to estimating the aerodynamic coefficients for such finite wings. Called *Prandtl's lifting line theory* (see, e.g., Ref. 16), this method allows, among other properties, the estimate of the lift slope $a = dC_L/d\alpha$ for a finite

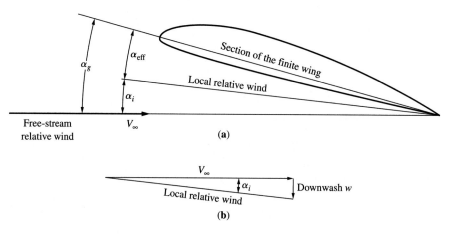

Figure 2.22   Illustration of induced and effective angles of attack, and downwash w.

wing in terms of the lift slope of the airfoil section $a_0 = dc_l/d\alpha$ as

for AR = 4 or 7

$$a = \frac{a_0}{1 + a_0/(\pi e_1 \text{AR})}$$

High-aspect-ratio straight wing
(incompressible)

[2.15]

where a and a_0 are the lift slope *per radian* and e_1 is a factor that depends on the geometric shape of the wing, including the aspect ratio and taper ratio. (The taper ratio is defined in Fig. 2.20. It is the ratio of the tip chord c_t to the root chord c_r.) Values of e_1 are typically on the order of 0.95.

The results from Eq. (2.15) show that the lift slope for a finite wing decreases as the aspect ratio decreases. This is a general result—as the aspect ratio decreases, the induced flow effects over the wing due to the tip vortices are stronger, and hence at a given angle of attack, the lift coefficient is decreased. This is clearly seen in Fig. 2.23. Here, experimentally measured lift curves are shown for seven different finite wings with the same airfoil cross section but with different aspect ratios. Note that the angle of attack for zero lift, denoted $\alpha_{L=0}$ is the same for all seven wings; at zero lift the induced effects theoretically disappear. At any given angle of attack larger than $\alpha_{L=0}$, say, α_1 in Fig. 2.23, the value of C_L becomes smaller as the aspect ratio is decreased.

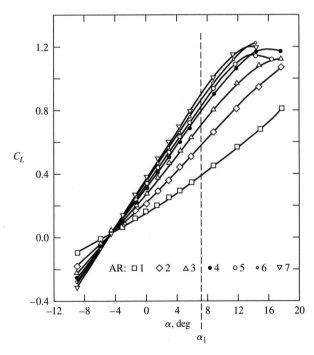

Figure 2.23   Effect of aspect ratio on the lift curve.

Prandtl's lifting line theory, hence Eq. (2.15), does not apply to low-aspect-ratio wings. Equation (2.15) holds for aspect ratios of about 4 or larger. Also, the lifting line theory does not predict the influence of AR on $C_{L_{\max}}$, which is governed by viscous effects. As sketched in Fig. 2.23, experiments show that as AR is reduced, $C_{L_{\max}}$ is also reduced, and that maximum lift occurs at higher angles of attack.

Equation (2.15) for the lift slope of the finite wing applies to incompressible flow, which limits its use to low-speed aircraft. However, during World War II, the flight velocities of many straight-wing subsonic airplanes penetrated well into the compressible flow regime (flow Mach numbers of 0.3 and higher). Today most straight-wing turboprop-powered civil transports and business airplanes fly routinely in the subsonic compressible flow regime. For this flight regime, Eq. (2.15) must be modified by an appropriate *compressibility correction*. Historically, the first and simplest compressibility correction was derived independently by Ludwig Prandtl in Germany and by Hermann Glauert in England in the 1920s for the case of subsonic compressible flow over airfoils. Called the *Prandtl–Glauert rule*, it allowed the incompressible lift slope for an airfoil to be modified for compressibility effects; we have already discussed this modification in conjunction with the trends shown in Figs. 2.10 and 2.12. For a high-aspect-ratio straight wing, we will also use the Prandtl–Glauert rule for a compressibility correction.

This is carried out as follows. Consider a thin *airfoil* at small to moderate angle of attack. Denote the low-speed, incompressible lift slope for this airfoil by a_0, as in Eq. (2.15). Denote the high-speed compressible value of the lift slope for the same airfoil at a free-stream Mach number M_∞ by $a_{0,\text{comp}}$. The *Prandtl–Glauert rule* is

$$a_{0,\text{comp}} = \frac{a_0}{\sqrt{1 - M_\infty^2}}$$

Let us assume that Eq. (2.15), which is obtained from Prandtl's lifting line theory, also holds for subsonic compressible flow, that is,

$$a_{\text{comp}} = \frac{a_{0,\text{comp}}}{1 + a_{0,\text{comp}}/(\pi e_1 \text{AR})} \qquad \textbf{[2.15a]}$$

where a_{comp} is the compressible lift slope for the finite wing. Replacing $a_{0,\text{comp}}$ in Eq. (2.15a) with the Prandtl–Glauert rule, we have

$$a_{\text{comp}} = \frac{a_0/\sqrt{1 - M_\infty^2}}{1 + a_0/(\pi e_1 \text{AR}\sqrt{1 - M_\infty^2})}$$

or

$$\boxed{a_{\text{comp}} = \frac{a_0}{\sqrt{1 - M_\infty^2} + a_0/(\pi e_1 \text{AR})}} \qquad \begin{array}{c}\text{Subsonic high-aspect-ratio} \\ \text{straight wing} \\ \text{(compressible)}\end{array} \qquad \textbf{[2.16]}$$

where M_∞ is the free-stream Mach number. Equation (2.16) gives a quick, but approximate correction to the lift slope; because it is derived from linear subsonic flow theory (see, e.g., Ref. 16), it is not recommended for use for M_∞ greater than 0.7. Figure 2.24 is an illustration of the variation of lift slope with free-stream Mach

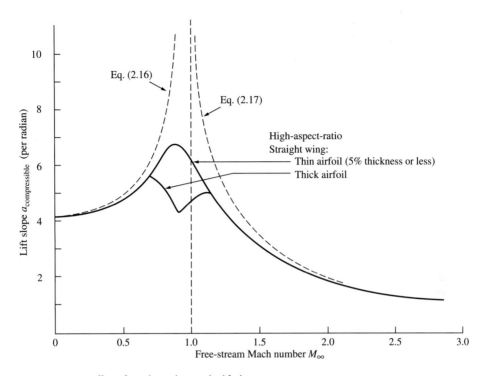

Figure 2.24 Effect of Mach number on the lift slope.

number for a high-aspect-ratio straight wing. Results obtained From Eq. (2.16) are shown as the dashed curve at the left in Fig. 2.24. The solid curve in Fig. 2.24 is representative of actual experimental data for a high-aspect-ratio straight wing. Note that Fig. 2.24 (for a finite wing) is similar to Fig. 2.10 (for an airfoil).

For Mach numbers closer to 1—the transonic regime—there are compressibility corrections for pressure coefficient that attempt to take into account the nonlinear nature of transonic flow (see, e.g., Ref. 16), and which, when integrated over the wing surface, lead to predictions of the lift slope that are more accurate at higher subsonic values of M_∞ than Eq. (2.16). However, today the preferred method of calculating the transonic lift coefficient is to use computational fluid dynamics (see, e.g., Ref. 21) to numerically solve the appropriate nonlinear Euler or Navier-Stokes equations for the transonic flow field over the wing, and then to integrate the calculated surface pressure distribution to obtain the lift.

For supersonic flow over a high-aspect-ratio straight wing, the lift slope can be approximated from supersonic linear theory (see Ref. 16) as

$$a_{comp} = \frac{4}{\sqrt{M_\infty^2 - 1}} \qquad \text{Supersonic high-aspect-ratio straight wing} \qquad \textbf{[2.17]}$$

where a_{comp} is per radian. The variation of lift slope predicted by Eq. (2.17) is shown as the dashed curve at the right in Fig. 2.24.

Example 2.5

Consider a straight wing of aspect ratio 6 with an NACA 2412 airfoil. Assuming low-speed flow, calculate the lift coefficient at an angle of attack of 6°. For this wing, the span effectiveness factor $e_1 = 0.95$.

Solution

From Fig. 2.6a, $a_0 = 0.105$ per degree and $\alpha_{L=0} = -2.2°$. The lift slope is given by Eq. (2.15).

$$a = \frac{a_0}{1 + a_0/(\pi e_1 AR)}$$

where a and a_0 are *per radian*.

$$a_0 = 0.105 \text{ per degree} = (0.105)(57.3) = 6.02 \text{ per radian}$$

Hence, from Eq. (2.15),

$$a = \frac{6.02}{1 + 6.02/[\pi(0.95)(6)]} = 4.51 \text{ per radian}$$

or

$$a = \frac{4.51}{57.3} = 0.079 \text{ per degree}$$

$$C_L = a(\alpha - \alpha_{L=0}) = 0.079[6 - (-2.2)] = \boxed{0.648}$$

Note: Comparing the above result for a finite wing with that for the airfoil as obtained in Example 2.3, we have for $\alpha = 6°$

Airfoil: $c_l = 0.85$

Finite wing: $C_L = 0.648$

As expected, the finite aspect ratio reduces the lift coefficient; in this case, for AR = 6, the reduction is by 24%—a nontrivial amount. For lower aspect ratios, the reduction will be even greater.

Example 2.6

What is the lift coefficient for the same wing at the same angle of attack as in Example 2.5, but for a free-stream Mach number of 0.7?

Solution

From Eq. (2.16),

$$a_{comp} = \frac{a_0}{\sqrt{1 - M_\infty^2} + a_0/(\pi e_1 AR)}$$

where a_0 is the *incompressible* lift slope for the *airfoil* and a_{comp} is the *compressible* lift slope for the *finite wing*. From Fig. 2.6a, $a_0 = 0.105$ per degree = 6.02 per radian. Hence,

$$a_{comp} = \frac{6.02}{\sqrt{1 - (0.7)^2} + 6.02/[\pi(0.95)(6)]} = 5.73 \text{ per radian}$$

Note: The finite aspect ratio reduces the lift slope, but the effect of compressibility increases the lift slope. In this example, the two effects almost compensate each other, and the compressible value of the finite-wing lift slope is almost the same as the incompressible value of the airfoil lift slope. The lift coefficient is given by

$$C_L = a_{comp}(\alpha - \alpha_{L=0})$$

where $a_{comp} = 5.73/57.3 = 0.1$ per degree. Hence,

$$C_L = 0.1[6 - (-2.2)] = \boxed{0.82}$$

Example 2.7

Calculate the lift coefficient for a high-aspect-ratio straight wing with a thin symmetric airfoil at an angle of attack of 6° in a supersonic flow in Mach 2.5.

Solution

From Eq. (2.17),

$$a_{comp} = \frac{4}{\sqrt{M_\infty^2 - 1}} = \frac{4}{\sqrt{(2.5)^2 - 1}} = 1.746 \text{ per radian}$$

or

$$a_{comp} = \frac{1.746}{57.3} = 0.0305 \text{ per degree}$$

Hence,

$$C_L = a_{comp}\alpha = 0.0305(6) = \boxed{0.183}$$

Note: Comparing this lift coefficient at Mach 2.5 with those obtained in Examples 2.5 and 2.6, we see that the magnitude of the supersonic lift coefficient is considerably smaller than that of the subsonic values (even taking into account the different zero-lift angles of attack).

Low-Aspect-Ratio Straight Wings   When applied to straight wings at AR < 4, Eq. (2.15) progressively yields poorer results as the aspect ratio is reduced. Why? The reason is that Eq. (2.15) is derived from a theoretical model which represents the finite wing with a single lifting line across the span of the wing. This is a good model when the aspect ratio is large; by examining the sketch in Fig. 2.25a, it is intuitively clear that a long, narrow wing planform might be reasonably modeled by a single lifting line from one wing tip to the other. However, when the aspect ratio is small, such as sketched in Fig. 2.25b, the same intuition leads to some misgivings—how can a short, stubby wing be properly modeled by a single lifting line? The fact is—it cannot. Instead of a single spanwise lifting line, the low-aspect-ratio wing must be modeled by a large number of spanwise vortices, each located at a different chordwise station, such as sketched in Fig. 2.25c. This is the essence of *lifting surface theory*. Today, the general concept of a lifting surface is the basis for a large number of *panel codes*—elaborate computer programs which numerically solve for the inviscid aerodynamic wing properties—lift slope, zero-lift angle of attack, moment coefficients, and induced

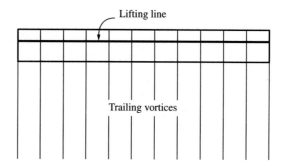

(a) High-aspect-ratio wing. Lifting line is a
reasonable representation of the wing.

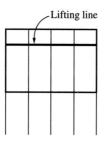

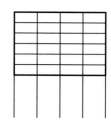

(b) Low-aspect-ratio wing. Lifting line
is a poor representation of the wing.

(c) Low-aspect-ratio wing. Lifting surface
is a better representation of the wing.

Figure 2.25   Contrast of lifting line and lifting surface models.

drag coefficients (to be discussed shortly). Modern panel methods can quickly and accurately calculate the inviscid flow properties of low-aspect-ratio straight wings, and every aerospace company and laboratory have such panel codes in their "numerical tool box." There is an extensive literature on panel methods; for a basic discussion see Ref. 16, and for a more thorough presentation, especially for three-dimensional panel codes, see Ref. 22.

An approximate relation for the lift slope for low-aspect-ratio straight wings was obtained by H. B. Helmbold in Germany in 1942 (Ref. 23). Based on a lifting surface solution for elliptic wings, Helmbold's equation is

$$a = \frac{a_0}{\sqrt{1 + [a_0/(\pi\,\text{AR})]^2} + a_0/(\pi\,\text{AR})}$$

Low-aspect-ratio
straight wing   **[2.18a]**
(incompressible)

where a and a_0 are per radian. Equation (2.18a) is remarkably accurate for wings with AR < 4. This is shown in Fig. 2.26, which gives experimental data for the lift slope for rectangular wings as a function of AR from 0.5 to 6; these data are compared with the predictions from Prandtl's lifting line theory, Eq. (2.15), and Helmbold's

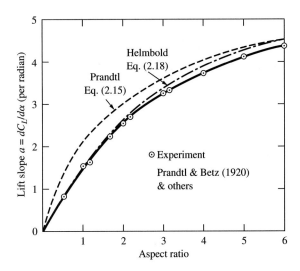

Figure 2.26 Lift slope for rectangular wings as a function of the aspect ratio.

equation, Eq. (2.18*a*). For subsonic compressible flow, Eq. (2.18*a*) is modified as follows (the derivation is given later, in our discussion of swept wings):

$$a_{\text{comp}} = \frac{a_0}{\sqrt{1 - M_\infty^2 + [a_0/(\pi\,\text{AR})]^2} + a_0/(\pi\,\text{AR})}$$

Subsonic
low-aspect-ratio **[2.18*b*]**
straight wing
(compressible)

where a_{comp} and a_0 are per radian.

In the case of supersonic flow over a low-aspect-ratio straight wing, Eq. (2.17) is not appropriate. At low aspect ratios, the Mach cones from the wing tips cover a substantial portion of the wing, hence invalidating Eq. (2.17). Instead, Hoerner and Borst (Ref. 18) suggest the following equation, obtained from supersonic linearized theory for three-dimensional wings:

$$a_{\text{comp}} = \frac{4}{\sqrt{M_\infty^2 - 1}}\left(1 - \frac{1}{2\text{AR}\sqrt{M_\infty^2 - 1}}\right)$$

Supersonic
low-aspect-ratio **[2.18*c*]**
straight wing

where a_{comp} is per radian. This equation is valid as long as the Mach cones from the two wing tips do not overlap.

In airplane design, when are we concerned with low-aspect-ratio straight wings? The answer is, not often. Just scanning the pictures of the airplanes discussed in Chapter 1, the only airplane we see with a very low-aspect-ratio straight wing is the Lockheed F-104, shown in Fig. 1.30. A three-view of the F-104 is given in Fig. 2.27; the wing aspect ratio is 2.97. At subsonic speeds, a low-aspect-ratio wing

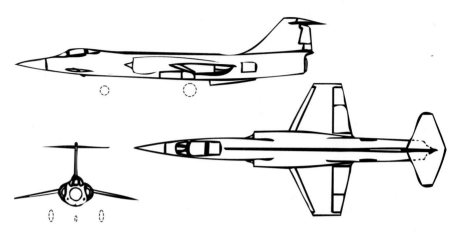

Figure 2.27   Three-view of the Lockheed F-104 Starfighter. Aspect ratio = 2.94.

is plagued by large induced drag, and hence subsonic aircraft (since World War I) do not have low-aspect-ratio wings. On the other hand, a low-aspect-ratio straight wing has low supersonic wave drag, and this is why such a wing was used on the F-104—the first military fighter designed for sustained Mach 2 flight. At subsonic speeds, and especially for takeoff and landing, the low-aspect-ratio wings were a major liability to the F-104. Fortunately, there are two other wing planforms that reduce wave drag without suffering nearly as large a penalty at subsonic speeds, namely, the swept wing and the delta wing. Hence, we will now shift our attention to these planforms.

Example 2.8

Helmbold's equation for low-aspect-ratio straight wings, Eq. (2.18a), in the limit as the aspect ratio becomes very large, reduces to Eq. (2.15) for high-aspect-ratio straight wings. Indeed, Eq. (2.18a) can be viewed as a higher approximation that holds for both low- and high-aspect-ratio straight wings, providing even greater accuracy than Eq. (2.15) for the high-aspect-ratio case, albeit the differences are small for high aspect ratios. To illustrate this, calculate the lift coefficient for the wing described in Example 2.5 at 6° angle of attack, using Helmbold's equation, and compare the results with those from Example 2.5 using Eq. (2.15).

Solution

From Example 2.5, $a_0 = 6.02$ radian and $AR = 6$. Hence,

$$\frac{a_0}{\pi AR} = \frac{6.02}{\pi(6)} = 0.319$$

From Eq. (2.18a),

$$a = \frac{a_0}{\sqrt{1 + [a_0/(\pi AR)]^2} + a_0/(\pi AR)} = \frac{6.02}{\sqrt{1 + (0.319)^2} + 0.319}$$

$$= 4.4 \text{ per radian}$$

or

$$a = \frac{4.4}{57.3} = 0.077 \text{ per degree}$$

$$C_L = a(\alpha - \alpha_{L=0}) = 0.077[6 - (-2.2)] = \boxed{0.629}$$

Compared to the result of $C_L = 0.648$ obtained in Example 2.5, the results obtained from Eqs. (2.15) and (2.18a) differ by only 3% for an aspect ratio of 6.

Consider a straight wing of aspect ratio 2 with an NACA 2412 airfoil. Assuming low-speed flow, calculate the lift coefficient at an angle of attack of 6°. Assume $e_1 = 0.95$.

Example 2.9

Solution
This is the same set of conditions as in Example 2.5, except for a much smaller aspect ratio. We have

$$\frac{a_0}{\pi R} = \frac{6.02}{\pi(2)} = 0.955$$

From Eq. (2.18a),

$$a = \frac{a_0}{\sqrt{1 + [a_0/(\pi \text{AR})]^2} + a_0/(\pi \text{AR})} = \frac{6.02}{\sqrt{1 + (0.955)^2} + 0.955}$$

$$= 2.575 \text{ per radian}$$

or

$$a = \frac{2.575}{57.3} = 0.0449 \text{ per degree}$$

$$C_L = a(\alpha - \alpha_{L=0}) = 0.0449[6 - (-2.2)] = \boxed{0.368}$$

This result is to be compared with that from Example 2.5, where $C_L = 0.648$. In reducing the aspect ratio from 6 to 2, the lift coefficient is reduced by 43%—a dramatic decrease.

Calculate the lift coefficient for a straight wing of aspect of ratio 2 at an angle of attack of 6° in a supersonic flow at Mach 2.5. Assume a thin, symmetric airfoil section.

Example 2.10

Solution
From Eq. (2.18c),

$$a_{\text{comp}} = \frac{4}{\sqrt{M_\infty^2 - 1}}\left(1 - \frac{1}{2\text{AR}\sqrt{M_\infty^2 - 1}}\right)$$

$$= \frac{4}{\sqrt{(2.5)^2 - 1}}\left[1 - \frac{1}{2(2)\sqrt{(2.5)^2 - 1}}\right] = 1.555 \text{ per radian}$$

$$= \frac{1.555}{57.3} = 0.027 \text{ per degree}$$

$$C_L = a\alpha = 0.0027(6) = \boxed{0.163}$$

Note: This result for a low-aspect-ratio wing at $M_\infty = 2.5$ is only 10% less than that obtained in Example 2.7 for a high-aspect-ratio wing at the same Mach number. The aspect ratio effect on lift coefficient for supersonic wings is substantially less than that for subsonic wings.

Swept Wings   The main function of a swept wing is to reduce wave drag at transonic and supersonic speeds. Since the topic of this subsection is lift, let us examine the lifting properties of swept wings.

Simply stated, a swept wing has a lower lift coefficient than a straight wing, everything else being equal. An intuitive explanation of this effect is as follows. Consider a straight wing and a swept wing in a flow with a free-stream velocity V_∞, as sketched in Fig. 2.28a and **b**, respectively. Assume that the aspect ratio is high for both wings, so that we can ignore tip effects. Let u and w be the components of V_∞ perpendicular and parallel to the leading edge, respectively. The pressure distribution over the airfoil section oriented perpendicular to the leading edge is mainly governed by the chordwise component of velocity u; the spanwise component of velocity w has little effect on the pressure distribution. For the straight wing in Fig. 2.28a, the chordwise velocity component u is the full V_∞; there is no spanwise component, that is, $w = 0$. However, for the swept wing in Fig. 2.28**b**, the chordwise component of velocity u is smaller than V_∞, that is $u = V_\infty \cos \Lambda$, where Λ is the sweep angle shown in Fig. 2.28**b**. For the swept wing, the spanwise component of velocity w is a finite value, but it has little effect on the pressure distribution over the airfoil section. Since u for the swept wing is smaller than u for the straight wing, the difference in pressure between the top and bottom surfaces of the swept wing will be less than the difference in pressure between the top and bottom surfaces of the straight wing. Since lift is generated by these differences in pressure, the lift on the swept wing will be

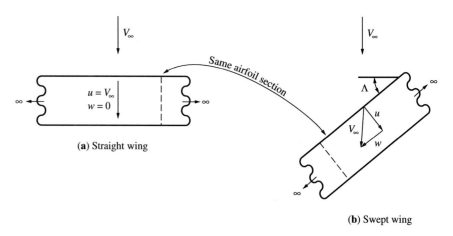

(a) Straight wing

(b) Swept wing

Figure 2.28   Effect of sweeping a wing.

less than that on the straight wing. Although this explanation is a bit naive because it ignores the details of the flow fields over both wings, it captures the essential idea.

The geometry of a tapered swept wing is illustrated in Fig. 2.29. The wingspan b is the straight-line distance between the wing tips, the wing planform area is S, and the aspect ratio and the taper ratio are defined as before, namely, $AR \equiv b^2/S$ and taper ratio $\equiv c_t/c_r$. For the tapered wing, the sweep angle Λ is referenced to the half-chord line, as shown in Fig. 2.29. (In some of the literature, the sweep angle is referenced to the quarter-chord line; however, by using the half-chord line as reference, the lift slope for a swept wing becomes independent of taper ratio, as discussed below.)

Just as in the case of low-aspect-ratio straight wings, Prandtl's lifting line theory does not apply directly to swept wings. Hence, Eq. (2.15) does not apply to swept wings. Instead, the aerodynamic properties of swept wings at low speeds must be calculated from lifting surface theory (i.e., numerical panel methods) in the same spirit as in our discussion on low-aspect-ratio straight wings. However, for an approximate calculation of the lift slope for a swept finite wing, Kuchemann (Ref. 24) suggests the following approach. From the discussion associated with Fig. 2.28, the lift slope for an infinite swept wing should be $a_0 \cos \Lambda$, where a_0 is the lift slope for the airfoil section perpendicular to the leading edge. Replacing a_0 in Helmbold's equation, Eq. (2.18a), with $a_0 \cos \Lambda$, we have

$$a = \frac{a_0 \cos \Lambda}{\sqrt{1 + [(a_0 \cos \Lambda)/(\pi AR)]^2} + (a_0 \cos \Lambda)/(\pi AR)}$$

Swept wing
(incompressible)

[2.19]

where a and a_0 are per radian. Equation (2.19) is an approximation for the incompressible lift slope for a finite wing of aspect ratio AR and sweep angle Λ (referenced to the half-chord line). The subsonic compressibility effect is added to Eq. (2.19) by

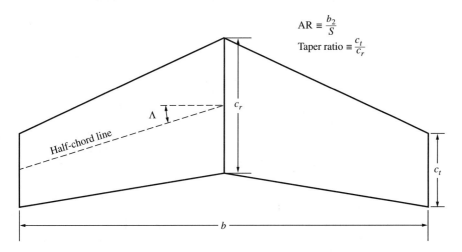

$$AR \equiv \frac{b_2}{S}$$

$$\text{Taper ratio} \equiv \frac{c_t}{c_r}$$

Figure 2.29 Swept-wing geometry.

replacing a_0 with $a_0/\sqrt{1 - M_{\infty,n}}$, where $M_{\infty,n}$ is the component of the free-stream Mach number perpendicular to the half-chord line of the swept wing, or $M_{\infty,n} = M_\infty \cos \Lambda$. Letting $\beta = \sqrt{1 - M_\infty^2 \cos^2 \Lambda}$, we replace a_0 in Eq. (2.19) with a_0/β, obtaining

$$a_{\text{comp}} = \frac{(a_0 \cos \Lambda)/\beta}{\sqrt{1 + [(a_0 \cos \Lambda)/(\pi \text{AR} \beta)]^2} + (a_0 \cos \Lambda)/(\pi \text{AR} \beta)} \qquad [2.20]$$

Multiply both numerator and denominator in Eq. (2.20) by β, we have

$$a_{\text{comp}} = \frac{a_0 \cos \Lambda}{\sqrt{\beta^2 + [(a_0 \cos \Lambda)/(\pi \text{AR})]^2} + (a_0 \cos \Lambda)/(\pi \text{AR})} \qquad [2.21]$$

Recalling that $\beta = \sqrt{1 - M_\infty^2 \cos^2 \Lambda}$, we can write Eq. (2.22) as

$$a_{\text{comp}} = \frac{a_0 \cos \Lambda}{\sqrt{1 - M_\infty^2 \cos^2 \Lambda + [(a_0 \cos \Lambda)/(\pi \text{AR})]^2} + (a_0 \cos \Lambda)/(\pi \text{AR})}$$

Subsonic swept wing
(compressible) $\qquad$ [2.22]

where a_{comp} and a_0 are per radian. Note that Eq. (2.22) reduces to Eq. (2.18b) when $\Lambda = 0°$; hence, the above derivation also constitutes a derivation of Eq. (2.18b).

The previous discussion on swept wings pertains to subsonic flow. For a swept wing moving at supersonic speeds, the aerodynamic properties depend on the location of the leading edge relative to a Mach wave emanating from the apex of the wing. For example, consider Fig. 2.30, which shows two wings with different leading-edge

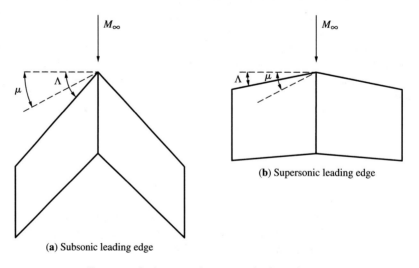

(a) Subsonic leading edge

(b) Supersonic leading edge

Figure 2.30   Illustration of subsonic and supersonic leading edges.

sweep angles in a flow with the same supersonic free-stream Mach number. The Mach angle μ is given by $\mu = \text{Arcsin}\,(1/M_\infty)$. In Fig. 2.30a, the wing leading edge is swept inside the Mach cone, that is, $\Lambda > \mu$. For this case, the component of M_∞ perpendicular to the leading edge is subsonic; hence, the swept wing is said to have a *subsonic leading edge.* For the wing in supersonic flight, there is a weak shock that emanates from the apex, but there is *no* shock attached elsewhere along the wing leading edge. In contrast, in Fig. 2.30b, the wing leading edge is swept outside the Mach cone, that is, $\Lambda < \mu$. For this case, the component of M_∞ perpendicular to the leading edge is supersonic; hence the swept wing is said to have a *supersonic leading edge.* For this wing in supersonic flight, there will be a shock wave attached along the entire leading edge. A swept wing with a subsonic leading edge behaves somewhat as a wing at subsonic speeds, although the actual free-stream Mach number is supersonic. That is, the top and bottom surfaces of the wing can communicate with each other in the vicinity of the leading edge, just as occurs in a purely subsonic flow. A swept wing with a supersonic leading edge, with its attached shock along the leading edge, behaves somewhat as a supersonic flat plate at the angle of attack. That is, the top and bottom surfaces of the wing do not communicate with each other. For these reasons, the aerodynamic properties of the two swept wings shown in Fig. 2.30 are different.

There is no convenient engineering formula for the rapid calculation of the lifting properties of a swept wing in supersonic flow. Most companies and laboratories use computational fluid dynamic techniques to calculate the pressure distribution over the wing, and then they find the lift by integrating the pressure distribution over the surface, taking the component of the resultant force perpendicular to the relative wind. In lieu of such detailed numerical calculations, Raymer (Ref. 25) suggests the use of a series of charts prepared by the U.S. Air Force for quick, design-oriented calculations for swept wings. A sampling of these charts is given in Fig. 2.31, one each for the six different wing planforms shown in the figure. Each planform corresponds to a different taper ratio, denoted by λ at the top of each chart. In Fig. 2.31, Λ_{LE} is the leading-edge swept angle, $\beta = \sqrt{M_\infty^2 - 1}$, and C_{N_α} is the slope of the *normal force coefficient* with angle of attack α. For ordinary supersonic cruising flight, we can readily assume that the normal force coefficient C_N, is representative of the lift coefficient C_L, that is, $C_L \approx C_N$. The reason for this is as follows. The dynamic pressure is given by

$$q_\infty \equiv \frac{1}{2}\rho_\infty V_\infty^2 = \frac{1}{2}\gamma p_\infty \left(\frac{\rho_\infty}{\gamma p_\infty}\right) V_\infty^2$$

$$= \frac{1}{2}\gamma p_\infty \frac{V_\infty^2}{a_\infty^2} = \frac{1}{2}\gamma p_\infty M_\infty^2$$

[2.23]

Equation (2.23) shows that $q_\infty \propto M_\infty^2$, and hence the dynamic pressure can be large at supersonic Mach numbers. For an airplane cruising at supersonic speeds in steady level flight, the lift is equal to the weight

$$L = W = q_\infty S C_L$$

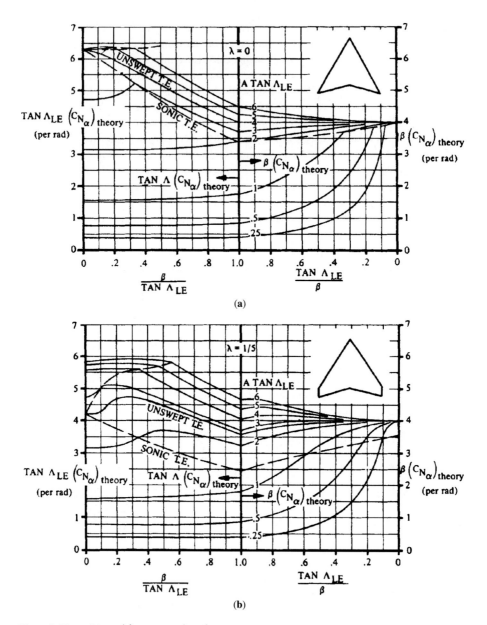

Figure 2.31   Normal-force-curve slope for supersonic wings. (From USAF DATCOM, Air Force Flight Dynamics Lab, Wright-Patterson AFB, Ohio.) (*continued*)

or

$$C_L = \frac{W}{q_\infty S}$$ **[2.24]**

From Eq. (2.24), when q_∞ is large, C_L is small. In turn, C_L is small when the angle

(continued)

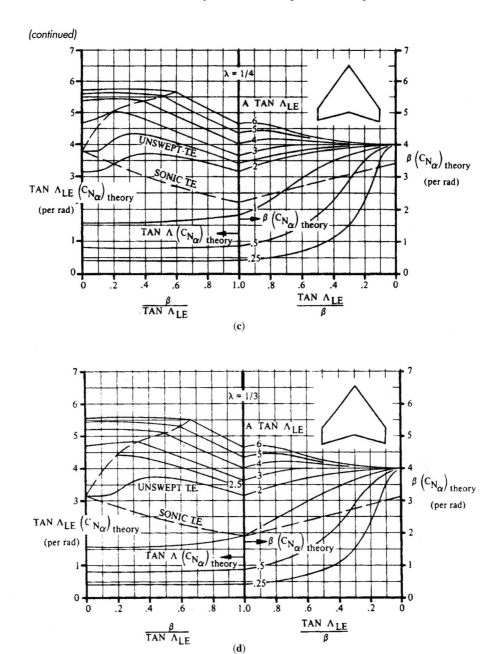

(c)

(d)

of attack is small. And when α is small, then $C_L \approx C_N$. Therefore, for normal design purposes, we can assume that C_{N_α} in Fig. 2.31 is the same as the lift slope $dC_L/d\alpha$. Finally, in each chart in Fig. 2.31, the different curves shown are for different values

(concluded)

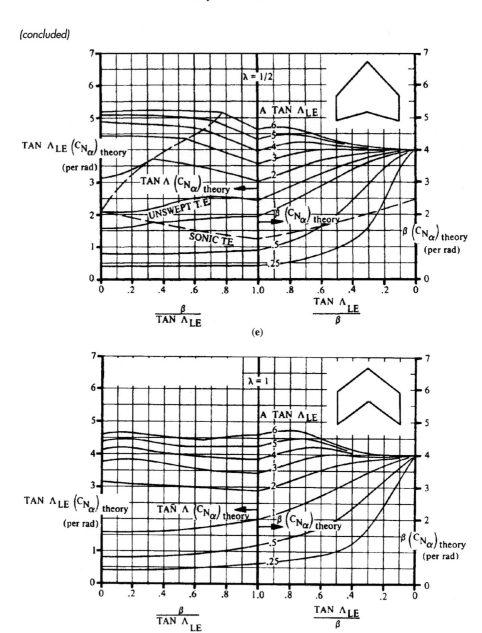

(e)

(f)

of the parameter AR Tan Λ_{LE}. To use Fig. 2.31 to find the lift slope for a given swept wing, carry out the following steps:

1. For the given wing, calculate $\beta/$ Tan Λ_{LE}. This is the abscissa on the left side of the charts. If this number is less than 1, use the left side. If the number is

greater than 1, invert it, and use the right side of the charts, where the abscissa is $(\text{Tan}\Lambda_{LE})/\beta$.

2. Pick the chart corresponding to the taper ratio λ of the given wing. If λ is in between the values shown in the charts, interpolation between charts will be necessary.

3. Calculate AR Tan Λ_{LE} for the given wing. This is a parameter in the charts. Find the curve in the chart corresponding to the value of this parameter. Most likely, interpolation between two curves will be needed.

4. Read the corresponding value from the ordinate; this value will correspond to Tan $\Lambda_{LE}(C_{N_\alpha})$ if the left side of the chart is being used, and it will correspond to βC_{N_α} if the right side is used.

5. Extract C_{N_α} dividing the left ordinate by Tan Λ_{LE}, or by dividing the right ordinate by β, as the case may be.

6. We assume that the supersonic swept wing is thin, to minimize wave edge. Hence, to calculate lift, assume a flat surface wing, where $L = 0$ at $a = 0°$. Recalling our assumption that $C_L \approx C_N$, calculate C_L from

$$C_L = C_{N_\alpha}\alpha \qquad (\alpha \text{ in radians})$$

Consider a swept wing with a taper ratio of 0.5, leading edge sweep angle of 45°, and an aspect ratio of 3. Calculate the lift coefficient at Mach 2 at an angle of attack of 2°.

Example 2.11

Solution

For taper ratio λ equal to 0.5, use chart (e) in Fig. 2.31.

$$\beta = \sqrt{M_\infty^2 - 1} = \sqrt{2^2 - 1} = 1.732$$
$$\text{Tan } \Lambda_{LE} = \text{Tan } 45° = 1$$

Since $\beta > \text{Tan } \Lambda_{LE}$, we will use the right side of chart (e).

$$\frac{\text{Tan } \Lambda_{LE}}{\beta} = \frac{1}{1.732} = 0.577$$

Also, the parameter AR Tan $\Lambda_{LE} = (3)(1) = 3$. In chart (e), find the curve corresponding to AR Tan $\Lambda_{LE} = 3$. The point on this curve corresponding to the abscissa of 0.577 has the ordinate $\beta(C_{N_\alpha}) = 4$. Hence,

$$C_{N_\alpha} = \frac{4}{\beta} = \frac{4}{1.732} = 2.31 \text{ per radian}$$

Since $\alpha = 2° = 0.0349$ rad,

$$C_L = C_{N_\alpha}\alpha = (2.31)(0.0349) = \boxed{0.0806}$$

To go further with this calculation, assume the wing area is 3,900 ft², which is about that for the Concorde supersonic transport. Assume Mach 2 flight at a standard altitude of 50,000 ft, where $p_\infty = 243.6$ lb/ft². Let us calculate the lift generated by the wing for an angle of attack of 2°. From Eq. (2.23),

$$q_\infty = \frac{1}{2}\gamma p_\infty M_\infty^2 = \frac{1.4}{2}(243.6)(2)^2 = 682 \text{ lb/ft}^2$$

Hence,

$$L = q_\infty S C_L = (682)(3{,}900)(0.0806) = 214{,}400 \text{ lb}$$

Note: The maximum fuel-empty weight of the Concorde is 200,000 lb. Although we are by no means making a direct comparision here, the above calculation of the lift for our example wing for our example conditions shows that supersonic wings can produce a lot of lift at low angles of attack (hence with low values of the lift coefficient).

For more details on the aerodynamics of supersonic wings, see the extensive discussion in chapter 11 of Ref. 26.

Example 2.12 | Consider the wing described in Example 2.5, except with a sweep angle of 35°. Calculate the low-speed lift coefficient at 6° angle of attack and compare with the straight-wing results from Example 2.5.

Solution

$$\frac{a_0 \cos \Lambda}{\pi \text{AR}} = \frac{6.02 \cos 35°}{\pi(6)} = 0.262 \text{ per radian}$$

From Eq. (2.19),

$$a = \frac{a_0 \cos \Lambda}{\sqrt{1 + [(a_0 \cos \Lambda)/(\pi \text{AR})]^2} + (a_0 \cos \Lambda)/(\pi \text{AR})}$$

$$= \frac{6.02 \cos 35°}{\sqrt{1 + (2.62)^2} + 0.262} = 3.8057 \text{ per radian}$$

or

$$a = \frac{3.8057}{57.3} = 0.0664 \text{ per degree}$$

$$C_L = a(\alpha - \alpha_{L=0}) = 0.0664[6 - (-2.2)] = \boxed{0.544}$$

Note: The straight wing result from Example 2.5 is $C_L = 0.648$. For this case, sweeping the wing by 35° decreases the lift coefficient by 16%.

DESIGN CAMEO

A swept wing is utilized in airplane design to reduce the transonic and supersonic wave drag—it is a design feature that is associated with high-speed airplanes. However, it is important for the designer to recognize that wing sweep is usually a detriment at low speeds. In the above example, we have seen that the low-speed lift coefficient is reduced by sweeping the wing. For the designer, this complicates the design of the airplane for good landing and takeoff performance. To compensate, swept-wing airplanes are frequently designed with elaborate high-lift devices (multielement trailing-edge flaps, leading-edge flaps and slats, etc.). Such high-lift devices are discussed in Chapter 5.

Delta Wings Swept wings that have planforms such as shown in Fig. 2.32 are called *delta wings.* Interest in delta wings for airplanes goes as far back as the early work done by Alexander Lippisch in Germany during the 1930s. Delta wings are employed on many aircraft designed for supersonic flight, for example, the F-102 (Fig. 1.31) and the Concorde (Fig. 1.35). The supersonic lifting characteristics of delta wings are essentially given by the data in Fig. 2.31, which have already been discussed in the previous section. In this section we concentrate on the subsonic flow over delta wings.

The flow field over a low-aspect-ratio delta wing at low speeds is completely different from that for a straight wing or a high-aspect-ratio swept wing. A qualitative sketch of the flow field over a delta wing at angle of attack is given in Fig. 2.33. The

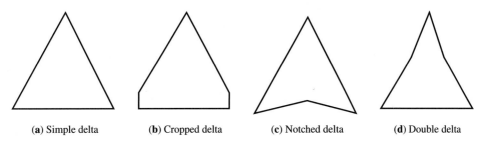

(a) Simple delta   (b) Cropped delta   (c) Notched delta   (d) Double delta

Figure 2.32   Four versions of a delta-wing planform. (After Loftin, Ref. 13.)

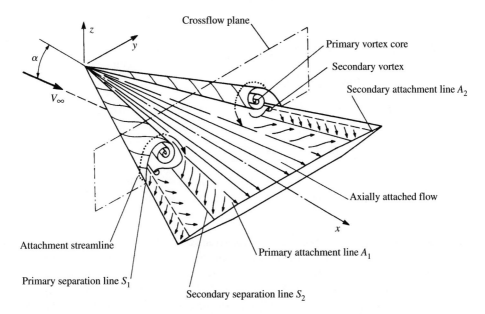

Figure 2.33   Schematic of the subsonic flow over the top of a delta wing at angle of attack. (Courtesy of John Stollery, Cranfield Institute of Technology, England.)

dominant aspect of this flow is the two vortices that are formed along the highly swept leading edges, and that trail downstream over the top of the wing. This vortex pattern is created by the following mechanism. The pressure on the bottom surface of the wing is higher than the pressure on the top surface. Thus, the flow on the bottom surface in the vicinity of the leading edge tries to curl around the leading edge from the bottom to the top. If the leading edge is relatively sharp, the flow will separate along its entire length. This separated flow curls into a primary vortex above the wing just inboard of each leading edge, as sketched in Fig. 2.33. The stream surface which has separated at the leading edge (the primary separation line S_1 in Fig. 2.33) loops above the wing and then reattaches along the primary attachment line (line A in Fig. 2.33). The primary vortex is contained within this loop. A secondary vortex is formed underneath the primary vortex, with its own separation line, denoted by S_2 in Fig. 2.33, and its own reattachment line A_2.

Unlike many separated flows in aerodynamics, the vortex pattern over a delta wing shown in Fig. 2.33 is a friendly flow in regard to the production of lift. The vortices are strong and generally stable. They are a source of high energy, relatively high vorticity flow, and the local static pressure in the vicinity of the vortices is small. Hence, the vortices create a lower pressure on the top surface than would exist if the vortices were not there. This increases the lift compared to what it would be without the vortices. The portion of the lift due to the action of the leading-edge vortices is called the *vortex lift*. A typical variation of C_L for a delta wing as a function of angle of attack is shown in Fig. 2.34 (after Ref. 18). Here, low-speed experimental data are plotted for a delta wing with an aspect ratio of 1.46. Also shown is a theoretical calculation which assumes potential flow without the leading-edge vortices; this is identified as *potential flow lift* in Fig. 2.34. The difference between the experimental data and the potential flow lift is the *vortex lift*. The vortex lift is a major contributor to the overall lift; note that in Fig. 2.34 the vortex lift is about equal to the potential flow lift in the higher angles of attack.

The lift curve in Fig. 2.34 illustrates three important characteristics of the lift of low-aspect-ratio delta wings:

1. The lift slope is small, on the order of 0.05 per degree.
2. The lift, however, continues to increase over a large range of angle of attack. In Fig. 2.34, the stalling angle of attack is about 35°. The net result is a reasonable value of $C_{L,\max}$, on the order of 1.35.
3. The lift curve is *nonlinear*, in contrast to the linear variation exhibited by conventional wings for subsonic aircraft. The vortex lift is mainly responsible for this nonlinearity.

The next time you have an opportunity to watch a delta-wing aircraft take off or land, for example, the televised landing of the space shuttle, note the large angle of attack of the vehicle. Also, this is why the Concorde supersonic transport, with its low-aspect-ratio deltalike wing, lands at a high angle of attack. In fact, the angle of attack is so high that the front part of the fuselage must be mechanically drooped upon landing in order for the pilots to see the runway.

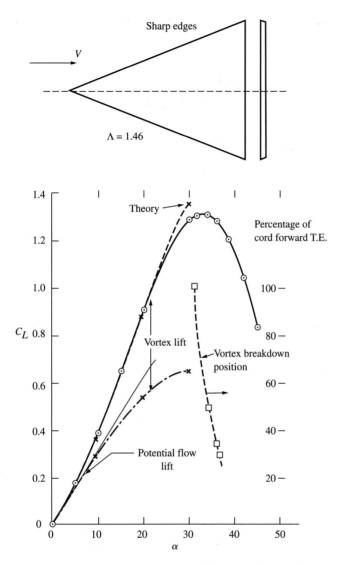

Figure 2.34   Lift coefficient curve for a delta wing in low-speed subsonic flow. (After Hoerner and Borst, Ref. 18.)

Kuchemann (Ref. 24) describes an approximate calculation for the normal force coefficient C_N for slender delta wings at low speeds. Defining the length l and the semispan s as shown in Fig. 2.35, the quantity $\alpha/(s/l)$ becomes a type of similarity parameter which allows normal force data for delta wings of different aspect ratios to collapse approximately to the same curve. In Fig. 2.35, $C_N/(s/l)^2$ is plotted versus $\alpha/(s/l)$, and the several sets of experimental data shown in this figure follow the

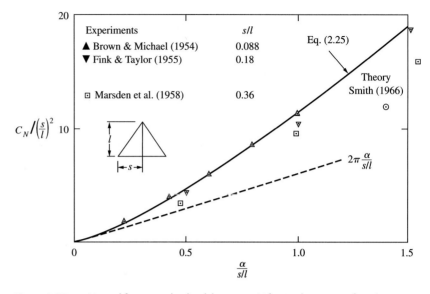

Figure 2.35   Normal forces on slender delta wings. (After Kuchemann, Ref. 24.)

same trend. Moreover, the experimental data are in reasonable agreement with the approximate analytical result of J. H. B. Smith, given by

$$\frac{C_N}{(s/l)^2} = 2\pi \left(\frac{\alpha}{s/l}\right) + 4.9 \left(\frac{\alpha}{s/l}\right)^{1.7} \qquad \text{Low-speed delta wing} \qquad \textbf{[2.25]}$$

where α is in radians. Note that Eq. (2.25) shows C_N as a *nonlinear* function of angle of attack, consistent with the experimental data for delta wings.

Example 2.13

Using Eq. (2.25), calculate the low-speed lift coefficient of a delta wing of aspect ratio 1.46 at an angle of attack of 20°. This is the same delta wing for which the experimental data in Fig. 2.34 apply. Compare the calculated result with the data shown in Fig. 2.34.

Solution

From the geometry of the triangular planform shown in Fig. 2.35, the planform area S is given by

$$S = \frac{1}{2}(2s)l = sl$$

The value of s/l is determined by the aspect ratio as follows.

$$AR \equiv \frac{b^2}{S} = \frac{(2s)^2}{sl} = \frac{4s^2}{sl} = 4\frac{s}{l}$$

Hence,

$$\frac{s}{l} = \frac{AR}{4} = \frac{1.46}{4} = 0.365$$

The angle of attack α, in radians, is

$$\alpha = \frac{20}{57.3} = 0.349 \text{ rad}$$

Hence,

$$\frac{\alpha}{s/l} = \frac{0.349}{0.365} = 0.956$$

From Eq. (2.25),

$$\frac{C_N}{(s/l)^2} = 2\pi \left(\frac{\alpha}{s/l}\right) + 4.9 \left(\frac{\alpha}{s/l}\right)^{1.7}$$

$$= 2\pi (0.956) + 4.9(0.956)^{1.7} = 10.57$$

Thus,

$$C_N = 10.57 \left(\frac{s}{l}\right)^2 = 10.57(0.365)^2 = 1.408$$

$$C_L = C_N \cos \alpha = 1.408 \cos 20° = \boxed{1.323}$$

The experimental data in Fig. 2.34 give a value of $C_L = 0.95$ at $\alpha = 20°$; the accuracy of Eq. (2.25) is within 39% for this case. Equation (2.25) is in better agreement with the different experimental data shown in Fig. 2.35.

2.8.2 Wing-Body Combinations

We normally think of wings as the primary source for lift for airplanes, and quite rightly so. However, even a pencil at an angle of attack will generate lift, albeit small. Hence, lift is produced by the fuselage of an airplane as well as the wing. The mating of a wing with a fuselage is called a *wing-body combination*. The lift of a wing-body combination is *not* obtained by simply adding the lift of the wing alone to the lift of the body alone. Rather, as soon as the wing and body are mated, the flow field over the body modifies the flow field over the wing, and vice versa—this is called the *wing-body interaction*.

There is no accurate analytical equation which can predict the lift of a wing-body combination, properly taking into account the nature of the wing-body aerodynamic interaction. Either the configuration must be tested in a wind tunnel, or a computational fluid dynamic calculation must be made. We cannot even say in advance whether the combined lift will be greater or smaller than the sum of the two parts.

However, for subsonic speeds, we can take the following approach for preliminary airplane performance and design considerations. Figure 2.36 shows data obtained from Hoerner and Borst (Ref. 18) for a circular fuselage-midwing combination, as sketched at the top of the figure. The diameter of the fuselage is d, and the wingspan

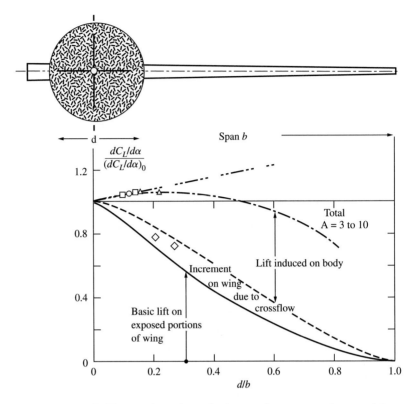

Figure 2.36   The lift-curve slope of wing-fuselage combinations as a function of the diameter ratio d/b. (After Hoerner and Borst, Ref. 18.)

is b. The lift slope of the wing-body combination, denoted by $dC_L/d\alpha$, divided by the lift slope of the wing alone, denoted by $(dC_L/d\alpha)_0$, is shown as a function of d/b. The magnitudes of the three contributions to the lift are identified in Fig. 2.36 as (1) the basic lift due to exposed portions of the wing, (2) the increase in lift on the wing due to crossflow from the fuselage acting favorably on the pressure distribution on the wing, and (3) the lift on the fuselage, taking into account the interaction effect with the wing flow field. The interesting result shown in this figure is that, for a range of d/b from 0 (wing only) to 6 (which would be an inordinately fat fuselage with a short, stubby wing), the *total* lift for the wing-body combination is essentially constant (within about 5%). Hence, the lift of the wing-body combination can be treated as simply the lift on the complete wing by itself, including that portion of the wing that is masked by the fuselage. This is illustrated in Fig. 2.37. In other words, the lift of the wing-body combination shown in Fig. 2.37**a** can be approximated by the lift on the wing of planform area S shown in Fig. 2.37**b**. This is the same as saying that the wing lift is effectively carried over by the fuselage for that part of the wing that is masked by the fuselage. For subsonic speeds, this is a reasonable approximation for preliminary airplane performance and design considerations. Hence, in all our future references to the planform area of a wing of an airplane, it will be construed

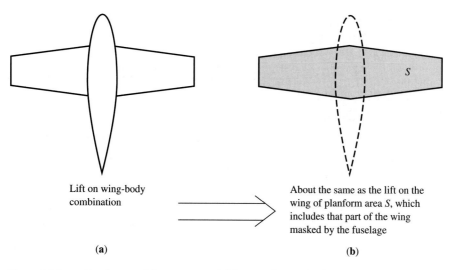

Lift on wing-body combination

About the same as the lift on the wing of planform area *S*, which includes that part of the wing masked by the fuselage

(a)   (b)

Figure 2.37   Significance of the conventional definition of wing planform area.

as the area *S* shown in Fig. 2.37**b**, and the lift of the wing-body combination will be considered as the lift on the wing alone of area *S*.

Wing-body interactions at supersonic speed can involve complex shock wave interactions and impingements on the surface. We will make no effort here to examine such interactions. In practice, we must usually depend on wind tunnel tests and/or computational fluid dynamic calculations for the aerodynamic properties of such supersonic configurations.

2.8.3   Drag

When you watch an airplane flying overhead, or when you ride in an airplane, it is almost intuitive that your first aerodynamic thought is about lift. You are witnessing a machine that, in straight and level flight, is producing enough aerodynamic lift to equal the weight of the machine. This keeps it in the air—a vital concern. Indeed, in Sections 2.8.1 and 2.8.2, we discuss the production of lift at some length. But this is only part of the role of aerodynamics. It is equally important to produce this lift as *efficiently as possible*, that is, with as little drag as possible. The ratio of lift to drag L/D is a good measure of aerodynamic efficiency. In Section 2.1 we mentioned that a barn door will produce lift at angle of attack, but it also produces a lot of drag at the same time—the L/D for a barn door is terrible. For such reasons, minimizing drag has been one of the strongest drivers in the historical development of applied aerodynamics. In airplane performance and design, drag is perhaps the most important aerodynamic quantity. The purpose of this section is to focus your thoughts on drag and to provide some methods for its estimation.

The subject of drag has been made confusing historically because so many different types of drag have been defined and discussed over the years. However, we can easily cut through this confusion by recalling the discussion in Section 2.2. There are only two sources of aerodynamic force on a body moving through a fluid—the pressure distribution and the shear stress distribution acting over the body surface. Therefore, *there are only two general types of drag:*

> *Pressure drag*—due to a net imbalance of surface pressure acting in the drag direction
>
> *Friction drag*—due to the net effect of shear stress acting in the drag direction

All the different types of drag that have been defined in the literature fall in one or the other of the above two categories. It is important to remember this.

It is also important to recognize that the analytical prediction of drag is much harder and more tenuous than that of lift. Drag is a different kind of beast—it is driven in large part by viscous effects. Closed-form analytical expressions for drag exist only for some special cases. Even computational fluid dynamics is much less reliable for drag predictions than for lift. Indeed, in a recent survey by Jobe (Ref. 27), the following comment is made:

> Except for the isolated cases of drag due to lift at small angle of attack and supersonic wave drag for smooth, slender bodies, drag prediction is beyond the capability of current numerical aerodynamic models.

However, faced with this situation, people responsible for airplane design and analysis have assimilated many empirical data on drag, and have synthesized various methodologies for drag prediction. About these methodologies, Jobe (Ref. 27) states:

> Each has its own peculiarities and limitations. Additionally each airframe manufacturer has compiled drag handbooks that are highly prized and extremely proprietary.

Hence, in this section we will be able to provide analytical formulas for only a few aspects of drag prediction. In lieu of such formulas, we will explore some of the empirical aspects of drag, and hopefully will give you some idea of what can be done to predict drag for purposes of preliminary performance analyses and conceptual design of airplanes.

We organized our discussion of lift in Sections 2.8.1 and 2.8.2 around different wing and body shapes. The effect of Mach number for each shape was dealt with in turn. However, the physical nature of drag, as well as its prediction, is more fundamentally affected by Mach number than is lift. Therefore, we will organize our discussion of drag around the different Mach-number regimes: subsonic, transonic, and supersonic.

Subsonic Drag

Airfoils   Let us first consider the case of drag on a two-dimensional airfoil shape in subsonic flow. We have already discussed this matter somewhat in Section 2.5; variations of the airfoil drag coefficient are shown in Figs. 2.6**b**, 2.9, 2.11, and 2.18. Return to Fig. 2.18, for example, where the drag coefficient for an NACA 64-212 airfoil is shown as a function of c_l, and hence as a function of α (due to the linear variations of c_l with α). The drag coefficient in this figure is labeled the section drag coefficient; it is also frequently called the *profile drag coefficient*. Profile drag is a combination of two types of drag:

$$\text{Profile drag} = \begin{bmatrix} \text{skin-friction} \\ \text{drag} \end{bmatrix} + \begin{bmatrix} \text{pressure drag due} \\ \text{to flow separation} \end{bmatrix}$$

Skin-friction drag is self-explanatory; it is due to the frictional shear stress acting on the surface of the airfoil. *Pressure drag due to flow separation* is caused by the imbalance of the pressure distribution in the drag direction when the boundary layer separates from the airfoil surface. (Note that, for an inviscid flow with no flow separation, theoretically the pressure distribution on the back portion of the airfoil creates a force pushing forward, which is exactly balanced by the pressure distribution on the front portion of the airfoil pushing backward. Hence, in a subsonic inviscid flow over a two-dimensional body, there is no net pressure drag on the airfoil—this phenomenon is called *d'Alembert's paradox* after the eighteenth-century mathematician who first obtained the result. In contrast, when the flow separates from the airfoil, the integrated pressure distribution becomes unbalanced between the front and back parts of the airfoil, producing a net drag force. This is the pressure drag due to flow separation.) Frequently, the pressure drag due to flow separation is called simply the *form drag*. In coefficient form, we have

$$
\begin{array}{ccccc}
c_d & = & c_f & + & c_{d,p} \\[6pt]
\begin{pmatrix} \text{Profile} \\ \text{drag coefficient} \end{pmatrix} & = & \begin{pmatrix} \text{skin-friction} \\ \text{drag coefficient} \end{pmatrix} & + & \begin{pmatrix} \text{form drag coefficient,} \\ \text{or pressure drag coefficient} \\ \text{due to flow separation} \end{pmatrix}
\end{array}
\qquad \textbf{[2.26]}
$$

For relatively thin airfoils and wings, c_f can be approximated by formulas for a flat plate. But even here there are major uncertainties in regard to the transition of laminar flow to turbulent flow in the boundary layer. Turbulence is still a major unsolved problem in classical physics, and the prediction of where on a surface transition occurs is uncertain. For a purely laminar flow, c_f for a flat plate in incompressible flow is given by

$$c_f = \frac{1.328}{\sqrt{\text{Re}}} \qquad \text{laminar} \qquad \textbf{[2.27]}$$

where $c_f = D_f/(q_\infty S)$, $\text{Re} = \rho_\infty V_\infty c/\mu_\infty$, D_f is the friction drag on *one* side of the flat plate, S is the planform area of the plate, c is the length of the plate in the flow direction (the chord length for an airfoil), and ρ_∞, V_∞, and μ_∞ are the free-stream density, velocity, and viscosity coefficient, respectively. Equation (2.27) is an *exact* theoretical relation for laminar incompressible flow over a flat plate. No such exact result exists for turbulent flow. Instead, a number of different approximate relations have been developed over the years. The results of various empirical flat-plate formulas for incompressible turbulent flow are shown in Fig. 2.38, where c_f is plotted versus Re. For reference, the Karman-Schoenherr curve shown in Fig. 2.38

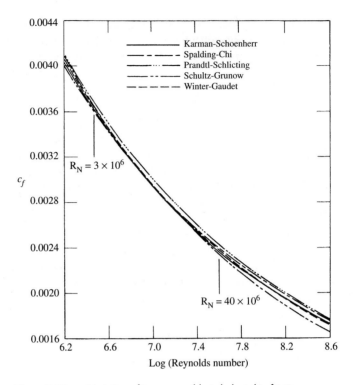

Figure 2.38   Variation of incompressible turbulent skin-friction coefficient for a flat plate as a function of Reynolds number.

is obtained from the relation

$$(c_f)^{-1/2} = 4.13 \log(\text{Re } c_f) \qquad \text{turbulent} \qquad \textbf{[2.28]}$$

which is one of the most widely used formulas for estimating turbulent flat-plate skin friction. The calculation of c_f from Eq. (2.28) must be done implicitly. Jobe (Ref. 27) recommends an alternate formula developed by White and Christoph in which c_f is more easily calculated in an explicit manner from

$$c_f = \frac{0.42}{\ln^2(0.056 \text{ Re})} \qquad \textbf{[2.29]}$$

Equation (2.29) is claimed to be accurate to ±4% in the Reynolds number range from 10^5 to 10^9. However, there remains the question as to *where* to apply the above formulas, which is a matter of where transition occurs. Equation (2.27) is valid as long as the flow is completely laminar. Equations (2.28) and (2.29) are applicable as long as the flow is completely turbulent. The latter is a reasonable assumption for most conventional airplanes in subsonic flight; the flow starts out laminar at the leading edge, but at the high Reynolds numbers normally encountered in flight, the extent of laminar flow is very small, and transition usually occurs very near the leading edge—so close that we can frequently assume that the surface is completely covered with a turbulent boundary layer. The location at which transition actually occurs on the surface is a function of a number of variables; suffice it to say that the transition Reynolds number is

$$\text{Re}_{\text{trans}} = \frac{\rho_\infty V_\infty x_{\text{tr}}}{\mu_\infty} \approx 350{,}000 \text{ to } 1{,}000{,}000$$

for low-speed flows, where x_{tr} is the distance of the transition point along the surface measured from the leading edge. Generally, a predicted value of x_{tr} is quite uncertain. For this reason, many preliminary drag estimates simply assume that the boundary layer is turbulent starting right at the leading edge.

To return to Eq. (2.26), the analytical prediction of $c_{d,p}$, the form drag coefficient, is still a current research question. No simple equations exist for the estimation of $c_{d,p}$, nor does computational fluid dynamics always give the right answer. Instead, $c_{d,p}$ is usually found from experiment. [What really happens is that the net profile drag coefficient c_d in Eq. (2.26) is measured, such as given in Fig. 2.18, and then $c_{d,p}$ can be backed out of Eq. (2.26) if a reasonable estimate of c_f exists.]

At subsonic speeds below the drag-divergence Mach number, the variation of c_d with Mach number is very small; indeed, for a first approximation it is reasonable to assume that c_d is relatively constant across the subsonic Mach number range. This is reflected in the left-hand side of Fig. 2.11.

Finite Wings   Consider the subsonic drag on a finite wing. This drag is more than just the profile drag. The same induced flow effects due to the wing-tip vortices that were discussed in Section 2.8.1 result in an extra component of drag on a three-dimensional lifting body. This extra drag is called *induced drag*. Induced drag is purely a pressure drag. It is caused by the wing tip vortices which generate an induced, perturbing flow field over the wing, which in turn perturbs the pressure distribution over the wing surface in such a way that the integrated pressure distribution yields an increase in drag—the induced drag D_i. For a high-aspect-ratio straight wing, Prandtl's lifting line theory shows that the induced drag coefficient, defined by

$$C_{D_i} \equiv \frac{D_i}{q_\infty S}$$

is given by

$$C_{D_i} = \frac{C_L^2}{\pi e \text{AR}} \qquad \qquad \textbf{[2.30]}$$

where e is the span efficiency factor, given by

$$e = \frac{1}{1 + \delta} \qquad \qquad \textbf{[2.31]}$$

In Eq. (2.31), δ is calculated from lifting line theory. It is a function of aspect ratio and taper ratio and is plotted in Fig. 2.39. Note that $\delta \geq 1$, so that $e \leq 1$. Examining

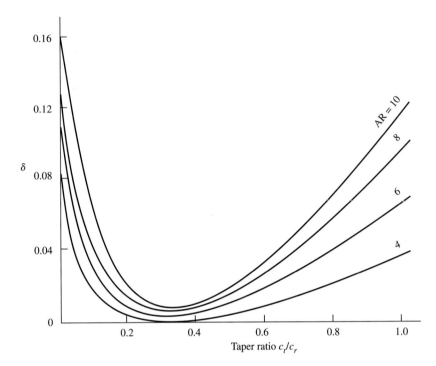

Figure 2.39   Induced drag factor as a function of taper ratio for wings of different aspect ratios.

Eq. (2.30), we see that it makes physical sense that C_{D_i} should be a function of the lift coefficient (and a strong function, at that, varying as the *square* of C_L). This is because the generation of wing-tip vortices is associated with a higher pressure over the bottom of the wing and a lower pressure over the top of the wing—the same mechanism that produces lift. Indeed, it would be naive for us to assume that lift is free. The induced drag is the penalty that is paid for the production of lift. Imagine, for example, a Boeing 747 weighing 500,000 lb in a straight and level flight. The airplane is producing 500,000 lb of lift. This costs money—the money to pay for the extra fuel consumed by the engines in producing the extra thrust necessary to overcome the induced drag.

   If our objective is to reduce the induced drag, Eq. (2.30) shows us how to do it. First, we want e to be as close to unity as possible. The value of e is always less than 1 except for a wing that has a spanwise lift distribution that varies elliptically over the span, for which $e = 1$. However, as seen in Fig. 2.39, δ is usually on the order of 0.05 or smaller for most wings, which means that e varies between 0.95 and 1.0—a relatively minor effect. Therefore, trying to design a wing that will have a spanwise lift distribution that is as close to elliptical as possible may not always be an important feature. Rather, from Eq. (2.30), we see that the *aspect ratio* plays a strong role; if we can double the aspect ratio, then we can reduce the induced drag by a factor of 2. The fact that increasing the aspect ratio reduces the induced drag also makes physical sense. Since $AR \equiv b^2/S$, for a wing of fixed area, increasing the aspect ratio moves the wing tips farther from the center of the wing. Since the strength of the induced flow due to the wing-tip vortices decays with lateral distance from each vortex, the farther removed the vortices, the weaker the overall induced flow effects and hence the smaller the induced drag. Thus, the clear message from Eq. (2.30) is that increasing the aspect ratio is the major factor in reducing the induced drag.

If aerodynamics were the only consideration in the design of an airplane, all subsonic aircraft would have wings with extremely large aspect ratios in order to reduce the induced drag—the wings would look like slats from a venetian blind. However, in order to make such a long, narrow wing structurally sound, the weight of the internal wing structure would be prohibitive. As a design compromise, the aspect ratios of most subsonic aircraft range between 6 and 9. The following is a list of the aspect ratios of some classic subsonic airplanes.

Airplane	Aspect Ratio
Lockheed Vega (Fig. 1.19)	6.11
Douglas DC-3 (Fig. 1.22)	9.14
Boeing 747 (Fig. 1.34)	7.0

Some special-purpose aircraft have larger aspect ratios. Sailplanes have aspect ratios that range from 10 to about 30. For example, the Schweizer SGS 1-35 has an aspect ratio of 23.3. The Lockheed U-2 reconnaissance aircraft (Fig. 2.40) has as aspect ratio of 14.3 and is capable of flying as high as 90,000 ft. [Reducing the induced drag for the U-2 was of paramount importance. At very high altitudes, where the

Figure 2.40   Lockheed U-2.

air density is low, the U-2 generates its lift by flying at high values of C_L. From Eq. (2.30), the induced drag is going to be large. To minimize this effect, the designers of the U-2 exerted every effort to make the aspect ratio as large as possible.]

We end this discussion about induced drag by noting that, in England, induced drag is usually called *vortex drag*. For some reason, this terminology has not been picked up in the United States. The term *induced drag* was coined by Ludwig Prandtl and Max Munk at Gottingen University in Germany in 1918, and we have carried on with this tradition to the present. This author feels that the descriptor *vortex drag* is much more explicit as to its source and is therefore preferable. However, in this book we continue with tradition and use the label *induced drag*.

Example 2.14

Consider the wing described in Example 2.5. For low-speed flow, calculate the lift-to-drag ratio for this wing at 6° angle of attack. Assume the span efficiency factor e is 0.95.

Solution

The induced drag coefficient is given by Eq. (2.30). From Example 2.5, at $\alpha = 6°$, $C_L = 0.648$. Hence, from Eq. (2.30),

$$C_{D_i} = \frac{C_L^2}{\pi e \mathrm{AR}} = \frac{(0.648)^2}{\pi (0.95)(6)} = 0.0234$$

The sum of the skin friction and form drag (pressure drag due to flow separation for the wing) is approximately given by the airfoil profile drag coefficient, plotted in Fig. 2.6b. From these data, when the airfoil is at 6° angle of attack ($c_l = 0.85$), the value of c_d is 0.0076 (assuming a Reynolds number on the order of 9×10^6). Hence, for the finite wing, the total drag coefficient is given by

$$C_D = c_d + C_{D_i} = 0.0076 + 0.0234 = 0.0312$$

The lift-to-drag ratio is

$$\frac{L}{D} = \frac{C_L}{C_D} = \frac{0.648}{0.031} = \boxed{20.9}$$

Note: Recall from Example 2.3 that for the *airfoil* at $\alpha = 6°$, $L/D = 111.8$, much higher than that for the finite wing. The dramatic reduction of L/D between the airfoil value and the finite-wing value is completely due to the finite-wing induced drag.

DESIGN CAMEO

Aspect ratio is one of the most important design features of an airplane. For subsonic airplane design, it is a major factor in determining the maximum value of L/D at cruise conditions, which in turn has a major impact on the maximum range of an airplane (discussed in Chapter 5). Everything else being equal, the higher the aspect ratio, the higher the maximum L/D. Of course, in any airplane design process, not everything else is equal. As noted earlier, as the design aspect ratio is increased, the wing structure must be made stronger. This increases the weight of the airplane, which is an undesirable feature. So the airplane designer is faced

(*continued*)

with a compromise—one of many in the airplane design process (as discussed in Chapters 7 and 8). However, the point made here is that, during the interactive design process, if it becomes important to increase the design value of the maximum L/D, then one of the powerful tools available to the designer is an increase in aspect ratio.

Fuselages   The fuselage by itself experiences substantial drag—a combination of skin-friction drag and pressure drag due to flow separation. The skin-friction drag is a direct function of the *wetted surface area* S_w, which is the area that would get wet if the fuselage were immersed in water. This makes physical sense because the shear stress is tugging at every square inch exposed to the airflow. The reference area used to define the drag coefficient is usually *not* the wetted surface area, which is fine because the reference area is just that—a reference quantity. But for some of our subsequent discussions it is useful to realize that the actual value of the aerodynamic skin-friction drag physically depends on the actual wetted surface area.

When the fuselage is mated to a wing and other appendages, the net drag is usually *not* the direct sum of the individual drags for each part. For example, the presence of the wing affects the airflow over the fuselage, and the fuselage affects the airflow over the wing. This sets up an interacting flow field over both bodies which changes the pressure distribution over both bodies. The net result is usually an increase in the pressure drag; this increase is called *interference drag*. Interference drag is almost always positive—the net drag of the combined bodies is almost always greater than the sum of the drags of the individual parts.

The prediction of interference drag is primarily based on previous experimental data. There are no analytical, closed-form expressions for such drag.

Summary   For subsonic drag, the following definitions for different contributions to the total drag are summarized below.

Skin-friction drag: Drag due to frictional shear stress integrated over the surface.

Pressure drag due to flow separation (form drag): The drag due to the pressure imbalance in the drag direction caused by separated flow.

Profile drag: The sum of skin friction drag and form drag. (The term *profile drag* is usually used in conjunction with two-dimensional airfoils; it is sometimes called *section drag*.)

Interference drag: An additional pressure drag caused by the mutual interaction of the flow fields around each component of the airplane. The total drag of the combined body is usually greater than that of the sum of its individual parts; the difference is the interference drag.

Parasite drag: The term used for the profile drag for a complete airplane. It is that portion of the total drag associated with skin friction and pressure drag due to flow separation, integrated over the complete airplane surface. It *includes* interference drag. We have more to say about parasite drag in Section 2.9.

Induced drag: A pressure drag due to the pressure imbalance in the drag direction caused by the induced flow (downwash) associated with the vortices created at the tips of finite wings.

Zero-lift drag: (Usually used in conjunction with a complete airplane configuration.) The parasite drag that exists when the airplane is at its zero-lift angle of attack, that is, when the lift of the airplane is zero. We elaborate in Section 2.9.

Drag due to lift: (Usually used in conjunction with a complete airplane.) That portion of the total airplane drag measured above the zero-lift drag. It consists of the *change* in parasite drag when the airplane is at an angle of attack different from the zero-lift angle, plus the induced drag from the wings and other lifting components of the airplane. We elaborate in Section 2.9.

The items summarized above are the main categories of drag. They need not be confusing as long as you keep in mind their physical source; each one is due to either skin friction or a pressure imbalance in the drag direction. As you begin to look at the airplane in greater detail, the above categories are sometimes broken down into more detailed subcategories. Here are a few such examples:

External store drag: An increase in parasite drag due to external fuel tanks, bombs, rockets, etc., carried as payload by the airplane, but mounted externally from the airframe.

Landing gear drag: An increase in parasite drag when the landing gear is deployed.

Protuberance drag: An increase in parasite drag due to "aerodynamic blemishes" on the external surface, such as antennas, lights, protruding rivets, and rough or misaligned skin panels.

Leakage drag: An increase in parasite drag due to air leaking into and out of holes and gaps in the surface. Air tends to leak in where the external pressure distribution is highest and to leak out where the external pressure distribution is lowest.

Engine cooling drag: An increase in parasite drag due to airflow through the internal cooling passages for reciprocating engines.

Flap drag: An increase in both parasite drag and induced drag due to the deflection of flaps for high-lift purposes.

Trim drag: The induced drag of the tail caused by the tail lift necessary to balance the pitching moments about the airplane's center of gravity. In a conventional rear-mounted tail, the lift of the tail is frequently downward to achieve this balance. When this is the case, the wing must produce extra lift to counter the downward lift on the tail; the resulting increase in the wing induced drag is then included in the trim drag.

This list can go on almost indefinitely. A good example of the drag buildup on a typical subsonic airplane is shown in Fig. 2.41. Here, we start with a completely

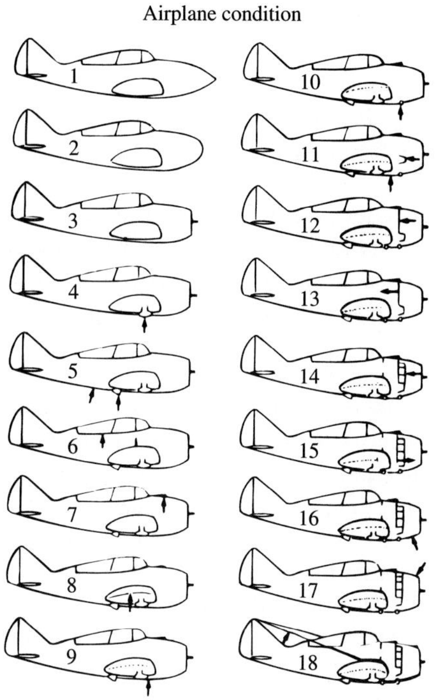

Airplane condition	Condition number	Description	C_D ($C_L = 0.15$)	ΔC_D	ΔC_D, %[a]
	1	Completely faired condition, long nose fairing	0.0166		
	2	Completely faired condition, blunt nose fairing	0.0169		
	3	Original cowling added, no airflow through cowling	0.0186	0.0020	12.0
	4	Landing-gear seals and fairing removed	0.0188	0.0002	1.2
	5	Oil cooler installed	0.0205	0.0017	10.2
	6	Canopy fairing removed	0.0203	−0.0002	−1.2
	7	Carburetor air scoop added	0.0209	0.0006	3.6
	8	Sanded walkway added	0.0216	0.0007	4.2
	9	Ejector chute added	0.0219	0.0003	1.8
	10	Exhaust stacks added	0.0225	0.0006	3.6
	11	Intercooler added	0.0236	0.0011	6.6
	12	Cowling exit opened	0.0247	0.0011	6.6
	13	Accessory exit opened	0.0252	0.0005	3.0
	14	Cowling fairing and seals removed	0.0261	0.0009	5.4
	15	Cockpit ventilator opened	0.0262	0.0001	0.6
	16	Cowling venturi installed	0.0264	0.0002	1.2
	17	Blast tubes added	0.0267	0.0003	1.8
	18	Antenna installed	0.0275	0.0008	4.8
		Total		0.0109	

[a]Percentages based on completely faired condition with long nose fairing.

Figure 2.41 The breakdown of various sources of drag on a late 1930s airplane, the Seversky XP-41. [Experimental data from Paul J. Coe, "Review of Drag Cleanup Tests in the Langley Full-Scale Tunnel (from 1935 to 1945) Applicable to Current General Aviation Airplanes," NASA TN-D-8206, 1976.]

streamlined basic configuration (condition 1 in Fig. 2.41), where the drag coefficient (for $C_L = 0.15$) is 0.0166. Conditions 2 through 18 progressively add various practical aspects to the basic configuration, and the change in drag coefficient for each addition as well as the running total drag are tabulated at the right in Fig. 2.41. For the complete configuration (condition 18), the total drag coefficient is 0.0275.

Transonic Drag Shock waves—that is the difference between transonic flow and purely subsonic flow. In a transonic flow, even though the free-stream Mach number is less than 1, local regions of supersonic flow occur over various parts of the airplane, and these local supersonic pockets are usually terminated by the presence of shock waves. This phenomenon has already been discussed in conjunction with airfoils and sketched in Fig. 2.11. Return to Fig. 2.11; we see the qualitative variation of c_d versus M_∞, and the prominent transonic drag rise near Mach 1. This drag rise is due to the presence of shock waves, as shown in Fig. 2.11; it is exclusively a *pressure* *drag* effect. It occurs in two ways. First, and primarily, the strong adverse pressure gradient across the shock causes the boundary layer to separate from the surface—this creates pressure drag due to flow separation. Second, even if the boundary layer did not separate, there is a loss of total pressure across the shock which ultimately would cause a net static pressure imbalance in the drag direction—also a pressure drag. The net effect of these combined phenomena is the large drag rise near Mach 1 shown in Fig. 2.11.

Although Fig. 2.11 is for an airfoil, the same qualitative effect occurs for complete airplanes. For example, Fig. 2.42 shows the transonic drag rise for the Northrop T-38 jet trainer. Here, the zero-lift drag coefficient $C_{D,0}$ is plotted versus free-stream Mach number; note that $C_{D,0}$ experiences about a factor-of-3 increase in the transonic regime.

No closed-form analytical formulas exist to predict the transonic drag rise. Even computational fluid dynamics, which has been applied to the computation of transonic flows for more than 25 years, does not always give the right answer, principally due to uncertainties in the calculation of the shock-induced separated flow. Jobe (Ref. 27) states: "The numerous authors in the field of numerical transonic aerodyanmics have reached a consensus: Transonic drag predictions are currently unreliable by any method." The burden of transonic drag prediction falls squarely on empirical data from wind tunnel tests and flight experiments. However, in spite of the difficulty of predicting the transonic drag rise, there are two principal design features that have been developed in the last half of the twentieth century which serve to reduce the drag rise itself, or to delay its effect: the *transonic area rule* and the *supercritical airfoil*. Let us briefly examine these features.

Area Rule   We first mentioned the area rule in conjunction with the F-102 delta wing fighter shown in Fig. 1.31. The essence of the area rule is sketched in Figs. 2.43 and 2.44. In Fig. 2.43**a**, the top view of a non-area-ruled airplane is shown; here, the variation of the cross-sectional area with the longitudinal distance is not smooth, that is, it has some discontinuities in it, particularly where the cross-sectional area of the wing is added to that of the fuselage. Prior to the early 1950s, aircraft designers did not realize that the kinks in the cross-sectional area distribution caused a large transonic drag rise. However, in the mid-1950s, principally based on the highly intuitive experimental work of Richard Whitcomb, an aerodynamicist at NACA Langley Aeronautical Laboratory, it became evident that the cross-sectional area distribution for transonic and supersonic airplanes should be smooth—no kinks. This can be achieved in part by decreasing the cross-sectional area of the fuselage in the wing region to compensate for the cross-sectional area increase due to the wings. Such an area-ruled airplane is sketched in Fig. 2.43**b**. The area ruling causes the fuselage to have a "Coke bottle" shape. The effect of area ruling is to reduce the peak transonic drag rise, as sketched in Fig. 2.44. The actual drag data for the F-102 before and after area ruling are given in Fig. 2.45. The minimum drag coefficient is plotted versus the free-stream Mach number for (**a**) the original, non-area-ruled prototype (solid curve) and (**b**) the modified, area-ruled airplane (labeled *revised* in Fig. 2.45 and given by the dashed line). Note the decrease in peak drag coefficient for the

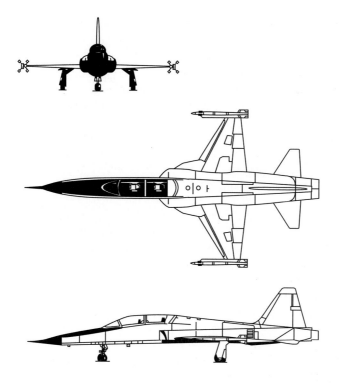

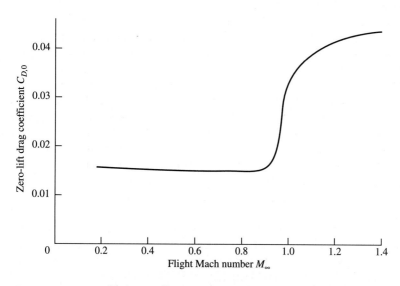

Figure 2.42 Zero-lift drag coefficient variation with Mach number, and three-view, for the Northrop T-38 jet trainer (U.S. Air Force).

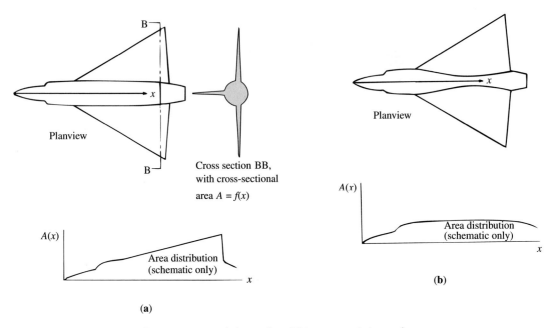

Figure 2.43   Schematics of (**a**) a non-area-ruled aircraft and (**b**) an area-ruled aircraft.

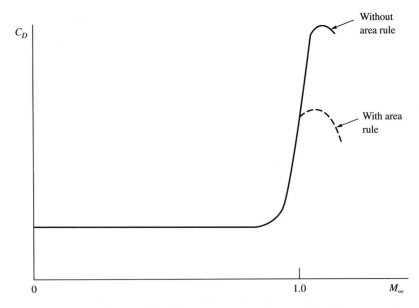

Figure 2.44   A schematic of the drag-rise properties of area-ruled and non-area-ruled aircraft.

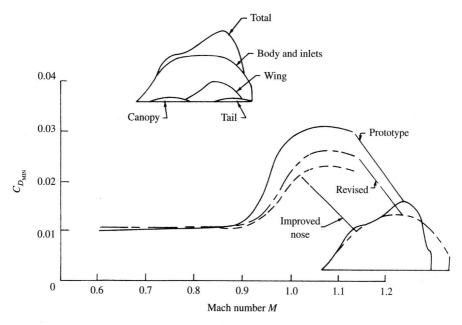

Figure 2.45 Minimum drag coefficient as a function of Mach number for the F-102; comparison of cases with and without area rule. (After Loftin, Ref. 13.)

area-ruled airplane. To the bottom right of Fig. 2.45, the cross-sectional area distributions of the two aircraft are shown; note the smoother, more regular variation for the area-ruled aircraft (the dashed curve). For the sake of reference, the area buildup of the original, non-area-ruled prototype is shown at the upper left of Fig. 2.45, illustrating the area contributions from various parts of the aircraft. For additional reference, the cross-sectional area buildup of a generic high-speed, area-ruled transport airplane is shown in Fig. 2.46, patterned after Refs. 26 and 28.

Supercritical Airfoil Return again to Fig. 2.11. Note that the drag-divergence Mach number M_{DD} occurs slightly above the critical Mach number M_{crit}. Conventional wisdom after World War II was that M_{DD} could be increased only by increasing M_{crit}. Indeed, the NACA laminar-flow airfoil series, particularly the NACA 64-series airfoils, were found to have relatively high values of M_{crit}. This is why the NACA 64-series airfoil sections found wide application on high-speed airplanes for several decades after World War II. This was not because of any possibility of laminar flow, as was the original intent of the airfoil design, but rather because, after the fact, these airfoil shapes were found to have values of M_{crit} higher than those for the other standard NACA airfoil families.

In 1965, Richard Whitcomb (of area-rule fame) developed a high-speed airfoil shape using a different rationale than that described above. Rather than increasing the value of M_{crit}, Whitcomb designed and tested a new family of airfoil shapes intended to increase the *increment* between M_{DD} and M_{crit}. The small increase of free-stream

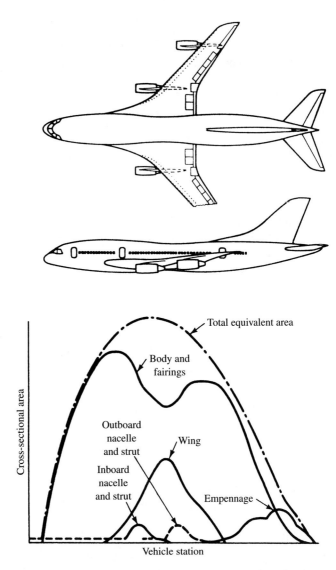

Figure 2.46 Cross-sectional area distribution breakdown for a typical, generic high-speed subsonic transport. (After Goodmanson and Gratzer, Ref. 28.)

Mach number above M_{crit} but before drag divergence occurs is like a "grace period"; Whitcomb worked to increase the magnitude of this grace period. This led to the design of the *supercritical airfoil* as discussed below.

The intent of supercritical airfoils is to increase the value of M_{DD}, not necessarily M_{crit}. This is achieved as follows. The supercritical airfoil has a relatively flat top, thus encouraging a region of supersonic flow with *lower* local values of M than those of the NACA 64 series. In turn, the terminating shock is weaker, thus creating less drag. The shape of a supercritical airfoil is compared with an NACA 64-series airfoil in Fig. 2.47. Also shown are the variations of the pressure coefficient C_p, for both airfoils. Figure 2.47**a** and **b** pertains to the NACA 64-series airfoil at Mach 0.69, and Fig. 2.47**c** and **d** is for the supercritical airfoil at Mach 0.79. In spite of the fact that the 64-series airfoil is at a lower M_∞, the extent of the supersonic flow reaches farther above the airfoil, the local supersonic Mach numbers are higher, and the terminating shock wave is stronger. Clearly, the supercritical airfoil shows more desirable flow field characteristics; namely, the extent of the supersonic flow is closer to the surface, the local supersonic Mach numbers are lower [as evidenced by smaller (in magnitude) negative values of C_p], and the terminating shock wave is weaker. As a result, the value of M_{DD} is higher for the supercritical airfoil. This is verified by the experimental data given in Fig. 2.48, taken from Ref. 29. Here, the value of M_{DD} is 0.79 for the supercritical airfoil in comparison with 0.67 for the NACA 64 series.

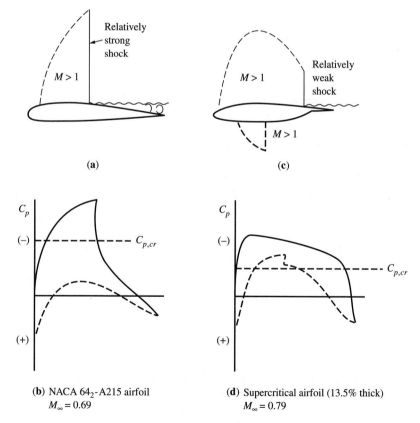

Figure 2.47   Standard NACA 64-series airfoil compared with a supercritical airfoil at cruise lift conditions.

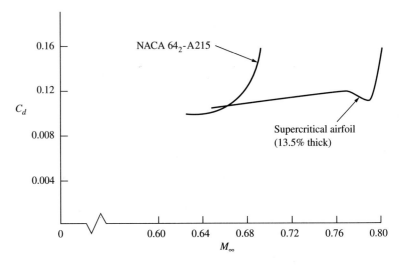

Figure 2.48   The drag-divergence properties of a standard NACA 64-series airfoil
and a supercritical airfoil.

Because the top of the supercritical airfoil is relatively flat, the forward 60%
of the airfoil has negative camber, which lowers the lift. To compensate, the lift is
increased by having extreme positive camber on the rearward 30% of the airfoil. This
is the reason for the cusplike shape of the bottom surface near the trailing edge.

A detailed description of the rationale as well as some early experimental data
for supercritical airfoils is given by Whitcomb in Ref. 29, which should be consulted
for more details.

Supersonic Drag   Shock waves are the dominant feature of the flow field around an
airplane flying at supersonic speeds. The presence of shock waves creates a pressure
pattern around the supersonic airplane which leads to a strong pressure imbalance in
the drag direction, and which integrated over the surface gives rise to *wave drag*.

Supersonic wave drag is a pressure drag. This is best seen in the supersonic flow
over a flat plate at angle of attack, as shown in Fig. 2.49. The shock and expansion
wave pattern creates a constant pressure on the bottom surface of the plate that is
larger than the free-stream pressure p_∞ and a constant pressure over the top surface
of the plate that is smaller than p_∞. This pressure distribution creates a resultant
aerodynamic force perpendicular to the plate, which is resolved into lift and drag, as
shown in Fig. 2.49. The drag is called *wave drag* D_w, because it is a ramification of
the supersonic wave pattern on the body. For small angles of attack, the lift slope is
expressed by Eq. (2.17), discussed earlier, which gives for the lift coefficient

$$c_l = \frac{4\alpha}{\sqrt{M_\infty^2 - 1}} \qquad\qquad \textbf{[2.32]}$$

The corresponding expression for the wave drag coefficient $c_{d,w}$ is

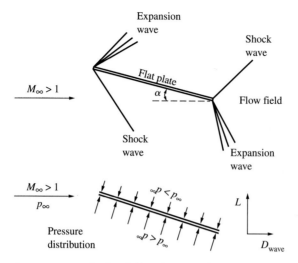

Figure 2.49   The flow field and pressure distribution for a flat plate at angle of attack in a supersonic flow.

$$c_{d,w} = \frac{4\alpha^2}{\sqrt{M_\infty^2 - 1}}$$ **[2.33]**

Since both lift and wave drag occur at angle of attack for the flat plate, and both are zero at $\alpha = 0$, the wave drag expressed by Eq. (2.33) is *wave drag due to lift*. This is in contrast to a body with thickness, such as the supersonic wedge at zero angle of attack, shown in Fig. 2.50. The pressure increase across the shock leads to a constant pressure along the two inclined faces that is greater than p_∞. The pressure decrease across the expansion waves at the corners of the base leads to a base pressure that is generally less than p_∞. Examining the pressure distribution over the wedge, as shown in Fig. 2.50, we clearly see that a net drag is produced. This is again called *wave drag*. But we also see from the surface pressure distribution in Fig. 2.50 that the lift will be zero. Hence, D_w in Fig. 2.50 is an example of *zero-lift wave drag*.

The above examples are just for the purpose of introducing the concept of supersonic wave drag, and to indicate that it consists of two parts:

(Wave drag) = (zero-lift wave drag) + (wave drag due to lift)   **[2.34]**

There exist various computer programs, based on small-perturbation linearized supersonic theory, for the calculation of supersonic wave drag. In fact, Jobe (Ref. 27) states:

Linear supersonic aerodynamic methods are the mainstay of the aircraft industry and are routinely used for preliminary design because of their simplicity and versatility despite their limitations to slender configurations at low lift coefficients. Not surprisingly most successful supersonic designs to date have adhered to the theoretical and geometrical limitations of these analysis methods.

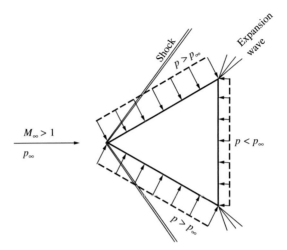

Figure 2.50   The flow field and pressure distribution for a wedge at 0° angle of attack in a supersonic flow.

At subsonic and transonic speeds, we ignored the effect of Mach number on the friction drag coefficient. However, at supersonic speeds, the effects of compressibility and heat transfer should be taken into account. Such matters are the subject of classical compressible boundary layer theory (e.g., see chapter 6 of Ref. 30). Here we will simply present some results for flat-plate skin-friction coefficients that can be used for preliminary design estimates. Figure 2.51, obtained from Ref. 30, gives the variation of the laminar skin-friction coefficient as a function of Mach number and wall–free-stream temperature ratio T_w/T_e. The Mach number variation accounts for compressibility effects, and the variation with T_w/T_e accounts for heat transfer at the surface. Figure 2.52, also obtained from Ref. 30, gives the variation of turbulent skin friction for an adiabatic wall as a function of Mach number. In Fig. 2.52, c_f is the compressible turbulent flat-plate skin-friction coefficients, and $c_{f_{inc}}$ is the incompressibile value, obtained from Eq. (2.28) or (2.29).

2.8.4   Summary

In this section on the buildup of lift and drag, we have dissected the aerodynamics of the airplane from the point of view of the properties of various components of the airplane, as well as the effects of different speed regimes—subsonic, transonic, and supersonic. In the process, we have presented

1.   Some physical explanations to help you better understand the nature of lift and drag, and to sort out the myriad definitions associated with our human efforts to understand this nature

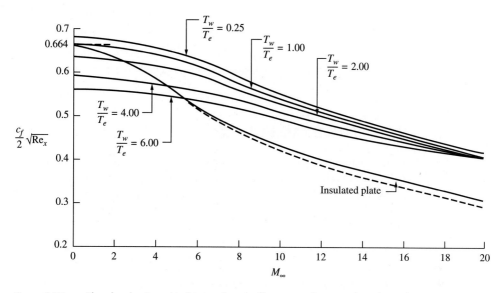

Figure 2.51 Flat-plate laminar skin-friction drag coefficient as a function of Mach number.

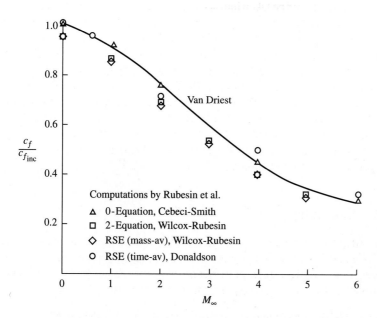

Figure 2.52 Flat-plate turbulent skin-friction drag coefficient as a function of Mach number: adiabatic wall, $Re_L = 10^7$.

2. Some equations, graphs, and approaches for the estimation of lift and drag for various components of the airplane, and how they fit together

We now move on to the concept of overall airplane lift and drag, and how it is packaged for our future discussions on airplane performance and design.

2.9   THE DRAG POLAR

In this section we treat the aerodynamics of the complete airplane, and we focus on a way in which the aerodynamics can be wrapped in a single, complete package—the drag polar. Indeed, the drag polar is the culmination of our discussion of aerodynamics in this chapter. Basically, all the aerodynamics of the airplane is contained in the drag polar. What is the drag polar? How can we obtain it? Why is it so important? These questions are addressed in this section.

2.9.1   More Thoughts on Drag

As a precursor to this discussion, and because drag is such a dominant consideration in airplane aerodynamics, it is interesting to compare the relative percentages for the various components of drag for typical subsonic and supersonic airplanes. This is seen in the bar charts in Fig. 2.53; the data are from Jobe (Ref. 27). These bar charts illustrate relative percentages; they do not give the actual magnitudes. A generic subsonic jet transport is treated in Fig. 2.53**a**; both cruise at Mach 0.8 and takeoff

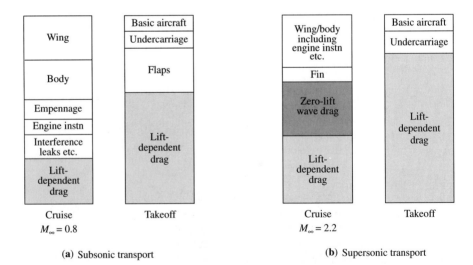

(**a**) Subsonic transport

(**b**) Supersonic transport

Figure 2.53   Comparison of cruise and takeoff drag breakdowns for (**a**) a generic subsonic transport and (**b**) a generic supersonic transport.

conditions are shown. Similarly, a generic slender, delta-wing, supersonic transport is treated in Fig. 2.53**b**; both cruise at Mach 2.2 and takeoff conditions are shown. Note the following aspects, shown in Fig. 2.53:

1. For the subsonic transport in Fig. 2.53**a**, the elements labeled wing, body, empennage, engine installations, interference, leaks, undercarriage, and flaps are the contributors to the zero-lift parasite drag; that is, they stem from friction drag and pressure drag (due to flow separation). The element labeled lift-dependent drag (drag due to lift) stems from the increment of parasite drag associated with the change in angle of attack from the zero-lift valve, and the induced drag. Note that most of the drag at cruise is parasite drag, whereas most of the drag at takeoff is lift-dependent drag, which in this case is mostly induced drag associated with the high lift coefficient at takeoff.

2. For the supersonic transport in Fig. 2.53**b**, more than two-thirds of the cruise drag is wave drag—a combination of zero-lift wave drag and the lift-dependent drag (which is mainly wave drag due to lift). This dominance of wave drag is the major aerodynamic characteristic of supersonic airplanes. At takeoff, the drag of the supersonic transport is much like that of the subsonic transport, except that the supersonic transport experiences more lift-dependent drag. This is because the low-aspect-ratio delta wing increases the induced drag, and the higher angle of attack required for the delta wing at takeoff (because of the lower lift slope) increases the increment in parasite drag due to lift.

Elaborating on the breakdown of subsonic cruise drag shown in Fig. 2.53**a**, we note that, of the total parasite drag at cruise, about two-thirds is usually due to skin friction, and the rest is form drag and interference drag. Since friction drag is a function of the total *wetted* surface area of the airplane (as noted in Section 2.4), an estimate of the parasite drag of the whole airplane should involve the wetted surface area. The wetted surface area S_{wet} can be anywhere between 2 and 8 times the reference planform area of the wing S. At the conceptual design stage of an airplane, the wetted surface area can be estimated based on historical data from previous airplanes. For example, Fig. 2.54 gives the ratio S_{wet}/S for a number of different types of aircraft, ranging from a flying wing (the B-2) to a large jumbo jet (the Boeing 747). Although not very precise, Fig. 2.54 can be used in the conceptual design stage to estimate S_{wet} for the given S and aircraft type. In turn, the zero-lift parasite drag D_0 can be expressed in terms of an equivalent skin friction coefficient C_{fe} and S_{wet} as follows:

$$D_0 = q_\infty S_{wet} C_{fe} \qquad \text{[2.35]}$$

In Eq. (2.35), C_{fe} is a function of Reynolds number based on mean chord length, as given in Fig. 2.55, after Jobe (Ref. 27). The equivalent skin-friction coefficient includes form drag and interference drag as well as friction drag. The more conventional zero-lift drag coefficient $C_{D,0}$ is defined in terms of the planform area S

$$C_{D,0} = \frac{D_0}{q_\infty S} \qquad \text{[2.36]}$$

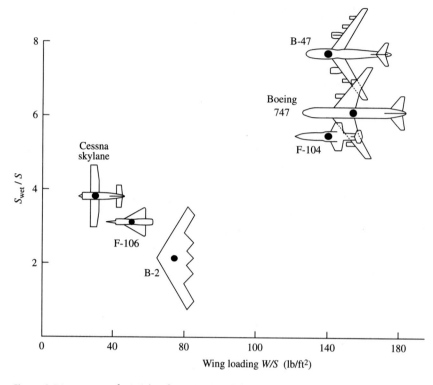

Figure 2.54   Ratio of wetted surface area to reference area for a number of different airplane configurations.

Substituting Eq. (2.35) into (2.36), we have

$$C_{D,0} = \frac{q_\infty S_{\text{wet}} C_{\text{fe}}}{q_\infty S} = \frac{S_{\text{wet}}}{S} C_{\text{fe}} \qquad \textbf{[2.37]}$$

Equation (2.37) can be used to obtain an estimate for $C_{D,0}$ by finding S_{wet}/S from Fig. 2.54 and C_{fe} from Fig. 2.55.

Example 2.15 | Estimate the zero-lift drag coefficient of the Boeing 747.

Solution

From Fig. 2.54, for the Boeing 747

$$S_{\text{wet}}/S = 6.3$$

From Fig. 2.55, given the assumption that the Boeing 747 and the Lockheed C-5 are comparable airplanes in size and flight conditions,

$$C_{\text{fe}} = 0.0027$$

Hence, from Eq. (2.37),

$$C_{D,0} = \frac{S_{\text{wet}}}{S} C_{\text{fe}} = (6.3)(0.0027) = \boxed{0.017}$$

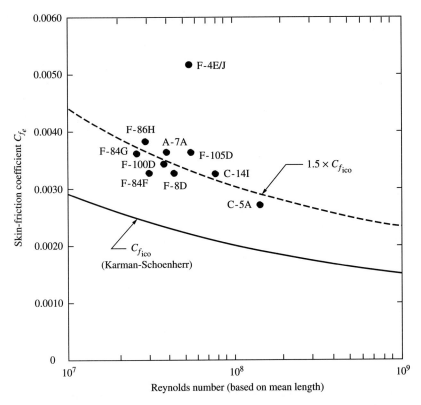

Figure 2.55 Equilvalent skin-friction drag for a variety of airplanes. (After Jobe, Ref. 27.)

2.9.2 The Drag Polar: What Is It and How Is It Used?

For every aerodyamic body, there is a relation between C_D and C_L that can be expressed as an equation or plotted on a graph. Both the equation and the graph are called the *drag polar*. Virtually all the aerodynamic information about an airplane necessary for a performance analysis is wrapped up in the drag polar. We examine this matter further and construct a suitable expression for the drag polar for an airplane.

From Section 2.8.3 on drag, we can write the total drag for an airplane as the following sum:

$$(\text{Total drag}) = (\text{parasite drag}) + (\text{wave drag}) + (\text{induced drag}) \qquad \textbf{[2.38]}$$

In coefficient form, Eq. (2.38) becomes

$$C_D = C_{D,e} + C_{D,w} + \frac{C_L^2}{\pi e \text{AR}} \qquad \textbf{[2.39]}$$

The parasite drag coefficient $C_{D,e}$ can be treated as the sum of its value at zero lift $C_{D,e,0}$ and the increment in parasite drag $\Delta C_{D,e}$ due to lift. Another way to look at the source of $\Delta C_{D,e}$ is to realize that lift is a function of angle of attack α and that $\Delta C_{D,e}$ is due to the change in orientation of the airplane, that is, the change in

α required to produce the necessary lift. That is, the skin-friction drag (to a lesser extent) and the pressure drag due to flow separation (to a greater extent) change when α changes; the sum of these changes creates $\Delta C_{D,e}$. Moreover, if we return to Fig. 2.6b, which is plot of c_d and c_l for an airfoil, we note that the change in c_d, denoted Δc_d, measured above its minimum value seems to vary approximately as the square of c_l. The source of c_d is friction drag and pressure drag due to flow separation (form drag). These physical phenomena are exactly the same source of $C_{D,e}$. Since Δc_d varies approximately as c_l^2, we can reasonably assume that $\Delta C_{D,e}$ varies as C_L^2. Indeed, we assume

$$C_{D,e} = C_{D,e,0} + \Delta C_{D,e} = C_{D,e,0} + k_1 C_L^2 \qquad [2.40]$$

where k_1 is a suitable proportionality constant.

Next, we can dissect the wave drag coefficient $C_{D,w}$ in a similar fashion; that is, $C_{D,w}$ is the sum of the zero-lift wave drag coefficient $C_{D,w,0}$ and the change $\Delta C_{D,w}$ due to lift. Recalling our discussion of supersonic drag in Section 2.8.3, we note that, for a flat plate at angle of attack, the substitution of Eq. (2.32) into (2.33) yields

$$c_{d,w} = \frac{4\alpha^2}{\sqrt{M_\infty^2 - 1}} = \frac{4}{\sqrt{M_\infty^2 - 1}} \left(\frac{c_l \sqrt{M_\infty^2 - 1}}{4} \right)^2$$

$$= \frac{c_l^2 \sqrt{M_\infty^2 - 1}}{4} \qquad [2.41]$$

Since $c_{d,w}$ is simply the wave drag coefficient due to lift, and since Eq. (2.41) shows that $c_{d,w}$ varies as c_l^2, we are comfortable with the assumption that $\Delta C_{D,w}$ varies as C_L^2. Hence,

$$C_{D,w} = C_{D,w,0} + \Delta C_{D,w} = C_{D,w,0} + k_2 C_L^2 \qquad [2.42]$$

where k_2 is an appropriate proportionality constant.

Substituting Eqs. (2.40) and (2.42) into Eq. (2.39), we have

$$C_D = C_{D,e,0} + C_{D,w,0} + k_1 C_L^2 + k_2 C_L^2 + \frac{C_L^2}{\pi e \text{AR}} \qquad [2.43]$$

In Eq. (2.43), define $k_3 \equiv 1/(\pi e \text{AR})$. Then Eq. (2.43) becomes

$$C_D = C_{D,e,0} + C_{D,w,0} + (k_1 + k_2 + k_3) C_L^2 \qquad [2.44]$$

The sum of the first two terms is simply the zero-lift drag coefficient $C_{D,0}$

$$C_{D,e,0} + C_{D,w,0} \equiv C_{D,0} \qquad [2.45]$$

Also, let

$$k_1 + k_2 + k_3 \equiv K \qquad [2.46]$$

Substituting Eqs. (2.45) and (2.46) into Eq. (2.44), we have for the complete airplane

$$\boxed{C_D = C_{D,0} + K C_L^2} \qquad [2.47]$$

Equation (2.47) is the *drag polar* for the airplane. In Eq. (2.47), C_D is the total drag coefficient, $C_{D,0}$ is the zero-lift parasite drag coefficient (usually called just the zero-lift drag coefficient), and KC_L^2 is the drag due to lift. The form of Eq. (2.47) is valid for both subsonic and supersonic flight. At supersonic speeds, $C_{D,0}$ contains the wave drag at zero lift, along with the friction and form drags, and the effect of wave drag due to lift is contained in the value used for K.

A graph of C_L versus C_D is sketched in Fig. 2.56. This is simply a plot of Eq. (2.47), hence the curve itself is also called the *drag polar*. The label *drag polar* for this type of plot was coined by the Frenchman Gustave Eiffel in 1909 (see Section 2.10). The origin of this label is easily seen in the sketch shown in Fig. 2.57. Consider an airplane at an angle of attack α, as shown in Fig. 2.57**a**. The resultant aerodynamic force R makes an angle θ with respect to the relative wind. If R and θ are drawn on a piece of graph paper, they act as polar coordinates which locate point a in Fig. 2.57**b**. If α is changed in Fig. 2.51**a**, then new values of R and θ are produced; these new values locate a second point, say point b, in Fig. 2.57**b**. The locus of all such points for all values of α forms the drag polar in Fig. 2.57**b**. Thus, the drag polar is nothing more than the resultant aerodynamic force plotted in polar coordinates—hence the name *drag polar*. Note that each point on the drag polar corresponds to a different angle of attack for the airplane. Also, note that a plot of L versus D, as shown in Fig. 2.57**b**, yields the same curve as a plot of C_L versus C_D, as shown in Fig. 2.56. In most cases, the drag polar is plotted in terms of the aerodynamic coefficients rather than the aerodynamic forces.

Another feature of the drag polar diagram, very closely related to that shown in Fig. 2.57**b**, is sketched in Fig. 2.58. Consider a straight line (the dashed line) drawn from the origin to point 1 on the drag polar. The length and angle of this line

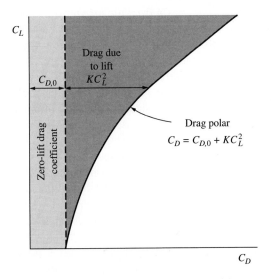

Figure 2.56 Schematic of the components of the drag polar.

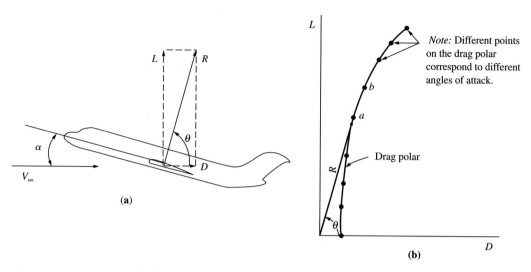

Figure 2.57   Construction for the resultant aerodynamic force on a drag polar.

correspond to the resultant force coefficient C_R and its orientation relative to the free-stream direction θ, as discussed above. Also, point 1 on the drag polar corresponds to a certain angle of attack α_1 of the airplane. The slope of the line 0–1 is equal to C_L/C_D, that is, lift-to-drag ratio. Now imagine that we ride up the polar curve shown in Fig. 2.58. The slope of the straight line from the origin will first increase, reach a maximum at point 2, and then decrease such as shown by line 0–3. Examining Fig. 2.58, we see that the line 0–2 is *tangent* to the drag polar. *Conclusion: The tangent line to the drag polar drawn from the origin locates the point of maximum lift-to-drag ratio for the airplane.* Moreover, the angle of attack associated with the tangent point α_2 corresponds to that angle of attack for the airplane when it is flying at $(L/D)_{\text{max}}$. Sometimes this tangent point (point 2 in Fig. 2.58) is called the *design point* for the airplane, and the corresponding value of C_L is sometimes called the *design lift coefficient* for the airplane. Also note from Fig. 2.58 that the maximum lift-to-drag ratio clearly does *not* correspond to the point of minimum drag.

There has been a subtlety in our discussion of the drag polar. In all our previous sketches and equations for the drag polar, we have tacitly assumed that the zero-lift drag is also the minimum drag. This is reflected in the vertex of each parabolically shaped drag polar in Figs. 2.56 to 2.58 being on the horizontal axis for $C_L = 0$. However, for real airplanes, this is usually not the case. When the airplane is pitched to its zero-lift angle-of-attack $\alpha_{L=0}$, the parasite drag may be slightly higher than the minimum value, which would occur at some small angle of attack slightly above $\alpha_{L=0}$. This situation is sketched in Fig. 2.59. Here, the drag polar in Fig. 2.56 has simply been translated vertically a small distance; the shape, however, stays the same. The equation for the drag polar in Fig. 2.59 is obtained directly from Eq. (2.47) by translating the value of C_L; that is, in Eq. (2.47), replace C_L with $C_L - C_{L_{\text{min drag}}}$. Hence, for the type of drag polar sketched in Fig. 2.59, the analytical equation is

$$C_D = C_{D_{\text{min}}} + K(C_L - C_{L_{\text{min drag}}})^2 \qquad \textbf{[2.48]}$$

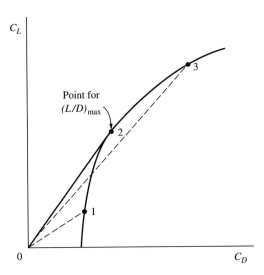

Figure 2.58   Construction for maximum lift-to-drag point on a drag polar.

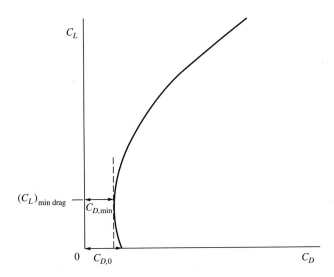

Figure 2.59   Illustration of minimum drag and drag at zero lift.

For airplanes with wings of moderate camber, the difference between $C_{D,0}$ and $C_{D_{min}}$ is very small and can be ignored. We make this assumption in this book, and hence we treat Eq. (2.47) as our analytical equation for the drag polar in the subsequent chapters.

For purposes of instruction, let us examine the drag polars for several real airplanes. The low-speed ($M_\infty < 0.4$) drag polar for the Lockheed C-141 military jet

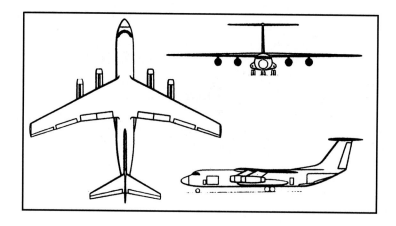

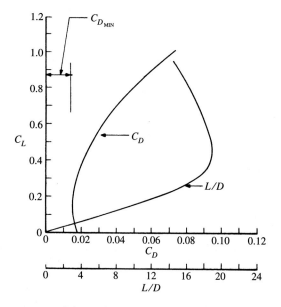

Figure 2.60   Low-speed drag polar and L/D variation for the Lockheed C-141A (shown in three-view).

transport is given in Fig. 2.60, and the drag polar at $M_\infty = 0.8$ for the McDonnell F4C jet fighter is given in Fig. 2.61. It is worthwhile studying these drag polars, just to obtain a feeling for the numbers for C_L and C_D. Also, note that each of the drag polars in Figs. 2.60 and 2.61 is for a given Mach number (or Mach number range). It is important to remember that C_L and C_D are functions of the Mach number; hence the same airplane will have different drag polars for different Mach numbers. At low subsonic Mach numbers, the differences will be small and can be ignored. However, at high subsonic Mach numbers, especially above the critical Mach number, and for supersonic Mach numbers, the differences will be large. This trend is illustrated in

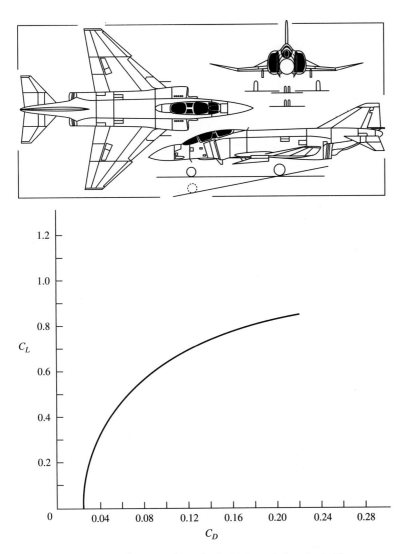

Figure 2.61 Drag polar at Mach 0.8 for the McDonnell–Douglas F4 Phantom (shown in three-view).

Fig. 2.62, which gives the drag polars for the McDonnell–Douglas F-15 jet fighter at 30,000-ft altitude for a range of free-stream Mach numbers. Subsonic and transonic drag polars are shown in Fig. 2.62**a**. Note the large increases in the minimum drag coefficient as the Mach number is increased through the transonic regime, and how this translates the entire drag polar to the right. This increase in $C_{D,\min}$ is to be expected; it is due to the drag-divergence effects illustrated, for example, in Fig. 2.11. Supersonic drag polars are shown in Fig. 2.62**b**. Here, we note a progressive decrease in $C_{D,\min}$ as M_∞ is increased, consistent with the supersonic trend illustrated in Fig. 2.11. Also note that the magnitude of C_L decreases as M_∞ is increased, consistent with the supersonic trend illustrated in Fig. 2.10. Hence, in Fig. 2.62**b**, as M_∞

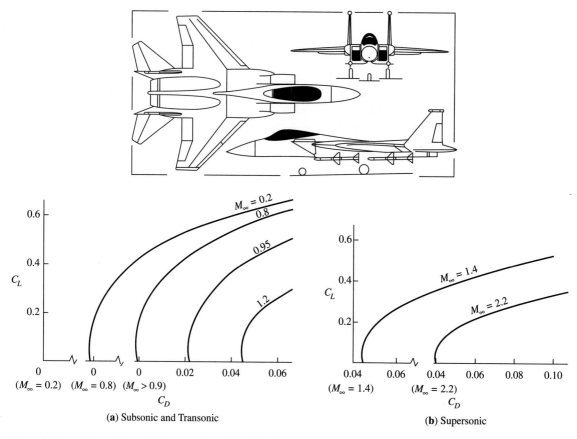

Figure 2.62   Drag polars at different Mach numbers for the McDonnell–Douglas F-15 (shown in three-view). **(a)** Subsonic and transonic speeds. **(b)** Supersonic speeds. Please note in parts **(a)** and **(b)** that the origin for C_D is different for different Mach numbers, as indicated by the broken abscissa.

increases, the supersonic drag polar shifts toward the left and gets "squashed down" closer to the horizontal axis.

Design Cameo

An accurate drag polar is essential to good airplane design. At the beginning of the preliminary design process (Chapters 7 and 8), every effort (theoretical and experimental) is made to obtain a good approximation for the drag polar. As the airplane design goes through iteration and refinement, the prediction of the drag polar also goes through a similar

iteration and refinement. With this in mind, let us reflect again on the two drag polars sketched in Figs. 2.56 and 2.59. In Fig. 2.56, the drag polar is for an airplane that has the minimum drag coefficient at zero lift. This would be the case, for example, for an airplane with a symmetric fuselage, a wing with a symmetric airfoil, and zero incidence angle

(continued)

between the wing chord and the axis of symmetry of the fuselage. Such an airplane would have zero lift at $0°$ angle of attack, and the drag would be a minimum at the same $0°$ angle of attack. In contrast, the drag polar sketched in Fig. 2.59, where the zero-lift drag coefficient is not the same as the minimum drag coefficient, applies to an airplane with some effective camber; the zero-lift drag coefficient $C_{D,0}$ is obtained at some angle of attack different from zero. This is the case for most airplanes, such as that shown in Fig. 2.60.

In the remainder of this book, we assume that the difference between $C_{D,0}$ and $C_{D,min}$ is small, and we will deal with the type of drag polar shown in Fig. 2.56. This has the advantage of leading to relatively straightforward analytical formulas for the various airplane performance characteristics discussed in Chapters 5 and 6—analytical formulas which are very useful to the designer in the preliminary design process. Moreover, many of the airplane performance characteristics are relatively insensitive to whether the form of the drag polar is given by Fig. 2.56 or 2.59, within reason. However, when the stage in the design process is reached where design *optimization* is carried out, it is important to deal with a more accurate drag polar as sketched in Fig. 2.59. Otherwise, the optimization process may converge to a misleading configuration. The reader is cautioned about this effect on the design. However, all the educational goals of the subsequent chapters are more readily achieved by assuming a drag polar of the form shown in Fig. 2.56, and hence we continue with this assumption.

2.10 HISTORICAL NOTE: THE ORIGIN OF THE DRAG POLAR—LILIENTHAL AND EIFFEL

The first drag polar in the history of aerodynamics was constructed by Otto Lilienthal in Germany toward the end of the nineteenth century. Lilienthal played a pivotal role in the development of aeronautics, as discussed in Chapter 1. Among his many contributions was a large bulk of aerodynamic lift and drag measurements on flat plates and thin, cambered airfoils, which he published in 1889 and in his classic book entitled *Birdflight as the Basis of Aviation* (Ref. 31). Later, these results were tabulated by Lilienthal; this became the famous *Lilienthal table* used by the Wright brothers in their early flying machine work. However, of interest in the present section is that in Ref. 31 Lilienthal also plotted his data in the form of drag polars.

Before we pursue this matter further, let us expand on our earlier discussion of Lilienthal in Chapter 1, and take a closer look at the man himself.

Otto Lilienthal was born in Anklam, Germany, on May 23, 1848, to middle-class parents. His mother was an educated woman, interested in artistic and cultural matters, and was a trained singer. His father was a cloth merchant who died when Otto was only 13 years old. Lilienthal was educated in Potsdam and Berlin; in 1870, he graduated with a degree in mechanical engineering from the Berlin Trade Academy (now the respected Technical University of Berlin). A photograph of Lilienthal is shown in Fig. 2.63. After serving in the Franco–Prussian War, Lilienthal married and went into business for himself. He obtained a patent for a compact, efficient, low-cost boiler, and in 1881 he opened a small factory in Berlin to manufacture his boilers. This boiler factory became his lifelong source of monetary income. However, throughout his adulthood, Lilienthal lived a simultaneous "second life," namely, that of an aerodynamic researcher and aeronautical enthusiast. As early as 1866, with the help of his brother Gustav, Lilienthal began a series of protracted

Figure 2.63   Otto Lilienthal (1848–1896).

aerodynamic experiments to measure the lift and drag on a variety of different-shaped lifting surfaces. In Lilienthal's words, these experiments continued "with some long interruptions until 1889." Lilienthal's measurements fell into two categories—those obtained with a whirling arm device and later those obtained outside in the natural wind. In 1889, Lilienthal finally gathered together his data and published them in Ref. 31, which has become one of the classics of pre-twentieth-century aeronautics. For a lengthy description and evaluation of Lilienthal's aerodynamics, see Ref. 8.

Figure 2.64 is one of many similar charts found in Lilienthal's book. It is a plot of the measured resultant aerodynamic force (magnitude and direction) for a range of angle of attack for a flat plate. The arrows from the origin (in the lower left corner) to the solid curve are the resultant force vectors; each arrow corresponds to a different angle of attack for the flat plate. The vertical and horizontal components of each arrow are the lift and drag, respectively. The solid curve is clearly a *drag polar*. Moreover, if we take the drag coefficient for a flat plate oriented perpendicular to the flow to be $C_D = 1$ (approximately true), then the length of the arrow at 90° can be considered a *unit* length, and relative to this unit length, the vertical and horizontal lengths of each arrow are equal to C_L and C_D, respectively. (See Ref. 8 for a full explanation.) In any event, Fig. 2.64 and the dozens of other similar plots for curved airfoils in Lilienthal's book represent the first drag polars in the history of aerodynamics.

Lilienthal's contributions to pre-twentieth-century aerodynamics were seminal. However, he is much more widely known for his development of the hang glider, and for his more than 2,000 successful glider flights during 1891 to 1896. Lilienthal developed the first successful, human-carrying gliders in the history of aeronautics. With these, he advanced the progress in aeronautics to a new height; he was the first person to find out what it takes to operate a flying machine in the air, even though an engine was not involved. Unfortunately, on the morning of August 9, 1896, during a flight in one of his gliders, Lilienthal encountered an unexpected thermal eddy which stalled his aircraft, and he crashed to the ground from a height of 50 ft. With a broken spine, Lilienthal died the next day in a Berlin clinic. As we first discussed in Chapter 1, at that time he had been working on an engine for his gliders, and there are some historians who feel that, had Lilienthal lived, he might have beaten the Wright brothers to the punch, and he might have been the first to fly a successful airplane. However, it was not to be.

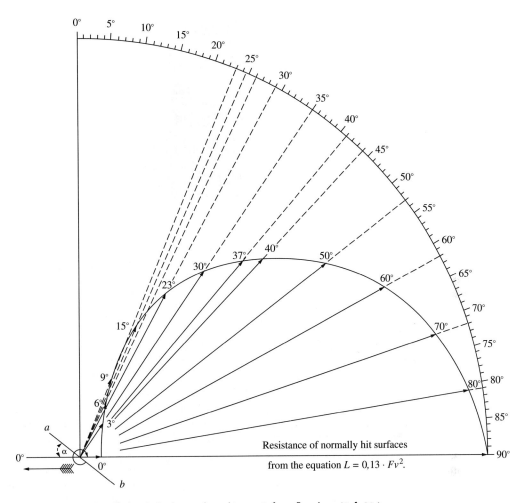

Figure 2.64 One of Lilienthal's drag polars; this one is for a flat plate. (Ref. 31.)

Even though Lilienthal was the first to construct a drag polar, he did not identify the plot as such. The name *drag polar* was coined about two decades later by Gustave Eiffel in Paris. Eiffel was a distinguished civil engineer who specialized in metal structures, and who is perhaps best known for the construction of the Eiffel Tower in Paris. A photograph of Eiffel is shown in Fig. 2.65. In the later years of his life, Eiffel became very active in aerodynamics. Beginning in 1902, he conducted a series of experiments by dropping various aerodynamic shapes from the Eiffel Tower and measuring their drag. In early 1909, he constructed a wind tunnel within the shadow of the tower, where he carried out extensive measurements of aerodynamic forces and pressure distributions on various wings and airplane models. The results of these tests were published in Ref. 32. Among the many plots in Ref. 32 are drag polars, which he referred to as *polar diagrams*. One such drag polar measured by Eiffel was for a model of the wing of the *Wright Flyer*; this drag polar is shown in Fig. 2.66 as the solid curve. Figure 2.66 is reproduced directly from Eiffel's book (Ref. 32).

Figure 2.65   Gustave Eiffel (1832–1923).

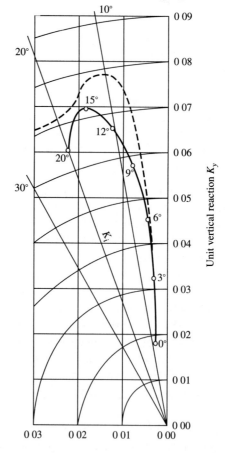

Unit horizontal reaction K_x

Figure 2.66   A drag polar for a wind tunnel model of the wing of the *Wright Flyer*, measured and published by Eiffel in 1910. (Ref. 32.)

From that time on to the present, all such diagrams have been called drag polars in the aerodynamic literature.

Eiffel contributed much more to the discipline of aerodynamics than that discussed above. For example, he designed a style of subsonic wind tunnel called the *Eiffel-type tunnel*; Eiffel-type tunnels are still widely used all across the world. He designed and tested airfoils; many of the French-built World War I airplanes used Eiffel airfoils. Eiffel continued an intensive program of aerodynamic research and development throughout the war and until his death in 1923 at the age of 91. For unexplained reasons, the fact that the builder of the Eiffel Tower was also the leading aerodynamicist in France during the period from 1902 to 1923 has become almost forgotten by modern aerodynamicists. Yet this intellectually powerful man contributed greatly to the historical development of aerodynamics after the turn of the century, and his legacy lives on in the way we do business in modern aerodynamics, especially in regard to experimental aerodynamics. For an extensive discussion of Eiffel and his contributions to aerodynamics, see Ref. 8.

2.11 SUMMARY

This has been a chapter on applied aerodynamics—aerodynamic concepts, formulas, and data to be applied to our discussions of airplane performance and design in subsequent chapters. Even though we have limited ourselves to applications of aerodynamics, we still have covered a wide range of topics. We have concentrated on the following aspects:

1. The sources of any aerodynamic force and moment on a body are the surface pressure distribution and the surface skin-friction distribution, integrated over the complete exposed surface of the body. Pressure distribution and skin-friction distribution—these are the two hands nature uses to reach out and grab a moving body immersed in a fluid.

2. Dimensionless coefficients are used to quantify these forces and moments. For a given shaped body, the lift, drag, and moment coefficients are functions of the angle of attack, Mach number, and Reynolds number. The question as to *how* C_L, C_D, and C_M vary with α, M_∞, and Re was examined.

3. There exists an aerodynamic center on a body, that is, that point about which moments may be finite, but do not vary with angle of attack. We set up a short procedure for calculating the location of the aerodynamic center.

4. There is an existing body of airfoil nomenclature. We looked at it and explained it.

5. Lift and drag on an airplane can be viewed as built up from those on various parts of the airplane—wing, fuselage, etc. However, the total lift and drag are *not* equal to the sum of the parts, due to aerodynamic interference effects.

6. Wing aerodynamics is a function of the wing shape. High-aspect-ratio straight wing, high-aspect-ratio swept wing, low-aspect-ratio wing, and a delta wing were the typical planform shapes considered here.

7. The drag polar, a plot of C_L versus C_D (or vice versa), contains almost all the necessary aerodynamics for an airplane performance analysis, and hence for a preliminary design of an airplane.

Look over the above list again. If the important details associated with each item do not readily come to mind, return to the pertinent section and refresh your memory. It is important that you have a comfortable feel for the applied aerodynamics discussed here. When you are ready, proceed to the next chapter, where we will examine some of the applied aspects of propulsion necessary for our subsequent airplane performance and design analyses.

PROBLEMS

2.1   We wish to design a wind tunnel test to accurately measure the lift and drag coefficients that pertain to the Boeing 777 in actual flight at Mach 0.84 at an altitude of 35,000 ft. The wingspan of the Boeing 777 is 199.9 ft. However, to fit in the wind tunnel test section, the wingspan of the wind tunnel model of the Boeing 777 is 6 ft. The pressure of the airstream in the test section of the wind tunnel is 1 atm. Calculate the necessary values of the airstream velocity, temperature, and density in the test section. Assume that the viscosity coefficient varies as the square root of the temperature. *Note:* The answer to this problem leads to an *absurdity*. Discuss the nature of this absurdity in relation to the real world of wind tunnel testing.

2.2   Consider an NACA 2412 airfoil (data given in Fig. 2.6) with chord of 1.5 m at an angle of attack of 4°. For a free-stream velocity of 30 m/s at standard sea-level conditions, calculate the lift and drag per unit span. *Note:* The viscosity coefficient at standard sea-level conditions is 1.7894×10^{-5} kg/(m·s).

2.3   For the airfoil and conditions in Problem 2.2, calculate the lift-to-drag ratio. Comment on its magnitude.

2.4   For the NACA 2412 airfoil, the data in Fig. 2.6a show that, at $\alpha = 6°$, $c_l = 0.85$ and $c_{m_{c/4}} = -0.037$. In Example 2.4, the location of the aerodynamic center is calculated as $x_{\text{a.c.}}/c = -0.0053$, where $x_{\text{a.c.}}$ is measured relative to the quarter-chord point. From this information, calculate the value of the moment coefficient about the aerodynamic center, and check your result with the measured data in Fig. 2.6b.

2.5   Consider a finite wing of aspect ratio 4 with an NACA 2412 airfoil; the angle of attack is 5°. Calculate (a) the lift coefficeint at low speeds (incompressible flow) using the results of Prandtl's lifting line theory, and (b) the lift coefficient for $M_\infty = 0.7$. Assume that the span efficiency factor for lift is $e_1 = 0.90$.

2.6   Using Helmbold's relation for low-aspect-ratio wings, calculate the lift coefficient of a finite wing of aspect ratio 1.5 with an NACA 2412 airfoil section. The wing is at an angle of attack of 5°. Compare this result with that obtained from Prandtl's lifting line theory for high-aspect-ratio wings. Comment on the different between the two results. Assume a span efficiency factor $e_1 = 1.0$.

2.7 Consider the wing described in Problem 2.5, except now consider the wing to be swept at 35°. Calculate the lift coefficient at an angle of attack of 5° for $M_\infty = 0.7$. Comparing this with the result of Problem 2.5b, comment on the effect of wing sweep on the lift coefficient.

2.8 Consider a wing with a thin, symmetric airfoil section in a Mach 2 airflow at an angle of attack of 1.5°. Calculate the lift cofficient
(*a*) For the airfoil section.
(*b*) For the wing if it is a straight wing with an aspect ratio of 2.56.
(*c*) For the wing if it is swept at an angle of 60°, with an aspect ratio of 2.56 and a taper ratio of unity. [*Note:* These are approximately the characteristics of the wing for the BAC (English Electric) Lightning supersonic fighter designed and built in England during the 1960s.]

2.9 The Anglo-French Concorde supersonic transport has an ogival delta wing with as aspect ratio of 1.7. Assuming a triangular planform shape, estimate the low-speed lift coefficient for this wing at an angle of attack of 25°.

2.10 Consider inviscid supersonic flow over a two-dimensional flat plate.
(*a*) What is the value of the maximum lift-to-drag ratio?
(*b*) At what angle of attack does it occur?

2.11 Consider viscous supersonic flow over a two-dimensional flat plate.
(*a*) Derive an expression for the maximum lift-to-drag ratio.
(*b*) At what angle of attack does it occur?
In parts (*a*) and (*b*), couch your results in terms of the skin-friction drag coefficient, $c_{d,f}$ and free-stream Mach number. Assume, that $c_{d,f}$ is independent of the angle of attack.

2.12 Estimate the zero-lift drag coefficient of the General Dynamics F-102.

3

Some Propulsion Characteristics

The chief obstacle (to successful powered flight) has hitherto been the lack of a sufficiently light motor in proportion to its energy; but there has recently been such marked advance in this respect, that a partial success with screws is even now almost in sight.

Octave Chanute, U.S. aeronautical pioneer;
from his *Progress in Flying Machines*, 1894

Since the beginning of powered flight, the evolutions of both the aero-vehicle and aeropropulsion systems are strongly interrelated, and are governed by a few major thrusts, namely: demands for improving reliability, endurance and lifetime; improvements in flight performance, such as speed, range, altitude maneuverability; and in more recent time, strongest emphasis on overall economy. Under these thrusts the technologies of aero-vehicle and propulsion system advanced continuously.

Hans von Ohain, German inventor
of the jet engine; comments made
in 1979 during a reflection of
the fortieth anniversary of the first
flight of a jet-propelled airplane

3.1   INTRODUCTION

Thrust and the way it is produced are the subjects of this chapter. In keeping with the spirit of Chapter 2 on aerodynamics, this chapter emphasizes only those aspects of flight propulsion that are necessary for our subsequent discussions of airplane performance and design. We examine in turn the following types of aircraft propulsion mechanisms:

1. Reciprocating engine/propeller
2. Turbojet
3. Turbofan
4. Turboprop

In each case, we are primarily concerned with two characteristics: thrust (or power) and fuel consumption. These are the two propulsion quantities that directly dictate the performance of an airplane. Also, note that missing from the above list is rocket engines. The use of rockets as the primary propulsion mechanism for airplanes is very specialized; the Bell X-1, the first airplane to fly faster than the speed of sound, and the North American X-15, the first airplane to fly at hypersonic speeds, are examples of aircraft powered by rocket engines. Rockets are also sometimes used for assisted takeoffs; JATO, which is an acronym for jet-assisted takeoff, is a bundle of small rockets mounted externally to the airplane, and it was used during and after World War II as a means of shortening the takeoff distance for some airplanes. However, we will not focus on rocket propulsion as a separate entity in this chapter.

Why do different aircraft propulsion devices exist? We have listed above four different devices, ranging from propellers connected to reciprocating engines or gas turbines, to pure turbojet engines. Of course, there is an historical, chronological thread. Beginning with Langley's *Aerodrome* and the *Wright Flyer*, the first airplanes were driven by propellers connected to internal combustion reciprocating engines. Then the invention of the jet engine in the late 1930s revolutionized aeronautics and allowed the development of transonic and supersonic airplanes. But this historical thread is not the answer to the question. For example, many airplanes today are still powered by the classical propeller/reciprocating engine combination, a full 50 years after the jet revolution. Why? There is a rather general, sweeping answer to these questions, having to do with the compromise between thrust and efficiency. This is the subject of the next section.

3.2   THRUST AND EFFICIENCY—THE TRADEOFF

In an elementary fashion, we can state that a propeller/reciprocating engine combination produces comparably low thrust with great efficiency, a turbojet produces considerably higher thrust with less efficiency, and a rocket engine produces tremendous thrust with poor efficiency. In this sense, there is a tradeoff—more thrust means less efficiency in this scenario. This tradeoff is the reason why all four propulsion mechanisms listed in Section 3.1 are still used today—the choice of a proper power plant for an airplane depends on what you want that airplane to do.

What is the technical reason for this tradeoff—thrust versus efficiency? First, let us consider the fundamental manner in which thrust is produced. (For a more detailed and elaborate discussion and derivation of the thrust equation, see, e.g., chapter 9 of Ref. 3, or any book on flight propulsion, such as Refs. 33 and 34.) Consider Fig. 3.1**a**, which shows a stream tube of air flowing from left to right through a generic propulsive device; this device may be a propeller, a jet engine, etc. The function of the propulsive device is to produce thrust T, acting toward the left, as sketched in Fig. 3.1**b**. No matter what type of propulsive device is used, the thrust is exerted on the device via the net resultant of the pressure and shear stress distributions acting on the exposed surface areas, internal and/or external, at each point where the air contacts any part of the device. This is consistent with our discussion of aerodynamic force in Chapter 2. The pressure and shear stress distributions are the two hands of nature that reach out and grab hold of any object immersed in an airflow. These two hands of

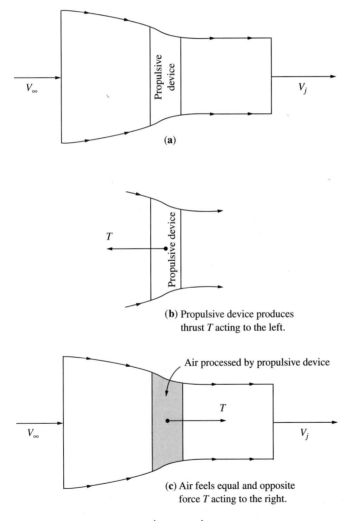

(a)

(b) Propulsive device produces
thrust T acting to the left.

(c) Air feels equal and opposite
force T acting to the right.

Figure 3.1   Reaction principle in propulsion.

nature grab the propulsive device and exert a force on it, namely, the thrust T, shown in Fig. 3.1**b**. The air exerts thrust on the device. However, from Newton's third law—namely, that for every action, there is an equal and opposite reaction—the propulsion device will exert on the air an equal and opposite force T, acting towards the right, as sketched in Fig. 3.1**c**. Now imagine that you are the air, and you experience the force T acting toward the right. You will accelerate toward the right; if your initial velocity is V_∞ far ahead of the propulsion device, you will have a larger velocity V_j downstream of the device, as sketched in Fig. 3.1**c**. We call V_j the *jet velocity*. The change in velocity $V_j - V_\infty$ is related to T through Newton's second law, which states that the force on an object is equal to the time rate of change of momentum of that object. Here, the "object" is the air flowing through the propulsion device, and the force on the air is T, as shown in Fig. 3.1**c**. Momentum is mass times velocity. Let $\dot{m}$ be the *mass flow* (for example, kg/s or slug/s) through the stream tube in Fig. 3.1**c**. We are assuming steady flow, so $\dot{m}$ is the same across any cross section of the stream tube. Hence, the momentum *per unit time* entering the stream tube at the left is $\dot{m}V_\infty$, and that leaving the stream tube at the right is $\dot{m}V_j$. Thus, the time rate of change of momentum of the air flowing through the propulsion device is simply the momentum flowing out at the right minus the momentum flowing in at the left, namely, $\dot{m}V_j - \dot{m}V_\infty$, or $\dot{m}(V_j - V_\infty)$. From Newton's second law, this time rate of change of momentum is equal to the force T. That is,

$$T = \dot{m}(V_j - V_\infty)$$ [3.1]

Equation (3.1) is the *thrust equation* for our generic propulsion device. (We note that a more detailed derivation of the thrust equation takes into account the additional force exerted by the pressure acting on the "walls" of the stream tube; for our analysis here, we are assuming this effect to be small, and we are ignoring it. For a more detailed derivation, see the control volume analysis in chapter 9 of Ref. 3.)

Let us now consider the matter of efficiency, which has a lot to do with the "wasted" kinetic energy left in the exhaust jet. In Fig. 3.1 we have visualized the situation when the propulsive device is stationary, and the air is moving through the device, with an initial upstream air velocity of V_∞. Clearly, velocities V_∞ and V_j are relative to the device. If we are sitting in the laboratory with the stationary device, we see the air moving both in front of and behind the device with velocities V_∞ and V_j, respectively. However, consider the equivalent situation where the propulsive device *moves* with a velocity V_∞ into stationary air, as shown in Fig. 3.2. This is the usual case in practice; the propulsive device is mounted on an airplane, and the airplane flies with velocity V_∞ into still air. *Relative to the device*, the flow picture is *identical* to that sketched in Fig. 3.1, with an upstream velocity relative to the device equal to V_∞ and a downstream velocity relative to the device equal to V_j. However, for us sitting in the laboratory, we do not see velocities V_∞ and V_j at all; rather, we see stationary air in front of the device, we see the device hurtling by us at a velocity V_∞, and we see the air behind the device moving in the opposite direction with a velocity (relative to the laboratory) of $V_j - V_\infty$, as shown in Fig. 3.2. In essence, before the moving device enters the laboratory, the air in the room is stationary, hence it has no kinetic

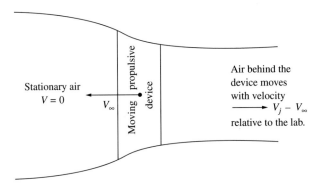

Figure 3.2   Sketch of the propulsive device moving into stationary air with velocity V_∞.

energy. After the device flies through the room, the air in the laboratory is no longer stationary; rather, it is moving in the opposite direction with velocity $V_j - V_\infty$. This moving air, which is left behind after the device has passed through the laboratory, has a kinetic energy per unit mass of $\frac{1}{2}(V_j - V_\infty)^2$. This kinetic energy is totally wasted; it performs no useful service. It is simply a loss mechanism associated with the generation of thrust. It is a source of inefficiency.

We can now define a propulsive efficiency as follows. Recall from basic mechanics that when you exert a force on a body moving at some velocity, the *power* generated by that force is

$$\text{Power} = \text{force} \times \text{velocity} \qquad \textbf{[3.2]}$$

See, for example, chapter 6 of Ref. 3 for a derivation of Eq. (3.2). Consider an airplane moving with velocity V_∞ being driven by a propulsion device with thrust T. The *useful* power, called the *power available* P_A provided by the propulsive device, is

$$P_A = T V_\infty \qquad \textbf{[3.3]}$$

However, the propulsive device is actually putting out more power than that given by Eq. (3.3) because the device is also producing the wasted kinetic energy in the air left behind. Power is energy per unit time. The wasted kinetic energy per unit mass of air is $\frac{1}{2}(V_j - V_\infty)^2$, as described above. Since $\dot{m}$ is the mass flow of air through the propulsive device (mass per unit time), then $\frac{1}{2}\dot{m}(V_j - V_\infty)^2$ is the *power wasted* in the air jet behind the device. Hence,

$$\text{Total power generated by propulsive device} = T V_\infty + \frac{1}{2}\dot{m}(V_j - V_\infty)^2 \qquad \textbf{[3.4]}$$

The propulsive efficiency, denoted by η_p, can be defined as

$$\eta_p = \frac{\text{useful power available}}{\text{total power generated}} \qquad \textbf{[3.5]}$$

Substituting Eqs. (3.3) and (3.4) into Eq. (3.5), we have

$$\eta_p = \frac{T V_\infty}{T V_\infty + \frac{1}{2}\dot{m}(V_j - V_\infty)^2} \qquad \text{[3.6]}$$

Substituting the thrust equation, Eq. (3.1), into Eq. (3.6), we have

$$\eta_p = \frac{\dot{m}(V_j - V_\infty)V_\infty}{\dot{m}(V_j - V_\infty)V_\infty + \frac{1}{2}\dot{m}(V_j - V_\infty)^2} \qquad \text{[3.7]}$$

By dividing numerator and denominator by $\dot{m}(V_j - V_\infty)V_\infty$, Eq. (3.7) becomes

$$\eta_p = \frac{1}{1 + \frac{1}{2}(V_j - V_\infty)/V_\infty} = \frac{1}{\frac{1}{2}(1 + V_j/V_\infty)}$$

or

$$\boxed{\eta_p = \frac{2}{1 + V_j/V_\infty}} \qquad \text{[3.8]}$$

The nature of the tradeoff between thrust and efficiency is now clearly seen by examining Eq. (3.1) with one eye and Eq. (3.8) with the other eye. From Eq. (3.8), maximum (100%) propulsive efficiency is obtained when $V_j = V_\infty$; for this case, $\eta_p = 1$. This makes sense. In this case, when the propulsion device hurtles through the laboratory at velocity V_∞ into the stationary air ahead of it, and the air is exhausted from the device with a velocity V_j relative to the device which is equal to the velocity of the device itself ($V_j = V_\infty$), then relative to the laboratory, the air simply appears to plop out of the back end of the device with no velocity. In other words, since the air behind the device is not moving in the laboratory, there is no wasted kinetic energy. On the other hand, if $V_j = V_\infty$, Eq. (3.1) shows that $T = 0$. Here is the compromise; we can achieve a maximum propulsive efficiency of 100%, but with no thrust—a self-defeating situation.

In this compromise, we can find the reasons for the existence of the various propulsion devices listed in Section 3.1. A propeller, with its relatively large diameter, processes a large mass of air, but gives the air only a small increase in velocity. In light of Eq. (3.1), a propeller produces thrust by means of a large $\dot{m}$ with a small $V_j - V_\infty$, and therefore in light of Eq. (3.8), η_p is high. The propeller is inherently the most efficient of the common propulsive devices. However, the thrust of a propeller is limited by the propeller tip speed; if the tip speed is near or greater than the speed of sound, shock waves will form on the propeller. This greatly increases the drag on the propeller, which increases the torque on the reciprocating engine, which reduces the rotational speed (rpm) of the engine, which reduces the power obtained from the engine itself, and which is manifested in a dramatic reduction of thrust. In addition, the shock waves reduce the lift coefficient of the affected airfoil sections making up the propeller, which further decreases thrust. The net effect is that, at high speeds, a propeller becomes ineffective as a good thrust-producing device. This is why there are no propeller-driven transonic or supersonic airplanes.

In contrast to a propeller, a gas-turbine jet engine produces its thrust by giving a comparably smaller mass of air a much larger increase in velocity. Reflecting on Eq. (3.1), we see that $\dot{m}$ may be smaller than that for a propeller, but $V_j - V_\infty$ is much larger. Hence, jet engines can produce enough thrust to propel airplanes to transonic and supersonic flight velocities. However, because V_j is much larger than V_∞, from Eq. (3.8) the propulsive efficiency of a jet engine will be less than that for a propeller.

Because of the tradeoffs discussed above, in modern aeronautics we see low-speed airplanes powered by the reciprocating engine/propeller combination, because of the increased propulsive efficiency, and we see high-speed airplanes powered by jet engines, because they can produce ample thrust to propel aircraft to transonic and supersonic speeds. We also see the reason for a turbofan engine—a large multiblade fan driven by a turbojet core—which is designed to generate the thrust of a jet engine but with an efficiency that is more reflective of propellers. An even more direct combination is a propeller driven by a gas-turbine engine—the turboprop—which has a nice niche with airplanes in the 300 to 400 mi/h range.

In summary, the purpose of this section has been to give you an overall understanding of the fundamental tradeoffs associated with different flight propulsion devices. This understanding is helpful for studies of airplane performance and discussions about airplane design. In the subsequent sections, we briefly examine those aspects of each class of propulsive device which are directly relevant to our considerations of airplane performance and design.

3.3 THE RECIPROCATING ENGINE/PROPELLER COMBINATION

The basic operation of a four-stroke spark-ignition engine is illustrated in Fig. 3.3. Illustrated here is a piston-cylinder arrangement, where the translating, up-and-down movement of the piston is converted to rotary motion of the crankshaft via a connecting rod. On the intake stroke (Fig. 3.3**a**), the intake valve is open, the piston moves down, and fresh fuel-air mixture is sucked into the cylinder. During the compression stroke (Fig. 3.3**b**), the valves are closed, the piston moves up, and the gas in the cylinder is compressed to a higher pressure and temperature. Combustion is initiated approximately at the top of the compression stroke; as a first approximation, the combustion is fairly rapid, and is relatively complete before the piston has a chance to move very far. Hence, the combustion is assumed to take place at constant volume. During combustion, the pressure increases markedly. This high pressure on the face of the piston drives the piston down on the power stroke (Fig. 3.3**c**). This is the main source of power from the engine. Finally, the exhaust valve opens, and the piston moves up on the exhaust stroke, pushing most of the burned fuel-air mixture out of the cylinder. Then the four-stroke cycle is repeated. This four-stroke internal combustion engine concept has been in existence for more than a century; it was developed by Nikolaus Otto in Germany in 1876 and patented in 1877. (Strangely enough, although Otto worked in Germany, his 1877 patent was taken out in the

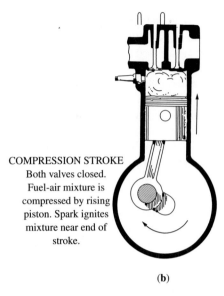

INTAKE STROKE
Intake valve opens,
thus admitting charge
of fuel and air. Exhaust
valve closed for most of
stroke.

Intake

Spark plug

Cylinder

Piston

Crank
(and crankshaft)

Connecting
rod

(a)

COMPRESSION STROKE
Both valves closed.
Fuel-air mixture is
compressed by rising
piston. Spark ignites
mixture near end of
stroke.

(b)

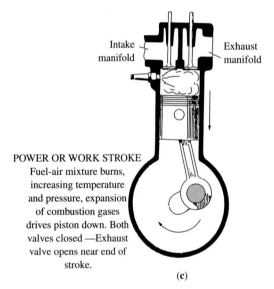

Intake
manifold

Exhaust
manifold

POWER OR WORK STROKE
Fuel-air mixture burns,
increasing temperature
and pressure, expansion
of combustion gases
drives piston down. Both
valves closed —Exhaust
valve opens near end of
stroke.

(c)

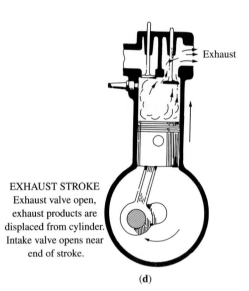

Exhaust

EXHAUST STROKE
Exhaust valve open,
exhaust products are
displaced from cylinder.
Intake valve opens near
end of stroke.

(d)

Figure 3.3   Diagram of the four-stroke Otto cycle for internal combustion spark-ignition engines. (*After Edward F. Obert, Internal Combustion Engines and Air Pollution, Intext, 1973.*)

United States.) Appropriately, the four-stroke process illustrated in Fig. 3.3 is called the Otto cycle.

The business end of the reciprocating engine is the rotating crankshaft—this is the means by which the engine's power is transmitted to the outside world—a wheel axle in the case of an automobile, or a propeller in the case of an airplane. On what characteristics of the engine does this power depend? The answer rests on three primary features. First, there is the shear size of the engine, as described by the *displacement.* On its travel from the top of a stroke (top dead center) to the bottom of the stroke (bottom dead center), the piston sweeps out a given volume, called the *displacement* of the cylinder. The total displacement of the engine is that for a cylinder, multiplied by the number of cylinders; we denote the displacement by d. The larger the displacement, the larger the engine power output, everything else being the same. Second, the number of times the piston moves through its four-stroke cycle per unit time will influence the power output. The more power strokes per minute, the greater the power output of the engine. Examining Fig. 3.3, we note that the shaft makes 2 revolutions (r) for each four-stroke cycle. Clearly, the more revolutions per minute (rpm), the more power will be generated. Hence, the power output of the engine is directly proportional to the *rpm*. Third, the amount of force applied by the burned gas on the face of the piston after combustion will affect the work performed during each power stroke. Hence, the higher the pressure in the cylinder during the power stroke, the larger will be the power output. An average pressure which is indicative of the pressure level in the cylinder is defined as the *mean effective pressure* p_e. Therefore, we can state that the power output from the engine to the crankshaft, called the *shaft brake power P*, is

$$\boxed{P \propto d p_e \text{RPM}} \qquad \text{[3.9]}$$

A typical internal combustion reciprocating engine is shown in Fig. 3.4.

Figure 3.4   Textron Lycoming TIO 540-AE2A turbocharged piston engine.

The *specific fuel consumption* is a technical figure of merit for an engine which reflects how efficiently the engine is burning fuel and converting it to power. For an internal combustion reciprocating engine, the specific fuel consumption c is defined as

$$c = \text{weight of fuel burned per unit power per unit time}$$

or

$$c = \frac{\text{weight of fuel consumed for given time increment}}{(\text{power output})(\text{time increment})} \qquad \textbf{[3.10]}$$

In this book, we will always use consistent units in our calculations, either the English engineering system or the international system (SI). (See chapter 2 of Ref. 3 for a discussion of the significance of consistent units.) Hence, c is expressed in terms of the units

$$[c] = \frac{\text{lb}}{(\text{ft·lb/s}) \, (\text{s})}$$

or

$$[c] = \frac{\text{N}}{\text{W·s}}$$

However, over the years, conventional engineering practice has quoted the specific fuel consumption in the inconsistent units of pounds of fuel consumed per horsepower per hour; these are the units you will find in most specifications for internal combustion reciprocating engines. To emphasize this difference, we will denote the specific fuel consumption in these inconsistent units by the symbol SFC. Hence, by definition,

$$[\text{SFC}] = \frac{\text{lb}}{\text{hp·h}}$$

Before making a calculation which involves specific fuel consumption, we always convert the inconsistent units of SFC to the consistent units of c.

3.3.1 Variations of Power and Specific Fuel Consumption with Velocity and Altitude

In Eq. (3.9), P is the power that comes from the engine shaft; it is sometimes called *shaft power*. Consider the engine mounted on an airplane. As the airplane velocity V_∞ is changed, the only variable affected in Eq. (3.9) is the pressure of the air entering the engine manifold, due to the stagnation of the airflow in the engine inlet. (Sometimes this is called a *ram effect*.) In effect, as V_∞ increases, this "ram pressure" is increased; it is reflected as an increase in p_e in Eq. (3.9), which in turn increases P via Eq. (3.9). For the high-velocity propeller-driven fighter airplanes of World War II, this effect had some significance. However, today reciprocating engines are used only on low-speed general aviation aircraft, and the ram effect can be ignored. Hence, we assume in this book that

$$\boxed{P \text{ is reasonably constant with } V_\infty}$$

For the same reason, the specific fuel consumption is also assumed to be independent of V_∞:

SFC is constant with V_∞

In the United States, two principal manufacturers of aircraft reciprocating engines are Teledyne Continental and Textron Lycoming. The horsepower ratings at sea level for these engines generally range from 75 to 300 hp. For these engines, a typical value of SFC is 0.4 lb of fuel consumed per horsepower per hour.

As the airplane's altitude changes, the engine power also changes. This is easily seen in Eq. (3.9). The air pressure (also air density) decreases with an increase in altitude; in turn this reduces p_e in Eq. (3.9), which directly reduces P. The variation of P with altitude is usually given as a function of the local air density. To a first approximation, we can assume

$$\frac{P}{P_0} = \frac{\rho}{\rho_0} \qquad \textbf{[3.11]}$$

where P and ρ are the shaft power output and density, respectively, at a given altitude and P_0 and ρ_0 are the corresponding values at sea level. There is also a temperature effect on mean effective pressure p_e in Eq. (3.9). An empirical correlation given by Torenbeck (Ref. 35) for the altitude variation of P is

$$\frac{P}{P_0} = 1.132\frac{\rho}{\rho_0} - 0.132 \qquad \textbf{[3.12]}$$

The specific fuel consumption is relatively insensitive to changes in altitude, at least for the altitude range for general aviation aircraft. Hence, we assume

SFC is constant with altitude

The decrease in power with increasing altitude, as indicated by Eqs. (3.11) and (3.12), is for engines *without superchargers*. As early as World War I, it was fully recognized that this decrease in power could be eliminated, or at least delayed to a higher threshold altitude, by compressing the manifold pressure to values above ambient pressure. This compression is carried out by a compressor geared to the engine shaft (a supercharger) or driven by a small turbine mounted in the engine exhaust (a turbosupercharger). These devices tend to maintain a constant value of p_e for the engine as the altitude increases, and hence from Eq. (3.9) the power is essentially constant with altitude. Supercharging was important for the high-performance military and civil transport airplanes of the 1930s and 1940s (the era of the mature, propeller-driven aircraft described in Section 1.2.3). However, supercharging adds expense to the engine, and most general aviation airplanes today are not supercharged. For example, of the 45 different piston engines manufactured by Textron Lycoming as listed in *Jane's All the World's Aircraft* (Ref. 36), only one is supercharged.

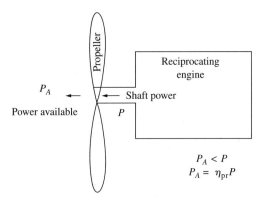

Figure 3.5   Schematic illustrating shaft power P and power available P_A from a propeller/reciprocating engine combustion.

3.3.2   The Propeller

Wilbur Wright in 1902 was the first person to recognize that a propeller is essentially a twisted wing oriented vertically to the longitudinal axis of the airplane, and that the forward thrust generated by the propeller is essentially analogous to the aerodynamic lift generated on a wing. And like a wing, which also produces friction drag, form drag, induced drag, and wave drag, a rotating propeller experiences the same sources of drag. This propeller drag is a loss mechanism; that is, it robs the propeller of some useful power. This power loss means that the net power output of the engine/propeller combination is always less than the shaft power transmitted to the propeller through the engine shaft. Hence, the power available P_A from the engine/propeller combination is always less than P. This is illustrated schematically in Fig. 3.5. The propeller efficiency η_{pr} is defined such that

$$P_A = \eta_{pr} P \qquad\qquad \textbf{[3.13]}$$

where $\eta_{pr} < 1$.

The propeller efficiency is a function of the *advance ratio J*, defined as

$$J = \frac{V_\infty}{ND}$$

where V_∞ is the free-stream velocity, N is the number of propeller revolutions per second, and D is the propeller diameter. This makes sense when you examine the local airflow velocity relative to a given cross section of the propeller, as sketched in Fig. 3.6. Here the local relative wind is the vector sum of V_∞ and the translational motion of the propeller airfoil section due to the propeller rotation, namely, $r\omega$, where r is the radial distance of the airfoil section from the propeller hub and ω is the angular velocity of the propeller. The angle between the airfoil chord line and the plane of

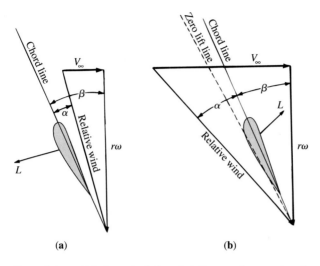

(a)   (b)

Figure 3.6   Velocity and relative wind diagrams for a section of a revolving propeller: **(a)** Case for low V_∞ and **(b)** case for high V_∞.

rotation is the *pitch angle* β. The angle of attack α is the angle between the chord line and the local relative wind. The angle of attack clearly depends on the relative values of V_∞ and $r\omega$. In Fig. 3.6a, V_∞ is small, and α is a fairly large positive value, producing an aerodynamic "lift" L acting in the general thrust direction. In Fig. 3.6b, the value of V_∞ has greatly increased; all other parameters remain the same. Here, the relative wind has moved to the other side of the airfoil section, giving rise to a negative α and an aerodynamic lift force L pointing in the opposite direction of positive thrust. Conclusion: The local angle of attack, and hence the thrust generated by the propeller, depends critically on V_∞ and $r\omega$. Note that $r\omega$ evaluated at the propeller tip is $(D/2)(2\pi N)$, or

$$(r\omega)_{\text{tip}} = \pi ND$$

Hence, the ratio $V_\infty/r\omega$, which sets the direction of the local relative wind (see Fig. 3.6), is given by

$$\frac{V_\infty}{r\omega} = \frac{V_\infty}{r(2\pi N)}$$ [3.14]

Evaluated at the propeller tip, Eq. (3.14) gives

$$\left(\frac{V_\infty}{r\omega}\right)_{\text{tip}} = \frac{V_\infty}{(D/2)(2\pi N)} = \frac{V_\infty}{\pi ND} = \frac{J}{\pi}$$ [3.15]

Clearly, from Eq. (3.15), the advance ratio J, a dimensionless quantity, plays a strong role in propeller performance; indeed, dimensional analysis shows that J is a similarity parameter for propeller performance, in the same category as the Mach number and Reynolds number. Hence, we should intuitively feel more comfortable with the opening statement of this paragraph—propeller efficiency is indeed a function of J.

A typical variation of η_{pr} with J is given in Fig. 3.7, obtained from the experimental measurements of Hartman and Biermann (Ref. 37) for an NACA propeller with a Clark Y airfoil section and three blades. Seven separate propeller efficiency curves are shown in Fig. 3.7, each one for a different propeller pitch angle β, measured at the station 75% of the blade length from the propeller hub. Examining this figure, we see that η_{pr} for a given, fixed pitch angle is 0 at $J = 0$, increases as J increases, goes through a maximum value at some value of J, and then goes to 0 at some higher value of J. Let us examine why the propeller efficiency curve has this shape. First, the reason why $\eta_{pr} = 0$ at $J = 0$ is seen from Eqs. (3.3) and (3.13). Consider an airplane at zero velocity (standing motionless on the ground) with the engines running, producing thrust (this is called the *static* thrust). From Eq. (3.3), $P_A = 0$ when $V_\infty = 0$; no power is produced at zero velocity, even though the propulsive mechanism is generating thrust. When Eq. (3.13) is applied in this case, P_A is 0 but P is finite—P is the shaft power coming from the internal combustion reciprocating engine and is not a direct function of V_∞. Hence, for $V_\infty = 0$, Eq. (3.13) dictates that $\eta_{pr} = 0$. Also, when $V_\infty = 0$, then $J = V_\infty/(ND) = 0$. This is why $\eta_{pr} = 0$ at $J = 0$ for all the curves shown in Fig. 3.7.

The shape of the propeller efficiency curve as J is increased above 0 is explained as follows. For clarity, the variation of η_{pr} with J for a given fixed-pitch propeller is sketched generically in Fig. 3.8**a**. Also, the variation of the lift-to-drag ratio for a given propeller airfoil cross section versus angle of attack is sketched generically in Fig. 3.8**b**. We will discuss the phenomena shown in Fig. 3.8, keeping in mind the geometry shown in Fig. 3.6. For a fixed-pitch propeller, β for a given propeller cross section is fixed, by definition. (Keep in mind that a propeller blade is twisted; hence β is different for each cross section; a fixed-pitch propeller is one where the value

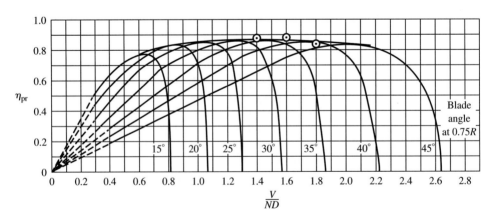

Figure 3.7   Propeller efficiency as a function of advance ratio for various pitch angles. Three-bladed propeller with Clark Y sections. (*After McCormick, Ref. 50.*)

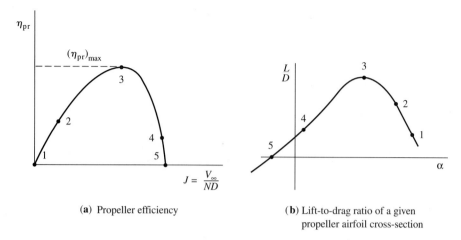

(a) Propeller efficiency

(b) Lift-to-drag ratio of a given propeller airfoil cross-section

Figure 3.8 Effect of section lift-to-drag ratio on propeller efficiency.

of β at any given cross section is essentially "locked in" mechanically, i.e., the pilot cannot change β during flight.) Examine Fig. 3.6. For a given N, $r\omega$ is constant. However, as the airplane changes its velocity, V_∞ will change, and consequently the angle of attack α will change, as shown in Fig. 3.6. At $V_\infty = 0$, the angle of attack is the same as the angle between the propeller airfoil chord line and the plane of rotation; that is, the angle of attack is also the pitch angle (for this case only, where $V_\infty = 0$). For a pitch angle of, say, $30°$, the angle of attack is also $30°$; for this case the airfoil section most likely would be stalled. This situation is labeled in Fig. 3.8a and b by point 1. In Fig. 3.8a, $J = 0$ when $V_\infty = 0$, hence point 1 is at the origin. The angle of attack is large; this is indicated in Fig. 3.8b by point 1 being far out on the right-hand side of the L/D curve. Returning to Fig. 3.6, we imagine that V_∞ is increased, keeping N constant; this gives us point 2, illustrated in Fig. 3.8. Note from Fig. 3.8b that L/D is increased; that is, the given airfoil section is now operating with an improved aerodynamic efficiency. Let us continue to increase V_∞, say, to a value such that the angle of attack corresponds to the peak value of L/D; this is shown as point 3 in Fig. 3.8b. Also, if all the other propeller airfoil cross sections are designed to simultaneously have α correspond to the point of $(L/D)_{max}$, then the net efficiency of the propeller will be maximum, as shown by point 3 in Fig. 3.8a. Let us continue to increase V_∞, keeping everything else the same. The angle of attack will continue to decrease, say, to point 4 in Fig. 3.8b. However, this corresponds to a very low value of L/D, and hence will result in poor propeller efficiency, as indicated by point 4 in Fig. 3.8a. Indeed, if V_∞ is increased further, the local relative wind will eventually flip over to a direction below the airfoil chord line, as shown in Fig. 3.6b, and the direction of the local lift vector will flip also, acting in the negative thrust direction. When this happens, the propeller efficiency is totally destroyed, as indicated by point 5 in Fig. 3.8a and b.

In summary, we have explained why the curve of η_{pr} versus J first increases as J is increased, then peaks at a value $(\eta_{pr})_{max}$, and finally decreases abruptly. This is why the propeller efficiency curves shown in Figs. 3.7 and 3.8**a** look the way they do.

In Section 1.2.3, we mentioned that a technical milestone of the era of the mature propeller-driven airplane was the development of the variable-pitch propeller, and subsequently the constant-speed propeller. Please return to Section 1.2.3, and review the discussion surrounding these propeller developments. This review will help you to better understand and appreciate the next two paragraphs.

For fixed-pitch propellers, which were used exclusively on all airplanes until the early 1930s, the maximum η_{pr} is achieved at a specific value of J (hence a specific value of V_∞). This value of J was considered the design point for the propeller, and it could correspond to the cruise velocity, or velocity for maximum rate of climb, or whatever condition the airplane designer considered most important. However, whenever V_∞ was different from the design speed, η_{pr} decreased precipitously, as reflected in Fig. 3.8**a**. The off-design performance of a fixed-pitch propeller caused a degradation of the overall airplane performance that became unacceptable to airplane designers in the 1930s. However, the solution to this problem is contained in the data shown in Fig. 3.7, where we see that maximum η_{pr} for different pitch angles occurs at different values of J. Indeed, for the propeller data shown in Fig. 3.7, the locus of the points for maximum η_{pr} forms a relatively flat envelope over a large range of J (hence V_∞), at a value of approximately $(\eta_{pr})_{max} = 0.85$. Clearly, if the pitch of the propeller could be changed by the pilot during flight so as to ride along this flat envelope, then high propeller efficiency could be achieved over a wide range of V_∞. Thus, the *variable-pitch propeller* was born; here the entire propeller blade is rotated by a mechanical mechanism located in the propeller hub, and the degree of rotation is controlled by the pilot during flight. The improvement in off-design propeller performance brought about by the variable-pitch propeller was so compelling that this design feature is ranked as one of the major aeronautical technical advances of the 1930s.

However, the variable-pitch propeller per se was not the final answer to propeller design during the era of the mature propeller-driven airplane; rather, an improvisation called the *constant-speed propeller* eventually supplanted the variable-pitch propeller in most high-performance propeller-driven airplanes. To understand the technical merit of a constant-speed propeller, first return to Eq. (3.13). The power available from a reciprocating engine/propeller combination depends not only on propeller efficiency η_{pr}, but also on the shaft power P coming from the engine. In turn, P is directly proportional to the rotational speed (rpm) of the engine, as shown in Eq. (3.9). For a given throttle setting, the rpm of a piston engine depends on the load on the crankshaft. (For example, in your automobile, with the gas pedal depressed a fixed amount, the engine rpm actually slows down when you start climbing a hill, and hence your automobile starts to slow down; the load on the engine while climbing the hill is increased, and hence the engine rpm decreases for a fixed throttle setting.) For an airplane, the load on the shaft of the piston engine comes from the aerodynamic torque created on the propeller; this torque is generated by the component of aerodynamic force exerted on the propeller in the plane of rotation, acting through a moment arm to the shaft. This aerodynamic component is a resistance force, tending to retard the rotation of the propeller. In the case of the variable-pitch propeller, as the pilot changed the pitch angle, the torque changed, which in turn caused a change in the engine rpm away from the optimum value for engine operation. This was partially self-defeating; in the quest to obtain maximum η_{pr} by varying the propeller pitch, the engine power P was frequently degraded by the resulting change in rpm. Thus, the *constant-speed propeller* was born. The constant-speed propeller is a variant of the variable-pitch propeller wherein the pitch of the propeller is automatically varied by a governer mechanism so as to maintain a constant rpm for the engine. Although the constant-speed propeller is not always operating at maximum efficiency, the product $\eta_{pr}P$ in Eq. (3.13) is optimized. Also, the automatic feature of the constant-speed propeller frees the pilot to concentrate on other things—something especially important in combat.

The use of variable-pitch and constant-speed propellers greatly enhances the rate of climb for airplanes, compared to that for a fixed-pitch propeller. (Rate of climb is one of the important airplane performance characteristics discussed in detail in Chapter 5.) This advantage is illustrated in Fig. 3.9 from Carter (Ref. 38) dating from 1940. Figure 3.9 is shown as much for historical value as for technical edification; Carter's book was a standard text in practical airplane aerodynamics during the 1930s, and Fig. 3.9 shows clearly how much the advantage of a constant-speed propeller was appreciated by that time. In Fig. 3.9, the altitude versus horizontal distance climb path of a representative airplane is shown for three different propellers—fixed-pitch, two-position controllable (a kind of variable-pitch propeller with only two settings), and constant-speed. Tick marks at various points along each flight path give the time required from take off to reach that point. Clearly, the constant-speed propeller yields much better climb performance, that is, reaches a given altitude in less time

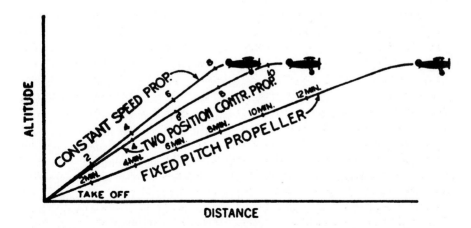

Figure 3.9   Comparison of airplane climb performance for three types of propellers: fixed-pitch, two-position (controllable), and constant-speed. Historic diagram by Carter (Ref. 38).

and over a shorter horizontal distance—a characteristic particularly important for the high-performance airplanes that characterize the era of the mature propeller-driven airplane.

Finally, we note another advantage of being able to vary the propeller pitch, namely, feathering of the propeller. A propeller is feathered when its pitch is adjusted so that the drag is minimized, and there is little or no tendency for autorotation when the engine is turned off but the airplane is still moving. The propeller is feathered when an engine failure occurs in flight, and sometimes when a multiengine airplane is taxiing on the ground with one or more engines turned off.

3.4   THE TURBOJET ENGINE

The basic components of a turbojet engine are illustrated schematically in Fig. 3.10**a**, and the generic variations (averaged over a local cross section) of static pressure p, static temperature T, and flow velocity V with axial distance through the engine are shown in Fig. 3.10**b**, **c**, and **d**, respectively. Flow enters the inlet diffuser with essentially the free-stream velocity V_∞. (In reality, the velocity entering the inlet is usually slightly slower or faster than V_∞, depending on the engine operating conditions; nature takes care of the adjustment to an inlet velocity different from V_∞ in that portion of the stream tube of air which enters the engine, but upstream of the entrance to the inlet.) In the diffuser (1–2), the air is slowed, with a consequent increase in p and T. It then enters the compressor (2–3), where work is done on the air by the rotating compressor blades, hence greatly increasing both p and T. After discharge from the compressor, the air enters the burner (or combustor), where it is mixed with fuel and burned at essentially constant pressure (3–4). The burned fuel-air mixture then expands through a turbine (4–5) which extracts work from the gas; the turbine is connected to the compressor by a shaft, and the work extracted from the turbine is transmitted via the shaft to operate the compressor. Finally, the gas expands through a nozzle (5–6) and is exhausted into the air with the jet velocity V_j.

The thrust generated by the engine is due to the net resultant of the pressure and shear stress distributions acting on the exposed surface areas, internal and external, at each point at which the gas contacts any part of the device, as described in Section 3.2. Figure 3.10**e** illustrates how each component of the turbojet contributes to the thrust; this figure is essentialy a picture of the "thrust buildup" for the engine. The internal duct of the diffuser and compressor has a component of surface area that faces in the thrust direction (toward the left in Fig. 3.10). The high pressure in the diffuser and especially in the compressor, acting on this forward-facing area, creates a large force in the thrust direction. Note in Fig. 3.10**e** that the accumulated thrust T grows with distance along the diffuser (1–2) and the compressor (2–3). This high pressure also acts on a component of forward-facing area in the burner, so that the accumulated value of T continues to increase with distance through the burner (3–4), as shown in Fig. 3.10**e**. However, in the turbine and nozzle, the net surface area has a component that faces in the rearward direction, and the pressure acting on this rearward-facing area creates a force in the negative thrust direction (to the right in Fig. 3.10). Thus, the accumulated thrust F *decreases* through the turbine (4–5) and nozzle (5–6), as

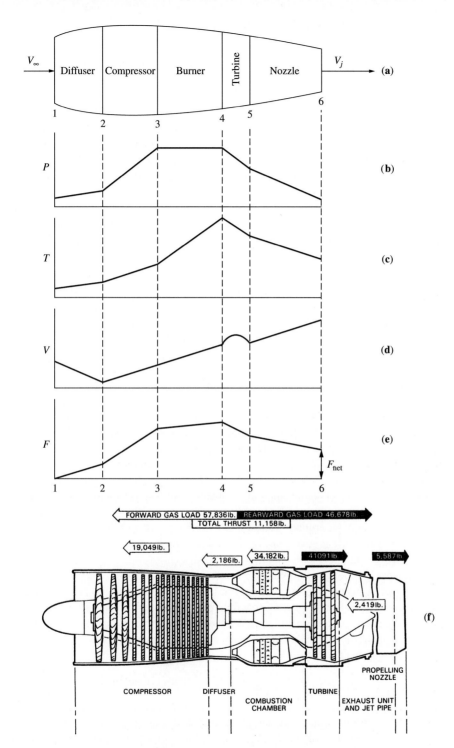

Figure 3.10 Distribution of (**a**) components, (**b**) pressure, (**c**) temperature, (**d**) velocity, and (**e**) local thrust; (**f**) integrated thrust through a generic turbojet engine.

shown in Fig. 3.10e. However, by the time the nozzle exit is reached (point 6), the *net* accumulated thrust F_{net} is still a positive value, as shown in Fig. 3.10e. This is the net thrust produced by the engine, that is, $T = F_{net}$. A more diagrammatic illustration of the thrust distribution exerted on a turbojet is shown in Fig. 3.10f.

The detailed calculation of the pressure and shear stress distributions over the complete internal surface of the engine would be a herculean task, even in the present day of the sophisticated computational fluid dynamics (CFD). (See Ref. 39 for an introductory book on CFD, written for beginners in the subject). However, the major jet engine manufacturers are developing the CFD expertise that will eventually allow such a calculation. Fortunately, the calculation of jet engine thrust is carried out infinitely more simply by drawing a control volume around the engine, looking at the time rate of change of momentum of the gas flow through the engine, and using Newton's second law to obtain the thrust. To a certain extent, we have already carried out this control volume analysis in Section 3.2, obtaining Eq. (3.1) for the thrust. However, in that derivation we simplified the analysis by not including the pressure acting on the front and back free surfaces of the control volume, and by not considering the extra mass due to the fuel added. A more detailed derivation (see, e.g., chapter 9 of Ref. 3) leads to a thrust equation which is slightly more refined than Eq. (3.1), namely,

$$T = (\dot{m}_{air} + \dot{m}_{fuel})\, V_j - \dot{m}_{air} V_\infty + (p_e - p_\infty)\, A_e \qquad \text{[3.16]}$$

where $\dot{m}_{air}$ and $\dot{m}_{fuel}$ are the mass flows of the air and fuel, respectively, p_e is the gas pressure at the exit of the nozzle, p_∞ is the ambient pressure, and A_e is the exit area of the nozzle. The first two terms on the right side of Eq. (3.16) are the time rate of change of momentum of the gas as it flows through the engine; these terms play the same role as the right-hand side of Eq. (3.1). The pressure term $(p_e - p_\infty)A_e$ in Eq. (3.16) is usually much smaller than the momentum terms. As a first approximation, it can be neglected, just as we did in obtaining Eq. (3.1).

A typical turbojet engine is shown in the photograph in Fig. 3.11. A cutaway drawing of a turbojet is given in Fig. 3.12, showing the details of the compressor, burner, turbine, and nozzle.

The specific fuel consumption for a turbojet is defined differently than that for a reciprocating piston engine given by Eq. (3.10). The measurable primary output from a jet engine is thrust, whereas that for a piston engine is power. Therefore, for a turbojet the specific fuel consumption is based on thrust rather than power; to make this clear, it is frequently called the *thrust* specific fuel consumption. We denote it by c_t, and it is defined as

$$c_t = \text{weight of fuel burned per unit thrust per unit time}$$

or

$$c_t = \frac{\text{weight of fuel consumed for given time increment}}{(\text{thrust output})\,(\text{time increment})} \qquad \text{[3.17]}$$

Figure 3.11   Rolls-Royce Conway RCo.10 turbojet engine. (Courtesy of Rolls Royce.)

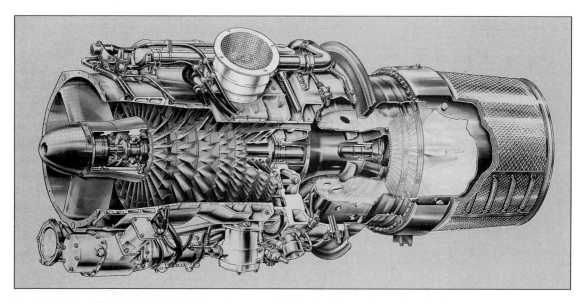

Figure 3.12   Rolls-Royce Viper 632 turbojet. (Courtesy of Rolls Royce.)

Consistent units for c_t are

$$[c_t] = \frac{\text{lb}}{\text{lb·s}} = \frac{1}{\text{s}}$$

or

$$[c_t] = \frac{N}{N \cdot s} = \frac{1}{s}$$

However, analogous to the case of the piston engine, the thrust specific fuel consumption (TSFC) has been conventionally defined using the inconsistent time unit of hour instead of second. To emphasize this difference, we will use the symbol TSFC for the thrust specific fuel consumption in inconsistent units. Hence, by definition,

$$[\text{TSFC}] = \frac{\text{lb}}{\text{lb} \cdot \text{h}} = \frac{1}{\text{h}}$$

3.4.1   Variations of Thrust and Specific Fuel Consumption with Velocity and Altitude

The thrust generated by a turbojet is given by Eq. (3.16). Questions: When the engine is mounted on an airplane flying through the atmosphere, how does the thrust vary with flight velocity? With altitude? Some hints regarding the answers can be obtained by examining the thrust equation given by Eq. (3.16). First, consider the mass flow of air $\dot{m}_{\text{air}}$. The mass flow of air entering the inlet (location 1 in Fig. 3.10a) is $\rho_\infty A_1 V_\infty$, where A_1 is the cross-sectional area of the inlet. As V_∞ is increased, V_j stays essentially the same (at least to first order); the value of V_j is much more a function of the internal compression and combustion processes taking place inside the engine than it is of V_∞. Hence, the difference $V_j - V_\infty$ tends to decrease as V_∞ increases. From Eq. (3.16), with V_∞ increasing but V_j staying about the same, the value of T is decreased. These two effects tend to cancel in Eq. (3.16), and therefore we might expect the thrust generated by a turbojet to be only a weak function of V_∞. This is indeed the case, as shown in Fig. 3.13 based on data from Hesse and Mumford (Ref. 40). Here the thrust for a typical small turbojet is given as a function of flight Mach number for two altitudes, sea level and 40,000 ft, and for three different throttle settings (denoted by different compressor rpm values) at each altitude. Note that, especially at altitude, T is a very weak function of Mach number. Hence, to a first approximation, in this book we consider that, for a turbojet flying at *subsonic speeds*,

$$\boxed{T \text{ is reasonably constant with } V_\infty}$$

A typical variation of TSFC for the same small turbojet is given in Fig. 3.14, also based on data from Ref. 40. Here we see a general trend where TSFC increases monotonically with flight Mach number. Note that, at low speed, the TSFC is about 1 lb of fuel/(lb of thrust/h)—an approximate value used frequently in airplane performance analyses. However, at high velocities, the increase in TSFC should be taken into account. Based on the data shown in Fig. 3.14, we write as a reasonable approximation, *for $M_\infty < 1$,*

$$\boxed{\text{TSFC} = 1.0 + k M_\infty} \qquad \textbf{[3.18]}$$

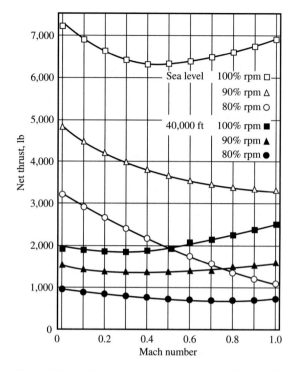

Figure 3.13 Typical results for the variation of thrust with subsonic Mach number for a turbojet.

where k is a function of altitude and throttle setting (engine rpm). For example, the data in Fig. 3.14 show that for an altitude of 40,000 ft, k is about 0.5 and is relatively insensitive to rpm.

There is a strong altitude effect on thrust, as can be seen by examining Eq. (3.16). Again, we note that $\dot{m}_{\text{air}} = \rho_\infty A_1 V_\infty$; hence $\dot{m}_{\text{air}}$ is directly proportional to ρ_∞. As the altitude increases, ρ_∞ decreases. In turn, from Eq. (3.16) where T is almost directly proportional to $\dot{m}_{\text{air}}$, thrust also decreases with altitude. Indeed, it is reasonable to express the variation of T with altitude in terms of the density ratio ρ/ρ_0, where ρ is the density at a given altitude and ρ_0 is sea-level density. Hence,

$$\boxed{\frac{T}{T_0} = \frac{\rho}{\rho_0}} \qquad \textbf{[3.19]}$$

where T_0 is the sea-level thrust.

In regard to the altitude effect on thrust specific fuel consumption, comparing the results in Fig. 3.14 for full throttle (100% rpm) at sea level and at 40,000 ft, we see little difference. It is reasonable to ignore this weak altitude effect and to assume that

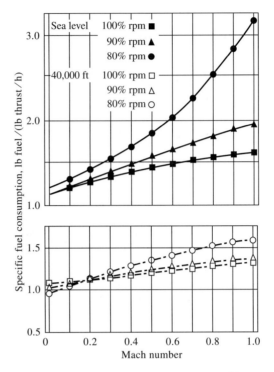

Figure 3.14   Typical results for the variation of thrust specific fuel consumption with subsonic Mach number for a turbojet.

TSFC is constant with altitude

The above discussion pertains to the performance of turbojets at subsonic speeds. Let us extend this discussion to the supersonic regime. One of the most important supersonic airplanes of the last quarter-century has been the Concorde supersonic transport (see Fig. 1.35). The Concorde is powered with four Rolls-Royce/SNECMA Olympus 593 engines—pure turbojets. The choice of turbojet engines for the Concorde instead of turbofan engines (to be discussed in the next section) keyed on the better thrust specific fuel consumption of a turbojet at the design cruise Mach number of 2.2. The variations of both T and thrust specific fuel consumption with supersonic Mach number for the Olympus 593 are shown in Fig. 3.15, after Mair and Birdsall (Ref. 41). Here, $\delta = p/p_0$, where p and p_0 are the pressures at altitude and sea level, respectively. In Fig. 3.15, T is given in units of kilonewtons, and c_t is expressed in terms of the *mass* of fuel consumed per newton of thrust per second. The results are shown for flight in the stratosphere, that is, for altitudes above 11 km, or 36,000 ft. What is important in Fig. 3.15 is the *variation* of T and c_t with Mach number. As M_∞ increases, T at constant altitude increases almost linearly, more than doubling

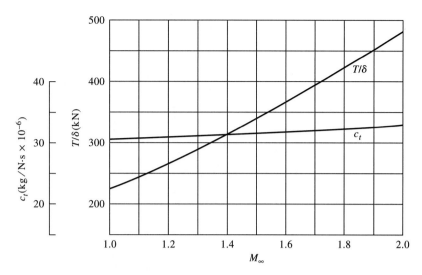

Figure 3.15 Typical results for the variation of thrust and thrust specific fuel consumption with supersonic Mach number for a turbojet.

from Mach 1 to Mach 2. This increase in thrust is presaged by the data in Fig. 3.13 where, for most of the curves, T is seen to be on the increase as the Mach number increases near Mach 1.

Why does T increase with M_∞ in the supersonic regime, whereas it is relatively constant in the subsonic regime? The answer lies in part in the large total pressures recovered in the supersonic inlet diffuser as M_∞ increases. Recall from gas dynamics (see, e.g., Ref. 42) that the ratio of total to static pressure is given by the isentropic relation

$$\frac{p_{\text{total}}}{p_{\text{static}}} = \left(1 + \frac{\gamma - 1}{2} M^2\right)^{\gamma/(\gamma-1)}$$
[3.20]

Although the compression process in a supersonic inlet diffuser is not precisely isentropic, Eq. (3.20) gives a reasonable first estimate of the total pressure recovered at the exit of the diffuser. Note from Eq. (3.20) that as M_∞ increases, particularly for supersonic values, p_{total} becomes quite large. This is essentially the pressure of the flow as it enters the compressor, through which the pressure is further increased considerably. The net effect of these higher pressure levels inside the engine for supersonic flight is that V_j in Eq. (3.16) is greatly increased; indeed, the nozzles for turbojets designed for supersonic flight are convergent-divergent supersonic nozzles rather than the purely convergent subsonic nozzles for turbojets used for subsonic airplanes. That is, for the conditions shown in Fig. 3.15, V_j is supersonic. Examining Eq. (3.16) as M_∞ is increased in supersonic flight, we see that *both* $\dot{m}_{\text{air}}$ and V_j are increased substantially, hence both combine to increase T.

In regard to the thrust specific fuel consumption at supersonic speeds, Fig. 3.15 shows only a small increase with M_∞. This is presaged in Fig. 3.14 where c_t is seen to bend over and becomes more constant near Mach 1. Hence, at supersonic speeds, we can assume that c_t is essentially constant.

Therefore, *for $M_\infty > 1$*, we can assume from the data in Fig. 3.15 that for the Olympus 593 turbojet,

$$\boxed{\frac{T}{T_{\text{Mach 1}}} = 1 + 1.18(M_\infty - 1)} \qquad \text{[3.21]}$$

We will take this result as a model for our subsequent analyses for supersonic, turbojet-powered aircraft. Also, we will assume that for *supersonic flight*,

$$\boxed{\text{TSFC is constant with} M_\infty}$$

3.5   THE TURBOFAN ENGINE

Recall our discussion in Section 3.2 of the tradeoff between thrust and efficiency, and how a propeller produces less thrust but with more efficiency, whereas a jet engine produces more thrust but with less efficiency. The *turbofan* engine is a propulsive mechanism the design of which strives to combine the high thrust of a turbojet with the high efficiency of a propeller. A schematic of a turbofan is shown in Fig. 3.16. Basically, a turbojet engine forms the *core* of the turbofan; the core contains the diffuser, compressor, burner, turbine, and nozzle. However, in the turbofan engine, the turbine drives not only the compressor, but also a large fan *external* to the core. The fan itself is contained in a shroud that is wrapped around the core, as shown in Fig. 3.16. The flow through a turbofan engine is split into two paths. One passes through the fan and flows externally over the core; this air is processed only by the fan, which is acting in the manner of a sophisticated, shrouded propeller. Hence,

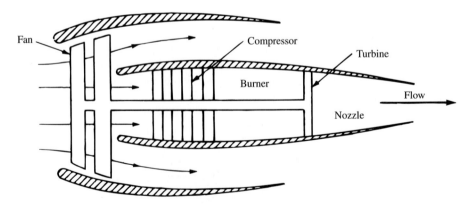

Figure 3.16   Schematic of a turbofan engine.

the propulsive thrust obtained from this flow through the fan is generated with an efficiency approaching that of a propeller. The second air path is through the core itself. The propulsive thrust obtained from the flow through the core is generated with an efficiency associated with a turbojet. The overall propulsive efficiency of a turbofan is therefore a compromise between that of a propeller and that of a turbojet. This compromise has been found to be quite successful—the vast majority of jet-propelled airplanes today are powered by turbofan engines.

An important parameter of a turbofan engine is the *bypass ratio*, defined as the mass flow passing through the fan, externally to the core (the first path described above), divided by the mass flow through the core itself (the second path described above). Everything else being equal, the higher the bypass ratio, the higher the propulsive efficiency. For the large turbofan engines that power airplanes such as the Boeing 747 (see Fig. 1.34), for example, the Rolls-Royce RB211 and the Pratt & Whitney JT9D, the bypass ratios are on the order of 5. Typical values of the thrust specific fuel consumption for these turbofan engines are 0.6 lb/(lb·h)—almost half that of a conventional turbojet engine. This is why turbofan engines are used on most jet-propelled airplanes today.

A photograph of a typical turbofan engine is shown in Fig. 3.17. A cutaway illustrating the details of the fan and the core is shown in Fig. 3.18.

Figure 3.17   **(a)** Rolls-Royce Tay Turbofan. *(Courtesy of Rolls-Royce.) (continued)*

(concluded)

(b) Pratt & Whitney PW 4084 turbofan. *(Courtesy of Pratt & Whitney.)*

(c) CFM56-5C high-bypass turbofan engine. CFM56 engines are produced by CFM International, a 50/50 joint company of General Electric of the United States and SNECMA of France. *(Courtesy of CFM International.)*

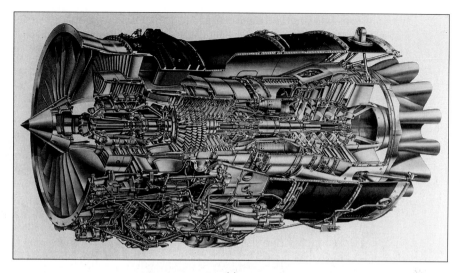

(a)

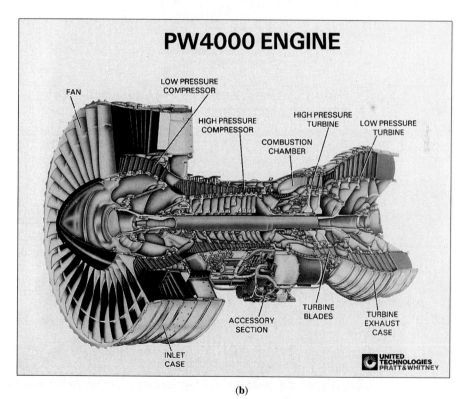

(b)

Figure 3.18 (**a**) Cutaway of the Rolls-Royce Tay. *(Courtesy of Rolls-Royce.)* (**b**) Cutaway of the Pratt & Whitney PW 4000 turbofan. *(Courtesy of Pratt & Whitney.)* *(continued)*

(concluded)

(c) Cutaway of the CFM56-5C high-bypass turbofan. *(Courtesy of CFM International.)*

3.5.1   Variations of Thrust and Specific Fuel Consumption with Velocity and Altitude

We first discuss the characteristics of high-bypass-ratio turbofans—those with bypass ratios on the order of 5. These are the class of turbofans that power civil transports. The performance of these engines seems to be closer to that of a propeller than that of a turbojet in some respects.

The thrust of a civil turbofan engine has a strong variation with velocity; thrust decreases as V_∞ increases. Let $T_{V=0}$ be the thrust at standard sea level and at zero flight velocity. A typical variation of $T/T_{V=0}$ with V_∞ for a range of velocities associated with takeoff is shown in Fig. 3.19; the data are for the Rolls-Royce RB211-535E4 turbofan found in Ref. 41. These data fit the curve

$$\frac{T}{T_{V=0}} = 1 - 2.52 \times 10^{-3} V_\infty + 4.34 \times 10^{-6} V_\infty^2 \qquad \textbf{[3.22]}$$

where V_∞ is in meters per second and holds for $V_\infty < 130$ m/s. *Caution*: Equation (3.22) holds for takeoff velocities only. The variation of $T/T_{V=0}$ for the same engine at higher subsonic velocities is shown in Fig. 3.20 for various altitudes from sea level to 11 km. For each altitude, two curves are given, the upper curve for the higher thrust used during climb and the lower curve for the lower thrust setting for cruise. The data are from Ref. 41. For a given, constant altitude, the decrease in thrust with Mach number can be correlated by

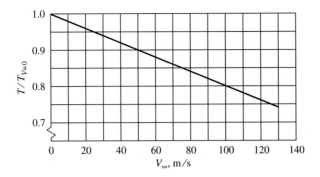

Figure 3.19 Maximum takeoff thrust as a function of velocity at sea level for the Rolls-Royce RB211-535E4 turbofan.

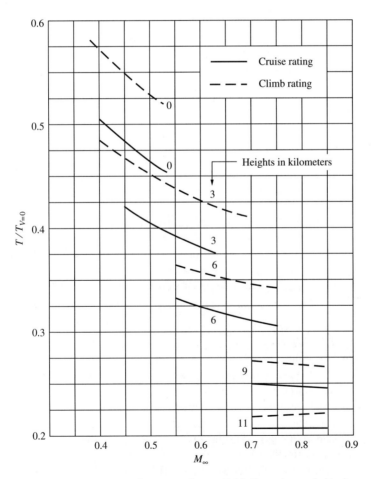

Figure 3.20 Variation of maximum thrust with Mach number and altitude for the Rolls-Royce RB211-535E4 turbofan. Note that $T_{V=0}$ is the thrust at zero velocity at sea level.

$$\boxed{\frac{T}{T_{V=0}} = AM_\infty^{-n}}$$ **[3.23]**

where A and n are functions of altitude. For example, at an altitude of 3 km, a reasonable correlation for the climb-rating thrust is

$$\frac{T}{T_{V=0}} = 0.369 M_\infty^{-0.305}$$ **[3.24]**

Keep in mind that Eqs. (3.22) to (3.24), as well as the curves in Figs. 3.19 and 3.20, are for a Rolls-Royce RB211 engine (designed for use with the Boeing 747 and similar large transport aircraft). They are given here only to illustrate the general trends.

Although the variation of T for a civil turbofan is a strong function of V_∞ (or M_∞) at lower altitudes, note from Fig. 3.20 that at the relatively high altitude of 11 km, T is relatively constant for the narrow Mach number range from 0.7 to 0.85. This corresponds to normal cruise Mach numbers for civil transports such as the Boeing 747. Hence, for the analysis of airplane performance in the cruise range, it appears reasonable to assume $T = $ constant.

The variation of T with altitude is approximated by

$$\boxed{\frac{T}{T_0} = \left(\frac{\rho}{\rho_0}\right)^m}$$ **[3.25]**

as given by Mattingly et al. (Ref. 43) and Mair and Birdsall (Ref. 41). Equation (3.25) is an empirical relation which holds for a large number of civil turbofan engines. The value of m depends on the engine design; it is usually near 1, but could be less than or greater than 1.

The variation of thrust specific fuel consumption with both altitude and Mach number is shown in Fig. 3.21 for the Rolls-Royce RB211-535E4 turbofan. Here, c_t/c_{t_∞} is the ratio of the thrust specific fuel consumption at the specified altitude and Mach number, denoted by c_t, to the value of c_t at zero velocity and at sea level, denoted by c_{t_∞}. The variation of c_t with velocity at a given altitude follows the relation

$$c_t = B(1 + kM_\infty)$$ **[3.26]**

where B and k are empirical constants found by correlating the data. Equation (3.26) is valid only for a limited range of M_∞ around the cruise value $0.7 < M_\infty < 0.85$. A glance at Fig. 3.21 shows why turbofans were not used on the Concorde supersonic transport, with its cruising Mach number of 2.2. As mentioned in Section 3.4, the thrust specific fuel consumption of a turbojet engine is almost constant with speed in the supersonic regime. However, for a turbofan, c_t increases markedly with increases in M_∞, as shown in Fig. 3.21. For this reason, a turbojet is more fuel-efficient than a turbofan is at the design Mach number of 2.2 for the Concorde.

The ordinate in Fig. 3.21 is expanded. Hence, the altitude effect on c_t looks larger than it really is. For example, at $M_\infty = 0.7$, there is about an 11% reduction of c_t when the altitude is increased from 3 to 11 km. Therefore, to first order, we assume the c_t is constant with altitude.

For low-bypass-ratio turbofans—those with bypass ratios between 0 and 1—the performance is somewhat different from that for the high-bypass-ratio case discussed above. The performance of low-bypass-ratio turbofans is much closer to that of a turbojet than that of a propeller, in contrast to the civil turbofan discussed earlier. Low-bypass-ratio turbofans are used on many high-performance jet fighter planes of today, such as the McDonnell-Douglas F-15. Typical generic variations of $T/T_{V=0}$ and c_t/c_{t_∞} versus M_∞ for a military, low-bypass-ratio turbofan are given in Fig. 3.22.

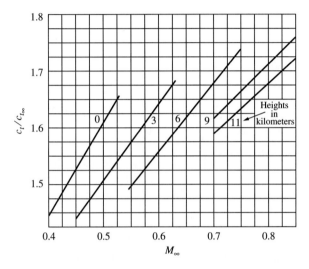

Figure 3.21 Variation of thrust specific fuel consumption with subsonic Mach number and altitude for the Rolls-Royce RB211-535E4 turbofan. Note that c_{t_∞} is the thrust specific fuel consumption at zero velocity at sea level.

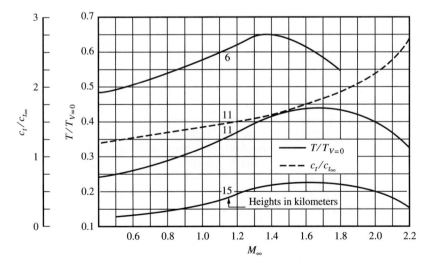

Figure 3.22 Variation of thrust and thrust specific fuel consumption with subsonic and supersonic Mach number and altitude for a generic military turbofan.

In contrast to the civil turbofan, here we see that after a small initial decrease at low subsonic Mach numbers, the thrust increases for increasing Mach number well above Mach 1. The typical decrease of thrust with altitude is also indicated in Fig. 3.22, where thrust curves are shown for altitudes of 6, 11, and 15 km.

The dashed line in Fig. 3.22 gives the variation of thrust specific fuel consumption versus Mach number for a military turbofan. Note that c_t for the low-bypass-ratio turbofan gradually increases as M_∞ increases for subsonic and transonic speeds, and begins to rapidly increase at Mach 2 and beyond. This is unlike the variation of c_t for a pure turbojet engine, which is relatively constant in the low supersonic regime (see Fig. 3.15).

3.6   THE TURBOPROP

The turboprop is essentially a propeller driven by a gas-turbine engine. Therefore, of all the gas-turbine devices described in this chapter, the turboprop is closest to the reciprocating engine/propeller combination discussed in Section 3.3. A schematic of a turboprop engine is shown in Fig. 3.23. Here, similar to the turbojet, the inlet air is compressed by an axial-flow compressor, mixed with fuel and burned in the combustor, expanded through a turbine, and then exhausted through a nozzle. However, unlike the turbojet, the turbine powers not only the compressor but also the propeller. In Fig. 3.23 a twin-spool arrangement is shown; the compressor is divided into two stages—low-pressure and high-pressure—where each stage is driven by a separate turbine—the low-pressure turbine and high-pressure turbine. The high-pressure turbine drives the high-pressure compressor. The low-pressure turbine drives both the low-pressure compressor and the propeller. By design, most of the available work in the flow is extracted by the turbines, leaving little available for jet thrust. For most turboprops, only about 5% of the total thrust is associated with the jet exhaust, and

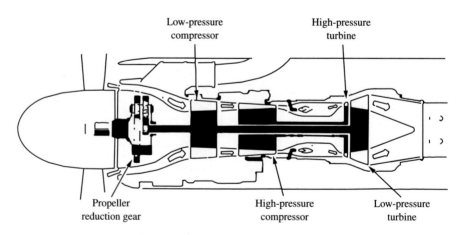

Figure 3.23   Schematic of a turboprop engine.

the remaining 95% comes from the propeller. In regard to the thrust and efficiency tradeoff discussed in Section 3.2, the turboprop falls in between the reciprocating engine/propeller combination and the turbofan or turbojet. The turboprop generates more thrust than a reciprocating engine/propeller device, but less than a turbofan or turbojet. On the other hand, the turboprop has a specific fuel consumption higher than that of the reciprocating engine/propeller combination, but lower than that of a turbofan or turbojet. (Keep in mind that the above are broad statements and are made only to give you a feeling for these tradeoffs. Definitive statements can only be made by comparing specific real engines with one another.) Also, the maximum flight speed of a turboprop-powered airplane is limited to that at which the propeller efficiency becomes seriously degraded by shock wave formation on the propeller—usually around $M_\infty = 0.6$ to 0.7. A photograph of a turboprop engine is shown in Fig. 3.24**a**, and a cutaway of the same engine is given in Fig. 3.24**b**.

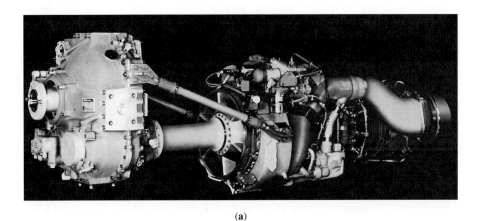

(**a**)

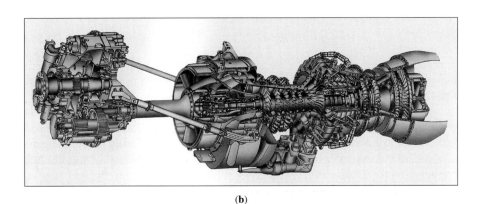

(**b**)

Figure 3.24 (**a**) The CT7 turboprop engine. *(Courtesy of General Electric.)* (**b**) A cutaway of the CT7 turboprop engine. *(Courtesy of General Electric.)*

As noted above, the thrust generated by the turboprop is the sum of the propeller thrust T_p and the jet thrust T_j. For the engine in flight at velocity V_∞, the power available from the turboprop is

$$P_A = (T_p + T_j)\, V_\infty \qquad\qquad \textbf{[3.27]}$$

Because of its closeness to the reciprocating engine/propeller mechanism, where the rating of engine performance is in terms of power rather than thrust, the performance of a turboprop is frequently measured in terms of power. The main business end of a turboprop is the shaft coming from the engine to which the propeller is attached via some type of gearbox mechanism. Hence the *shaft power P_s* coming from the engine is a meaningful quantity. Because of losses associated with the propeller as described in Section 3.3.2, the power obtained from the propeller/shaft combination is $\eta_{pr} P_s$. Hence, the net power available, which includes the jet thrust, is

$$P_A = \eta_{pr} P_s + T_j V_\infty \qquad\qquad \textbf{[3.28]}$$

Sometimes manufacturers rate their turboprops in terms of the *equivalent shaft power P_{es}* which is an overall power rating that *includes* the effect of the jet thrust. Here, we *imagine* that *all* the power from the engine is being delivered through the shaft (although we know that a part of it—about 5%—is really due to jet thrust). The equivalent shaft power is defined to be analogous to the shaft power coming from a reciprocating engine. Analogous to Eq. (3.13), P_{es} is *defined* by

$$P_A = \eta_{pr} P_{es} \qquad\qquad \textbf{[3.29]}$$

Combining Eqs. (3.28) and (3.29), we have

$$\eta_{pr} P_{es} = \eta_{pr} P_s + T_j V_\infty \qquad\qquad \textbf{[3.30]}$$

Solving Eq. (3.30) for P_{es}, we obtain

$$P_{es} = P_s + \frac{T_j V_\infty}{\eta_{pr}} \qquad\qquad \textbf{[3.31]}$$

Equation (3.31) shows how the defined equivalent shaft power is related to the actual shaft power P_s and the jet thrust T_j.

Turboprop engines clearly have an ambivalence—is thrust or power more germane? There is no definitive answer to this question, nor should there be. Once you become comfortable with Eqs. (3.27) to (3.31), you can easily accept this ambivalence. Of course, this ambivalence carries over to the definition of specific fuel consumption for a turboprop. Let $\dot{w}_{fuel}$ be the weight flow rate of the fuel (say, in pounds per second, or newtons per second). Also let T be the total thrust from the turboprop, $T = T_p + T_j$. Then the thrust specific fuel consumption can be defined as

$$c_t \equiv \frac{\dot{w}_{fuel}}{T} \qquad\qquad \textbf{[3.32]}$$

The specific fuel consumption can also be based on power, but because power can be treated as net power available P_A, shaft power P_s, or equivalent shaft power P_{es}, we have three such specific fuel consumptions, defined as

$$c_A \equiv \frac{\dot{w}_{\text{fuel}}}{P_A} \tag{3.33}$$

$$c_s \equiv \frac{\dot{w}_{\text{fuel}}}{P_s} \tag{3.34}$$

$$c_{es} \equiv \frac{\dot{w}_{\text{fuel}}}{P_{es}} \tag{3.35}$$

When you examine the manufacturer's specifications for specific fuel consumption for a turboprop, it is important to make certain which definition is being used.

Finally, we note a useful rule of thumb (Ref. 44) that, at static conditions (engine operating with the airplane at zero velocity on the ground), a turboprop produces about 2.5 lb of thrust per shaft horsepower.

3.6.1 Variations of Power and Specific Fuel Consumption with Velocity and Altitude

A typical variation of power available P_A from a turboprop (note that P_A includes the propeller efficiency) with Mach number and altitude is given in Fig. 3.25, based on data from Ref. 40. Keep in mind that as Mach 1 is approached, there is a serious degradation of power because of shock formation on the propeller. In Fig. 3.25, P_A at a given altitude first increases, reaches a maximum, and then decreases as M_∞ increases. Keeping in mind that $P_A = T_A V_\infty$, the decreasing P_A at the higher

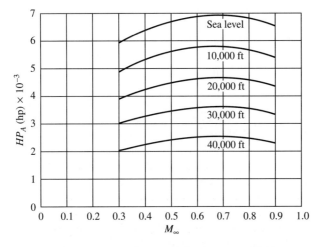

Figure 3.25 Variation of maximum horsepower available HP_A as a function of Mach number and altitude for a typical turboprop engine.

subsonic Mach numbers is due to a sharp degradation in T_A. The maximum P_A occurs in a Mach number range around 0.6 to 0.7—the upper limit for turboprop-powered airplanes. However, the net effect of the combined variation of thrust and V_∞ in Fig. 3.25 is to yield, to a first approximation, a relatively flat variation of P_A with M_∞. Hence, we can make the assumption that

$$\boxed{P_A \text{ is constant with } M_\infty}$$

for a turboprop. In regard to the altitude variation, the data in Fig. 3.25 are reasonably correlated by

$$\boxed{\frac{P_A}{P_{A,0}} = \left(\frac{\rho}{\rho_0}\right)^n \quad n = 0.7} \tag{3.36}$$

For other turboprop engines, the value of n in Eq. (3.36) will be slightly different.

Typical variations of the specific fuel consumption as a function of M_∞ and altitude are shown as the upper set of curves in Fig. 3.26, obtained from the data of Ref. 40. The specific fuel consumption shown here is c_A defined by Eq. (3.33). For all practical purposes, the results in Fig. 3.26 show that

$$\boxed{c_A \text{ is constant with both velocity and altitude}}$$

The lower set of curves shown in Fig. 3.26 gives the ratio of jet power to total power

$$\frac{T_j V_\infty}{P_A} = \frac{T_j V_\infty}{(T_p + T_j) V_\infty} = \frac{T_j}{T_p + T_j}$$

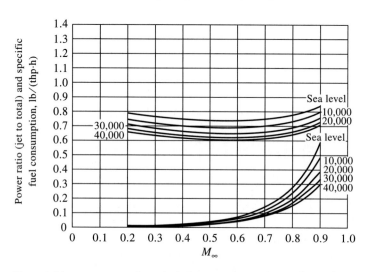

Figure 3.26   Variation of specific fuel consumption and ratio of jet to the total thrust horsepower with Mach number and altitude for a typical turboprop engine. Altitude is given in feet.

as a function of M_∞ and altitude. Note that this ratio is less than 0.05 (or 5%) for $M_\infty < 0.6$—the range for most turboprop-powered airplanes. The ratio increases rapidly above Mach 0.6, mainly because T_p is degraded due to shock formation on the propeller.

3.7 MISCELLANEOUS COMMENTS: AFTERBURNING AND MORE ON SPECIFIC FUEL CONSUMPTION

In a turbojet or turbofan engine, the fuel-air mixture in the combustion chamber is lean, and hence there is plenty of oxygen left in the exhaust gas that can be used for additional burning. A device that takes advantage of this situation is the *afterburner*, wherein extra fuel is injected into the exhaust gas and burned downstream of the turbine. A diagram of an afterburner is shown in Fig. 3.27. The afterburner is essentially a long duct downstream of the turbine into which fuel is sprayed and burned. At the exit of the afterburner duct is a variable-area nozzle; the variable-area feature is required by the different nozzle flow conditions associated with the afterburner turned on or off.

The afterburner is used for short periods of greatly increased thrust. The Concorde supersonic transport uses afterburners for rapid climb and acceleration after takeoff. Military fighter airplanes use afterburners for a fast takeoff and for bursts of speed during combat. With the afterburner operating, the weight flow of fuel increases markedly, so the pilot has to be careful to use the afterburner only when needed. The performance of a typical turbofan with afterburner is illustrated in Fig. 3.28. The solid curve gives the ratio of the thrust with afterburner on T_{AB} to the thrust without afterburner as a function of Mach number; clearly the afterburner is an effective device

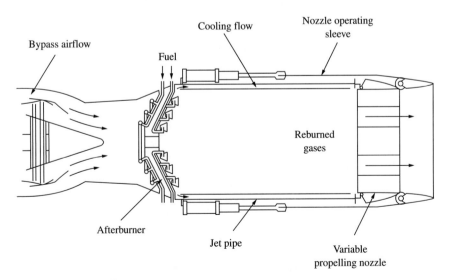

Figure 3.27 Schematic of an afterburner.

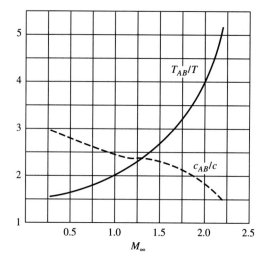

Figure 3.28   Effect of afterburning on thrust and
thrust specific fuel consumption for a
typical military turbofan at 11-km
altitude.

for thrust augmentation. The large increase in the ratio T_{AB}/T at the higher values of M_∞ in Fig. 3.28 is mainly due to the fact that T decreases at high Mach numbers. The dashed curve in Fig. 3.28 is the ratio of the thrust specific fuel consumption with and without afterburning c_{AB}/c. The use of the afterburner causes a dramatic increase in the thrust specific fuel consumption well above that for the afterburner off. As M_∞ increases, the ratio decreases, but it still remains substantially above unity.

An international note: In the British aeronautical literature, afterburning is called *reheat*.

The second miscellaneous comment in this section has to do with specific fuel consumption, which we have already seen may be couched in terms of thrust or power depending on the type of engine. Sometimes, in comparing the performance of a variety of engines, it is useful to quote the specific fuel consumption uniformly in terms of one or the other. It is easy to transform the specific fuel consumption c, defined in terms of power, to the thrust specific fuel consumption c_t, defined in terms of thrust, and vice versa, as follows. By definition,

$$c \equiv \frac{\dot{w}_{\text{fuel}}}{P}$$

[3.37]

and

$$c_t \equiv \frac{\dot{w}_{\text{fuel}}}{T}$$

[3.38]

Combining Eqs. (3.37) and (3.38), we have

$$c_t = \frac{cP}{T}$$ [3.39]

For the reciprocating engine/propeller combination, c is defined with P as the engine shaft power, as given by Eq. (3.10). In turn, P is related to the power available from the engine/propeller combination P_A, via Eq. (3.13), as

$$P = \frac{P_A}{\eta_{pr}}$$ [3.40]

Moreover,

$$P_A = T V_\infty$$ [3.41]

Combining Eqs. (3.40) and (3.41), we have

$$P = \frac{T V_\infty}{\eta_{pr}}$$ [3.42]

Substituting Eq. (3.42) into (3.39), we obtain

$$\boxed{c_t = \frac{c V_\infty}{\eta_{pr}}}$$ [3.43]

Equation (3.43) allows us to couch the specific fuel consumption for a reciprocating engine c in terms of an equivalent "thrust" specific fuel consumption c_t. The same relation can be used to couch the specific fuel consumption of a turboprop based on the equivalent shaft power c_{es}, defined by Eq. (3.35), in terms of an equivalent "thrust" specific fuel consumption c_t.

3.8 SUMMARY

The basic operation of various flight propulsion systems has been discussed in this chapter. In particular, the variations of thrust, power, and specific fuel consumption with flight velocity and altitude have been examined for each of these systems; this information is particularly relevant to the airplane performance and design concepts to be discussed in the remainder of this book. To help sort out these variations for different types of engines, a block diagram is shown in Fig. 3.29. Examine this block diagram carefully, and return to the pertinent sections of this chapter if you are not clear about any of the entries in the diagram. Please note that the velocity and altitude variations shown in Fig. 3.29 are approximate, first-order results, as explained throughout this chapter. They are useful for our purposes of estimating airplane performance and for the conceptual design of an airplane; they should not be taken too literally for any detailed analyses where more precise engine data are needed.

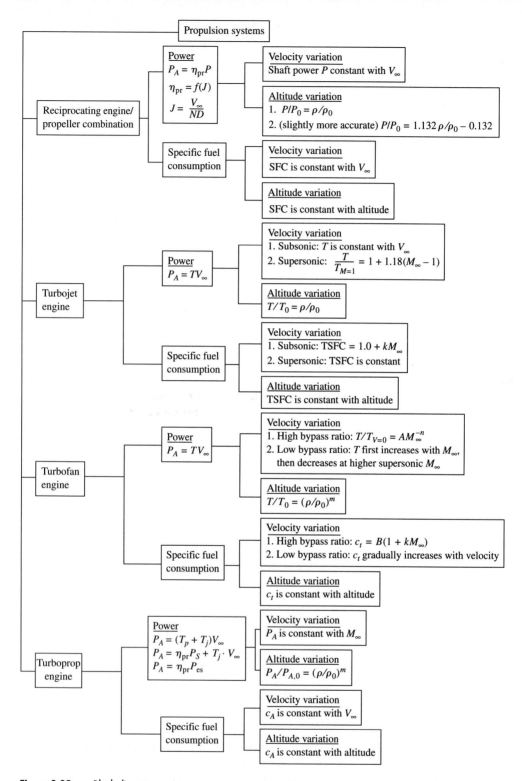

Figure 3.29 Block diagram summary.

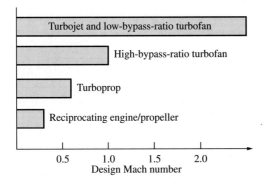

Figure 3.30   Mach number regimes for various power plants.

The different propulsive mechanisms discussed in this chapter—the reciprocating engine/propeller, the turboprop, the turbofan, and the turbojet—find their best flight applications in different parts of the Mach number spectrum. The flight Mach number regions most pertinent to each type of engine are summarized in the bar chart in Fig. 3.30 above.

With this chapter, we come to the end of Part 1, dealing with preliminary considerations. We are now in a position to use the aerodynamic and propulsion aspects discussed in Part 1 to address the performance of an airplane. Hence we move on to Part 2, which deals with such airplane performance.

PROBLEMS

3.1 The General Electric J79 turbojet produces a thrust of 10,000 lb at sea level. The inlet diameter is 3.19 ft. If an airplane equipped with the J79 is flying at standard sea level with a velocity of 1,000 ft/s, estimate (*a*) the jet velocity relative to the airplane and (*b*) the propulsive efficiency.

3.2 The Cessna model 310 twin-engine propeller-driven airplane is powered by two Continental IO-520-M engines rated at 285 hp each at 2,700 rpm at sea level. McCauley three-blade propellers are used, with a diameter of 6.27 ft. The maximum speed of the airplane at sea level is 238 mi/h. Assume the performance of the propeller is given by the propeller efficiency curves in Fig. 3.7, and that the propellers are variable-pitch so as to obtain the maximum efficiency. Calculate the maximum horsepower available from the engine-propeller combination at sea level.

3.3 Consider the design of turbojet engine intended to produce a thrust of 25,000 lb at a takeoff velocity of 220 ft/s at sea level. At takeoff, the gas velocity at the exit of the engine (relative to the engine) is 1,700 ft/s. The fuel-air ratio by mass is 0.03. The exit pressure is equal to the ambient pressure. Calculate the area of the inlet to the engine necessary to obtain this thrust.

3.4   A turbofan engine on a test stand in the laboratory operates continuously at a thrust level of 60,000 lb with a thrust specific fuel consumption of 0.5 h^{-1}. The fuel reservoir feeding the engine holds 1,000 gal of jet fuel. If the reservoir is full at the beginning of the test, how long can the engine run before the fuel reservoir is empty? *Note*: A gallon of fuel weighs 6.7 lb.

3.5   The thrust of a turbofan engine decreases as the flight velocity increases. The maximum thrust of the Rolls-Royce RB211 turbofan at zero velocity at sea level is 50,000 lb. Calculate the thrust at an altitude of 3 km at Mach 0.6.

3.6   The Allison T56 turboprop engine is rated at 4,910 equivalent shaft horsepower at zero velocity at sea level. Consider an airplane with this engine flying at 500 ft/s at sea level. The jet thrust is 250 lb, and the propeller efficiency is 0.9. Calculate the equivalent shaft horsepower at this flight condition.

3.7   The specific fuel consumption for the Teledyne Continental Voyager 200 liquid-cooled reciprocating engine is 0.375 lb/(bhp·h). When installed in an airplane which is flying at 200 mi/h with a propeller efficiency of 0.85, calculate the equivalent thrust specific fuel consumption.

$$0.375 \left[\frac{lb}{hp\, hr} \right] \left(\frac{1\, hp}{550\, \frac{ft\, lb}{s}} \times \frac{1\, hr}{3600\, s} \right) = 189.34 \times 10^{-9}\, \frac{lb}{ft - lb}$$

2

AIRPLANE PERFORMANCE

An airplane in motion through the atmosphere is responding to the "four forces of flight"—lift, drag, thrust, and weight. Just how it responds to these four forces determines how fast it flies, how high it can go, how far it can fly, and so forth. These are some of the elements of the study of *airplane performance*, a sub-speciality under the general discipline of flight mechanics (or flight dynamics). Airplane performance is the subject of Part 2 of this book. Here we will use our knowledge of the lift, drag, and thrust of an airplane, as discussed in Part 1, to analyze how a given airplane responds to the four forces of flight, and how this response determines its performance. In some respects, such a study helps to reinforce an appreciation for the "magic" of flight, and helps us to better understand the "mystery" of the flying machine.

chapter

4

The Equations of Motion

The power of knowledge, put it to the task,
No barrier will be able to hold you back,
It will support you even in flight!
It cannot be your Creator's desire
To chain his finest in the muck and the mire,
To eternally deny you flight!

Poem by Otto Lilienthal, 1889.
The last lines of this poem are engraved
into a commemorative stone which
marks the site of Lilienthal's fatal crash
at Gollenberg, near Stolln, Germany

4.1 INTRODUCTION

In Part 1 we have discussed some preliminary considerations—aspects of aeronautical engineering history, applied aerodynamics, and the generation of propulsive thrust and power—all intended to provide the background against which we will examine the major subjects of this book, namely, airplane performance and design. We are now ready to move into the first of these subjects—airplane performance. Here we are not concerned about the details of aerodynamics or propulsion; rather, we make use of aerodynamics mainly through the drag polar for a given airplane, and we consider the propulsive device simply in terms of thrust (or power) available and the specific fuel consumption. Our major concern is with the *movement* of a given airplane through the atmosphere, insofar as it is responding to the four forces of flight. This movement is governed by a set of equations called the *equations of motion*, which is the subject of this chapter.

4.2   THE FOUR FORCES OF FLIGHT

The four forces of flight—lift, drag, weight, and thrust, denoted by L, D, W, and T, respectively—are sketched in Fig. 4.1 for an airplane in level flight. The free-stream velocity V_∞ is always in the direction of the local flight of the airplane; in Fig. 4.1 the flight path is horizontal, and hence V_∞ is also along the horizontal. The airplane is moving from left to right, hence V_∞ is drawn pointing toward the left since it is a flow velocity relative to the airplane. By definition, the airplane lift and drag are perpendicular and parallel, respectively, to V_∞, as shown in Fig. 4.1. Lift and drag are aerodynamic forces; in Fig. 4.1, L and D represent the lift and drag, respectively, of the *complete airplane*, including the wing, tail, fuselage, etc. The weight always acts toward the center of the earth; for the level-flight case shown in Fig. 4.1, W is perpendicular to V_∞. The thrust is produced by whatever flight propulsion device is powering the airplane. In general, T is not necessarily in the free-stream direction; this is shown in Fig. 4.1 where T is drawn at an angle ϵ relative to the flight path. For the level-flight case shown in Fig. 4.1, all four forces are in the same plane, namely, the plane of the paper. This is also the longitudinal plane of symmetry for the airplane; the plane of symmetry splits the airplane into two symmetric halves.

The completely level-flight case shown in Fig. 4.1 is by far the simplest orientation of the airplane to analyze. Consider next the case of the airplane climbing (or descending) along a flight path that is angled to the horizontal, as shown in Fig. 4.2. In general, the flight path is curved, as shown. Let us consider the case where the curve

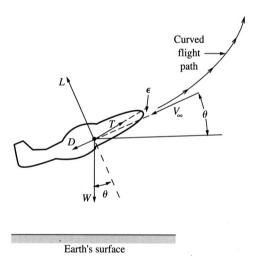

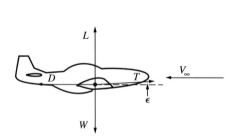

Figure 4.1   Four forces of flight—lift, drag, thrust, and weight. Illustration shows the case of a horizontal flight path. *Note:* For ordinary flight, lift and weight are much larger than thrust and drag; that is, for typical airplanes, $L/D \approx 10$ to 15.

Figure 4.2   Climbing flight.

of the flight path lies entirely in the plane of the page, that is, in the vertical plane perpendicular to the earth's surface. At any given instant as the airplane moves along this path, the local, instantaneous angle of the flight path, relative to the horizontal, is θ. Hence V_∞ is inclined at angle θ, which is called the local *climb angle* of the airplane. As before, L and D are perpendicular and parallel to V_∞. Weight W, acts toward the center of the earth, and hence is perpendicular to the earth's surface. For the airplane in climbing flight, the direction of W is inclined at the angle θ relative to the lift, as shown in Fig. 4.2. The vertical plane (page of the paper) is still the plane of symmetry for the airplane.

Starting with the airplane in the orientation shown in Fig. 4.2, we now rotate it about the longitudinal axis—the axis along the fuselage from the nose to the tail. That is, let us roll (or bank) the airplane through the roll angle ϕ shown in Fig. 4.3. This figure shows a more general orientation of the airplane in three-dimensional space, at an instantaneous climb angle of θ and an instantaneous roll angle ϕ. Examine Fig. 4.3 closely. The side view shows, in perspective, the airplane rolled toward you, the reader. Hence, the page is no longer the symmetry plane of the aircraft. Instead, the plane of symmetry is as shown in the head-on front view at the right in Fig. 4.3. This front view is a projection of the airplane and the forces on plane AA taken perpendicular to the local free-stream velocity V_∞. In this head-on front view, the plane of symmetry of the airplane is inclined to the local vertical through the roll angle ϕ.

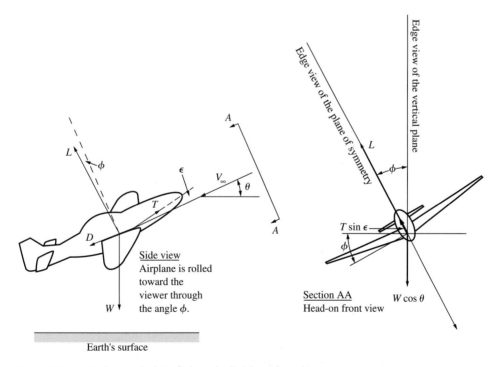

Figure 4.3 Airplane in climbing flight and rolled through angle ϕ.

Now consider the four forces of flight as they appear in Fig. 4.3. In the side view, the lift is shown, in perspective, rotated away from the local vertical through the angle ϕ; that is, the lift is inclined to the page at the roll angle ϕ. In the head-on front view, the lift L is clearly shown inclined to the vertical at angle ϕ. The thrust T, which is inclined to the flight path direction through the angle ϵ, is also rotated out of the plane of the page in the side view. In the head-on front view, T projects as the component $T \sin \epsilon$; this component is also rotated away from the vertical through angle ϕ. The weight W is always directed downward in the local vertical direction. Hence, in the side view, W is in the plane of the page. In the head-on front view, the weight projects as the component $W \cos \theta$, directed downward along the vertical. Finally, in the side view, the drag D, which is parallel to the local relative wind, is in the plane of the page. In the head-on view, since D is parallel to V_∞, the drag does not appear; its component projected on plane AA is zero.

4.3   THE EQUATIONS OF MOTION

The equations of motion for an airplane are simply statements of Newton's second law, namely,

$$\mathbf{F} = m\mathbf{a} \qquad \text{[4.1]}$$

Equation (4.1) is a vector equation, where the force $\mathbf{F}$ and the acceleration $\mathbf{a}$ are vector quantities. However, Eq. (4.1) can also be written in scalar form in terms of scalar components of $\mathbf{F}$ and $\mathbf{a}$. For example, if we choose an arbitrary direction in space, denoted by s, and we let F_s and a_s be the components of $\mathbf{F}$ and $\mathbf{a}$, respectively, in the s direction, then Eq. (4.1) gives

$$F_s = ma_s \qquad \text{[4.2]}$$

At this stage in our discussion, we have two choices. We could choose to develop the equations of motion in a very general, formalistic manner, dealing with a rotating, spherical earth and taking into account the acceleration of gravity with distance from the center of the earth. Such a development can be found in intermediate or advanced books on dynamics. A nice discussion of the general equations of motion is given by Vinh in Ref. 45. Our other choice is to assume a flat, stationary earth and to develop the equations of motion from a less formalistic, more physically motivated point of view. Since the flat-earth equations are all we need for the present book, and since our purpose is not to cover general dynamics, we make the latter choice.

Return to Fig. 4.3, and visualize the motion of the airplane along its curved flight path in three-dimensional space. Since we are interested in the *translational* motion of the airplane only, let us replace the airplane in Fig. 4.3 with a point mass at its center of gravity, with the four forces of flight acting through this point, as sketched in Fig. 4.4. The sketch in Fig. 4.4 is drawn so that the plane of the page is the plane formed by the local free-stream velocity V_∞ and the local vertical. Hence, in Fig. 4.4, both D and W are in the plane of the page. The component of lift in this plane is $L \cos \phi$.

The thrust is represented by its components in this plane, $T \cos \epsilon$ and $T \sin \epsilon \cos \phi$, parallel and perpendicular, respectively, to the local free-stream velocity V_∞.

The curvilinear motion of the airplane along the curved flight path, projected into the plane of Fig. 4.4, can be expressed by Newton's second law, by first taking components parallel to the flight path and then taking components perpendicular to the flight path. The component of force parallel to the flight path is, from Fig. 4.4,

$$F_\parallel = T \cos \epsilon - D - W \sin \theta \qquad \textbf{[4.3]}$$

The acceleration parallel to the flight path is

$$a_\parallel = \frac{dV_\infty}{dt} \qquad \textbf{[4.4]}$$

Hence, Newton's second law, taken parallel to the flight path, is

$$ma_\parallel = F_\parallel$$

or

$$m\frac{dV_\infty}{dt} = T \cos \epsilon - D - W \sin \theta \qquad \textbf{[4.5]}$$

In the direction perpendicular to the flight path, the component of force is

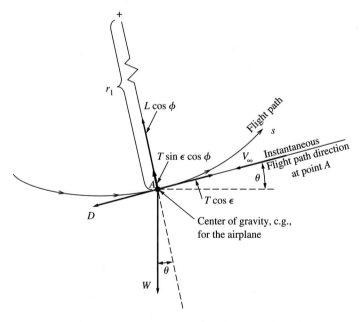

Figure 4.4   Forces projected into the plane formed by the local free-stream velocity V_∞ and the vertical (perpendicular to the surface of the earth).

$$F_\perp = L \cos \phi + T \sin \epsilon \cos \phi - W \cos \theta$$

The radial acceleration of the curvilinear motion, perpendicular to the flight path, is

$$a_\perp = \frac{V_\infty^2}{r_1}$$

where r_1 is the local radius of curvature of the flight path in the plane of the page in Fig. 4.4. Hence, Newton's second law, taken perpendicular to the flight path, is

$$ma_\perp = F_\perp$$

or

$$m\frac{V_\infty^2}{r_1} = L \cos \phi + T \sin \epsilon \cos \phi - W \cos \theta \qquad \textbf{[4.6]}$$

Return to Fig. 4.3, and visualize a horizontal plane—a plane parallel to the flat earth. The projection of the curved flight path on this horizontal plane is sketched in Fig. 4.5. The plane of the page in Fig. 4.5 is the horizontal plane. The instantaneous location of the airplane's center of gravity (c.g.) is shown as the large dot; the velocity vector of the airplane projects into this horizontal plane as the component $V_\infty \cos \theta$, tangent to the projected flight path at the c.g. location. The local radius of curvature of the flight path in the horizonatal plane is shown as r_2. The projection of the lift vector in the horizontal plane is $L \sin \phi$, and is perpendicular to the flight path, as shown in Fig. 4.5. The components of the thrust vector in the horizontal plane are $T \sin \epsilon \sin \phi$ and $T \cos \epsilon \cos \theta$ perpendicular and parallel, respectively, to the projected flight path in Fig. 4.5. The component of drag in this plane is $D \cos \theta$. Since the weight acts perpendicular to the horizontal, its component is zero in Fig. 4.5. If you are not quite clear about the force components shown in Fig. 4.5, go back and reread this paragraph,

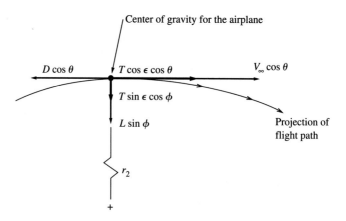

Figure 4.5   Forces projected into the horizontal plane parallel to the flat earth.

flipping back and forth between Figs. 4.3 and 4.5, until you feel comfortable with the sketch shown in Fig. 4.5.

Consider the force components in Fig. 4.5 that are perpendicular to the flight path at the instantaneous location of the center of gravity. The sum of these forces, denoted by F_2, is

$$F_2 = L \sin \phi + T \sin \epsilon \sin \phi$$

The instantaneous radial acceleration along the curvilinear path in Fig. 4.5 is

$$a_2 = \frac{(V_\infty \cos \theta)^2}{r_2}$$

From Newton's second law taken along the direction perpendicular to the flight path in the horizontal plane shown in Fig. 4.5, we have

$$m\frac{(V_\infty \cos \theta)^2}{r_2} = L \sin \phi + T \sin \epsilon \sin \phi \qquad \textbf{[4.7]}$$

Equations (4.5) to (4.7) are three equations which describe the translational motion of an airplane through three-dimensional space over a flat earth. They are called the *equations of motion* for the airplane. (There are three additional equations of motion that describe the rotational motion of the airplane about its axes; however, we are not concerned with the rotational motion here. Also, we have assumed no yaw of the airplane; i.e., the free-stream velocity vector has been treated as always parallel to the symmetry plane of the aircraft.) These equations of motion are simply statements of Newton's second law.

4.4   SUMMARY AND COMMENTS

In this short chapter we have discussed the four forces of flight—lift, drag, thrust, and weight. The translational motion of the airplane—its flight path and the instantaneous velocities and accelerations—is determined by these forces. The equations which relate the forces to the motion are obtained from Newton's second law. The resulting equations are called the *equations of motion* for the airplane. For the assumption of a flat earth and no yaw, the equations of motion are given by Eqs. (4.5) to (4.7).

Our discussion of airplane performance for the remainder of Part 2 of this book is based on various applications of the equations of motion. We will find that, to answer some questions about the performance of an airplane, Eqs. (4.5) to (4.7) can be greatly simplified. However, to address other aspects of performance, Eqs. (4.5) to (4.7) need to be used in almost their full form. In any event, with the equations of motion in our mind, we are now ready to examine these performance questions.

5

Airplane Performance: Steady Flight

We have the aerodynamic knowledge, the structural materials, the power plants, and the manufacturing capacities to perform any conceivable miracle in aviation. But miracles must be planned, nurtured, and executed with intelligence and hard work.

Glenn L. Martin, aviation
pioneer and manufacturer, 1954

5.1   INTRODUCTION

A three-view of the Gulfstream IV twin-turbofan executive transport is shown in Fig. 5.1. This airplane is considered one of the most advanced executive jet transports in existence today. The first flight of the prototype took place on September 19, 1985. On June 12, 1987, a regular production model took off from le Bourget Airport in Paris (the same airport at which Charles Lindbergh touched down on May 21, 1927, at the end of his famous transatlantic solo flight); 45 hours 25 minutes later, the same Gulfstream IV landed at le Bourget, setting a new world record for a westbound around-the-world flight (with four refueling stops). This length of time to fly around the world was only 12 h longer than it took Lindbergh to fly the Atlantic in 1927. The Gulfstream IV has a normal cruising speed of 528 mi/h at an altitude of 45,000 ft, which yields a cruising Mach number of 0.80. Its maximum range at cruising conditions with a maximum payload of 4,000 lb is 4,254 mi. The Gulfstream has a maximum rate of climb at sea level of 4,000 ft/min. Its stalling speed with flaps up is 141 mi/h; with the flaps down, the stalling speed reduces to 124 mi/h.

The facts and figures given above are a partial description of the *performance* of the airplane. They pertain to the airplane in *steady flight*; that is, the airplane is experiencing *no acceleration*. Such performance for unaccelerated flight is called

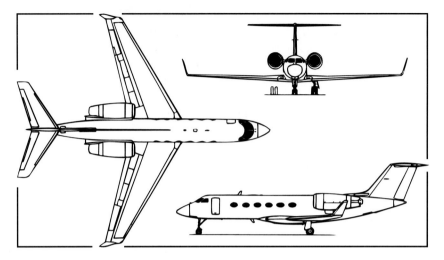

Figure 5.1   Three-view of the Gulfstream Aerospace Gulfstream IV executive jet transport.

static performance. In this chapter, we focus on aspects of the static performance of an airplane.

How do we know the static performance characteristics of the Gulfstream IV itemized above? One answer is that they can be measured in flight after the airplane is designed and built. But how can we calculate and analyze the performance of the airplane *before* it first flies? Indeed, how can we estimate the performance of a given airplane design before the airplane is actually built? The purpose of this chapter is to answer these and other related questions. In this chapter we develop analytical and graphical techniques to predict the static performance of an airplane. We see how to obtain the type of performance figures discussed earlier for the Gulfstream IV, and for any other type of conventional airplane as well.

Parenthetical note: The worked examples sprinkled throughout this chapter deal with an airplane patterned after the Gulfstream IV. The Gulfstream IV is powered by *turbofan* engines, which, as we have discussed in Chapter 3, experience a *decrease* in thrust as the flight velocity increases. This is in contrast to typical turbojet engines which, for subsonic speeds, have a relatively constant thrust with velocity. Nevertheless, for a pedagogical reason, we assume in the present worked examples that the thrust from the jet engines remains constant with velocity, as opposed to the actual situation of decreasing thrust. The pedagogical reason is this: in this chapter we highlight both graphical and analytical solutions of airplane performance. In the worked examples, both graphical and analytical approaches are used, and the answers from both approaches are compared with one another. If the engine thrust is a function of velocity, the analytical solutions, although still possible, become much more cumbersome. From a pedagogical point of view, making the analysis more cumbersome detracts from the fundamental ideas being presented. Therefore, we avoid this situation by assuming in the worked examples a constant thrust from the jet engines.

Please be aware that in some cases this will lead to results that are much too optimistic. The actual Gulfstream IV is already a high-performance airplane (a "hot" airplane); in some of the worked examples in this chapter, it will appear to be even "hotter." However, the purpose of the worked examples is to illustrate the basic concepts, and so nothing is lost, and indeed much is gained, by the simplicity in assuming a constant thrust with velocity. Some of the problems at the end of this chapter deal with the more realistic case of a variation of turbofan thrust with velocity. The results of these problems, compared with the corresponding worked examples in the text, give some idea of the differences obtained.

5.2 EQUATIONS OF MOTION FOR STEADY, LEVEL FLIGHT

Return to Fig. 4.1, which shows an airplane with a horizontal flight path. This airplane is in *level flight*; that is, the climb angle θ and roll angle ϕ are zero. Moreover, by definition, *steady flight* is flight with no acceleration. Hence, the governing equations of motion for steady, level flight are obtained from Eqs. (4.5) and (4.6) by setting $\theta, \phi, dV_\infty/dt$, and V_∞^2/r_1 equal to zero. (The normal acceleration V_∞/r_1 is zero by definition of steady flight, i.e., no acceleration; this is also consistent with the flight path being a straight line, where the radius of curvature r_1 is infinitely large.) The resulting equations are, from Eq. (4.5),

$$0 = T \cos \epsilon - D \qquad \text{[5.1]}$$

and from Eq. (4.6),

$$0 = L + T \sin \epsilon - W \qquad \text{[5.2]}$$

Although the engine thrust line is inclined at angle ϵ to the free-stream direction, this angle is usually small for conventional airplanes and can be neglected. Hence, for this chapter we assume that the thrust is aligned with the flight direction, that is, $\epsilon = 0$. For this case, Eqs. (5.1) and (5.2) reduce to, respectively,

$$\boxed{T = D} \qquad \text{[5.3]}$$

$$\boxed{L = W} \qquad \text{[5.4]}$$

Equations (5.3) and (5.4) can be obtained simply by inspection of Fig. 5.2, which illustrates an airplane in steady, level flight. In the simple force balance shown in Fig. 5.2, lift equals weight [Eq. (5.4)] and thrust equals drag [Eq. (5.3)]. Although we could have written these equations directly by inspection of Fig. 5.2 rather than derive them as special cases of the more general equations of motion, it is instructional to show that Eqs. (5.3) and (5.4) are indeed special cases of the general equations of motion—indeed, Eqs. (5.3) and (5.4) *are* the equations of motion for an airplane in steady, level flight.

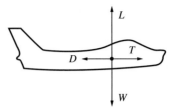

Figure 5.2   Force diagram for steady, level flight.

5.3   THRUST REQUIRED (DRAG)

Return again to Fig. 5.2. Imagine this airplane in steady, level flight at a given velocity and altitude, say, at 400 mi/h at 20,000 ft. To maintain this speed and altitude, enough thrust must be generated to exactly overcome the drag and to keep the airplane going—this is the *thrust required* to maintain these flight conditions. The thrust required T_R depends on the velocity, the altitude, and the aerodynamic shape, size, and weight of the airplane—it is an airframe-associated feature rather than anything having to do with the engines themselves. Indeed, the thrust required is simply equal to the *drag* of the airplane—it is the thrust required to overcome the aerodynamic drag.

A plot showing the variation of T_R with free-stream velocity V_∞ is called the *thrust required curve*; such a curve is shown in Fig. 5.3. It is one of the essential elements in the analysis of airplane performance. A thrust required curve, such as the one shown in Fig. 5.3, pertains to a given airplane at a given standard altitude. Keep in mind that the thrust required is simply the drag of the airplane, hence the thrust required cruve is nothing other than a plot of drag versus velocity for a given airplane at a given altitude. The thrust required curve in Fig. 5.3 is for the Northrop T-38 jet trainer (shown in Fig. 2.42) with a weight of 10,000 lb at an altitude of 20,000 ft.

Question: Why does the T_R curve in Fig. 5.3 look the way it does? Note that at the higher velocities, T_R increases with V_∞, which makes sense intuitively. However, at lower velocities, T_R *decreases* with V_∞, which at first thought is counterintuitive—it takes *less* thrust to fly faster? Indeed, there is some velocity at which T_R is a minimum value. What is going on here? Why is the thrust required curve shaped this way? We will address these questions in the next two subsections. First we examine the purely graphical aspects of the thrust required curve, showing how to calculate points on this curve. Then we follow with a theoretical analysis of the thrust required curve and associated phenomena.

5.3.1   Graphical Approach

Consider a given airplane flying at a given altitude in steady, level flight. For the given airplane, we know the following physical characteristics: weight W, aspect ratio AR, and wing planform area S. Equally important, we know the drag polar for

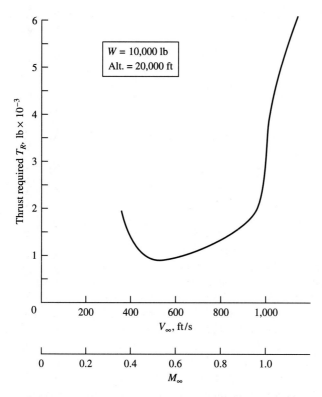

Figure 5.3 Thrust required curve for the Northrop T-38 jet trainer with a weight of 10,000 lb at an altitude of 20,000 ft.

the airplane, given by Eq. (2.47), repeated here:

$$C_D = C_{D,0} + K C_L^2 \qquad \text{[5.5]}$$

where $C_{D,0}$ and K are known for the given airplane. To calculate the thrust required curve, proceed as follows:

1. Choose a value of V_∞.
2. For the chosen V_∞, calculate C_L from the relation

$$L = W = \tfrac{1}{2}\rho_\infty V_\infty^2 S C_L$$

or

$$C_L = \frac{2W}{\rho_\infty V_\infty^2 S}$$

3. Calculate C_D from Eq. (5.5), repeated here.

$$C_D = C_{D,0} + K C_L^2$$

4. Calculate drag, hence T_R, from

$$T_R = D = \tfrac{1}{2}\rho_\infty V_\infty^2 S C_D$$

This is the value of T_R corresponding to the velocity chosen in step 1. This combination (T_R, V_∞) is one point on the thrust required curve.

5. Repeat steps 1 to 4 for a large number of different values of V_∞, thus generating enough points to plot the thrust required curve.

Example 5.1

Consider the Gulfstream IV twin-turbofan executive transport shown in Fig. 5.1. Calculate and plot the thrust required curve at an altitude of 30,000 ft, assuming a weight of 73,000 lb. Airplane data: $S = 950$ ft^2, $AR = 5.92$, $C_{D,0} = 0.015$, and $K = 0.08$. Hence the drag polar in the form given by Eq. (5.5) is

$$C_D = 0.015 + 0.08 C_L^2$$

Note: The above drag polar for the Gulfstream IV is only an educated guess by the author. Drag polar information for specific airplanes is sometimes difficult to find in the open literature because it is often proprietary to the manufacturer. The value of 0.015 chosen for $C_{D,0}$ is based on a generic value typical of streamlined, multiengine jet aircraft. The value of 0.08 chosen for K is estimated by first calculating k_3 in Eq. (2.44), where $k_3 = 1/(\pi e AR)$. Assuming a span efficiency factor $e = 0.9$, we have

$$k_3 = \frac{1}{\pi e AR} = \frac{1}{\pi(0.9)(5.92)} = 0.06$$

In Eq. (2.44), assume k_1 (associated with the increase in parasite drag due to lift) is about $\tfrac{1}{3}k_3$. Also, assume no wave drag, hence in Eq. (2.44), $k_2 = 0$. Thus, $K = k_1 + k_2 + k_3 = 0.02 + 0 + 0.06 = 0.08$. Because of these assumptions, the drag polar used in this calculation is only an approximation for the Gulfstream IV, and hence the computed results (and any of the related results to follow) are only an approximate representation of the performance of the Gulfstream IV as opposed to a precisely accurate result for the real airplane.

To calculate a point on the thrust required curve, let us follow the four-step procedure described earlier.

1. Choose $V_\infty = 500$ ft/s.

2. At a standard altitude of 30,000 ft (see Appendix B),

$$\rho_\infty = 8.9068 \times 10^{-4} \text{ slug/ft}^3$$

$$C_L = \frac{2W}{\rho_\infty V_\infty^2 S} = \frac{2(73,000)}{(8.9068 \times 10^{-4})(500)^2(950)} = 0.6902$$

3. $C_D = C_{D,0} + K C_L^2 = 0.015 + 0.08(0.69)^2 = 0.0531$

4. $T_R = D = \tfrac{1}{2}\rho_\infty V_\infty^2 S C_D = \tfrac{1}{2}(8.9068 \times 10^{-4})(500)^2(950)(0.053) = \boxed{5{,}617 \text{ lb}}$

Hence, to maintain straight and level flight at a velocity of 500 ft/s at an altitude of 30,000 ft, the airplane requires 5,617 lb of thrust. The calculation of other points on the thrust required curve for other velocities is tabulated in Table 5.1.

Table 5.1

V_∞ (ft/s)	C_L	C_D	T_R (lb)
300	1.9172	0.3090	11,768
400	1.0784	0.1080	7,313
500	0.6902	0.0531	5,617
600	0.4793	0.0334	5,084
700	0.3521	0.0249	5,166
800	0.2696	0.0208	5,636
900	0.2130	0.0186	6,384
1,000	0.1725	0.0174	7,354
1,100	0.1426	0.0166	8,512
1,200	0.1198	0.0161	9,838
1,300	0.1021	0.0158	11,321

The results are plotted in Fig. 5.4 as the solid curve.

Let us examine the trends shown in Table 5.1 and in Fig. 5.4. Keep in mind that the drag polar for this graph, namely $C_D = 0.015 + 0.08C_L^2$, does not account for the rapid drag divergence due to wave drag that would occur at a free-stream Mach number of about 0.85 (the maximum operating Mach number of the Gulfstream IV is 0.88, as listed in Ref. 36). Hence the portion of the T_R curve shown in Fig. 5.4 for $M_\infty > 0.85$ is more academic than real. However, this does not compromise the important points discussed below.

First, note the variation of C_L with V_∞ as tabulated in Table 5.1. At the lowest values of V_∞, C_L is very large; but as V_∞ increases, C_L decreases fairly rapidly. This is because for steady, level flight $L = W$ and

$$L = W = \tfrac{1}{2}\rho_\infty V_\infty^2 SC_L$$

At very low velocity, the necessary lift is generated by flying at a high lift coefficient, hence at a high angle of attack. However, as V_∞ increases, a progressively lower C_L is required to sustain the weight of the airplane because the necessary lift is generated progressively more by the increasing dynamic pressure $\tfrac{1}{2}\rho_\infty V_\infty^2$. Hence, as V_∞ increases, the angle of attack of the airplane progressively decreases, as sketched in Fig. 5.4.

With the above ideas in mind, we can now explain *why* the thrust required curve is shaped as it is—with T_R first *decreasing* with increasing velocity, reaching a minimum value, and then increasing as velocity further increases. To help us in this explanation, we write the drag as

$$D = \tfrac{1}{2}\rho_\infty V_\infty^2 SC_D$$

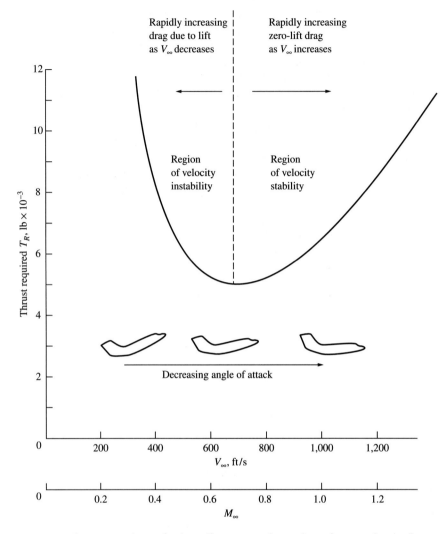

Figure 5.4   Thrust required curve for the Gulfstream IV at the conditions for Example 5.1, illustrating the regions of velocity instability and stability, and the direction of decreasing angle of attack with increasing velocity. Altitude = 30,000 ft; $W = 73,000$ lb.

where

$$C_D = C_{D,0} + K C_L^2$$

Hence

$$D = \tfrac{1}{2}\rho_\infty V_\infty^2 S C_D + \tfrac{1}{2}\rho_\infty V_\infty^2 S K C_L^2$$

$\underbrace{\qquad\qquad}_{\text{Zero-lift drag}}$ $\underbrace{\qquad\qquad}_{\text{drag due to lift}}$

[5.6]

At low velocity, where C_L is high, the total drag is dominated by the drag due to lift. Since the drag due to lift is proportional to the *square* of C_L, as seen in Eq. (5.6), and C_L decreases rapidly as V_∞ increases, the drag due to lift rapidly decreases, in spite of the fact that the dynamic pressure $\frac{1}{2}\rho_\infty V_\infty^2$ is increasing. This is why the T_R curve first *decreases* as V_∞ *increases*. This part of the curve is shown to the left of the vertical dashed line in Fig. 5.4—the region where the drag due to lift increases rapidly as V_∞ decreases. In contrast, as seen in Eq. (5.6), the zero-lift drag increases as the square of V_∞. At high velocity, the total drag is dominated by the zero-lift drag. Hence, as the velocity of the airplane increases, there is some velocity at which the increasing zero-lift drag exactly compensates for the decreasing drag due to lift; this is the velocity at which T_R is a minimum. At higher velocities, the rapidly increasing zero-lift drag causes T_R to increase with increasing velocity—this is the part of the curve shown to the right of the vertical dashed line in Fig. 5.4. These are the reasons why the T_R curve is shaped as it is—with T_R first decreasing with V_∞, passing through a minimum value, and then increasing with V_∞.

To reinforce the above discussion, Fig. 5.5 shows the individual variations of drag due to lift and zero-lift drag as functions of V_∞. Note that at the point of minimum T_R, the drag due to lift and the zero-lift drag are equal. From Eq. (5.6), this requires that $C_{D,0} = K C_L^2$. We will prove this result analytically in Section 5.4.1.

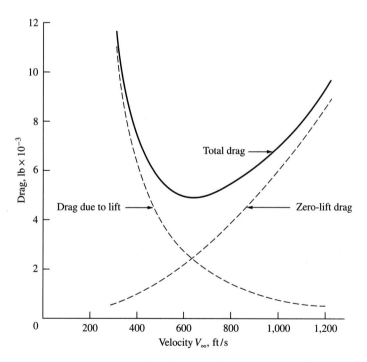

Figure 5.5   Drag versus velocity for the Gulfstream IV for the conditions of Example 5.1. Illustration of the variation of the drag due to lift and the zero-lift drag. Altitude = 30,000 ft; W = 73,000 lb.

It is undesirable to fly an airplane in the velocity range to the left of the vertical dashed line in Fig. 5.4. This is a region of *velocity instability*, as identified in Fig. 5.4. The nature of this velocity instability is as follows. Consider an airplane in steady, level flight at a velocity less than the velocity for minimum T_R, that is, to the left of the vertical dashed line in Fig. 5.4. This condition is sketched in Fig. 5.6**a**, where the airplane velocity is denoted by V_1. For steady flight, the engine throttle is adjusted such that the thrust from the engine exactly equals T_R. Now assume the airplane is perturbed in some fashion, say, by a horizontal gust, which momentarily decreases V_∞ for the airplane, say, to velocity V_2. This decrease in velocity $\Delta V = V_2 - V_1$ causes an increase in T_R (an increase in drag), denoted by $\Delta T_R = T_{R_2} - T_{R_1}$. But the engine throttle has not been touched, and momentarily the drag of the airplane is higher than the thrust from the engine. This *further slows down* the airplane and takes it even farther away from its original point, point 1 in Fig. 5.6**a**. *This is an unstable condition.* Similarly, if the perturbation momentarily increases V_∞ to V_3, where the increase in velocity is $\Delta V = V_3 - V_1$, then T_R (hence, drag) decreases, $\Delta T_R = T_{R_3} - T_{R_1}$. Again, the engine throttle has not been touched, and momentarily the thrust from the engine is higher than the drag of the airplane. This accelerates the airplane to an even higher velocity, taking it even farther away from its original point, point 1. Again, this is an *unstable condition*. This is why the region to the left of the vertical dashed line in Fig. 5.4 is a region of *velocity instability*.

The opposite occurs at velocities higher than that for minimum T_R, that is, to the right of the dashed vertical line in Fig. 5.4. As shown in Fig. 5.6**b**, a momentary increase in velocity $\Delta V = V_2 - V_1$ causes a momentary increase in T_R (hence drag). Since the throttle is not touched, momentarily the drag will be higher than the engine thrust, and the airplane will slow down; that is, it will tend to return back to its original point 1. This is a *stable condition*. Similarly, a momentary decrease in velocity $\Delta V = V_3 - V_1$ causes a momentary decrease in T_R (hence, drag). Since the throttle is not touched, momentarily the drag will be less than the engine thrust, and the airplane will speed up; that is, it will tend to return to its original point 1. Again, this is a *stable condition*. This is why the region to the right of the vertical dashed line in Fig. 5.4 is a region of *velocity stability*.

5.3.2 Analytical Approach

In this section we examine the thrust required curve from an analytical point of view, exploring the equations and looking for interesting relationships between the important parameters that dictate thrust required (drag).

For steady, level flight we have from Eqs. (5.3) and (5.4)

$$T_R = D = \frac{D}{W}W = \frac{D}{L}W$$

or

$$T_R = \frac{W}{L/D} \qquad \textbf{[5.7]}$$

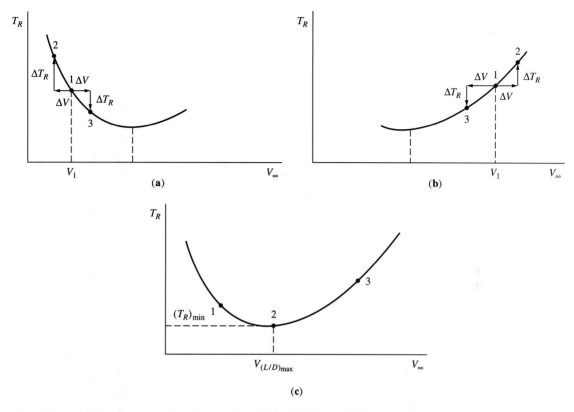

Figure 5.6 (**a**) The mechanism of velocity instability. (**b**) The mechanism of velocity stability. (**c**) Maximum T_R occurs at maximum lift-to-drag ratio, point 2. Points 1, 2, and 3 correspond to points 1, 2, and 3, respectively, in Fig. 5.7.

Examining Eq. (5.7), we see that for an airplane with fixed weight, T_R decreases as L/D increases. Indeed, *minimum T_R occurs when L/D is maximum.* This fact is noted on the thrust required curve sketched in Fig. 5.6c. *The lift-to-drag ratio is one of the most important parameters affecting airplane performance.* It is a direct measure of the aerodynamic efficiency of an airplane. The lift-to-drag ratio is the same as the ratio of C_L to C_D,

$$\frac{L}{D} = \frac{\frac{1}{2}\rho_\infty V_\infty^2 S C_L}{\frac{1}{2}\rho_\infty V_\infty^2 S C_D} = \frac{C_L}{C_D} \qquad \textbf{[5.8]}$$

Since C_L and C_D are both functions of the angle of attack of the airplane α, then L/D itself is a function of α. A generic variation of L/D with α for a given airplane is sketched in Fig. 5.7. Comparing the generic curves in both Figs. 5.6c and 5.7, we see that point 2 in both figures corresponds to the maximum value of L/D, denoted $(L/D)_{\max}$. The angle of attack of the airplane at this condition is denoted as $\alpha_{(L/D)_{\max}}$. The flight velocity at this condition is denoted by $V_{(L/D)_{\max}}$, which of course is the velocity at which T_R is a minimum. Imagine an airplane in steady, level flight at a

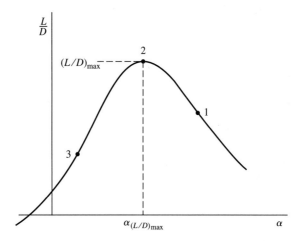

Figure 5.7   Schematic of the variation of lift-to-drag ratio for a given airplane as a function of angle of attack. Points 1, 3, and 3 correspond to points 1, 2, and 3, respectively, in Fig. 5.6c.

given altitude, with its thrust required curve given by the generic curve in Fig. 5.6c. If its velocity is high, say, given by point 3 in Fig. 5.6c, then its angle of attack is low, denoted by point 3 in Fig. 5.7. As seen in Fig. 5.7, this condition is far away from that for maximum L/D. As the airplane slows down, we move from right to left along the T_R curve in Fig. 5.6c and from left to right along the L/D curve in Fig. 5.7. As the airplane slows down, its angle of attack increases. Starting at point 3 in Fig. 5.7, L/D first increases, reaches a maximum (point 2), and then decreases. From Eq. (5.7), T_R correspondly first decreases, reaches a minimum (point 2 in Fig. 5.6c), and then increases. Point 1 in Figs. 5.6c and 5.7 corresponds to a low velocity, with a large angle of attack and with a value of L/D far away from its maximum value. When you are looking at T_R curve, it is useful to remember that each different point on the curve corresponds to a different angle of attack and a different L/D. To be more specific, consider the airplane in Example 5.1, with the corresponding data in Table 5.1. The variation of L/D with V_∞ can easily be found by dividing C_L by C_D, both found in Table 5.1. The results are plotted in Fig. 5.8, where the values of $(L/D)_{max}$ and $V_{(L/D)_{max}}$ are also marked.

The drag (hence T_R) for a given airplane in steady, level flight is a function of altitude (denoted by h), velocity, and weight:

$$D = f(h, V_\infty, W) \qquad [5.9]$$

This makes sense. When the altitude h changes, so does density ρ_∞; hence D changes. Clearly, as V_∞ changes, D changes. As W changes, so does the lift L; in turn, the induced drag (drag due to lift) changes, and hence the total drag changes. It is sometimes comfortable and useful to realize that drag for a given airplane depends

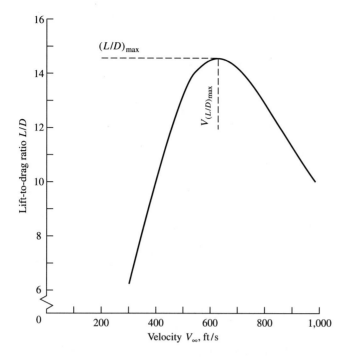

Figure 5.8 Variation of L/D with velocity for the Gulfstream IV at the conditions for Example 5.1. Altitude $= 30,000$ ft; $W = 73,000$ lb.

only on altitude, velocity, and weight. An expression for drag which explicitly shows this relationship is easily obtained from the drag polar:

$$D = q_\infty S C_D = q_\infty S(C_{D,0} + K C_L^2) \qquad \textbf{[5.10]}$$

From Eq. (5.4),

$$L = W = q_\infty S C_L = \tfrac{1}{2}\rho_\infty V_\infty^2 S C_L$$

we have

$$C_L = \frac{2W}{\rho_\infty V_\infty^2 S} \qquad \textbf{[5.11]}$$

Substituting Eq. (5.11) into (5.10), we obtain

$$D = \frac{1}{2}\rho_\infty V_\infty^2 S \left[C_{D,0} + 4K \left(\frac{W}{\rho_\infty V_\infty^2 S} \right)^2 \right]$$

or

$$D = \frac{1}{2}\rho_\infty V_\infty^2 S C_{D,0} + \frac{2KS}{\rho_\infty V_\infty^2}\left(\frac{W}{S}\right)^2 \qquad \textbf{[5.12]}$$

For a given airplane (with given S, $C_{D,0}$, and K), Eq. (5.12) explicitly shows the variation of drag with *altitude* (via the value of ρ_∞), *velocity* V_∞, and *weight* W.

Equation (5.12) can be used to find the flight velocities for a given value of T_R. Writing Eq. (5.12) in terms of the dynamic pressure $q_\infty = \frac{1}{2}\rho_\infty V_\infty^2$ and noting that $D = T_R$, we obtain

$$T_R = q_\infty S C_{D,0} + \frac{KS}{q_\infty}\left(\frac{W}{S}\right)^2 \qquad \textbf{[5.13]}$$

Multiplying Eq. (5.13) by q_∞, and rearranging, we have

$$q_\infty^2 S C_{D,0} - q_\infty T_R + KS\left(\frac{W}{S}\right)^2 = 0 \qquad \textbf{[5.14]}$$

Note that, being a quadratic equation in q_∞, Eq. (5.14) yields two roots, that is, two solutions for q_∞. Solving Eq. (5.14) for q_∞ by using the quadratic formula results in

$$q_\infty = \frac{T_R \pm \sqrt{T_R^2 - 4SC_{D,0}K(W/S)^2}}{2SC_{D,0}} \qquad \textbf{[5.15]}$$

$$= \frac{T_R/S \pm \sqrt{(T_R/S)^2 - 4C_{D,0}K(W/S)^2}}{2C_{D,0}}$$

By replacing q_∞ with $\frac{1}{2}\rho_\infty V_\infty^2$, Eq. (5.15) becomes

$$V_\infty^2 = \frac{T_R/S \pm \sqrt{(T_R/S)^2 - 4C_{D,0}K(W/S)^2}}{\rho_\infty C_{D,0}} \qquad \textbf{[5.16]}$$

The parameter T_R/S appears in Eq. (5.16); analogous to the wing loading W/S, the quantity T_R/S is sometimes called the *thrust loading*. However, in the hierarchy of parameters important to airplane performance, T_R/S is not quite as fundamental as the wing loading W/S or the thrust-to-weight ratio T_R/W (as will be discussed in the next section). Indeed, T_R/S is simply a combination of T_R/W and W/S via

$$\frac{T_R}{S} = \frac{T_R}{W}\frac{W}{S} \qquad \textbf{[5.17]}$$

Substituting Eq. (5.17) into (5.16) and taking the square root, we have our final expression for velocity:

$$V_\infty = \left[\frac{(T_R/W)(W/S) \pm (W/S)\sqrt{(T_R/W)^2 - 4C_{D,0}K}}{\rho_\infty C_{D,0}}\right]^{1/2} \qquad \textbf{[5.18]}$$

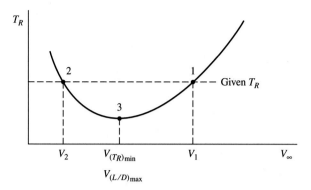

Figure 5.9   At a given T_R larger than the minimum value, there are two corresponding velocities, the low velocity V_2 and the high velocity V_1.

Equation (5.18) gives the two flight velocities associated with a given value of T_R. For example, as sketched in Fig. 5.9, for a given T_R there are generally two flight velocities which correspond to this value of T_R, namely, the higher velocity V_1 obtained from the positive discriminant in Eq. (5.18) and the lower velocity V_2 obtained from the negative discriminant in Eq. (5.18). It is important to note the characteristics of the airplane on which these velocities depend. From Eq. (5.18), V_∞ for a given T_R depends on

1. Thrust-to-weight ratio T_R/W
2. Wing loading W/S
3. The drag polar, that is, $C_{D,0}$ and K

Of course, V_∞ also depends on altitude via ρ_∞. As we progress in our discussion, we will come to appreciate that T_R/W, W/S, and the drag polar are the *fundamental parameters* that dictate airplane performance. Indeed, these parameters will be highlighted in Section 5.4.

When the discriminant in Eq. (5.18) equals zero, then only one solution for V_∞ is obtained. This corresponds to point 3 in Fig. 5.9, namely, the point of minimum T_R. That is, in Eq. (5.18) when

$$\left(\frac{T_R}{W}\right)^2 - 4C_{D,0}K = 0 \qquad \text{[5.19]}$$

then the velocity obtained from Eq. (5.18) is

$$V_{(T_R)_{\min}} = \left[\frac{1}{\rho_\infty C_{D,0}}\left(\frac{T_R}{W}\right)_{\min}\frac{W}{S}\right]^{1/2} \qquad \text{[5.20]}$$

The value of $(T_R/W)_{\min}$ is given by Eq. (5.19) as

$$\left(\frac{T_R}{W}\right)^2_{\min} = 4C_{D,0}K$$

or

$$\boxed{\left(\frac{T_R}{W}\right)_{\min} = \sqrt{4C_{D,0}K}} \qquad \textbf{[5.21]}$$

Substituting Eq. (5.21) into Eq. (5.20), we have

$$V_{(T_R)_{\min}} = \left(\frac{\sqrt{4C_{D,0}K}}{\rho_\infty C_{D,0}}\frac{W}{S}\right)^{1/2}$$

or

$$\boxed{V_{(T_R)_{\min}} = V_{(L/D)_{\max}} = \left(\frac{2}{\rho_\infty}\sqrt{\frac{K}{C_{D,0}}}\frac{W}{S}\right)^{1/2}} \qquad \textbf{[5.22]}$$

In Eq. (5.22), by stating that $V_{(T_R)_{\min}} = V_{(L/D)_{\max}}$, we are recalling that the velocity for minimum T_R is also the velocity for maximum L/D, as shown in Fig. 5.6. Indeed, since $T_R = D$ and $L = W$ for steady, level flight, Eq. (5.21) can be written as

$$\left(\frac{D}{L}\right)_{\min} = \sqrt{4C_{D,0}K} \qquad \textbf{[5.23]}$$

Since the minimum value of D/L is the reciprocal of the maximum value of L/D, then Eq. (5.23) becomes

$$\boxed{\left(\frac{L}{D}\right)_{\max} = \frac{1}{\sqrt{4C_{D,0}K}}} \qquad \textbf{[5.24]}$$

Surveying the results associated with minimum T_R (associated with point 3 on the curve in Fig. 5.9) as given by Eqs. (5.21), (5.22), and (5.24), we again see the role played by the parameters T_R/W, W/S, and the drag polar. From Eq. (5.21), we see that the value of $(T_R/W)_{\min}$ depends *only* on the drag polar, that is, the values of $C_{D,0}$ and K. From Eq. (5.22), the velocity for $(T_R)_{\min}$ depends on the altitude (via ρ_∞), the drag polar (via $C_{D,0}$ and K), and the wing loading W/S. Notice in Eqs. (5.21) and (5.22) that the airplane weight does not appear separately, but rather always appears as part of a ratio, namely T_R/W and W/S.

Looking more closely at Eqs. (5.21) and (5.22), we see that the value of $(T_R)_{\min}$ is independent of altitude, but that the velocity at which it occurs increases with increasing altitude (decreasing ρ_∞). This relationship is sketched in Fig. 5.10. Also, the effect of increasing the zero-lift drag coefficient $C_{D,0}$ is to increase $(T_R)_{\min}$ and to decrease the velocity at which it occurs. The effect of increasing the drag-due-to-lift factor K (say by decreasing the aspect ratio) is to increase $(T_R)_{\min}$ and increase the

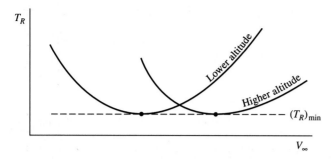

Figure 5.10 Effect of altitude on the point corresponding to minimum thrust required.

velocity at which it occurs. If the airplane's weight is increased, $(T_R)_{min}$ increases directly proportionally to W, given by Eq. (5.21), and the velocity at which it occurs increases as the square root of W, given by Eq. (5.22).

The maximum lift-to-drag ratio, as given by Eq. (5.24), is *solely* dependent on the drag polar. An *increase* in the zero-lift drag coefficient $C_{D,0}$ and/or an increase in the drag-due-to-lift factor K *decreases* the value of the maximum lift-to-drag ratio, which certainly makes sense. Here is where the wing aspect ratio plays a strong role. A higher aspect ratio results in a lower value of K and hence increases the lift-to-drag ratio.

Example 5.2

For the Gulfstream IV at the conditions stated in Example 5.1, calculate the minimum thrust required and the velocity at which it occurs. Compare the answers with the graphical results shown in Fig. 5.4.

Solution

From the data given in Example 5.1 we have $W = 73{,}000$ lb, $S = 950$ ft^2, $\rho_\infty = 8.9068 \times 10^{-4}$ slug/ft^3, $C_{D,0} = 0.015$, and $K = 0.08$. From Eq. (5.21),

$$\left(\frac{T_R}{W}\right)_{min} = \sqrt{4C_{D,0}K} = \sqrt{4(0.015)(0.08)} = 0.0693$$

Hence,

$$(T_R)_{min} = 0.0693W = 0.0693(73{,}000) = \boxed{5{,}058 \text{ lb}}$$

The wing loading is

$$\frac{W}{S} = \frac{73{,}000}{950} = 76.84 \text{ lb/ft}^2$$

Hence, the velocity for minimum T_R is, from Eq. (5.22),

$$V_{(T_R)_{min}} = \left(\frac{2}{\rho_\infty}\sqrt{\frac{K}{C_{D,0}}}\frac{W}{S}\right)^{1/2} = \left[\frac{2}{8.9068 \times 10^{-4}}\sqrt{\frac{0.08}{0.015}}(76.84)\right]^{1/2} = \boxed{631.2 \text{ ft/s}}$$

On a graph of T_R versus velocity, the above results state that the coordinates of the minimum point on the curve are $(T_R, V_\infty) = (5{,}058$ lb, 631.2 ft/s). Return to Fig. 5.4, and examine the thrust required curve. The results calculated above agree with the graphical results obtained in Section 5.2 for the location of the minimum point in Fig. 5.4.

5.3.3   Graphical and Analytical Approaches: Some Comments

For our study of thrust required, we have employed both a graphical approach (Section 5.3.1) and an analytical approach (Section 5.3.2). These two approaches complement each other. The graphical approach gives the global picture—an instantaneous visualization of how various characteristics vary over a range of, say, velocity. For example, Fig. 5.4 shows a complete T_R curve; we see at a glance how T_R varies with velocity, and in particular that a minimum point exists. In Fig. 5.5, we immediately see why the T_R curve in Fig. 5.4 is shaped the way it is—it is a sum of two components, one rapidly decreasing with V_∞ and the other rapidly increasing with V_∞. Also, it is instructive to be able to read from these curves the magnitudes of the variables. In subsequent sections we will illustrate yet another advantage of dealing with graphs for airplane performance, namely, the use of geometric constructions (such as drawing a line from the origin, tangent to the T_R curve) to identify certain specific aspects of airplane performance. For these reasons, we continue to use the graphical approach in our subsequent discussions.

The great advantage of the analytical approach is that it clearly delineates the fundamental parameters of the problem. For example, in Section 5.3.2 we have shown that most of the equations involve the thrust-to-weight ratio T_R/W, the wing loading W/S, the drag polar via $C_{D,0}$ and K, and the lift-to-drag ratio L/D. We discuss further the importance of these parameters in the next section. In contrast, the graphical approach, dealing mainly with numbers rather than relationships, does not always identify the fundamental parameters. For example, in constructing the graph shown in Fig. 5.4, we know the results depend on weight W. However, only through the analysis do we find out the more fundamental fact that W usually appears only in the form T_R/W or W/S [see, e.g., Eqs. (5.18), (5.21), and (5.22)]. Also, *how* one quantity varies with another quantity is shown by the equations. For example, from Eq. (5.24) we *know* that $(L/D)_{max}$ increases inversely proportionally to the square root of the zero-lift drag coefficient C_{D0} [see Eq. (5.24)]. For these reasons, we will continue to use the analytical approach (as well as the graphical approach) in our subsequent discussions.

5.4   THE FUNDAMENTAL PARAMETERS: THRUST-TO-WEIGHT RATIO, WING LOADING, DRAG POLAR, AND LIFT-TO-DRAG RATIO

In the equations derived in Section 5.3.2, the thrust required T_R rarely appears by itself; rather, it is usually found in combination with the weight T_R/W or the wing area T_R/S. Similarly, the weight does not occur in an isolated fashion in these equations; rather, it is always found in combination with the wing area W/S or the thrust-to-weight ratio T_R/W. Moreover, the *thrust loading* T_R/S can always be replaced with T_R/W because

$$\frac{T_R}{W} = \frac{T_R/S}{W/S}$$

Hence, *the thrust-to-weight ratio and the wing loading are fundamental parameters for airplane performance*, rather than just the thrust by itself and the weight by itself.

The equations in Section 5.3.2 also highlight the importance of $C_{D,0}$ and K, that is, the drag polar. These are the primary descriptors of the external aerodynamic properties of the airplane, and it stands to reason that they would appear prominently in the equations for airplane performance.

For steady, level flight, the lift-to-drag ratio is simply the reciprocal of the thrust-to-weight ratio:

$$\frac{L}{D} = \left(\frac{T_R}{W}\right)^{-1} \qquad \text{Steady, level flight}$$

Hence, for such a case, to discuss L/D and T_R/W is somewhat redundant. However, for accelerated flight (turning flight, takeoff, etc.) and climbing flight, T_R/W and L/D are different, and each one takes on its own significance. We have frequently emphasized the importance of L/D as a stand-alone indicator of aerodynamic efficiency. Let us examine further the implication of this ratio for airplane performance.

For the restricted case of a given airplane in steady, level flight, we have noted that the lift-to-drag ratio is a function of velocity; Fig. 5.8 is a plot of the variation of L/D with V_∞ for the Gulfstream IV in Example 5.1. The results shown in Fig. 5.8 are obtained directly from the tabulation and graphical approach described in Section 5.3.1. However, an equation for the curve shown in Fig. 5.8 is easily obtained by dividing Eq. (5.12) by the weight.

$$\frac{D}{W} = \frac{1}{2}\rho_\infty V_\infty^2 \frac{S}{W} C_{D,0} + \frac{2K}{\rho_\infty V_\infty^2} \frac{S}{W} \left(\frac{W}{S}\right)^2 \qquad \textbf{[5.25]}$$

Since $L = W$ for steady, level flight, Eq. (5.25) can be written as

$$\frac{D}{L} = \frac{\rho_\infty V_\infty^2 C_{D,0}}{2(W/S)} + \frac{2K}{\rho_\infty V_\infty^2} \frac{W}{S}$$

or

$$\boxed{\frac{L}{D} = \left(\frac{\rho_\infty V_\infty^2 C_{D,0}}{2(W/S)} + \frac{2K}{\rho_\infty V_\infty^2} \frac{W}{S}\right)^{-1}} \qquad \textbf{[5.26]}$$

Equation (5.26) is the analytical equation for the curve shown in Fig. 5.8. Note in Eq. (5.26) that W and S do not appear separately, but in the form of the wing loading W/S. Once again we are reminded of the fundamental nature of the wing loading.

Example 5.3

For the Gulfstream IV at the conditions given in Example 5.1, calculate the value of L/D for a velocity of 400 ft/s. Compare the calculated result with Fig. 5.8.

Solution

From Example 5.1, $W = 73,000$ lb, $S = 950$ ft^2, $C_{D,0} = 0.015$, $K = 0.08$, and $\rho_\infty = 8.9068 \times 10^{-4}$ slug/ft^3. The wing loading is

$$\frac{W}{S} = \frac{73,000}{950} = 76.84 \text{ lb/ft}^2$$

From Eq. (5.26)

$$\frac{L}{D} = \left(\frac{\rho_\infty V_\infty^2 C_{D,0}}{2W/S} + \frac{2K}{\rho_\infty V_\infty^2} \frac{W}{S} \right)^{-1}$$

$$= \left[\frac{(8.9068 \times 10^{-4})(400)^2(0.015)}{2(76.84)} + \frac{2(0.08)(76.84)}{(8.9068 \times 10^{-4})(400)^2} \right]^{-1} \qquad \boxed{\frac{L}{D} = 9.98}$$

Examining Fig. 5.8, we see that the value calculated above for L/D agrees with the value on the curve for $V_\infty = 400$ ft/s.

Consider the maximum value of L/D; clearly, from Fig. 5.8 we see that L/D goes through a maximum value $(L/D)_{max}$, and we know from Section 5.3.3 that this point corresponds to minimum T_R. Indeed, examining Eq. (5.26), we might assume that the value of $(L/D)_{max}$ would depend on the drag polar ($C_{D,0}$ and K), the wing loading, and the altitude (via ρ_∞). However, we have already shown that $(L/D)_{max}$ depends *only* on the drag polar, and not on the other parameters; this result is given by Eq. (5.24). On the other hand, the *velocity* at which maximum L/D is achieved *does* depend on altitude and wing loading, as shown in Eq. (5.22). Let us examine in a more general fashion these and other matters associated with maximum L/D.

5.4.1 Aerodynamic Relations Associated with Maximum C_L/C_D, $C_L^{3/2}/C_D$, and $C_L^{1/2}/C_D$

Equation (5.24) for $(L/D)_{max}$ was derived from a consideration of minimizing thrust required in steady, level flight. In reality, Eq. (5.24) is much more general, and the same result can be obtained by a simple consideration of the lift-to-drag ratio completely independent of any consideration of T_R, as follows. The lift-to-drag ratio is

$$\frac{L}{D} = \frac{C_L}{C_D} = \frac{C_L}{C_{D,0} + KC_L^2} \qquad \textbf{[5.27]}$$

For maximum C_L/C_D, differentiate Eq. (5.27) with respect to C_L and set the result equal to zero:

$$\frac{d(C_L/C_D)}{dC_L} = \frac{C_{D,0} + KC_L^2 - C_L(2KC_L)}{(C_{D,0} + KC_L^2)^2} = 0$$

Hence,

$$C_{D,0} + KC_L^2 - 2KC_L^2 = 0$$

or

$$\boxed{C_{D,0} = KC_L^2} \qquad \textbf{[5.28]}$$

From Eq. (5.28), *when L/D is a maximum value, the zero-lift drag equals the drag due to lift.* Furthermore, the value of $(L/D)_{\max}$ can be found by rewriting Eq. (5.28) as

$$C_L = \sqrt{\frac{C_{D,0}}{K}} \qquad \textbf{[5.29]}$$

and inserting Eqs. (5.28) and (5.29) into Eq. (5.27). [Keep in mind that since Eqs. (5.28) and (5.29) hold only for the condition of maximum L/D, then Eq. (5.27) with these insertions yields the value of maximum L/D.]

$$\left(\frac{L}{D}\right)_{\max} = \left(\frac{C_L}{C_{D,0} + KC_L^2}\right)_{\max} = \frac{\sqrt{C_{D,0}/K}}{C_{D,0} + C_{D,0}} = \frac{\sqrt{C_{D,0}/K}}{2C_{D,0}}$$

or

$$\boxed{\left(\frac{L}{D}\right)_{\max} = \left(\frac{C_L}{C_D}\right)_{\max} = \sqrt{\frac{1}{4C_{D,0}K}}} \qquad \textbf{[5.30]}$$

This result is the same as that obtained in Eq. (5.24). However, the above derivation made no assumptions about steady, level flight, and no consideration was given to minimum T_R. Equation (5.30) is independent of any such assumptions. It is a *general* result, having to do with the aerodynamics of the airplane via the drag polar. It is the same result whether the airplane is in climbing flight, turning flight, etc.

However, the *velocity* at which $(L/D)_{\max}$ is achieved *is* dependent on such considerations. This velocity will be different for climbing flight or turning flight compared to steady, level flight. Let us obtain the velocity at which maximum L/D is attained in steady, level flight. For this case, $L = W$, and hence

$$L = W = \tfrac{1}{2}\rho_\infty V_\infty^2 S C_L \qquad \textbf{[5.31]}$$

When L/D is a maximum, Eq. (5.29) holds. Substituting Eq. (5.29) into Eq. (5.31), and denoting, as before, the velocity at which L/D is a maximum by $V_{(L/D)_{\max}}$, we have

$$W = \frac{1}{2}\rho_\infty V_{(L/D)_{\max}}^2 S\sqrt{C_{D,0}/K} \qquad \textbf{[5.32]}$$

or

$$\frac{W}{S} = \frac{1}{2}\rho_\infty V^2_{(L/D)_{max}} \sqrt{C_{D,0}/K}$$

[5.33]

Solving Eq. (5.33) for the velocity, we obtain

$$V_{(L/D)_{max}} = \left(\frac{2}{\rho_\infty} \sqrt{\frac{K}{C_{D,0}}} \frac{W}{S} \right)^{1/2}$$

[5.34]

Equation (5.34) is *identical* to the result shown in Eq. (5.22). However, Eq. (5.22) was obtained from a consideration of minimum T_R whereas Eq. (5.34) was obtained strictly on the basis of the aerodynamic relationships that hold at maximum L/D, completely separate from any consideration of thrust required. The only restriction on Eqs. (5.22) and (5.34) is that they hold only for straight and level flight.

The value of $(L/D)_{max}$ and the flight velocity at which it is attained are important considerations in the analysis of range and endurance for a given airplane. Indeed, as we will show in Section 5.11, the maximum range for an airplane powered with a propeller/reciprocating engine combination is directly proportional to $(L/D)_{max}$. The maximum endurance for a jet-propelled airplane is also proportional to $(L/D)_{max}$. These matters will be made clear in Section 5.11. We mention them here to underscore the importance of the lift-to-drag ratio; L/D is clearly a measure of the aerodynamic efficiency of the airplane.

There are other aerodynamic ratios that play a role in airplane performance. For example, in Section 5.11 we will show that maximum endurance for a propeller/reciprocating engine airplane is proportional to the maximum value of $C_L^{3/2}/C_D$, and that the maximum range for a jet airplane is proportional to the maximum value of $C_L^{1/2}/C_D$. Because of the importance of these ratios, let us examine the aerodynamic relations associated with each.

First, consider $(C_L^{3/2}/C_D)_{max}$. By replacing C_D with the drag polar, this expression can be written as

$$\frac{C_L^{3/2}}{C_D} = \frac{C_L^{3/2}}{C_{D,0} + KC_L^2}$$

[5.35]

To find the conditions that hold for a maximum value of $C_L^{3/2}/C_D$, differentiate Eq. (5.35) with respect to C_L, and set the result equal to zero.

$$\frac{d\left(C_L^{3/2}/C_D\right)}{dC_L} = \frac{(C_{D,0} + KC_L^2)\left(\frac{3}{2}C_L^{1/2}\right) - C_L^{3/2}(2KC_L)}{C_{D,0} + KC_L^2} = 0$$

$$\frac{3}{2}C_{D,0}C_L^{1/2} + \frac{3}{2}KC_L^{5/2} - 2KC_L^{5/2} = 0$$

or

$$C_{D,0} = \tfrac{1}{3}KC_L^2$$

[5.36]

From Eq. (5.36), *when* $C_L^{3/2}/C_D$ *is a maximum value, the zero-lift drag equals one-third the drag due to lift.* Furthermore, the value of $(C_L^{3/2}/C_D)_{max}$ can be found by writing Eq. (5.36) as

$$C_L = \sqrt{3C_{D,0}/K} \qquad [5.37]$$

and substituting Eqs. (5.36) and (5.37) into Eq. (5.35). [Keep in mind that since Eqs. (5.36) and (5.37) hold only for the condition of maximum $C_L^{3/2}/C_D$, then Eq. (5.35) with these substitutions yields the value of maximum $C_L^{3/2}/C_D$.]

$$\left(\frac{C_L^{3/2}}{C_D}\right)_{max} = \left(\frac{C_L^{3/2}}{C_{D,0} + KC_L^2}\right)_{max} = \frac{(3C_{D,0}/K)^{3/4}}{C_{D,0} + 3C_{D,0}} = \frac{1}{4C_{D,0}}\left(\frac{3C_{D,0}}{K}\right)^{3/4}$$

or

$$\boxed{\left(\frac{C_L^{3/2}}{C_D}\right)_{max} = \frac{1}{4}\left(\frac{3}{KC_{D,0}^{1/3}}\right)^{3/4}} \qquad [5.38]$$

Note that the maximum value of $C_L^{3/2}/C_D$ is a function only of the drag polar, that is, $C_{D,0}$ and K.

In straight and level flight, where $L = W$, the velocity at which $(C_L^{3/2}/C_D)_{max}$ is achieved can be found as follows.

$$L = W - \tfrac{1}{2}\rho_\infty V_\infty^2 S C_L \qquad [5.39]$$

When $C_L^{3/2}/C_D$ is a maximum, Eq. (5.37) holds. Substituting Eq. (5.37) into (5.39), and denoting the velocity at which $C_L^{3/2}/C_D$ is a maximum by $V_{(C_L^{3/2}/C_D)_{max}}$, we have

$$W = \frac{1}{2}\rho_\infty V_{(C_L^{3/2}/C_D)_{max}}^2 S \sqrt{\frac{3C_{D,0}}{K}} \qquad [5.40]$$

Solving Eq. (5.40) for the velocity, we obtain

$$\boxed{V_{(C_L^{3/2}/C_D)_{max}} = \left(\frac{2}{\rho_\infty}\sqrt{\frac{K}{3C_{D,0}}}\frac{W}{S}\right)^{1/2}} \qquad (P_R)_{min} \qquad [5.41]$$

Comparing Eq. (5.41) with (5.22) for $V_{(L/D)_{max}}$, we see that

$$V_{(C_L^{3/2}/C_D)_{max}} = \left(\frac{1}{3}\right)^{1/4} V_{(L/D)_{max}}$$

or

$$\boxed{V_{(C_L^{3/2}/C_D)_{max}} = 0.76 V_{(L/D)_{max}}} \qquad [5.42]$$

Note from Eq. (5.42) that when the airplane is flying at $(C_L^{3/2}/C_D)_{max}$, it is flying *more slowly* than when it is flying at $(L/D)_{max}$; indeed, it is flying at a velocity 0.76 times that necessary for maximum L/D.

Consider $(C_L^{1/2}/C_D)_{max}$. Analogous to the above derivation, we find that for the maximum value of $C_L^{1/2}/C_D$,

$$C_{D,0} = 3KC_L^2 \qquad \textbf{[5.43]}$$

From Eq. (5.43), *when $C_L^{1/2}/C_D$ is a maximum value, the zero-lift drag equals 3 times the drag due to lift.* Furthermore, the value of $(C_L^{1/2}/C_D)_{max}$ is given by

$$\left(\frac{C_L^{1/2}}{C_D}\right)_{max} = \frac{3}{4}\left(\frac{1}{3KC_{D,0}^3}\right)^{1/4} \qquad \textbf{[5.44]}$$

The velocity at which $(C_L^{1/2}/C_D)_{max}$ is achieved is

$$V_{(C_L^{1/2}/C_D)_{max}} = \left(\frac{2}{\rho_\infty}\sqrt{\frac{3K}{C_{D,0}}}\frac{W}{S}\right)^{1/2} \qquad \textbf{[5.45]}$$

The derivation of Eqs. (5.43) to (5.45) is left to you as a homework problem. Comparing Eq. (5.45) with (5.22) for $V_{(L/D)_{max}}$, we see that

$$V_{(C_L^{1/2}/C_D)_{max}} = 3^{1/4}V_{(L/D)_{max}}$$

or

$$V_{(C_L^{1/2}/C_D)_{max}} = 1.32V_{(L/D)_{max}} \qquad \textbf{[5.46]}$$

Note from Eq. (5.46) that when the airplane is flying at $(C_L^{1/2}/C_D)_{max}$, it is flying *faster* than when it is flying at $(L/D)_{max}$; indeed, it is flying at a velocity 1.32 times that necessary for maximum L/D.

For the Gulfstream IV in Example 5.1, the variations of $C_L^{3/2}/C_D$ and $C_L^{1/2}/C_D$ with velocity are easily obtained from the individual values of C_L and C_D tabulated in Table 5.1. The graphical results are shown in Fig. 5.11, along with the previous results for C_L/C_D (which is the same as L/D). The various velocities at which $C_L^{3/2}/C_D$, C_L/C_D, and $C_L^{1/2}/C_D$ become maximum values are identified in Fig. 5.11. We can clearly see that

$$V_{(C_L^{3/2}/C_D)_{max}} < V_{(C_L/C_D)_{max}} < V_{(C_L^{1/2}/C_D)_{max}}$$

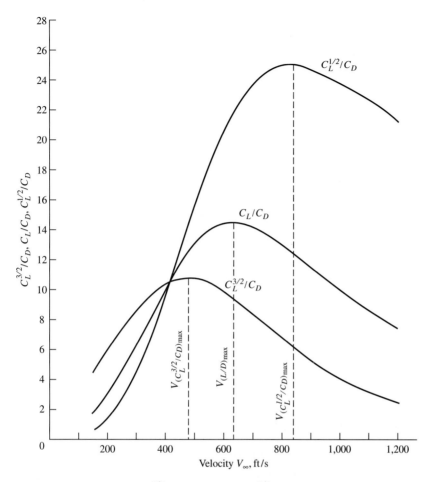

Figure 5.11 Variation of $C_L^{3/2}/C_D$, C_L/C_D, and $C_L^{1/2}/C_D$ versus velocity for the Gulfstream IV at the conditions set in Example 5.1. Altitude = 30,000 ft, $W = 73{,}000$ lb.

For the Gulfstream IV at the conditions given in Example 5.1, calculate the maximum values of $C_L^{3/2}/C_D$, C_L/C_D, and $C_L^{1/2}/C_D$, as well as the flight velocities at which they occur. | **Example 5.4**

Solution

The maximum values of $C_L^{3/2}/C_D$, C_L/C_D, and $C_L^{1/2}/C_D$ depend only on the drag polar, where $C_{D,0} = 0.015$ and $K = 0.08$. From Eq. (5.38),

$$\left(\frac{C_L^{3/2}}{C_D}\right)_{\max} = \frac{1}{4}\left(\frac{3}{KC_{D,0}^{1/3}}\right)^{3/4} = \frac{1}{4}\left[\frac{3}{(0.08)(0.015)^{1/3}}\right]^{3/4} = \boxed{10.83}$$

This value agrees with the graphical result shown in Fig. 5.11. The velocity at which this maximum occurs depends on the altitude and wing loading. At an altitude of 30,000 ft, $\rho_\infty = 8.9068 \times 10^{-4}$ slug/ft³. The wing loading is, from Example 5.2, $W/S = 76.84$ lb/ft². From Eq. (5.41)

$$V_{(C_L^{3/2}/C_D)\text{max}} = \left(\frac{2}{\rho_\infty} \sqrt{\frac{K}{3C_{D,0}}} \frac{W}{S} \right)^{1/2}$$

$$= \left[\frac{2}{8.9068 \times 10^{-4}} \sqrt{\frac{0.08}{3(0.015)}} (76.84) \right]^{1/2} = \boxed{479.6 \text{ ft/s}}$$

This value agrees with the graphical result shown in Fig. 5.11.
From Eq. (5.30),

$$\left(\frac{L}{D} \right)_\text{max} = \left(\frac{C_L}{C_D} \right)_\text{max} = \sqrt{\frac{1}{4C_{D,0}K}} = \sqrt{\frac{1}{4(0.015)(0.08)}} = \boxed{14.43} :$$

This value agrees with the graphical result shown in Fig. 5.11 and Fig. 5.8. The velocity at which this maximum occurs is given by Eq. (5.34):

$$V_{(L/D)\text{max}} = \left(\frac{2}{\rho_\infty} \sqrt{\frac{K}{C_{D,0}}} \frac{W}{S} \right)^{1/2} = \left[\frac{2}{8.9068 \times 10^{-4}} \sqrt{\frac{0.08}{0.015}} (76.84) \right]^{1/2} = \boxed{631.2 \text{ ft/s}}$$

This value agrees with the graphical result shown in Fig. 5.11 and Fig. 5.8.
From Eq. (5.44),

$$\left(\frac{C_L^{1/2}}{C_D} \right)_\text{max} = \frac{3}{4} \left(\frac{1}{3K C_{D,0}^3} \right)^{1/4} = \frac{3}{4} \left[\frac{1}{3(0.08)(0.015)^3} \right]^{1/4} = \boxed{25}$$

This value agrees with the graphical result shown in Fig. 5.11. The velocity at which this maximum occurs is given by Eq. (5.45):

$$V_{(C_L^{1/2}/C_D)\text{max}} = \left(\frac{2}{\rho_\infty} \sqrt{\frac{3K}{C_{D,0}}} \frac{W}{S} \right)^{1/2}$$

$$= \left[\frac{2}{8.9068 \times 10^{-4}} \sqrt{\frac{3(0.08)}{0.015}} (76.84) \right]^{1/2} = \boxed{830.8 \text{ ft/s}}$$

This value agrees with the graphical result shown in Fig. 5.11.
It is interesting to note that the velocities at which the maximums of the various aerodynamic ratios occur are in the ratio

$$V_{(C_L^{2/3}/C_D)\text{max}} : V_{(C_L/C_D)\text{max}} : V_{(C_L^{1/2}/C_D)\text{max}} = 479.6 : 631.2 : 830.8 = 0.76 : 1 : 1.32$$

This is precisely the velocity relationships indicated by Eqs. (5.42) and (5.46).

Example 5.5 | For the Gulfstream IV at the conditions given in Example 5.1, calculate and compare the zero-lift drag and the drag due to lift at (a) $(C_L^{3/2}/C_D)\text{max}$, (b) $(C_L/C_D)\text{max}$, and (c) $(C_L^{1/2}/C_D)\text{max}$.

Solution

(a) From Example 5.4, $V_{(C_L^{2/3}/C_D)\text{max}} = 479.6$ ft/s. The dynamic pressure is

$$q_\infty = \tfrac{1}{2}\rho_\infty V_\infty^2 = \tfrac{1}{2}(8.9068 \times 10^{-4})(479.6)^2 = 102.4 \text{ lb/ft}^2$$

The lift coefficient is, noting that $L = W$,

$$C_L = \frac{W}{q_\infty S} = \frac{73{,}000}{(102.4)(950)} = 0.7504$$

$$\text{Zero-lift drag} = q_\infty S C_{D,0} = (102.4)(950)(0.015) = \boxed{1{,}459.2 \text{ lb}}$$

$$\text{Drag due to lift} = q_\infty S K C_L^2 = (102.4)(950)(0.08)(0.7504)^2 = \boxed{4{,}382.3 \text{ lb}}$$

Comparing, we get

$$\frac{\text{Zero-lift drag}}{\text{Drag due to lift}} = \frac{1{,}459.2}{4{,}382.3} = 0.333 = \boxed{\frac{1}{3}}$$

This is precisely the prediction from Eq. (5.36), namely, that when $C_L^{3/2}/C_D$ is a maximum, the zero-lift drag equals one-third of the drag due to lift. This result is further reinforced in Fig. 5.12, which contains some of the same plots as given in Fig. 5.5 but illustrates the drag comparisons at the maxima of the various aerodynamic ratios.

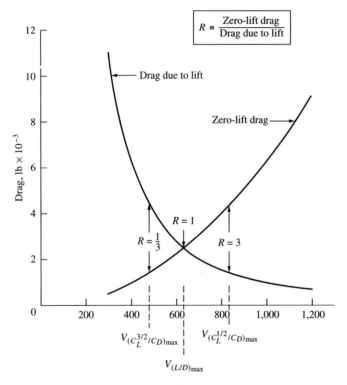

Figure 5.12 Comparison of zero-lift drag and drag due to lift for the Gulfstream IV at the conditions set in Example 5.1, emphasizing the relationships between these drag values for the maxima in $C_L^{3/2}/C_D$, C_L/C_D, and $C_L^{1/2}/C_D$.

(b) From Example 5.4, $V_{(L/D)_{max}} = 631.2$ ft/s.

$$q_\infty = \tfrac{1}{2}\rho_\infty V_\infty^2 = \tfrac{1}{2}(8.9068 \times 10^{-4})(631.2)^2 = 177.4 \text{ lb/ft}^2$$

$$C_L = \frac{W}{q_\infty S} = \frac{73,000}{(177.4)(950)} = 0.4332$$

Zero-lift drag $= q_\infty S C_{D,0} = (177.4)(950)(0.015) = \boxed{2,530 \text{ lb}}$

Drag due to lift $= q_\infty S K C_L^2 = (177.4)(950)(0.08)(0.4332)^2 = \boxed{2,530 \text{ lb}}$

Note: Since this calcuation is being done on a hand calculator, both drag values have been rounded to three significant figures, for comparison. Thus

$$\frac{\text{Zero-lift drag}}{\text{Drag due to lift}} = \frac{2,530}{2,530} = \boxed{1}$$

This is precisely the prediction from Eq. (5.28), namely, that when C_L/C_D is a maximum, the zero-lift drag equals the drag due to lift. This result is further reinforced in Fig. 5.12.

(c) From Example 5.4, $V_{(C_L^{1/2}/C_D)_{max}} = 830.8$ ft/s.

$$q_\infty = \tfrac{1}{2}\rho_\infty V_\infty^2 = \tfrac{1}{2}(8.9068 \times 10^{-4})(830.8)^2 = 307.4 \text{lb/ft}^2$$

$$C_L = \frac{W}{q_\infty S} = \frac{73,000}{(307.4)(950)} = 0.2500$$

Zero-lift drag $= q_\infty S C_{D,0} = (307.4)(950)(0.015) = \boxed{4,380 \text{ lb}}$

Drag due to lift $= q_\infty S K C_L^2 = (307.4)(950)(0.08)(0.25)^2 = \boxed{1,460 \text{ lb}}$

Comparing gives

$$\frac{\text{Zero-lift drag}}{\text{Drag due to lift}} = \frac{4,380}{1,460} = \boxed{3}$$

This is precisely the prediction from Eq. (5.43), namely, that when $(C_L^{1/2}/C_D)$ is a maximum, the zero-lift drag is 3 times the drag due to lift. This result is further reinforced in Fig. 5.12.

5.5   THRUST AVAILABLE AND THE MAXIMUM VELOCITY OF THE AIRPLANE

By definition, the thrust available, denoted by T_A, is the thrust provided by the power plant of the airplane. The various flight propulsion devices are described at length in Chapter 3. The single purpose of these propulsion devices is to reliably and efficiently provide *thrust* in order to propel the aircraft. Return to the force diagrams shown in Figs. 4.1 to 4.3 and in Fig. 5.2; the thrust T shown in these diagrams is what we are now labeling T_A and calling the thrust available. Unlike the thrust required T_R (discussed in Section 5.3), which has almost everything to do with the airframe (including the weight) of the airplane and virtually nothing to do with the power plant, the thrust available T_A has almost everything to do with the power plant and virtually nothing to do with the airframe. This statement is not completely true; there is always some aerodynamic interaction between the airframe and the power plant. The installation of the power plant relative to the airframe will set up an aerodynamic interaction that affects both the thrust produced by the power plant and the drag on the airframe. For conventional, low-speed airplanes, this interaction is usually small. However, for modern transonic and supersonic airplanes, it becomes more of a consideration. And for the hypersonic airplanes of the future, airframe and propulsion integration becomes a dominant design aspect. However, for this chapter, we do not consider such interactions; instead, we consider T_A to be completely associated with the flight propulsion device.

5.5.1 Propeller-Driven Aircraft

As described in Section 3.3.2, an aerodynamic force is generated on a propeller that is translating and rotating through the air. The component of this force in the forward direction is the thrust of the propeller. For a propeller/reciprocating engine combination, this propeller thrust is the thrust available T_A. For a turboprop engine, the propeller thrust is augmented by the jet exhaust, albeit by only a small amount (typically almost 5%), as described in Section 3.6. The combined propeller thrust and jet thrust is the thrust available T_A for the turboprop.

The qualitative variation of T_A with V_∞ for propeller-driven aircraft is sketched in Fig. 5.13. The thrust is highest at zero velocity (called the *static thrust*) and decreases with an increase in V_∞. The thrust rapidly decreases as V_∞ approaches sonic speed; this is because the propeller tips encounter compressibility problems at high speeds, including the formation of shock waves. It is for this reason (at least to the present) that propeller-driven aircraft have been limited to low to moderate subsonic speeds.

The propeller is attached to a rotating shaft which delivers *power* from a recipro-cating piston engine or a gas turbine (as in the case of the turboprop). For this reason, *power* is the more germane characteristic of such power plants rather than thrust. For example, in Ref. 36 the Teledyne Continental O-200-A four-cylinder piston engine is rated at 74.5 kW (or 100 hp) at sea level. Also in Ref. 36, the Allison T56-A-14 turboprop is rated at 3,661 ekW (equivalent kilowatts), or 4,910 ehp (equivalent horsepower); the concept of equivalent shaft power (which includes the effect of the jet thrust) is discussed in Section 3.6. What is important here is that for the analysis of the performance of a propeller-driven airplane, power is more germane than thrust. Therefore, we defer our discussion of propeller power plants to Section 5.7, which deals with power available.

However, should it be desired, the values of T_A for a propeller-driven airplane can be readily obtained from the power ratings as follows. The power available from a propeller/reciprocating engine combination is given by Eq. (3.13), repeated here:

$$P_A = \eta_{pr} P \qquad\qquad \textbf{[3.13]}$$

where η_{pr} is the propeller efficiency and P is the shaft power from the piston engine.

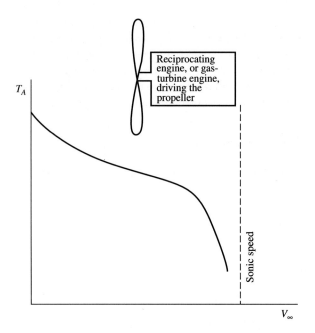

Figure 5.13   Sketch of the variation of thrust available versus velocity for a propellor-driven aircraft.

Since power is given by force times velocity (see Section 3.2), from Eq. (3.3) the power available from any flight propulsion device is

$$P_A = T_A V_\infty$$ [5.47]

Combining Eqs. (3.13) and (5.47) and solving for T_A, we get

$$T_A = \frac{\eta_{\mathrm{pr}} P}{V_\infty}$$ [5.48]

Similarly, for a turboprop, the power available is given by Eq. (3.29), repeated here:

$$P_A = \eta_{\mathrm{pr}} P_{\mathrm{es}}$$ [3.29]

Combining Eqs. (3.29) and (5.47) and solving for T_A, we have

$$T_A = \frac{\eta_{\mathrm{pr}} P_{\mathrm{es}}}{V_\infty}$$ [5.49]

Hence, for the given power ratings, the shaft power P for a piston engine and the equivalent shaft power P_{es} for a turboprop, Eqs. (5.48) and (5.49) give the thrust available for each type of power plant, respectively.

It is interesting to note that, as described in Chapter 3 and as summarized in Fig. 3.29, both P and P_{es} in Eqs. (5.48) and (5.49) are relatively constant with V_∞. By assuming a variable-pitch propeller such that the variation of η_{pr} with V_∞ is minimized, Eqs. (5.48) and (5.49) show that T_A decreases as V_∞ increases. This is consistent with the qualitative thrust available curve in Fig. 5.13, which shows maximum T_A at zero velocity and a decrease in T_A as V_∞ increases.

5.5.2 Jet-Propelled Aircraft

Turbojet and turbofan engines are rated in terms of thrust. Hence, for such power plants, T_A is the germane quantity for the analysis of airplane performance.

The turbojet engine is discussed in Section 3.4, where it was shown that, for *subsonic speeds*,

$$T_A \approx \text{constant with } V_\infty$$

and for *supersonic speeds*

$$\frac{T_A}{(T_A)_{\text{Mach 1}}} = 1 + 1.18(M_\infty - 1) \qquad \textbf{[3.21]}$$

The effect of altitude on T_A is given by Eq. (3.19)

$$\frac{T_A}{(T_A)_0} = \frac{\rho}{\rho_0} \qquad \textbf{[3.19]}$$

where $(T_A)_0$ is the thrust available at sea level and ρ_0 is the standard sea-level density.

The turbofan engine is discussed in Section 3.5. Unlike the turbojet, the thrust of a turbofan is a function of velocity. For the high-bypass-ratio turbofans commonly used for civil transports, thrust decreases with increasing velocity. (This is analogous to the thrust decrease with velocity for propellers sketched in Fig. 5.13, which makes sense because the large fan on a high-bypass-ratio turbofan is functioning much as a propeller.) Several relationships for the thrust variation with velocity (or Mach number) are given in Section 3.5. For example, Eq. (3.23) shows a functional relationship

$$\frac{T_A}{(T_A)_{V=0}} A M_\infty^{-n} \qquad \textbf{[3.23]}$$

where $(T_A)_{V=0}$ is the static thrust available (thrust at zero velocity) at standard sea level, and A and n are functions of altitude, obtained by correlating the data for a given engine. On the other hand, for a low-bypass-ratio turbofan, the thrust variation with velocity is much closer to that of a turbojet, essentially constant at subsonic speeds and increasing with velocity at supersonic speeds.

The altitude variation of thrust for a high-bypass-ratio civil turbofan is correlated in Eq. (3.25)

$$\frac{T_A}{(T_A)_0} = \left[\frac{\rho}{\rho_0} \right]^m \qquad \textbf{[3.25]}$$

where $(T_A)_0$ is the thrust available at sea level and ρ_0 is standard sea-level density.

For a performance analysis of a turbofan-powered airplane, the thrust available should be obtained from the engine characteristics provided by the manufacturer. The above discussion is given for general guidance only.

5.5.3 Maximum Velocity

Consider a given airplane flying at a given altitude, with a T_R curve as sketched in Fig. 5.14. For steady, level flight at a given velocity, say, V_1 in Fig. 5.14, the value of T_A is adjusted such that $T_A = T_R$ at that velocity. This is denoted by point 1 in Fig. 5.14. The pilot of the airplane can adjust T_A by adjusting the engine throttle in the cockpit. For point 1 in Fig. 5.14, the engine is operating at partial throttle, and the resulting value of T_A is denoted by $(T_A)_{\text{partial}}$. When the throttle is pushed all the way forward, maximum thrust available is produced, denoted by $(T_A)_{\text{max}}$. The airplane will accelerate to higher velocities, and T_R will increase, as shown in Fig. 5.14, until $T_R = (T_A)_{\text{max}}$, denoted by point 2 in Fig. 5.14. When the airplane is at point 2 in Fig. 5.14, any further increase in velocity requires more thrust than is available from the power plant. Hence, for steady, level flight, point 2 defines the maximum velocity V_{max} at which the given airplane can fly at the given altitude.

By definition, the thrust available curve is the variation of T_A with velocity at a given throttle setting and altitude. For the throttle full forward, $(T_A)_{\text{max}}$ is obtained. The maximum thrust available curve is the variation of $(T_A)_{\text{max}}$ with velocity at a given altitude. For turbojet and low-bypass-ratio turbofans, we have seen that at subsonic speeds, the thrust is essentially constant with velocity. Hence, for such power plants, the thrust available curve is a horizontal line, as sketched in Fig. 5.15. *In steady, level flight, the maximum velocity of the airplane is determined by the high-speed intersection of the thrust required and thrust available curves.* This is shown schmatically in Fig. 5.15.

Note that there is a low-speed intersection of the $(T_A)_{\text{max}}$ and T_R curves, denoted by point 3 in Fig. 5.15. At first glance, this would appear to define the minimum

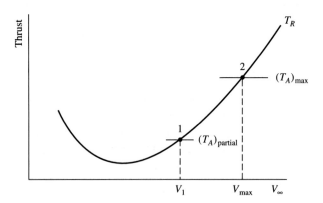

Figure 5.14   Partial- and full-throttle conditions; intersection of the thrust available and thrust required curves.

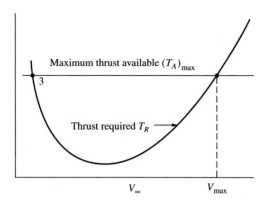

Figure 5.15 Thrust available curve for a turbojet and low-bypass-ratio turbofan is essentially constant with velocity at subsonic speeds. The high-speed intersection of the $(T_A)_{max}$ curve and the T_R curve determines the maximum velocity of the airplane.

velocity of the airplane in steady, level flight. However, what is more usual is that the minimum velocity of the airplane is determined by its stalling speed, which depends strongly on $C_{L_{max}}$ and wing loading. Such matters will be discussed in Section 5.9.

Finding V_{max} from the intersection of the thrust required and thrust available curves, as shown in Fig. 5.15, is a graphical technique. An analytical method for the direct solution of V_{max} follows from Eq. (5.18). For steady, level flight, $T_R = T_A$. For flight at V_{max}, the thrust available is at its maximum value. Hence,

$$T_R = (T_A)_{max}$$

In Eq. (5.18), replacing V_∞ with V_{max} and T_R with $(T_A)_{max}$, and taking the plus sign in the quadratic expression because we are interested in the highest velocity, we have

$$V_{max} = \left\{ \frac{[(T_A)_{max}/W](W/S) + (W/S)\sqrt{[(T_A)_{max}/W]^2 - 4C_{D,0}K}}{\rho_\infty C_{D,0}} \right\}^{1/2}$$

[5.50]

Equation (5.50) allows the direct calculation of the maximum velocity. Moreover, being an analytic equation, it clearly points out the parameters that influence V_{max}. Note in Eq. (5.50) that V_{max} depends on (1) the maximum thrust-to-weight ratio $(T_A)_{max}/W$, (2) wing loading W/S, (3) the drag polar via $C_{D,0}$ and K, and (4) altitude via ρ_∞. From this equation we see that

1. V_{max} increases as $(T_A)_{max}/W$ increases.
2. V_{max} increases as W/S increases.
3. V_{max} decreases as $C_{D,0}$ and/or K increases.

The altitude effect on V_{max} is also contained in Eq. (5.50). For example, for a turbojet-powered airplane with a thrust-altitude variation given by Eq. (3.19), namely, $T_A \propto \rho/\rho_0$, an analysis of Eq. (5.50) shows that V_{max} decreases as altitude increases. The proof of this statement is left for you as a homework problem.

Example 5.6	The Gulfstream IV in Example 5.1 is powered by two Rolls-Royce Tay 611-8 turbofans, each one rated at a maximum thrust at sea level of 13,850 lb. Calculate V_{max} at (a) sea level and (b) 30,000 ft. Assume that $m = 0.6$ in Eq. (3.25). We have noted that m can be less than, equal to, or greater than 1, depending on the particular turbofan engine. The assumption of $m = 0.6$ is for an engine with particularly good high-altitude performance; this will contribute to the airplane in these worked examples being a "hot" airplane.

Assume that the thrust is constant with velocity. (*Note:* As explained in Section 5.1, this assumption is made consistently for many of the worked examples in this chapter, although for an actual *turbofan* engine it is not the case. Please remind yourself of the rationale for this assumption, explained at the end of Section 5.1.)

Solution

(a) At sea level, $\rho_\infty = 0.002377$ slug/ft^3. From the given data in Example 5.1, $W = 73,000$ lb, $S = 950$ ft^2, $C_{D,0} = 0.015$, and $K = 0.08$. Hence,

$$\text{Wing loading} = \frac{W}{S} = \frac{73,000}{950} = 76.84 \text{ lb/ft}^2$$

$$\text{Thrust-to-weight ratio} = \frac{(T_A)_{max}}{W} = \frac{2(13,850)}{73,000} = 0.3795$$

From Eq. (5.50),

$$V_{max} = \left\{ \frac{[(T_A)_{max}/W](W/S) + (W/S)\sqrt{[(T_A)_{max}/W]^2 - 4C_{D,0}K}}{\rho_\infty C_{D,0}} \right\}^{1/2}$$

$$= \left[\frac{0.3795(76.84) + 76.84\sqrt{(0.3795)^2 - 4(0.015)(0.08)}}{(0.002377)(0.015)} \right]^{1/2} = \boxed{1{,}273.6 \text{ ft/s}}$$

Note: This result for V_{max} is slightly faster than the speed of sound at sea level, which is 1,117 ft/s. This result does not include the realistic drag-divergence phenomena near Mach 1, and hence is not indicative of the maximum velocity for the actual Gulfstream IV, which would be slightly less than the speed of sound.

(b) At 30,000 ft, $\rho_\infty = 8.9068 \times 10^{-4}$ slug/ft^3. From Eq. (3.25) for a civil turbofan,

$$(T_A)_{max} = (T_A)_0 \left[\frac{\rho}{\rho_0} \right]^{0.6} = (2)(13,850) \left[\frac{8.9068 \times 10^{-4}}{0.002377} \right]^{0.6} = 15,371 \text{ lb}$$

Hence,

$$\frac{(T_A)_{max}}{W} = \frac{15,371}{73,000} = 0.2106$$

From Eq. (5.50),

$$V_{max} = \left[\frac{0.2106(76.84) + 76.84\sqrt{(0.2106)^2 - 4(0.015)(0.08)}}{(8.9068 \times 10^{-4})(0.015)} \right]^{1/2} = \boxed{1{,}534.6 \text{ ft/s}}$$

Again we note that the drag polar assumed for this example does not include the large drag rise near Mach 1, and hence the V_{max} calculated above is unrealistically large. However, this example is intentionally chosen to emphasize two points, discussed below.

First, we have already noted (via a homework problem) that a turbojet-powered aircraft with $T_A \propto \rho/\rho_0$ will experience a decrease in V_{max} as altitude increases. This is a mathematical result obtained from Eq. (5.50). However, it is easily explained on a physical basis. The thrust decreases proportionally to the decrease in air density as the altitude increases. In contrast, the drag decreases slightly less than proportionally to the air density. Why? Even though $D = \frac{1}{2}\rho_\infty V_\infty^2 S C_D$, which would seem to indicate a decrease in drag proportional to the density decrease, keep in mind that (for a given velocity) the lift coefficient must increase with altitude in order for the lift to sustain the weight. Hence, the drag due to lift increases. Examining the drag equation

$$D = \frac{1}{2}\rho_\infty V_\infty^2 S C_D = \frac{1}{2}\rho_\infty V_\infty^2 S (C_{D,0} + K C_L^2)$$

we see that as ρ_∞ decreases and C_L increases as a result, D will decrease at a rate which is less than proportional to the air density. Hence, because the thrust drops off in direct proportion to density, we find that at altitude the thrust has decreased more than the drag, and hence V_{max} is smaller at altitude. The opposite is true for the turbofan-powered airplane in Example 5.6. Here, the thrust decreases more slowly than the drag decreases with altitude, and hence V_{max} grows larger as the altitude increases. Keep in mind that the discussion in this paragraph ignores the effect of drag divergence near Mach 1, hence it applies realistically to only those turbojet and turbofan aircraft flying below drag divergence.

This leads to the second point. Clearly the Gulfstream IV in Example 5.6 has plenty of thrust. The results of both parts (a) and (b) of the example show that if drag divergence did not occur, the airplane could fly at moderate supersonic speeds. Of course, the real Gulfstream IV does not go supersonic because it encounters drag divergence, and this large drag rise limits the Gulfstream IV to a maximum operating Mach number of 0.88 (see Ref. 36). This raises the question: Why does the Gulfstream IV have more thrust than it needs to achieve Mach 0.88? The answer is that considerations *other* than maximum flight velocity can dictate the design choice for maximum thrust for an airplane. For many cases, a large maximum thrust is necessary to achieve a reasonable takeoff distance along the ground. Also, maximum rate of climb and maximum turn rate are determined in part by maximum thrust. Rate of climb will be discussed in Section 5.9, and matters associated with takeoff and turn rate are considered in Chapter 6.

Historically, in the eras of the strut-and-wire biplanes and the mature propeller-driven airplane (see Chapter 1), maximum velocity *was* the primary consideration for sizing the engine—the more powerful the engine, the faster the airplane. However, in the era of jet-propelled airplanes, with engines that produce more than enough thrust for airplanes to bump up against the large drag divergence, the design considerations changed. For jet airplanes intended to be limited to subsonic flight, the sizing of the engine was influenced by other considerations, as mentioned above. On the other hand, for aircraft designed to fly at supersonic speeds and which have to penetrate the large transonic drag rise, engine size is still primarily driven by consideration of V_{max}.

DESIGN CAMEO

In Example 5.6 above, T_A was assumed to be constant with velocity—a reasonable assumption for a subsonic turbojet-powered airplane. However, the airplane treated in Example 5.6, indeed in most of the worked examples in this chapter, is patterned after the Gulfstream IV, which is powered by turbofan engines. The thrust available from a turbofan decreases with an increase in flight velocity of the airplane, as noted in Eq. (3.23). However, in the worked examples in this chapter, we assume that T_A is constant with V_∞ strictly for the purpose of simplicity and to allow us to highlight other aspects of airplane performance. This is not recommended for the preliminary design process for an airplane. During the design process, there are two general options for dealing with the engine:

1. The actual desired T_A to accomplish the design goals is determined through an iterative process—the "rubber engine" approach wherein the desired engine characteristics evolve along with the airframe characteristics. Then the engine manufacturers are approached for the design of a new engine to meet these characteristics. Considering the expense of designing a new engine, needless to say, this approach is used only in those few cases where the need and/or market for the new airplane is so compelling as to justify such a new engine.

2. Alternatively, the new airplane design is based on existing engines. In the iterative design process, T_A and other engine characteristics are known

parameters, and the design is optimized around these known values. This is the design option most often taken.

If an existing engine is to be used for a new airplane design, the known precise engine characteristics (variation of T_A and specific fuel consumption with velocity and altitude, etc.) should be used during the design process.

To illustrate the effect of more precise engine characteristics on airplane performance results, Problem 5.18 revisits worked Example 5.6 and assumes a variation of thrust available given by

$$\frac{T_A}{(T_A)_{V=0}} = 0.4M_\infty^{-0.6} \qquad \text{at sea level}$$

$$\frac{T_A}{(T_A)_{V=0}} = 0.222M_\infty^{-0.6} \qquad \text{at 30,000 ft}$$

where $(T_A)_{V=0}$ is the thrust available at sea level at zero velocity. The results for V_{max} at sea level and at 30,000 ft assuming the above variations for T_A are, for the answer to Problem 5.18,

At sea level: $V_{max} = 860$ ft/s

At 30,000 ft: $V_{max} = 945$ ft/s

Compare these results with $V_{max} = 1,273.6$ ft/s and 1,534.6 ft/s obtained earlier in worked Example 5.6, which assumed a constant T_A. Clearly, it is important to take into account the best available data for engine characteristics.

5.6   POWER REQUIRED

To begin, let us examine a general relation for power. Consider a force **F** acting on an object moving with velocity **V**, as sketched in Fig. 5.16**a**. Both **F** and **V** are vectors and may have different directions, as shown in Fig. 5.16**a**. At some instant, the object is located at a position given by the position vector **r**, as shown in Fig. 5.16**b**. Over a time increment dt, the object is displaced through the vector **dr**, shown in Fig. 5.16**b**. The work done on the object by the force **F** acting through the displacement **dr** is **F · dr**. Power is the time rate of doing work, or

$$\text{Power} = \frac{d}{dt}(\mathbf{F} \cdot \mathbf{dr}) = \mathbf{F} \cdot \frac{\mathbf{dr}}{dt}$$

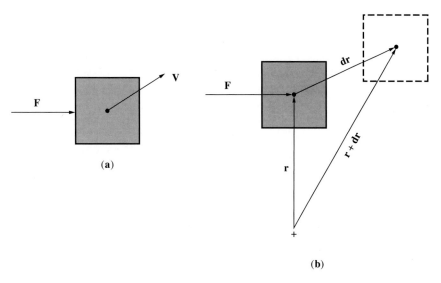

Figure 5.16 Force acting on a moving body. **(a)** Force and velocity vectors; **(b)** force and displacement vectors.

Since

$$\frac{d\mathbf{r}}{dt} = \mathbf{V}$$

then

$$\text{Power} = \mathbf{F} \cdot \mathbf{V} \qquad \textbf{[5.51]}$$

Equation (5.51) is the more general version of Eq. (3.2), which holds when the force and velocity are in the same direction.

Let us now apply Eq. (5.51) to an airplane in straight and level flight, as sketched in Fig. 5.2. The velocity of the airplane is V_∞. In Section 5.3, the concept of thrust required T_R was introduced, where $T_R = D$. In this section, we introduce the analogous concept of *power required*, denoted by P_R. Since in Fig. 5.2 both T and V_∞ are horizontal, the dot product in Eq. (5.51) gives for the power required

$$\boxed{P_R = T_R V_\infty} \qquad \textbf{[5.52]}$$

For some aspects of airplane performance, power rather than thrust is more germane, as we will soon see.

5.6.1 Graphical Approach

A graphical plot of P_R versus V_∞ for a given airplane at a given altitude is called the *power required curve*. The power required curve is easily obtained by multiplying thrust required by velocity via Eq. (5.52).

Example 5.7

Calculate the power required curve at 30,000 ft for the Gulfstream IV described in Example 5.1.

Solution

In Example 5.1 a tabulation of T_R versus V_∞ is made (see Table 5.1). This tabulation is repeated in Table 5.2 along with new entries for P_R obtained from Eq. (5.52). The values for P_R are first quoted in the consistent units of foot-pounds per second and then converted to the inconsistent unit of horsepower. Here we note that

$$1 \text{ hp} = 550 \text{ ft·lb/s} = 746 \ W$$

Table 5.2

V_∞ (ft/s)	T_R (lb)	P_R (ft·lb/s)	P_R (hp)
300	11,768	0.3530×10^7	6,419
400	7,313	0.2925×10^7	5,319
500	5,617	0.2809×10^7	5,107
600	5,084	0.3050×10^7	5,546
700	5,166	0.3616×10^7	6,575
800	5,636	0.4509×10^7	8,198
900	6,384	0.5746×10^7	10,447
1,000	7,354	0.7354×10^7	13,371
1,100	8,512	0.9363×10^7	17,023
1,200	9,838	0.1181×10^8	21,465
1,300	11,321	0.1472×10^8	26,759

The power required curve is plotted in Fig. 5.17.

The power required curve in Fig. 5.17 is qualitatively the same shape as the thrust required curve shown in Fig. 5.4; at low velocities, P_R first decreases as V_∞ increases, then goes through a minimum, and finally increases as V_∞ increases. The physical reasons for this shape are the same as discussed earlier in regard to the shape of the thrust required curve; that is, at low velocity, the drag due to lift dominates the power required, and at high velocity the zero-lift drag is the dominant factor. Quantitatively, the powered required curve is different from the thrust required curve. Comparison of Figs. 5.17 and 5.4 show that *minimum P_R occurs at a lower velocity than minimum T_R.*

5.6.2 Analytical Approach

A simple equation for P_R in terms of the aerodynamic coefficients is obtained as follows. From Eqs. (5.52) and (5.7), we have

$$P_R = T_R V_\infty = \frac{W}{C_L/C_D} V_\infty \qquad \textbf{[5.53]}$$

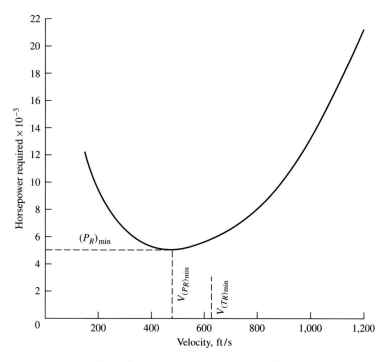

Figure 5.17   Calculated power required curve for the Gulfstream IV based on data in Example 5.1. Altitude = 30,000 ft, W = 73,000 lb.

Since $L = W$ for steady, level flight,

$$L = W = \tfrac{1}{2}\rho_\infty V_\infty^2 S C_L \qquad \text{[5.54]}$$

Solving Eq. (5.54) for V_∞, we have

$$V_\infty = \sqrt{\frac{2W}{\rho_\infty S C_L}} \qquad \text{[5.55]}$$

Substituting Eq. (5.55) into (5.53), we obtain

$$P_R = \frac{W}{C_L/C_D}\sqrt{\frac{2W}{\rho_\infty S C_L}}$$

or

$$P_R = \sqrt{\frac{2W^3 C_D^2}{\rho_\infty S C_L^3}} \qquad \text{[5.56]}$$

Examining Eq. (5.56), we see that

$$P_R \propto \frac{C_L^{3/2}}{C_D}$$

[5.57]

Hence, *minimum power required* occurs when the airplane is flying such that $C_L^{3/2}/C_D$ is a maximum value. In turn, all the characteristics associated with $(C_L^{3/2}/C_D)_{max}$ discussed in Section 5.4.1 hold for minimum P_R. In particular, at minimum P_R we have

1.

$$\left(\frac{C_L^{3/2}}{C_D}\right)_{max} = \frac{1}{4}\left(\frac{3}{KC_{D,0}^{3/2}}\right)^{3/4}$$

[5.38]

2. Zero-lift drag equals one-third of the drag due to lift.
3. The velocity at which P_R is a minimum occurs at

$$V_{(C_L^{3/2}/C_D)_{max}} = \left(\frac{2}{\rho_\infty}\sqrt{\frac{K}{3C_{D,0}}}\frac{W}{S}\right)^{1/2}$$

[5.41]

This velocity is less than that for minimum T_R, where C_L/C_D is a maximum. Indeed,

$$V_{(C_L^{3/2}/C_D)_{max}} = 0.76 V_{(L/D)_{max}}$$

[5.42]

Example 5.8

For the Gulfstream IV at the conditions given in Example 5.1, calculate the minimum power required and the velocity at which it occurs. Compare with the graphical results shown in Fig. 5.17.

Solution

The altitude is 30,000 ft, where $\rho_\infty = 8.9068 \times 10^{-4}$ slug/ft^3. Also, $W = 73,000$ lb and $S = 950$ ft^2. From the drag polar, $C_{D,0} = 0.015$ and $K = 0.08$. From Eq. (5.38), we obtained in Example 5.4

$$\left(\frac{C_L^{3/2}}{C_D}\right)_{max} = 10.83$$

Hence,

$$\left(\frac{C_D^2}{C_L^3}\right)_{min} = \left(\frac{1}{10.83}\right)^2 = 8.526 \times 10^{-3}$$

From Eq. (5.56),

$$P_R = \sqrt{\frac{2W^3 C_D^2}{\rho_\infty S C_L^3}}$$

Hence, minimum power required is given by

$$(P_R)_{\text{min}} = \sqrt{\frac{2W^3}{\rho_\infty S}\left(\frac{C_D^2}{C_L^3}\right)_{\text{min}}} = \sqrt{\frac{2(73,000)^3(8.526 \times 10^{-3})}{(8.9068 \times 10^{-4})(950)}}$$

$$= 2.8 \times 10^6 \text{ ft·lb/s} = \boxed{5{,}091 \text{ hp}}$$

The velocity at which minimum P_R occurs is that for flight at $(C_L^{3/2}/C_D)_{\text{max}}$; this velocity has already been calculated in Example 5.4 as

$$\boxed{V = 479.6 \text{ ft/s}}$$

The point of minimum P_R on the power required curve has the coordinates, from the above calculation, of $(P_R, V_\infty) = (5091 \text{ hp}, 479.6 \text{ ft/s})$. From Fig. 5.17, we see that these calculated coordinates agree with the graphical solution for the point of minimum P_R.

We note from the results of Example 5.2 that the flight velocity for minimum T_R occurs at $V_{(T_R)_{\text{min}}} = 631.2$ ft/s, which is greater than the velocity of 479.6 ft/s obtained above for minimum P_R. For the sake of comparison, the value of $V_{(T_R)_{\text{min}}}$ is shown in Fig. 5.17. This again emphasizes that the point of minimum power required occurs at a lower flight velocity than that for minimum thrust required.

5.7 POWER AVAILABLE AND MAXIMUM VELOCITY

By definition, the power available, denoted by P_A, is the power provided by the power plant of the airplane. As discussed in Section 3.2, the power available is given by Eq. (3.3), where

$$P_A = T_A V_\infty \qquad \textbf{[5.58]}$$

The maximum power available compared with the power required allows the calculation of the maximum velocity of the airplane. In this sense it is essentially an alternative to the method based on thrust considerations discussed in Section 5.5. We will examine the calculation of V_{max} by means of power considerations in Section 5.7.4. However, there are some aspects of airplane performance, rate of climb, for example, that depend more fundamentally on power than on thrust. Hence, the consideration of P_A in this section and P_R in the previous section is important in its own right.

5.7.1 Propeller-Driven Aircraft

Propellers are driven by reciprocating piston engines or by gas turbines (turboprop). The engines in both these cases are rated in terms of *power* (not thrust, as in the case of jet engines). Hence, for propeller-driven airplanes, *power available* is much more germane than thrust available, as discussed in Section 5.5.1.

Power available for a propeller/reciprocating engine combination is discussed in Section 3.3. In particular,

$$PA = \eta_{pr} P$$

[3.13]

where η_{pr} is the propeller efficiency and P is the shaft power from the reciprocating engine. Before you read further, review Section 3.3, paying particular attention to aspects of shaft power and propeller efficiency.

From the discussion in Section 3.3, we recall that the velocity and altitude effects on P for a piston engine are as follows:

1. Power P is reasonably constant with V_∞.

2. For an unsupercharged engine,

$$\frac{P}{P_0} = \frac{\rho}{\rho_0}$$

[3.11]

where P and ρ are the shaft power output and density, respectively, at altitude and P_0 and ρ_0 are the corresponding values at sea level. Taking into account the temperature effect, a slightly more accurate expression is

$$\frac{P}{P_0} = 1.132\frac{\rho}{\rho_0} - 0.132$$

[3.12]

3. For a supercharged engine, P is essentially constant with altitude up to the critical design altitude of the supercharger. Above this critical altitude, P decreases according to Eq. (3.11) or Eq. (3.12) with ρ_0 in these equations replaced by the density at the critical altitude, denoted by ρ_{crit}.

The power available for a turboprop is discussed in Section 3.6, which you should review before proceeding further. From that discussion, we have

$$PA = \eta_{pr} P_{es}$$

[3.29]

where P_{es} is the *equivalent* shaft power, which includes the effect of the jet thrust. Moreover, the velocity and altitude variations of P_A for a turboprop are as follows:

1. Power available P_A is reasonably constant with V_∞ (or M_∞).

2. The altitude effect is approximated by

$$\frac{P_A}{P_{A,0}} = \left(\frac{\rho}{\rho_0}\right)^n \qquad n = 0.7$$

[3.36]

5.7.2 Turbojet and Turbofan Engines

Turbojet and turbofan engines are rated in terms of thrust. Hence, to calculate the power available, simply use Eq. (5.58), repeated here:

$$P_A = T_A V_\infty$$

Turbojet engines are discussed in Section 3.4, which you should review at this point in your reading. The variation of P_A with velocity and altitude is reflected through the variation of T_A. Hence, for a turbojet engine:

1. At subsonic speeds, T_A is essentially constant. Hence, from Eq. (5.88), P_A is directly proportional to V_∞. For supersonic speeds, use Eq. (3.21) for T_A, that is,

$$\frac{T_A}{(T_A)_{\text{Mach 1}}} = 1 + 1.18(M_\infty - 1) \qquad \textbf{[3.21]}$$

   In this case, P_A for supersonic speeds is a nonlinear function of V_∞.

2. The effect of altitude on T_A is given by Eq. (3.19); the effect on P_A is the same.

$$\frac{P_A}{(P_A)_0} = \frac{\rho}{\rho_0} \qquad \textbf{[5.59]}$$

Turbofan engines are discussed in Section 3.5; you are encouraged to review this section before proceeding further. As in the case of the turbojet, the variation of P_A for a turbofan is reflected through the variation of T_A. Hence, for a turbofan:

1. The Mach number variation of thrust is given by Eq. (3.23), written as

$$T_A/(T_A)_{V=0} = A M_\infty^{-n} \qquad \textbf{[5.60]}$$

   Power available is then obtained from Eq. (5.58), $P_A = T_A V_\infty$. This will not, in general, be a linear variation for P_A. However, as noted in Section 3.5.1, for turbofans in the cruise range, T_A is essentially constant; hence in the cruise range, P_A varies directly with V_∞ via Eq. (5.58).

2. The altitude variation for turbofan thrust is approximated by Eq. (3.25), repeated here:

$$\frac{T_A}{(T_A)_0} = \left[\frac{\rho}{\rho_0}\right]^m \qquad \textbf{[5.61]}$$

Hence the variation of P_A with altitude is the same as given for T_A in Eq. (5.61), namely,

$$\frac{P_A}{(P_A)_0} = \left[\frac{\rho}{\rho_0}\right]^m \qquad \textbf{[5.62]}$$

5.7.3   Maximum Velocity

Consider a propeller-driven airplane. The power available P_A is essentially constant with velocity, as sketched in Fig. 5.18. The intersection of the maximum power available curve and the power required curve defines the maximum velocity for straight and level flight, as shown in Fig. 5.18.

Consider a jet-propelled airplane. Assuming T_A is constant with velocity, the power available at subsonic speeds varies linearly with V_∞ and is sketched in Fig. 5.19. The power required P_R is also sketched in Fig. 5.19. The high-speed intersection of the maximum power available curve and the power required curve defines the maximum velocity for straight and level flight, as shown in Fig. 5.19.

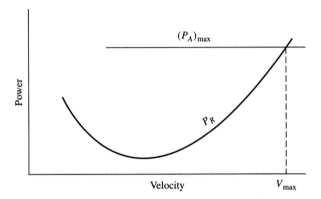

Figure 5.18   For a propellor-driven airplane, power available is essentially constant with velocity. The high-speed intersection of the maximum power available curve and the power required curve defines the maximum velocity of the airplane.

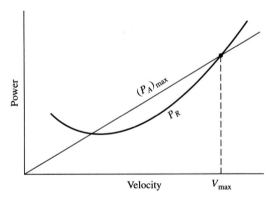

Figure 5.19   For a turbojet-powered airplane, power available varies essentially linearly with velocity. The high-speed intersection of the maximum power available curve and the power required curve defines the maximum velocity of the airplane.

Example 5.9

Using the graphical approach illustrated in Fig. 5.19, obtain the maximum velocity at 30,000 ft for the Gulfstream IV described in Example 5.1.

Solution

The power required curve is identical to that calculated in Example 5.7; it is plotted again in Fig. 5.20, extended to higher velocities. The maximum power available curve is obtained from Eqs. (5.58) and (5.61) as

$$P_A = T_A V_\infty = (T_A)_0 \left[\frac{\rho}{\rho_0} \right]^{0.6} V_\infty = 27{,}700 \left[\frac{8.9068 \times 10^{-4}}{0.002377} \right]^{0.6} V_\infty = 15{,}371 V_\infty$$

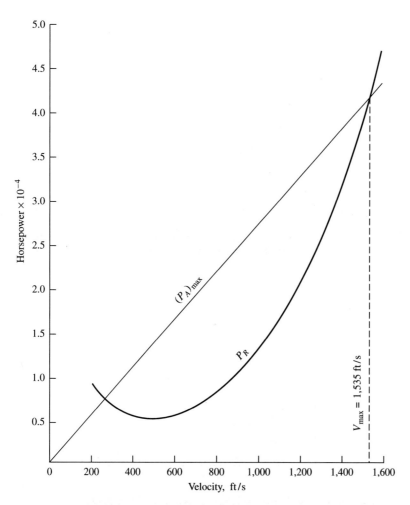

Figure 5.20 Calculated power required and power available curves for the Gulfstream IV at the conditions in Example 5.1. Altitude = 30,000 ft, W = 73,000 lb.

In terms of horsepower, where 550 ft·lb/s = 1 hp,

$$P_A = \frac{15{,}371}{550} V_\infty = 27.95 V_\infty \qquad \text{hp}$$

This linear relation for P_A is shown as the straight line in Fig. 5.20. The high-speed intersection of the $(P_A)_{max}$ and P_R curves occurs at a velocity of 1,535 ft/s. This is the maximum velocity attainable in straight and level flight at 30,000 ft, or

$$\boxed{V_{max} = 1{,}535 \text{ ft/s}}$$

This graphical solution for V_{max} using considerations of power required and power available gives the same result as that calculated from the consideration of thrust required and thrust available in Example 5.6, where the calculated value was shown to be $V_{max} = 1{,}534.6$ ft/s.

We emphasize again that the determination of V_{max} by means of power considerations (as carried out here) is simply an alternative to using the thrust considerations described earlier. Also, recall again that the value $V_{max} = 1{,}535$ ft/s at an altitude of 30,000 ft is unrealistically high because we have not included the transonic drag divergence effects in the drag polar.

5.8   EFFECT OF DRAG DIVERGENCE ON MAXIMUM VELOCITY

The purpose of the preceding sections has been to discuss the basic aspects of thrust required, thrust available, power required, and power available, and to show how these considerations can be used to calculate the maximum velocity of the airplane. In the examples used to illustrate these aspects, a drag polar was used that did not account for the large drag rise near Mach 1; rather, a conventional, subsonic drag polar was used for simplicity, because the major intent was to highlight the fundamental techniques. In reality, the drag polar used in the previous examples applies only below the drag divergence Mach number. In this section, we examine the effect of including the large transonic drag rise on our estimation of maximum velocity for the airplane.

Example 5.10

Consider the same Gulfstream IV highlighted in previous examples in this chapter. At 30,000 ft, estimate the magnitude of the transonic drag rise. Using this estimate, calculate the maximum velocity of the airplane at an altitude of 30,000 ft. Compare with the result obtained in Examples 5.6 and 5.9, where the drag-divergence effect was not included.

Solution

The flight Mach number at which the large transonic drag rise begins to occur is the drag-divergence Mach number M_{DD} discussed in Chapter 2. The precise value of M_{DD} for the Gulfstream IV is not readily available in the popular literature; for example, it is not quoted in Ref. 36. However, a reasonable value based on typical subsonic transports is 0.82. Hence, for this example, we assume that $M_{DD} = 0.82$.

To construct the slope of the drag curve in the drag-divergence region, we turn to Fig. 5.3 for guidance. Figure 5.3 shows actual drag-divergence data, albeit for a different airplane. Since we do not have the actual drag-divergence data for the Gulfstream IV, we assume that the trends are the same as those shown in Fig. 5.3. The data in Fig. 5.3, which are for the T-38, are repeated in Fig. 5.21. Although the ordinate in Fig. 5.3 is labeled *thrust required*, we recall that this is the same as the drag; hence the ordinate in Fig. 5.21 is labeled *drag*. Consider the two points 1 and 2 on the drag curve in Fig. 5.21. Point 1 is at $M_\infty = 0.9$ where $D = 1,750$ lb, and point 2 is at $M_\infty = 1.0$ where $D = 4,250$ lb. The drag rise is approximated by the straight line through points 1 and 2, shown as the dashed line in Fig. 5.21. The slope of this straight line, normalized by the value of drag at point 1, denoted D_1, is

$$\frac{d[D/D_1]}{dM_\infty} = \frac{(D_2 - D_1)/D_1}{M_2 - M_1} = \frac{(4,250 - 1,750)/1,750}{1.0 - 0.9} = 14.3$$

We will assume this same normalized slope holds for the Gulstream IV in this example.

At 30,000 ft, the standard air temperature is 411.86° R. The speed of sound is

$$a_\infty = \sqrt{\gamma R T_\infty} = \sqrt{(1.4)(1,716)(411.86)} = 994.7 \text{ ft/s}$$

Assuming $M_{DD} = 0.82$, the drag-divergence velocity is

$$V_{DD} = M_{DD}a_\infty = (0.82)(994.7) = 815.7 \text{ ft/s}$$

The drag curve *not* including drag-divergence effects for the Gulfstream IV at 30,000 ft is given in Fig. 5.5. This same drag curve is plotted in Fig. 5.22. The *modification* of this curve

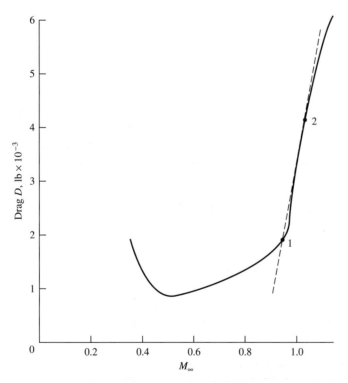

Figure 5.21 Drag versus Mach number for the T-38 jet trainer. Altitude = 20,000 ft, $W = 10,000$ lb.

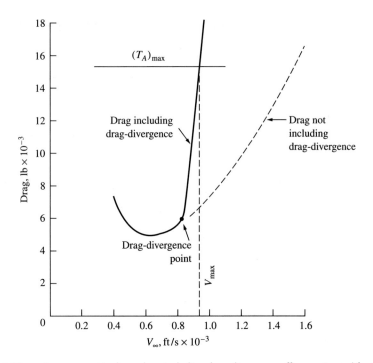

Figure 5.22   Drag versus Mach number, *including drag-divergence effects*, estimated for the Gulfstream IV. Altitude = 30,000 ft, W = 73,000 lb.

to account for drag-divergence effects starts at $V_\infty = 825.7$ ft/s, which is the velocity at drag divergence. Above this velocity, the drag is assumed to follow a sharp, linear increase with a slope given by the normalized value obtained from Fig. 5.21. Since Fig. 5.22 shows drag versus velocity, this normalized slope in terms of V_∞ is obtained as follows. Since $M = V/a$, then

$$dM = \frac{dV}{a}$$

Thus, the normalized slope in terms of V_∞ is

$$\frac{d[D/D_1]}{dV_\infty} = \frac{d[D/D_1]}{dM_\infty}\frac{dM_\infty}{dV_\infty} = \frac{d[D/D_1]}{dM_\infty}\frac{1}{a_\infty} = \frac{14.3}{994.7} = 0.01438 \ (\text{ft/s})^{-1}$$

The actual slope as shown in Fig. 5.22 is not normalized and is given by

$$\frac{dD}{dV_\infty} = D_1\frac{d[D/D_1]}{dV_\infty} = 0.01438D_1$$

The drag-divergence point shown in Fig. 5.22 is at $V_{DD} = 815.7$ ft/s, which corresponds to a drag $D_{DD} = 5{,}750$ lb. Let $D_1 = D_{DD}$. Then

$$\frac{dD}{dV_\infty} = 0.01438D_{DD} = 0.01438(5{,}750) = 82.7 \ \text{lb(ft/s)}^{-1}$$

Therefore, the equation of the straight line through the drag-divergence point in Fig. 5.22 is

$$D = D_{DD} + \frac{dD}{dV_\infty}(V - V_{DD})$$

or

$$D = 5,750 + 82.2(V - 815.7)$$

This is the straight line shown in Fig. 5.22.

In summary, the drag curve for the Gulfstream IV, taking into account the effects of drag divergence based on the above assumptions, is modeled for the purposes of this example by the solid curve shown in Fig. 5.22.

The maximum velocity of the airplane is readily obtained from Fig. 5.22 by the intersection of the maximum thrust available curve with the drag (i.e., thrust required) curve. From Example 5.6, $(T_A)_{max}$ at 30,000 ft is 15,371 lb, as shown in Fig. 5.22. The maximum velocity obtained from the intersection point is also shown in Fig. 5.22.

The value of V_{max} can also be obtained from the equation for the linear drag increase by replacing D with $(T_A)_{max}$ and V with V_{max}. That is, from

$$D = 5,750 + 82.2(V - 815.7)$$

we obtain

$$(T_A)_{max} = 5,750 + 82.2(V_{max} - 815.7)$$

Since $(T_A)_{max} = 15,371$ lb, we have

$$15,371 = 5750 + 82.2(V_{max} - 815.7)$$

Solving for V_{max} gives

$$\boxed{V_{max} = 933 \text{ ft/s}}$$

This corresponds to a maximum Mach number of

$$M_{max} = \frac{V_{max}}{a_\infty} = \frac{933}{994.7} = 0.938$$

Comparing the above results with those obtained in Examples 5.6 and 5.9, where the effects of drag divergence were not included, we see that the present result is much more realistic. Here the maximum velocity of the airplane is below sonic speed, as appropriate to the class of subsonic executive jet transports represented by the Gulfstream IV.

In Example 5.10, the estimation of the drag in the drag-divergence region was guided by flight data for a different airplane, namely, the T-38 supersonic trainer. How close the actual drag-divergence behavior of the Gulfstream IV is to the drag curve plotted in Fig. 5.22 is a matter of conjecture. An alternate method for the estimation of the drag coefficient in the drag-divergence region for a generic airplane configuration is given by Raymer in Ref. 25. Raymer's suggested procedure is as follows:

1. The zero-lift wave drag coefficient $C_{D,W,0}$ is discussed in Section 2.9.2 and is defined in conjunction with Eq. (2.42). An estimation of $C_{D,W,0}$ at or above

$M_\infty = 1.2$ can be obtained from the following equation:

$$C_{D,W,0} = E_{WD}\left[1 - 0.386(M_\infty - 1.2)^{0.57}\left(1 - \frac{\pi\Lambda^{0.77}}{100}\right)\right] \times [C_{D,W,0}]_{\text{Sears–Haack}}$$

[5.63]

where Λ is the sweep angle of the wing leading edge, E_{WD} is an empirical wave drag efficiency factor, and $(C_{D,W,0})_{\text{Sears–Haack}}$ is the theoretical zero-lift wave drag coefficient for the minimum drag Sears-Haack body of revolution, as described in Ref. 46, where

$$[C_{D,W,0}]_{\text{Sears–Haack}} = \frac{9\pi}{2S}\left[\frac{A_{\max}}{l}\right]^2$$

[5.64]

In Eq. (5.64), S is the wing area, $A_{\max}$ is the maximum cross-sectional area, and l is the longitudinal dimension of the body, reduced by the length of any portion of the body with constant cross-sectional area. Following Raymer's procedure (Ref. 25), Eqs. (5.63) and (5.64) are used to calculate $C_{D,W,0}$ at $M_\infty = 1.2$. This value is denoted by point A in Fig. 5.23. (Although Fig. 5.23 contains results for a specific

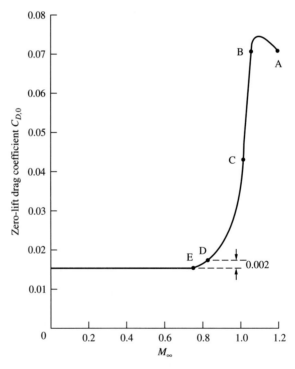

Figure 5.23   Drag-divergence curve for the Gulfstream IV as obtained by the procedure of Raymer (Ref. 25) in Example 5.11.

calculation, for the present discussion we treat Fig. 5.23 as a generic drag-divergence curve.)

2. The drag coefficient at $M_\infty = 1.05$, denoted by point B in Fig. 5.23, is assumed to be equal to that at $M_\infty = 1.2$.

3. The drag rise at $M_\infty = 1.0$ is assumed to be about one-half that at $M_\infty = 1.05$. This is denoted by point C on Fig. 5.23, which is located vertically halfway between points B and E. Point E is the drag coefficient at the critical Mach number; the drag coefficient at point E is the conventional subsonic value of $C_{D,0}$ before the drag rise sets in.

4. Point D denotes the zero-lift drag coefficient at the drag-divergence Mach number M_{DD}. It is assumed that the difference in drag coefficients between points D and E is about 0.002, as shown in Fig. 5.23.

Example 5.11

Using Raymer's procedure to obtain the drag-divergence curve as outlined above, calculate the maximum velocity of the Gulfstream IV at 30,000 ft. From Ref. 36, the maximum diameter of the fuselage is 7.83 ft, and the length is 78 ft. From the three-view in Fig. 5.1, the length of the constant-area section of the fuselage is measured as 43 ft.

Solution

From Eq. (5.63) evaluated at $M_\infty = 1.2$,

$$C_{D,W,0} = E_{WD}[C_{D,W,0}]_{\text{Sears–Haack}} \qquad \textbf{[5.65]}$$

In Eq. (5.64),

$$A_{\max} = \frac{\pi d^2}{4} = \frac{\pi (7.83)^2}{4} = 48.15 \text{ ft}^2$$

and

$$l = 78 - 43 = 35 \text{ ft}$$

Note that the Sears-Haack drag formula applies to bodies of revolution, and hence only the maximum fuselage cross-sectional area is used. Because the frontal areas of the wings and nacelles are not taken into account in this formula, our subsequent calculations will underpredict the drag. With this caveat, Eq. (5.64) yields

$$[C_{D,W,0}]_{\text{Sears–Haack}} = \frac{9\pi}{2S}\left[\frac{A_{\max}}{l}\right]^2 = \frac{9\pi}{2(950)}\left[\frac{48.15}{35}\right]^2 = 0.028$$

Based on the suggestion by Raymer (Ref. 25) that E_{WD} range from 1.4 to 2.0 for supersonic fighter, bomber, and transport designs, and can be between 2 and 3 for poor supersonic design, we choose (somewhat arbitrarily) a value of $E_{WD} = 2$ for the Gulfstream IV in this example.

Then from Eq. (5.65) we have

$$C_{D,W,0} = E_{WD}(C_{D,W,0})_{\text{Sears–Haack}} = 2(0.028) = 0.056$$

From Eq. (2.45), the total zero-lift drag coefficient, including the effect of wave drag, is

$$C_{D,0} = C_{D,e,0} + C_{D,W,0}$$

where $C_{D,e,0}$ is the purely subsonic value of zero-lift drag coefficient, which from Example 5.1 is 0.015. Thus

$$C_{D,0} = 0.015 + 0.0056 = 0.071$$

This is the estimated zero-lift drag coefficient at $M_\infty = 1.2$. This value is shown as point A in Fig. 5.23. It is also the same at point B. The value of $C_{D,0}$ at point C is then $(0.071 - 0.015)/2 + 0.015 = 0.043$. At this stage, the drag-divergence curve is faired through these points, as shown in Fig. 5.23.

Using the values of $C_{D,0}$ from Fig. 5.23, we calculate the drag at several Mach numbers. *Assume* $M_\infty = 0.9$. Then $C_{D,0} = 0.024$ (from Fig. 5.23). At 30,000 ft, $a_\infty = 994.7$ ft/s.

$$V_\infty = 0.9(994.7) = 895.23 \text{ ft/s}$$
$$q_\infty = \tfrac{1}{2}\rho_\infty V_\infty^2 = \tfrac{1}{2}(8.9068 \times 10^{-4})(895.23)^2 = 356.9 \text{ lb/ft}^2$$

To treat the drag due to lift, we need the value of C_L, obtained from

$$C_L = \frac{W}{q_\infty S} = \frac{73,000}{(356.9)(950)} = 0.215$$

From Eq. (2.47)

$$C_D = C_{D,0} + KC_L^2 = 0.024 + (0.08)(0.215)^2 = 0.0277$$
$$D = q_\infty S C_D = (356.9)(950)(0.02777) = 9,392 \text{ lb}$$

This value is denoted by point 1 in Fig. 5.24, which is a plot of drag in pounds as a function of M_∞.

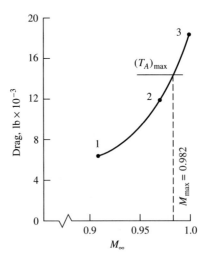

Figure 5.24   Drag curve for the conditions calculated in Example 5.11.

Assume $M_\infty = 1.0$. Then $C_{D,0} = 0.043$ (from Fig. 5.23). Since $V_\infty = 994.7$ ft/s,

$$q_\infty = \tfrac{1}{2}\rho_\infty V_\infty^2 = \tfrac{1}{2}(8.9068 \times 10^{-4})(994.7)^2 = 440.6 \text{ lb/ft}^2$$

$$C_L = \frac{W}{q_\infty S} = \frac{73{,}000}{(440.6)(950)} = 0.1744$$

$$C_D = C_{D,0} + KC_L^2 = 0.043 + (0.08)(0.1744)^2 = 0.0454$$

$$D = q_\infty S C_D = (440.6)(950)(0.0454) = 19{,}003 \text{ lb}$$

This value is denoted by point 3 in Fig. 5.24.

Assume $M_\infty = 0.97$. Then $C_{D,0} = 0.031$ (from Fig. 5.23), and

$$V_\infty = 0.97(994.7) = 964.9 \text{ ft/s}$$

$$q_\infty = \tfrac{1}{2}(8.9068 \times 10^{-4})(964.9)^2 = 414.6 \text{ lb/ft}^2$$

$$C_L = \frac{73{,}000}{(414.6)(950)} = 0.1853$$

$$C_D = 0.031 + (0.08)(0.1853)^2 = 0.0337$$

$$D = q_\infty S C_D = (414.6)(950)(0.0337) = 13{,}273 \text{ lb}$$

This value is denoted by point 2 in Fig. 5.24.

The drag curve in Fig. 5.24 is faired through points 1, 2, and 3. Also shown in Fig. 5.24 is the maximum thrust available, $(T_A)_{max} = 15{,}371$ lb. The intersection of these two curves yields a maximum flight Mach number $(M_\infty)_{max} = 0.982$. In turn, the predicted V_{max} is

$$V_{max} = a_\infty M_\infty = (994.7)(0.982) = \boxed{976.8 \text{ ft/s}}$$

Recall that the use of the Sears-Haack drag formula, Eq. (5.64), applies to bodies of revolution, and that the value of A_{max} used in this formula is the maximum cross-sectional area of the fuselage. The frontal area of the wings, nacelles, and other parts of the airplane is not included. This by itself will lead to an underprediction of drag and an overprediction of V_{max}. However, this aspect is masked to a certain extent by all the other uncertainties in the calculation procedure; the assumption that the empirical wave drag efficiency factor E_{WD} is 2 for this example is a case in point.

The result obtained in Example 5.11 is 4.7% higher than that obtained in Example 5.10. This is not a bad comparison considering the two totally different methods used in these two examples, as well as the difficulty in obtaining accurate estimates of the drag-divergence curve in the absence of precise experimental measurements for the actual airplane being analyzed. In conclusion to this section on the effect of drag divergence on maximum velocity of an airplane, we note the following:

1. For a transonic airplane such as the Gulfstream IV discussed here, taking into account the drag-divergence effects is absolutely essential in obtaining a realistic prediction of V_{max}.

2. For a preliminary performance analyses of such airplanes, the two rather empirical approaches discussed in this section for the estimation of drag-divergence effects are useful.

DESIGN CAMEO

An estimation of the variation of C_D with M_∞ through the transonic region, that is, a prediction of the drag-divergence curve, is one of the more imprecise aspects of the airplane preliminary design process. The approach illustrated in Example 5.11 is just one of several approximate techniques. Indeed, an examination of Eq. (5.63) indicates that as the sweep angle Λ increases, $C_{D,W,0}$ increases, which is not the proper physical effect; C_{DW0} should decrease as Λ increases. Another approximation is given by Steven Brandt et al. in their recent book entitled *Introduction of Aeronautics: A Design Perspective*, where Eq. (5.63) is replaced by

$$C_{D,W,0} = E_{WD}(0.74 + 0.37 \cos \Lambda)$$
$$\times \left(1 - 0.3\sqrt{M_\infty - M_{C,D,0,\max}}\right)(C_{D,W,0})_{\text{Sears–Haack}}$$

where

$$M_{C,D,0,\max} = \frac{1}{\cos^{0.2} \Lambda}$$

For the Gulfstream IV, $\Lambda = 27°40'$. With this sweep angle and $M_\infty = 1.2$, the above equation yields

$$C_{D,W,0} = 0.052$$

This is compared with $C_{D,W,0} = 0.056$ obtained in Example 5.11. The equation given by Raymer, as used in Example 5.11, yields a more conservative result by 8%. Note that the formula given by Brandt shows the proper qualitative variation of $C_{D,W,0}$ with sweep angle; that is, $C_{D,W,0}$ decreases as Λ increases.

From the point of view of the preliminary design process, an 8% difference in the prediction of $C_{D,W,0}$ is relatively small, and either of the results discussed above would be reasonable for a starter. As the design progresses and the drag polar is refined (including input from wind tunnel tests), the above calculations are replaced with more precise data.

5.9   MINIMUM VELOCITY: STALL AND HIGH-LIFT DEVICES

Return to Fig. 5.15 for a moment, where a schematic of the thrust required and maximum thrust available curves is shown. Note that there are two intersections of these curves—a high-speed intersection that determines $V_{\max}$ and a low-speed intersection (shown as point 3 in Fig. 5.15). *Question:* Does the low-speed intersection determine the minimum velocity of the airplane for steady, level flight? *Answer:* It may or may not. Indeed, it is more likely that the airplane, as it slows down, would encounter stall before it could ever reach the minimum velocity defined by point 3 in Fig. 5.15. This is not a hard-and-fast rule, and both possibilities should be examined for a given airplane in order to ascertain its allowable minimum velocity in steady, level flight.

5.9.1 Calculation of Stalling Velocity: Role of $(C_L)_{max}$

The variation of lift coefficient with angle of attack for an airplane is discussed in Section 2.5. In particular, this variation is sketched in Fig. 2.7, where the maximum lift coefficient $(c_l)_{max}$ is defined. Return to Fig. 2.7 and the related text in Section 2.5, and review the nature of $(c_l)_{max}$ before you go further.

For an airplane, the variation of total lift coefficient C_L is *qualitatively* the same as that for an airfoil. In particular, consider the sketch in Fig. 5.25, which shows a generic variation of lift coefficient versus angle of attack for a conventional airplane. As the airplane's angle of attack increases, the lift coefficient first increases, then goes through a maximum, and finally decreases. This local maximum is denoted by point 1 in Fig. 5.25. The value of the lift coefficient at point 1 is the maximum lift coefficient, denoted by $(C_L)_{max}$. The angle of attack at which $(C_L)_{max}$ is obtained is the stalling angle of attack, denoted α_{stall}. As α is increased above α_{stall}, the lift decreases precipitously, and the airplane is said to be stalled. Also as stall occurs, the drag coefficient increases rapidly; this combined with the loss in lift causes the lift-to-drag ratio of the airplane to plummet as the angle of attack increases beyond the stall. For these reasons, conventional airplanes are not flown in the stall region. Since the very beginning of manned flight, stall has been a common cause of aircraft crashes. Indeed, we discussed in Section 1.2.1 how Otto Lilienthal was killed in 1896 when his glider stalled and crashed to the ground.

Note that in the stall region, C_L does not go to zero. Indeed, at angles of attack well beyond the stall, C_L may actually recover and exceed the value given by the local maximum at point 1. This is illustrated in Fig. 5.26. However, the drag becomes prohibitive in the stall region, so a second local maximum of C_L at high angle of

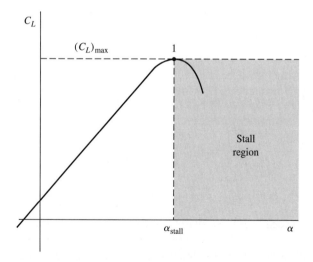

Figure 5.25 Schematic of lift coefficient versus angle of attack of an airplane.

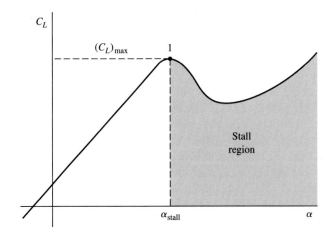

Figure 5.26   Schematic of lift coefficient versus angle of attack for an airplane, including the high angle of attack region far beyond the stall.

attack for a conventional airplane is of academic interest only. The operable value of $(C_L)_{max}$ used in an airplane performance analysis is that given by the first local maximum, namely, the value at point 1 sketched in Figs. 5.25 and 5.26.

At a given altitude, the velocity at which an airplane stalls is determined by both $(C_L)_{max}$ and the wing loading. For steady, level flight,

$$L = W = \tfrac{1}{2}\rho_\infty V_\infty^2 S C_L$$

Hence,

$$V_\infty = \sqrt{\frac{2}{\rho_\infty}\frac{W}{S}\frac{1}{C_L}} \qquad [5.66]$$

When $(C_L)_{max}$ is inserted into Eq. (5.66), the corresponding value of V_∞ is the stalling velocity V_{stall}.

$$V_{stall} = \sqrt{\frac{2}{\rho_\infty}\frac{W}{S}\frac{1}{(C_L)_{max}}} \qquad [5.67]$$

From Eq. (5.67), clearly V_{stall} depends on altitude (via ρ_∞), wing loading W/S, and maximum lift coefficient $(C_L)_{max}$. In particular, examining Eq. (5.67), we see that

1. Velocity V_{stall} increases with increasing altitude (decreasing ρ_∞). Reconnaissance airplanes that fly at extremely high altitudes, such as the Lockheed U-2, fly at velocities near stall because C_L must be large in order to generate enough lift to equal the weight in the rarefied air.
2. Velocity V_{stall} increases with increased wing loading W/S. Airplanes designed for high-speed cruise generally have high wing loadings, with correspondingly high

stalling velocity. Note again that W and S do not appear separately in Eq. (5.67), but rather in the combination W/S. This underscores again the significance of wing loading as a fundamental parameter in airplane performance.

3. Velocity V_{stall} decreases with increased $(C_L)_{\text{max}}$. This is the saving grace for airplanes with high wing loadings; such aircraft are usually designed with mechanical devices—high-lift devices such as flaps, slats, etc.—to give substantial increases in $(C_L)_{\text{max}}$ on takeoff and landing.

The stall velocity of an airplane is one of its most important performance characteristics. The takeoff and landing speeds are slightly higher than V_{stall}. Everything else being equal, it is advantageous to have a reasonably low value of V_{stall}. From Eq. (5.67), the *aerodynamic parameter* that controls V_{stall} is $(C_L)_{\text{max}}$. Starting with Otto Lilienthal and the Wright brothers about a century ago, aeronautical engineers have been concerned about the maximum lift coefficient obtainable with a given airfoil or wing. Then, in the late 1920s and early 1930s, as the speeds (and hence wing loadings) increased for new airplane designs, the natural $(C_L)_{\text{max}}$ for a given configuration was no longer sufficient. Artificial means became necessary to increase $(C_L)_{\text{max}}$ for landing and takeoff. This is the purpose of a variety of mechanical high-lift devices, discussed in Section 5.9.3. Before examining high-lift devices, we briefly examine the physical nature and cause of stall.

5.9.2   The Nature of Stall—Flow Separation

As shown in Fig. 2.7 and discussed in the related text, stall is caused by flow separation. See Ref. 3 for a discussion of the aerodynamic causes and effects of flow separation. We will only note here that flow separation usually occurs in the face of a strong adverse pressure gradient along the surface, such as downstream of the minimum pressure point on an airfoil at high angle of attack. The fluid elements deep within the aerodynamic boundary layer on the surface have been robbed of much of their original kinetic energy by the dissipative effects of friction, and they simply cannot make their way back uphill against a strong region of increasing pressure (strong adverse pressure gradient). As a consequence, the boundary layer separates from the surface. This changes the whole structure of the flow. In turn, the surface pressure distribution on the body is changed, and the changes are in such a fashion that lift is decreased and pressure drag is increased—hence stall.

Flow separation on a three-dimensional body is a particularly complex phenomenon. Just to underscore this statement, examine Fig. 5.27. Here, the low-speed flow is shown over a straight rectangular wing with an aspect ratio of 3.5 at an angle of attack of 20°. The photograph in Fig. 5.27**a** is from Winkelmann and Barlow (Ref. 47); it shows an oil flow pattern over the top surface of the wing. The rather complex three-dimensional flow aspects are self-evident. The drawing in Fig. 5.27**b** is the corresponding pattern of skin-friction lines on the surface, as conjectured by Peake and Tobak (Ref. 48). The drawing in Fig. 5.27**c** is the conjectured flow in the symmetry plane, that is, in the cut AA.

The complexity of flow separation makes its prediction on all but the simplest of shapes difficult and uncertain. Even the modern techniques of computational fluid

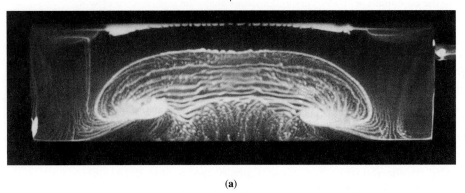

(a)

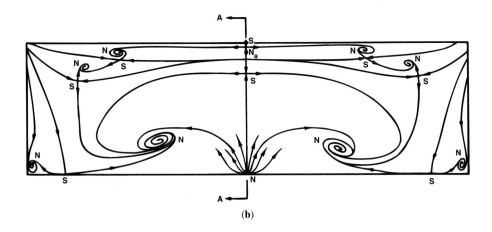

(b)

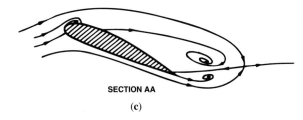

SECTION AA

(c)

Figure 5.27 (a) Oil flow pattern on the top surface of a Clark Y airfoil; aspect ratio = 3.5, $\alpha = 20°$, $R_e = 2.5 \times 10^5$. *(From Winkelmann and Barlow, Ref. 47.)* (b) Corresponding pattern of skin-friction lines as conjectured by Peake and Tobak, Ref. 48. (c) Conjectured flow pattern in the symmetry plane.

dynamics have difficulty in the accurate calculation of separated flows (Ref. 39). In turn, a reliable analytical prediction of $(C_L)_{max}$ for an airplane does not exist. Therefore, for the prediction of airplane performance, wind tunnel or flight test data are the only reasonable sources for values of $(C_L)_{max}$.

5.9.3 High-Lift Devices

If the natural value of $(C_L)_{max}$ for an airplane is not high enough for safe takeoff and landing, it can be increased by mechanical high-lift devices. The most common of these are shown schematically in Fig. 5.28, obtained from Loftin (Ref. 13). The airfoil configurations in Fig. 5.28 are arranged on a vertical scale that denotes the relative values of $(c_l)_{max}$. The following is a list of the configurations shown in Fig. 5.28, and the number of each item on the list corresponds to the identifying number in Fig. 5.28.

1. *The plain airfoil.* Here, $(c_l)_{max}$ is shown as about 1.4.

2. *The plain flap.* Here, the rear section of the airfoil is hinged so that it can be rotated downward. A plain flap is also illustrated in Fig. 5.29**a**. With a simple plain flap, $(c_l)_{max}$ can be almost doubled—Fig. 5.28 shows $(c_l)_{max}$ slightly above 2.4. A plain flap creates more lift simply by mechanically increasing the effective camber of the airfoil. It also increases the drag and pitching moment.

3. *The split flap.* Here, only the bottom surface of the airfoil is hinged, as also shown in Fig. 5.29**b**. In Fig. 5.28, the split flap is shown at a slightly higher $(c_l)_{max}$ than that for a plain flap. The split flap performs the same function as a plain flap, mechanically increasing the effective camber. However, the split flap produces more drag and less change in the pitching moment compared to a plain flap. The split flap was invented by Orville Wright in 1920, and it was employed, because of its

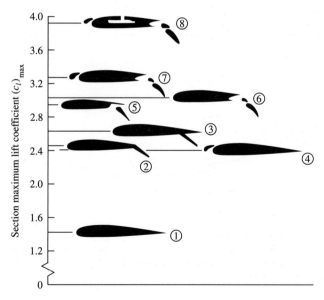

Figure 5.28 Typical values of airfoil maximum lift coefficient for various types of high-lift devices: (1) airfoil only, (2) plain flap, (3) split flap, (4) leading-edge slat, (5) single-slotted flap, (6) double-slotted flap, (7) double-slotted flap in combination with a leading-edge slat, (8) addition of boundary-layer suction at the top of the airfoil. *(From Loftin, Ref. 13.)*

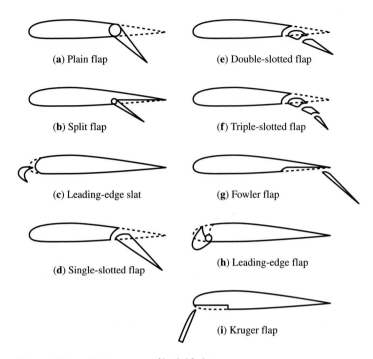

Figure 5.29   Various types of high-lift devices.

simplicity, on many of the 1930s and 40s airplanes. However, because of the higher drag associated with split flaps, they are rarely used on modern airplanes.

4.   *The leading-edge slat.* This is a small, highly cambered airfoil located slightly forward of the leading edge of the main airfoil. When deployed, a slat is essentially a flap at the leading edge, but with a gap between the flap and the leading edge, as shown in Fig. 5.29**c**. The function of a leading-edge slat is primarily to modify the pressure distribution over the top surface of the airfoil. The slat itself, being highly cambered, experiences a much lower pressure over its top surface; but the flow interaction results in a higher pressure over the top surface of the main airfoil section. This mitigates to some extent the otherwise strong adverse pressure gradient that would exist over the main airfoil section, hence delaying flow separation over the airfoil. In the process $(c_l)_{max}$ is increased with no significant increase in drag. In Fig. 5.28, the leading-edge slat is shown to produce about the same increase in $(c_l)_{max}$ as the plain flap.

5.   *The single-slotted flap.* Unlike the plain flap, which is sealed between the top and bottom surfaces, the single-slotted flap allows a gap between the top and bottom surfaces, as shown in Fig. 5.29**d**. The slot allows the higher-pressure air on the bottom surface of the airfoil to flow through the gap, modifying and stabilizing the boundary layer over the top surface of the airfoil. Indeed, flow through the slot creates a low pressure on the leading edge of the flap, and essentially a new boundary layer is formed over the flap which allows the flow to remain attached to very high flap deflections. Figure 5.28 indicates that a single-slotted flap generates a

considerably higher $(c_l)_{max}$ than a plain flap. Single-slotted flaps are in common use on light, general aviation airplanes.

6. *The double-slotted flap.* Here, the flap is divided into two segments, each with a slot, as sketched in Fig. 5.29e. If one slot is good, two are even better, as indicated by the slight increase in $(c_l)_{max}$ shown in Fig. 5.28. This benefit is achieved at the cost of increased mechanical complexity.

7. *The double-slotted flap in combination with a leading-edge slat.* There is a mutual benefit to be obtained by employing both leading- and trailing-edge devices in combination on the same airfoil. The corresponding increase in $(c_l)_{max}$ is shown in Fig. 5.28.

8. *Addition of boundary layer suction.* The low-energy boundary layer flow over the top surface of the airfoil is the culprit, in combination with the adverse pressure gradient, which causes flow separation and hence stall. By mechanically sucking away a portion of the boundary layer through small holes or slots in the top surface of the airfoil, flow separation can be delayed. This can lead to substantial increases in $(c_l)_{max}$, as shown in Fig. 5.28. However, the increased mechanical complexity and cost of this device, along with the power requirements on the pumps, diminish its attractiveness as a design option. Active boundary layer suction has not yet been used on standard, production airplanes. It remains in the category of an advanced technology item.

Several types of high-lift devices not shown in Fig. 5.28 are sketched in Fig. 5.29**f** to **i**. A triple-slotted flap is shown in Fig. 5.29**f**. This design is used on several commercial transports with high wing loadings; the Boeing 747 shown in Fig. 1.34 is a case in point. An airfoil equipped with leading-edge devices and a triple-slotted flap generates about the ultimate in high $(c_l)_{max}$ associated with purely mechanical high-lift systems. However, it is also almost the ultimate in mechanical complexity. For this reason, in the interest of lower design and production costs, recent airplane designs have returned to simpler mechanisms. For example, the Boeing 767 has single-slotted outboard flaps and double-slotted inboard flaps.

A Fowler flap is sketched in Fig. 5.29**g**. We have mentioned the Fowler flap before, in Fig. 1.23. The Fowler flap, when deployed, not only deflects downward, hence increasing the effective camber, but also translates or tracks to the trailing edge of the airfoil, hence increasing the exposed wing area with a further increase in lift. Today, the concept of the Fowler flap is combined with the double-slotted and triple-slotted flaps. The triple-slotted flaps on a Boeing 747, mentioned earlier, are also Fowler flaps.

A leading-edge flap is illustrated in Fig. 5.29**h**. Here, the leading edge pivots downward, increasing the effective camber. Unlike the leading-edge slat shown in Fig. 5.29**c**, the leading-edge flap is sealed, with no slot.

A Kruger flap is shown in Fig. 5.29**i**. This is essentially a leading-edge slat which is thinner, and which lies flush with the bottom surface of the airfoil when not deployed. Hence, it is suitable for use with thinner airfoils.

The effect of slats and flaps on the lift curve is shown schematically in Fig. 5.30. In Fig. 5.30**a** to **c**, the lift curve labeled *unflapped airfoil* pertains to no flap deflection at either the leading or trailing edge—it represents the lift curve for the basic airfoil itself, with no high-lift device included. The angle of attack α in Fig. 5.30**a** to **c**

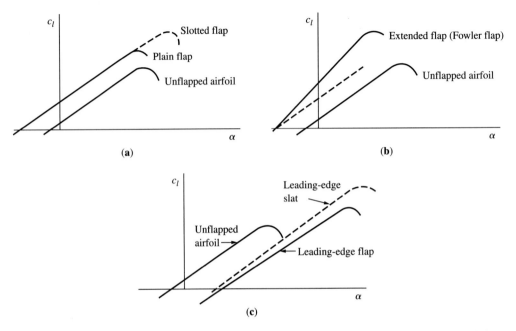

Figure 5.30   Effect of various high-lift devices on the lift curve. **(a)** Plain flap and slotted flap; **(b)** extended (Fowler) flap; **(c)** leading-edge flap and slat.

as usual represents the angle of attack of the basic airfoil, as shown in Fig. 5.31**a**. Now imagine that the basic airfoil in Fig. 5.31**a** has a plain flap, and that the flap is deflected through the angle δ as shown in Fig. 5.31**b**. Imagine that this flap deflection is locked in (i.e., fixed), and the flapped airfoil is pitched through a range of angle of attack α, where α is still defined as the angle between the original chord line and the free-stream direction, as shown in Fig. 5.31**b**. The resulting variation of c_l with α is given by the curve labeled *plain flap* in Fig. 5.30**a**. Note that the effect of the flap deflection is to shift the lift curve to the left. The lift slope of the flapped airfoil remains essentially the same as that for the basic airfoil; the zero-lift angle of attack is simply shifted to a lower value. The reason for this left shift of the lift curve has two components:

1. When the flap is deflected downward as shown in Fig. 5.31**b**, the effective camber of the airfoil is increased. A more highly cambered airfoil has a more negative zero-lift angle of attack.

2. If a line is drawn from the trailing edge of the flap through the airfoil leading edge (the dashed line in Fig. 5.31**b**) and this line is treated as a "virtual" chord line, then the flapped airfoil is at a "virtual" angle of attack which is larger than α, as sketched in Fig. 5.31**b**. That is, compared to the angle of attack of the unflapped airfoil α (which is the quantity on the abscissa of Fig. 5.30**a** to **c**), the flapped airfoil appears to the free stream to have a slightly higher angle of attack.

Also note in Fig. 5.30**a** that the deflected plain flap results in a larger $(c_l)_{max}$ than the unflapped airfoil, and that this maximum lift coefficient generally occurs at a smaller angle of attack than that for the unflapped airfoil.

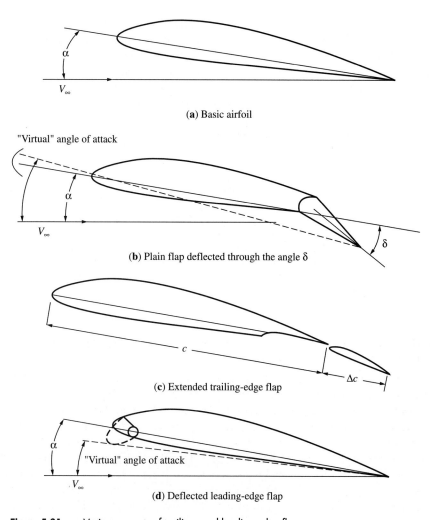

(a) Basic airfoil

"Virtual" angle of attack

(b) Plain flap deflected through the angle δ

(c) Extended trailing-edge flap

(d) Deflected leading-edge flap

Figure 5.31 Various aspects of trailing- and leading-edge flaps.

If the plain flap were replaced by a slotted flap, such as sketched in Fig. 5.29**d**, then the lift curve would be extended as indicated by the dashed portion shown in Fig. 5.30**a**. The high-energy flow through the slot from the bottom surface to the top surface delays flow separation; the airfoil can then be pitched to a higher angle of attack before stall occurs, with a consequent increase in $(c_l)_{max}$, as shown by the dashed curve in Fig. 5.30**b**. The presence of the slot does not materially change the lift slope or the zero-lift angle of attack.

The effect of a Fowler flap on the lift curve is sketched in Fig. 5.30**b**. As shown in Fig. 5.29**g**, the deployment of a Fowler flap not only increases the effective camber of the airfoil and "virtual" angle of attack (hence shifting the zero-lift angle of attack to the left, as shown in Fig. 5.30**b**), but also increases the effective planform area, which in turn increases the slope of the lift curve. This increased slope is easily seen in Fig. 5.30**b**, compared to the slope of the dashed line which is drawn parallel to the unflapped lift curve. The dashed line would hold for a given flap deflection with *no* area extension. The reason why an extended flap increases the lift slope is as follows. Consider the extended flap sketched in Fig. 5.31c. For simplicity, we show the case of extension only, with no flap deflection. For the basic airfoil with no extension, the lift per unit span is given by

$$L = q_\infty c c_l \qquad \text{[5.68]}$$

where c is the chord of the basic airfoil, as shown in Fig. 5.31c. Let L^* denote the lift per unit span with the flap extended by the distance Δc, as also sketched in Fig. 5.31c. For this case,

$$L^* = q_\infty(c + \Delta c)c_l \qquad \text{[5.69]}$$

Dividing by cq_∞, we see that Eq. (5.69) becomes

$$\frac{L^*}{q_\infty c} = \left(1 + \frac{\Delta c}{c}\right)c_l \qquad \text{[5.70]}$$

Basing the lift coefficient for the airfoil with an extended flap on the chord of the basic airfoil with no extension, and denoting this lift coefficient by c_l^*, where

$$c_l^* \equiv \frac{L^*}{q_\infty c}$$

we have from Eq. (5.70)

$$c_l^* = \left(1 + \frac{\Delta c}{c}\right)c_l \qquad \text{[5.71]}$$

Differentiating Eq. (5.71) with respect to α, we have

$$\frac{dc_l^*}{d\alpha} = \left(1 + \frac{\Delta c}{c}\right)\frac{dc_l}{d\alpha} \qquad \text{[5.72]}$$

In Eq. (5.72), $dc_l/d\alpha$ is the lift slope of the airfoil with no flap extension. Clearly, with a flap extension of Δc, the lift slope is increased by the amount $\Delta c/c$, as seen from Eq. (5.72). This increase in the lift slope is sketched in Fig. 5.30b.

Consider a leading-edge flap as sketched in Fig. 5.29h. The effect of the deflection of a leading-edge flap on the lift curve is shown in Fig. 5.30c. The lift curve is shifted to the right, with virtually no change in the lift slope. Why the shift to the right, in contrast to the left shift associated with a trailing-edge flap? The answer is illustrated in Fig. 5.31d, which shows an airfoil with a deflected leading-edge flap. When the flap is deflected, the effective camber is increased, which shifts the lift curve to the left. However, this is more than compensated by the influence of the "virtual" angle of attack. In Fig. 5.31d, the solid line drawn from the leading edge to the trailing edge of the basic airfoil is the chord line of the basic airfoil, and this chord line relative to the free-stream velocity defines the angle of attack α, as usual. When the leading-edge flap is deflected, the dashed line drawn through the leading edge of the flap and the trailing edge of the airfoil defines a *virtual* chord line, and this virtual chord line relative to the free-stream velocity defines the virtual angle of attack shown in Fig. 5.31d. This virtual angle of attack is smaller than α, which shifts the lift curve to the right. The net effect of the deflection of a leading-edge flap is usually a right shift of the lift curve, as sketched in Fig. 5.30c. Note also that a leading-edge flap results in a higher $(c_l)_{max}$ relative to the basic airfoil.

If the leading-edge flap is replaced by a leading-edge slat, as sketched in Fig. 5.29**c**, the same right shift of the lift curve takes place; but because the deployment of the leading-edge slat effectively increases the planform area by a small amount, there is a small increase in the lift slope, as indicated by the dashed curve in Fig. 5.30**c**. Because of the favorable influence of the flow through the slot between the slat and the basic airfoil, $(c_l)_{max}$ for the leading-edge slat is higher than that for the leading-edge flap, also sketched in Fig. 5.30**c**.

5.9.4   Interim Summary

Some type of high-lift device has been used on almost every airplane designed since the early 1930s. Its purpose is to increase the airplane's maximum lift coefficient $(C_L)_{max}$, hence reducing V_{stall} via Eq. (5.67). However, simply reducing V_{stall} to some small value is not the important story obtained from Eq. (5.67). The real story is told by Eq. (5.50) for the maximum velocity of an airplane. Return to Eq. (5.50) for a moment. Note that for a given maximum thrust-to-weight ratio $(T_A)_{max}/W$, V_{max} is directly proportional to $\sqrt{W/S}$. The higher the wing loading, the higher V_{max} is. This is why most high-speed airplanes are designed with high wing loadings. Now return to Eq. (5.67), which shows that V_{stall} is proportional to $\sqrt{W/S}$ also. If nothing else is done, a high-speed airplane will have an inordinately high V_{stall} because of the high wing loading. The solution to this problem is also embodied in Eq. (5.67)—increase $(C_L)_{max}$ sufficiently that, in spite of the large W/S, V_{stall} will be acceptable. In turn, high-lift devices are the means to obtain the sufficient increase in $(C_L)_{max}$. From this point of view, the argument can be made that high-lift devices make efficient high-speed flight possible.

On another note, we should mention that the values of $(C_L)_{max}$ achieved by high-lift devices on *finite wings* will be less than the typical airfoil values of $(c_l)_{max}$ shown in Fig. 5.28. This is especially true for swept wings. Detailed information for $(C_L)_{max}$ for a given wing configuration should be obtained from wind tunnel tests. For our purposes in this chapter, some guidelines are given in Table 5.3, where the $(C_L)_{max}$ values are those recommended by Torenbeck in Ref. 35. In Table 5.3, Λ is the sweep angle of the quarter-chord line.

Table 5.3

High-Lift Device		Typical Flap Angle		$(C_L)_{max}/\cos \Lambda$	
Trailing Edge	**Leading Edge**	**Takeoff**	**Landing**	**Takeoff**	**Landing**
Plain flap		20°	60°	1.4–1.6	1.7–2.0
Single-slotted flap		20°	40°	1.5–1.7	1.8–2.2
Fowler flap					
single-slotted		15°	40°	2.0–2.2	2.5–2.9
double-slotted		20°	50°	1.7–1.95	2.3–2.7
double-slotted	slat	20°	50°	2.3–2.6	2.8–3.2
triple-slotted	slat	20°	40°	2.4–2.7	3.2–3.5

Example 5.12   Calculate the minimum velocity of the Gulfstream IV at sea level based on (*a*) the low-speed intersection of the thrust available and the thrust required curves and (*b*) the stalling velocity. This airplane is equipped with single-slotted Fowler trailing-edge flaps. The wing sweep angle is 27°40′.

Solution

(*a*) The minimum velocity based on the low-speed intersection of the thrust available and thrust required curves (point 3 in Fig. 5.15) can be obtained from Eq. (5.18) by using the minus sign in the numerator, that is,

$$V_{\min} = \left[\frac{(T/W)(W/S) - (W/S)\sqrt{(T/W)^2 - 4C_{D,0}K}}{\rho_\infty C_{D,0}} \right]^{1/2}$$

In the above equation, from the data for the Gulfstream IV given in Example 5.6, namely, $W/S = 76.84$ lb/ft², $T/W = 0.3795$, $C_{D,0} = 0.015$, $K = 0.08$, and $\rho_\infty = 0.002377$ slug/ft³, we have

$$V_{\min} = \left[\frac{0.3795(76.84) - 76.84\sqrt{(0.3795)^2 - 4(0.015)(0.08)}}{0.002377(0.015)} \right]^{1/2} = \boxed{117 \text{ ft/s}}$$

(*b*) From Eq. (5.67),

$$V_{\text{stall}} = \sqrt{\frac{2}{\rho_\infty} \frac{W}{S} \frac{1}{(C_L)_{\max}}}$$

From Table 5.3, for a single-slotted Fowler flap in its most fully deployed configuration (that for landing), we choose

$$\frac{(C_L)_{\max}}{\cos \Lambda} = 2.7$$

Hence,

$$(C_L)_{\max} = 2.7 \text{ Cos } 27°40′ = 2.39$$

Thus,

$$V_{\text{stall}} = \sqrt{\frac{(2)(76.84)}{(0.002377)(2.39)}} = \boxed{164.5 \text{ ft/s}}$$

Clearly, the stalling velocity defines the minimum velocity for the Gulfstream IV in steady, level flight. The velocity calculated for the low-speed intersection of the T_A and T_R curves, namely, 117 ft/s, is of academic interest only.

It is interesting to note that the stalling velocity for the actual Gulfstream IV is given as 182 ft/s in Ref. 36. This value is quoted for both wheels and flaps down, and for a maximum landing weight of 58,500 lb. This weight is less than the maximum takeoff weight of 73,000 lb, which was the value of W used in the present worked example. In any event, our calculation of $V_{\text{stall}} = 164.5$ ft/s is a reasonable approximation.

5.10 RATE OF CLIMB

Imagine that you are flying an airplane, and you suddenly encounter a major obstacle ahead—a large building, a hill, or even a mountain. The ability of your airplane to fly up and over such obstacles depends critically on its climbing characteristics. Or, imagine that you encounter bad weather or turbulence at some altitude, and you want to get out of it by climbing quickly to a higher altitude. How fast you can do this depends on the climbing characteristics of your airplane. Or, imagine that you are a military fighter pilot, and you scramble to take off and intercept a target at some prescribed altitude. You need to get to that target as soon as possible; how soon you can do so depends on the climbing characteristics of your airplane. For these and other reasons, the climb performance of an airplane is an essential part of the overall performance scenario. Climb performance is the subject of this section.

The previous sections of this chapter have dealt with steady, *level* flight of an airplane. In this section we change our focus to an airplane in steady, unaccelerated *climbing* flight. This case was introduced in Chapter 4, and a sketch of an airplane in climbing flight is shown in Fig. 4.2. Return to Fig. 4.2 and study it carefully. Note that the climb angle θ, is defined as the angle between the instantaneous flight path direction (the direction of the relative wind V_∞) and the horizontal. (Please note that θ is *not* the angle of attack of the airplane—a misconception frequently held initially by students new to the subject. The angle of attack, not labeled in Fig. 4.2, is as usual the angle between the chord line and the relative wind.) The equations of motion for accelerated flight along a curved flight path are given by Eqs. (4.5) to (4.7). Please review the derivation and discussion of these equations in Chapter 4 before going on.

In this section we consider steady (unaccelerated) climb. Hence, in Eq. (4.5), $dV_\infty/dt = 0$; in Eq. (4.6), $V_\infty^2/r_1 = 0$; and in Eq. (4.7), $(V_\infty \cos\theta)^2/r_2 = 0$. The latter two statements imply $r_1 \to \infty$ and $r_2 \to \infty$, that is, flight along a straight path. This also implies that the bank angle ϕ is zero. The equations of motion for this case become, from Eqs. (4.5) and (4.6),

$$T \cos\epsilon - D - W \sin\theta = 0 \qquad \text{[5.73]}$$

$$L + T \sin\epsilon - W \cos\theta = 0 \qquad \text{[5.74]}$$

Furthermore, for simplicity, we assume the thrust line is in the direction of flight, that is, $\epsilon = 0$. Hence, Eqs. (5.73) and (5.74) become, respectively,

$$T - D - W \sin\theta = 0 \qquad \text{[5.75]}$$

$$L - W \cos\theta = 0 \qquad \text{[5.76]}$$

The force diagram consistent with Eqs. (5.75) and (5.76) is shown in Fig. 5.32, which is a specialized version of Fig. 4.2.

The inset in Fig. 5.32 is a vector diagram resolving the velocity of the airplane V_∞ into its horizontal and vertical components V_H and V_V, respectively. In particular, the vertical component is, by definition, the *rate of climb* of the airplane; we denote

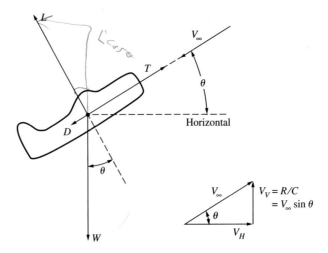

Figure 5.32   Force and velocity diagrams for climbing flight.

the rate of climb by R/C. From this diagram,

$$\boxed{R/C = V_\infty \sin \theta}$$

[5.77]

Multiplying Eq. (5.75) by V_∞/W, we have

$$\boxed{V_\infty \sin \theta = R/C = \frac{TV_\infty - DV_\infty}{W}}$$

[5.78]

In Eq. (5.78), TV_∞ is the power available, and DV_∞ is the power required to overcome the drag. We define

$$\boxed{TV_\infty - DV_\infty \equiv \text{excess power}}$$

[5.79]

Hence, from Eq. (5.78),

$$\boxed{R/C = \frac{\text{excess power}}{W}}$$

[5.80]

Clearly, rate of climb depends on raw power in combination with the weight of the airplane. The higher the thrust, the lower the drag, and the lower the weight, the better the climb performance—all of which makes common sense even without the benefit of the above equations.

At this stage in our discussion, it is important to note that, for steady climbing flight, lift is less than weight; indeed, from Eq. (5.76),

$$L = W \cos \theta$$

[5.81]

This is because, for climbing flight, part of the weight of the airplane is supported by the thrust, and hence less lift is needed than for level flight. In turn, this has an impact on drag; less lift means less drag due to lift. For a given velocity V_∞, the drag in climbing flight is less than that for level flight. Quantitatively, we can write for steady climbing flight,

$$C_L = \frac{L}{q_\infty S} = \frac{W \cos \theta}{q_\infty S} \qquad \textbf{[5.82]}$$

From the drag polar,

$$D = q_\infty S C_D = q_\infty S \left(C_{D,0} + K C_L^2\right) \qquad \textbf{[5.83]}$$

Substituting C_L from Eq. (5.82) into Eq. (5.83), we have

$$D = q_\infty S \left[C_{D,0} + K \left(\frac{W \cos \theta}{q_\infty S} \right)^2 \right]$$

or

$$D = q_\infty S C_{D,0} + \frac{K W^2 \cos^2 \theta}{q_\infty S} \qquad \textbf{[5.84]}$$

The value of D from Eq. (5.84) is the value that goes in Eq. (5.78) for rate of climb. Combining Eqs. (5.78) and (5.84), we have after some algebraic manipulations (the details are left for a homework problem)

$$V_\infty \sin \theta = V_\infty \left[\frac{T}{W} - \frac{1}{2} \rho_\infty V_\infty^2 \left(\frac{W}{S} \right)^{-1} C_{D,0} - \frac{W}{S} \frac{2K \cos^2 \theta}{\rho_\infty V_\infty^2} \right] \qquad \textbf{[5.85]}$$

Note that in Eq. (5.85) the weight does not appear separately, but rather in the form of the thrust-to-weight ratio T/W and the wing loading W/S. Once again we observe the importance of these two design parameters, this time in regard to climb performance.

Equation (5.85) is the key to the *exact* solution of the climb performance of an airplane. Unfortunately, it is unwieldly to solve. Note that V_∞ and θ appear on both sides of the equation. In principle, for a given V_∞, Eq. (5.85) can be solved by trial and error for θ, hence yielding $R/C = V_\infty \sin \theta$ for the given value of V_∞. Or, for a given value of θ, Eq. (5.85) can be solved by trial and error for V_∞, hence yielding $R/C = V_\infty \sin \theta$ for the given θ.

Fortunately, for a preliminary performance analysis, this hard work is usually not necessary. Let us make the assumption that *for the drag expression only*, $\cos \theta \approx 1$. For example, in Eq. (5.84), set $\cos \theta = 1$. This assumption leads to remarkably accurate results for climb performance for climb angles as large as $50°$ degrees. Indeed, in their elegant analysis in Ref. 41, Mair and Birdsall show that for a climb angle of $50°$, by making the assumption that $\cos \theta = 1$ in the drag expression, the error in the calculated climb angle is $2.5°$ or smaller, and the error in the calculated rate of climb is 3% or less. This is particularly fortuitous, because the normal climb angles of conventional airplanes are usually less than $15°$. Hence, in the remainder of this section, we assume $\cos \theta = 1$ in the drag expression. A more general energy-based method which can be applied to accelerated climb and which accurately treats the case for any climb angle (even $90°$) will be discussed in Chapter 6.

5.10.1 Graphical Approach

Return to Eq. (5.78) for the rate of climb. On the right-hand side, the term TV_∞ is the power available, discussed in Section 5.7, and the term DV_∞ *with the assumption of* $\cos\theta = 1$ is the power required for steady, level flight, discussed in Section 5.6. Hence the *excess power*, defined in Eq. (5.79) and used for the calculation of rate of climb in Eq. (5.80), is the difference between the power available and the power required curves, where the power required curve is for steady, level flight. In the sequence of most normal performance analyses, the P_R curve will already be available from the calculation of maximum velocity. The excess power is identified in Fig. 5.33 for both propeller-driven and jet-propelled aircraft; at a given V_∞, the excess power is simply the difference between the ordinates of the P_A and P_R curves. This directly leads to a graphical construction for the variation of R/C with V_∞, as sketched in Fig. 5.34. At any V_∞, measure the excess power from the difference between the P_A and P_R curves shown in Fig. 5.34**a**. Divide this excess power by the weight, obtaining the value of R/C at this velocity via Eq. (5.80). Carry out this process for a range of V_∞, obtaining the corresponding values of R/C. The locus of these values for R/C is sketched in Fig. 5.34**b**, which is a graph of R/C versus velocity for the airplane. Recall that the P_A and P_R curves sketched in Fig. 5.34**a** are for a *given altitude*, hence the variation of R/C versus velocity sketched in Fig. 5.34**b** is also for a given altitude. Also, note that at some velocity the difference between the P_A and P_R curves will be a maximum, as identified in Fig. 5.34**a**; in turn, this is the velocity at which R/C is a maximum value, as identified in Fig. 5.34**b**. Similarly, the velocity at which the P_A and P_R curves intersect is the maximum velocity for steady, level flight, as discussed in Section 5.7. No excess power exists at $V_{\max}$, and hence $R/C = 0$ at this velocity, as shown in Fig. 5.34**b**.

An even more useful graphical construction is the *hodograph diagram*, which is a plot of the aircraft's vertical velocity V_V versus its horizontal velocity V_H, as sketched in Fig. 5.35. The hodograph diagram is slightly different from the curve

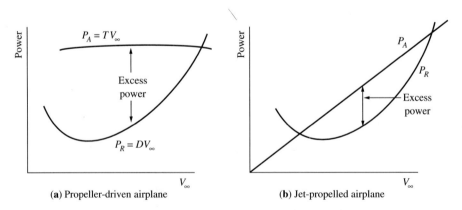

Figure 5.33   Illustration of excess power for **(a)** propellor-driven airplane and **(b)** jet-propelled airplane.

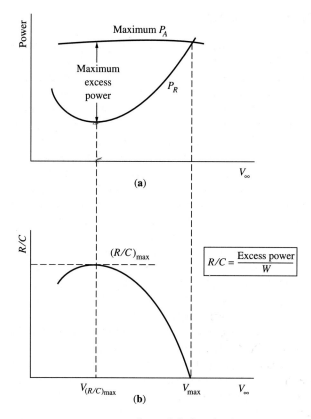

Figure 5.34 Variation of rate of climb with velocity at a given altitude.

shown in Fig. 5.34**b**. In both cases the ordinate is R/C, which by definition is the vertical component of velocity V_V. However, in Fig. 5.35 the abscissa is the horizontal component of velocity V_H, *not* the total velocity V_∞ which is the abscissa in Fig. 5.34**b**. The geometric relation among V_∞, V_H, V_V, and θ is also shown in Fig. 5.35, for convenience. Consider an arbitrary point on the hodograph curve, denoted by point 1 in Fig. 5.35. Draw a line from the origin to point 1. Geometrically, the length of the line is V_∞, and the angle it makes with the horizontal axis is the corresponding climb angle at that velocity. Point 2 in Fig. 5.35 denotes the maximum R/C; the length of the line from the origin to point 2 is the airplane velocity at maximum R/C, denoted by $V_{\max R/C}$, and the angle it makes with the horizontal axis is the climb angle for maximum R/C, or $\theta_{\max R/C}$. A line drawn through the origin and tangent to the hodograph curve locates point 3 in Fig. 5.35. The angle of this line relative to the horizontal defines the maximum possible climb angle $\theta_{\max}$, as shown in Fig. 5.35. The length of the line from the origin to the tangent point (point 3) is the velocity at the maximum climb angle. *Important:* Looking at Fig. 5.35, we see that the maximum rate of climb does *not* correspond to the maximum climb angle. The maximum climb angle $\theta_{\max}$ is important when you want to clear an obstacle

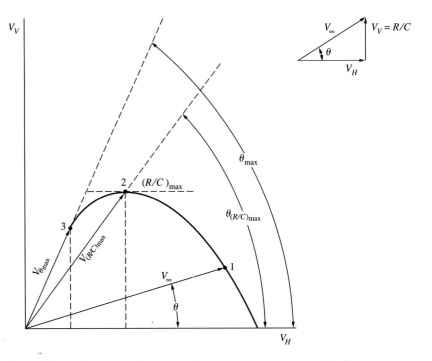

Figure 5.35   Hodograph diagram for climb performance at a given altitude.

while covering the minimum horizontal distance along the ground. The maximum rate of climb $(R/C)_{\max}$ is important when you want to achieve a certain altitude in a minimum amount of time. (The calculation of the time to climb to a given altitude is considered in Section 5.12.) Note that V_∞ is smallest at $\theta_{\max}$, and it increases as θ is made smaller. This is why $(R/C)_{\max}$ does not occur at $\theta_{\max}$; rather, since $R/C = V_\infty \sin\theta$, from point 3 to point 2 in Fig. 5.35 the increase in V_∞ exceeds the decrease in $\sin\theta$, leading to an increase in $V_\infty \sin\theta$.

5.10.2   Analytical Approach

By making the assumption in the drag relation that $\cos\theta = 1$, Eq. (5.85) becomes

$$V_\infty \sin\theta = R/C = V_\infty \left[\frac{T}{W} - \frac{1}{2}\rho_\infty V_\infty^2 \left(\frac{W}{S}\right)^{-1} C_{D,0} - \frac{W}{S}\frac{2K}{\rho_\infty V_\infty^2} \right] \qquad \textbf{[5.86]}$$

Given V_∞, the corresponding R/C can be calculated directly from Eq. (5.86). The corresponding climb angle can be found from

$$\sin\theta = \frac{R/C}{V_\infty} \qquad \textbf{[5.87]}$$

Alternately, we note that by dividing Eq. (5.86) by V_∞, we obtain

$$\sin \theta = \frac{T}{W} - \frac{1}{2}\rho_\infty V_\infty^2 \left(\frac{W}{S}\right)^{-1} C_{D,0} - \frac{W}{S}\frac{2K}{\rho_\infty V_\infty^2} \qquad \textbf{[5.88]}$$

Equation (5.86), with its counterpart Eq. (5.88), contains some useful information about climb performance and the design parameters of an airplane that dictate climb performance. In particular:

1. Equation (5.86) is simply an elaborate form of Eq. (5.78), repeated here in a slightly different form.

$$V_\infty \sin \theta = V_\infty \frac{T - D}{W} \qquad \textbf{[5.89]}$$

Clearly, from Eq. (5.89), more thrust, less drag, and smaller weight all work to increase the rate of climb. Equation (5.86) spells out more clearly the design parameters. For example, increasing the thrust-to-weight ratio increases R/C. The last two terms in Eq. (5.86) represent the zero-lift drag and the drag due to lift, respectively, both divided by the weight. A decrease in $C_{D,0}$ or K, or in both, increases R/C.

2. The effect of increasing altitude usually is to decrease R/C. All three terms in Eq. (5.86) are sensitive to altitude through ρ_∞. The effect of altitude on T depends on the type of power plant used. However, for turbojets, turbofans, and unsupercharged piston engines with propellers, thrust at a given V_∞ decreases with altitude (as discussed in Section 5.5 and in Chapter 3). For an airplane with any reasonable climb capacity, the dominant term in Eq. (5.86) is T/W; hence when T/W decreases with increasing altitude, R/C also decreases. However, for supercharged piston engines with variable-pitch, constant-speed propellers, the story may be different. Up to the critical altitude of the supercharged engine, power output is reasonably constant; hence at a given V_∞, the thrust output of the propeller can be maintained reasonably constant with increasing altitude by increasing the propeller pitch angle. The consequences of this on the altitude variation of R/C at a given V_∞ depend on how drag varies with altitude at the same V_∞. The drag is given by the last two terms in Eq. (5.86). The middle term shows that at a given V_∞ the zero-lift drag decreases with increasing altitude, whereas the last term shows that at a given V_∞ the drag due to lift increases with increasing altitude. If V_∞ is low, the drag due to lift dominates the total drag, and hence in this low-velocity range, drag increases with altitude at a given V_∞. If V_∞ is high, the zero-lift drag dominates the total drag, and hence in this high-velocity range, drag decreases with altitude at a given V_∞. Hence, in this high-velocity range, for an airplane with a supercharged piston engine, the R/C for a given V_∞ theoretically can increase with altitude. We repeat again that, in general, the dominant term in Eq. (5.86) is T/W, and the effect of altitude on T/W dominates the altitude variation of R/C.

3. From Eq. (5.86), wing loading also affects R/C. At a given arbitrary V_∞, this effect is a mixed bag. Note from the drag terms in Eq. (5.86) that increasing W/S decreases the zero-lift drag and increases the drag due to lift. Hence, in the

low-velocity range where drag due to lift is dominant, an increase in the design W/S results in a decrease in R/C at the same V_∞. However, in the high-velocity range where zero-lift drag is dominant, an increase in W/S results in an increase in R/C at the same V_∞.

The above considerations, gleaned from Eq. (5.86) for a given, arbitrary V_∞, are rather general, and in some cases the trends are somewhat mixed. More specific information on the airplane design parameters that *optimize* climb performance can be obtained by studying the cases for *maximum* climb angle θ_{max} and *maximum* rate of climb $(R/C)_{max}$. We now turn our attention to these two specific cases.

Maximum Climb Angle   Dividing Eq. (5.78) by V_∞, we have

$$\sin\theta = \frac{T}{W} - \frac{D}{W} \qquad [5.90]$$

From Eq. (5.76),

$$W = \frac{L}{\cos\theta} \qquad [5.91]$$

Replace W in the drag term of Eq. (5.90) by Eq. (5.91):

$$\sin\theta = \frac{T}{W} - \frac{\cos\theta}{L/D} \qquad [5.92]$$

By making the assumption that $\cos\theta = 1$, Eq. (5.92) becomes

$$\sin\theta = \frac{T}{W} - \frac{1}{L/D} \qquad [5.93]$$

Consider the case of a jet-propelled airplane where the thrust is essentially constant with velocity. Then Eq. (5.93) dictates that the maximum climb angle θ_{max} will occur when the lift-to-drag ratio is a maximum, that is, *for a jet-propelled airplane.*

$$\boxed{\sin\theta_{max} = \frac{T}{W} - \frac{1}{(L/D)_{max}}} \qquad [5.94]$$

Recalling Eq. (5.30) for $(L/D)_{max}$, we see that Eq. (5.94) can be written as

$$\boxed{\sin\theta_{max} = \frac{T}{W} - \sqrt{4C_{D,0}K}} \qquad [5.95]$$

The flight velocity corresponding to θ_{max} is obtained as follows. From Eq. (5.76)

$$L = W\cos\theta = \tfrac{1}{2}\rho_\infty V_\infty^2 SC_L \qquad [5.96]$$

For maximum L/D, Eq. (5.29) holds.

$$C_L = \sqrt{\frac{C_{D,0}}{K}} \qquad [5.29]$$

Substituting Eq. (5.29) into Eq. (5.96), we have, where θ and V_∞ in Eq. (5.96) now become θ_{max} and $V_{\theta_{max}}$, respectively,

$$W \cos \theta_{max} = \frac{1}{2} \rho_\infty V_{\theta_{max}}^2 S \sqrt{\frac{C_{D,0}}{K}} \qquad \text{[5.97]}$$

Solving Eq. (5.97) for $V_{\theta_{max}}$, we have *for a jet-propelled airplane,*

$$\boxed{V_{\theta_{max}} = \sqrt{\frac{2}{\rho_\infty} \left(\frac{K}{C_{D,0}}\right)^{1/2} \frac{W}{S} \cos \theta_{max}}} \qquad \text{[5.98]}$$

Finally, the rate of climb that corresponds to the maximum climb angle is given by

$$(R/C)_{\theta_{max}} = V_{\theta_{max}} \sin \theta_{max} \qquad \text{[5.99]}$$

where, in Eq. (5.99), $V_{\theta_{max}}$ is obtained from Eq. (5.98) and θ_{max} is obtained from Eq. (5.95). Note from Eq. (5.95) that θ_{max} does *not* depend on wing loading, but from Eq. (5.98), $V_{\theta_{max}}$ varies directly as $(W/S)^{1/2}$. Hence, everything else being equal, for flight at θ_{max}, the rate of climb is higher for higher wing loadings. Also, the effect of altitude is clearly seen from these results. Since $(L/D)_{max}$ does not depend on altitude, then from Eq. (5.94), θ_{max} decreases with altitude because T decreases with altitude. However, from Eq. (5.98) $V_{\theta_{max}}$ increases with altitude. These are competing effects in determining $(R/C)_{\theta_{max}}$ from Eq. (5.99). However, the altitude effect on θ_{max} usually dominates, and $(R/C)_{\theta_{max}}$ usually decreases with increasing altitude.

Caution: For a given airplane, it is possible for $V_{\theta_{max}}$ to be less than the stalling velocity. For such a case, it is not possible for that airplane to achieve the theoretical maximum climb angle.

Consider the case for a propeller-driven airplane. From Eq. (5.48),

$$T_A = \frac{\eta_{pr} P}{V_\infty} \qquad \text{[5.48]}$$

where η_{pr} is the propeller efficiency and P is the shaft power from the reciprocating piston engine (or the effective shaft power P_{es} for a turboprop). The product $\eta_{pr} P$ is the power available P_A, which we assume to be constant with velocity. The climb angle for the propeller-driven airplane is given by Eq. (5.88) with Eq. (5.48) inserted for the thrust, that is,

$$\sin \theta = \frac{\eta_{pr} P}{V_\infty W} - \frac{1}{2} \rho_\infty V_\infty^2 \left(\frac{W}{S}\right)^{-1} C_{D,0} - \frac{W}{S} \frac{2K}{\rho_\infty V_\infty^2} \qquad \text{[5.100]}$$

In Eq. (5.100), $\eta_{pr} P$ is assumed constant with velocity. Although Eq. (5.100) does not give useful information directly for θ_{max}, such information can be obtained by differentiating Eq. (5.100) and setting the derivative equal to zero, thus defining the conditions that will maximize $\sin \theta$.

Differentiating Eq. (5.100) with respect to V_∞, we have

$$\frac{d(\sin \theta)}{dV_\infty} = -\frac{\eta_{pr} P}{W V_\infty^2} - \rho_\infty V_\infty \left(\frac{W}{S}\right)^{-1} C_{D,0} + 2 \frac{W}{S} \frac{K}{\frac{1}{2} \rho_\infty V_\infty^3} \qquad \text{[5.101]}$$

Setting the right-hand side of Eq. (5.101) to zero, we obtain after a few algebraic steps (with V_∞ now representing $V_{\theta_{max}}$)

$$V_{\theta_{max}}^4 + \frac{\eta_{pr}(P/W)(W/S)}{\rho_\infty C_{D,0}} V_{\theta_{max}} - \frac{4(W/S)^2 K}{\rho_\infty^2 C_{D,0}} = 0 \qquad \textbf{[5.102]}$$

This author cannot find any analytical solution to Eq. (5.102), nor can he find any such solution in the existing literature. However, Hale in Ref. 49 has shown that for a typical propeller-driven airplane, the magnitudes of the last two terms in Eq. (5.102) are much larger than the magnitude of the first term, and hence a reasonable approximation can be obtained by dropping the $V_{\theta_{max}}^4$ term in Eq. (5.102), obtaining for $V_{\theta_{max}}$ *for a propeller-driven airplane,*

$$V_{\theta_{max}} \approx \frac{4(W/S)K}{\rho_\infty \eta_{pr}(P/W)} \qquad \textbf{[5.103]}$$

In turn, $V_{\theta_{max}}$ obtained from Eq. (5.103) can be inserted into Eq. (5.100) to obtain θ_{max}.

Caution: Once again we note that for a given airplane, it is possible for $V_{\theta_{max}}$ to be less than V_{stall}. For such a case, it is not possible for the airplane to achieve the theoretical maximum climb angle.

Maximum Rate of Climb   Consider the case of a jet-propelled airplane where T is relatively constant with V_∞. Rate of climb is given by Eq. (5.86). Conditions associated with maximum rate of climb can be found by differentiating Eq. (5.86) and settng the derivative equal to zero. Differentiating Eq. (5.86) with respect to V_∞, we have

$$\frac{d(R/C)}{dV_\infty} = \frac{T}{W} - \frac{3}{2}\rho_\infty V_\infty^2 \left(\frac{W}{S}\right)^{-1} C_{D,0} + \frac{W}{S}\frac{2K}{\rho_\infty V_\infty^2} \qquad \textbf{[5.104]}$$

If we set the right-hand side of Eq. (5.104) equal to zero and then divide it by $3/2\rho_\infty(W/S)^{-1}C_{D,0}$, we obtain

$$V_\infty^2 - \frac{2(T/W)(W/S)}{3\rho_\infty C_{D,0}} - \frac{4K(W/S)^2}{3\rho_\infty^2 C_{D,0}V_\infty^2} = 0 \qquad \textbf{[5.105]}$$

Recalling from Eq. (5.30) that $[L/D]_{max} = 1/\sqrt{4KC_{D,0}}$, we see that the last term in Eq. (5.105) can be expressed in terms of $(L/D)_{max}$. Also, multiplying by V_∞^2, Eq. (5.105) becomes

$$V_\infty^4 - \frac{2(T/W)(W/S)}{3\rho_\infty C_{D,0}}V_\infty^2 - \frac{(W/S)^2}{3\rho_\infty^2 C_{D,0}^2(L/D)_{max}^2} = 0 \qquad \textbf{[5.106]}$$

For simplicity, let

$$Q \equiv \frac{W/S}{3\rho_\infty C_{D,0}} \qquad \textbf{[5.107]}$$

$$x \equiv V_\infty^2 \qquad \textbf{[5.108]}$$

Then Eq. (5.106) can be written as

$$x^2 - 2\frac{T}{W}Qx - \frac{3Q^2}{(L/D)_{max}^2} = 0 \qquad \textbf{[5.109]}$$

Eq. (5.109) is a quadratic equation in terms of x (that is, in terms of V_∞^2). From the quadratic formula, we obtain

$$x = \frac{2(T/W)Q \pm \sqrt{4(T/W)^2 Q^2 + 12Q^2/(L/D)_{max}^2}}{2} \qquad \textbf{[5.110]}$$

By factoring $(T/W)Q$ out of the radical, Eq. (5.110) becomes

$$x = \frac{T}{W}Q \pm \frac{T}{W}Q\sqrt{1 + 3/(L/D)_{max}^2 (T/W)^2}$$

or

$$x = \frac{T}{W}Q\left\{1 \pm \sqrt{1 + \frac{3}{(L/D)_{max}^2 (T/W)^2}}\right\} \qquad \textbf{[5.111]}$$

In Eq. (5.111), the minus sign gives a negative value of x; this is nonphysical, hence we will use only the plus sign. Finally, replacing Q and x in Eq. (5.111) with their definitions given in Eqs. (5.107) and (5.108), respectively, and noting that V_∞ represents $V_{(R/C)_{max}}$, we have *for a jet-propelled airplane*,

$$V_{(R/C)_{max}} = \left\{\frac{(T/W)(W/S)}{3\rho_\infty C_{D,0}}\left[1 + \sqrt{1 + \frac{3}{(L/D)_{max}^2 (T/W)^2}}\right]\right\}^{1/2} \qquad \textbf{[5.112]}$$

An equation for the maximum rate of climb is obtained by substituting $V_{(R/C)_{max}}$ from Eq. (5.112) into Eq. (5.86). To simplify the resulting expression, let

$$Z \equiv 1 + \sqrt{1 + \frac{3}{(L/D)_{max}^2 (T/W)^2}} \qquad \textbf{[5.113]}$$

Then Eq. (5.112) becomes

$$V_{(R/C)_{max}} = \left[\frac{(T/W)(W/S)Z}{3\rho_\infty C_{D,0}}\right]^{1/2} \qquad \textbf{[5.114]}$$

Substituting Eq. (5.114) into Eq. (5.86), we have

$$(R/C)_{max} = \left[\frac{(T/W)(W/S)Z}{3\rho_\infty C_{D,0}}\right]^{1/2}$$

$$\times \left[\frac{T}{W} - \frac{1}{2}\rho_\infty \frac{(T/W)(W/S)ZC_{D,0}}{3\rho_\infty C_{D,0}(W/S)} - \frac{2(W/S)K(3\rho_\infty C_{D,0})}{\rho_\infty(T/W)(W/S)Z}\right]$$

or

$$(R/C)_{max} = \left[\frac{(T/W)(W/S)Z}{3\rho_\infty C_{D,0}}\right]\left[\frac{T}{W} - \frac{Z}{6}\frac{T}{W} - \frac{6KC_{D,0}}{(T/W)Z}\right] \qquad \textbf{[5.115]}$$

The last term in Eq. (5.115) can be written as follows, by recalling that $(L/D)^2_{\max} = 1/(4KC_{D,0})$.

$$\frac{6KC_{D,0}}{(T/W)Z} = \frac{(3/2)(T/W)4KC_{D,0}}{(T/W)^2 Z} = \frac{3T/W}{2(T/W)^2(L/D)^2_{\max}Z}$$

Hence, Eq. (5.115) becomes *for a jet-propelled airplane*

$$(R/C)_{\max} = \left[\frac{(W/S)Z}{3\rho_\infty C_{D,0}}\right]^{1/2} \left(\frac{T}{W}\right)^{3/2} \left[1 - \frac{Z}{6} - \frac{3}{2(T/W)^2(L/D)^2_{\max}Z}\right]$$

[5.116]

Equation (5.116) demonstrates that the thrust-to-weight ratio plays a powerful role in determining $(R/C)_{\max}$. Also note from Eqs. (5.112) and (5.116) that increasing the wing loading, everything else being equal, increases both $V_{(R/C)_{\max}}$ and $(R/C)_{\max}$. Indeed both $V_{(R/C)_{\max}}$ and $(R/C)_{\max}$ are proportional to $\sqrt{W/S}$. Finally, the effect of increasing altitude on $V_{(R/C)_{\max}}$ can be seen from Eq. (5.112). Assuming that T decreases with increasing altitude according to $T \propto \rho_\infty$ for a turbojet or $T \propto \rho_\infty^{0.6}$ for a turbofan, Eq. (5.112) shows that $V_{(R/C)_{\max}}$ is increased by increasing altitude. However, Eq. (5.116) clearly shows that $(R/C)_{\max}$, being dominated by the thrust-to-weight ratio, decreases with an increase in altitude.

Consider the case of a propeller-driven airplane with the power available $\eta_{\mathrm{pr}}P$ essentially constant. From Eq. (5.80) we can write

$$(R/C)_{\max} = \frac{\text{maximum excess power}}{W}$$

[5.117]

For a propeller-driven airplane with power available reasonably constant with velocity, the condition for maximum rate of climb is clearly seen in Fig. 5.36, which is an elaboration of Fig. 5.33a. For this case, maximum excess power, hence $(R/C)_{\max}$, occurs at the flight velocity for minimum power required. The conditions for minimum power required are discussed in Section 5.6.2. We have seen from Eq. (5.57) that minimum power required occurs when the airplane is flying at $(C_L^{3/2}/C_D)_{\max}$, and the flight velocity at which this occurs is given by Eq. (5.41). Hence, from Eq. (5.41), the flight velocity for maximum rate of climb is, *for a propeller-driven airplane,*

$$V_{(R/C)_{\max}} = \left(\frac{2}{\rho_\infty}\sqrt{\frac{K}{3C_{D,0}}\frac{W}{S}}\right)^{1/2}$$

[5.118]

An expression for the maximum rate of climb can be obtained by inserting Eq. (5.118) into Eq. (5.86), and noting that $TV_\infty = P_A = \eta_{\mathrm{pr}}P$ for a propeller-driven airplane, as follows. From Eq. (5.86),

$$R/C = \frac{\eta_{\mathrm{pr}}P}{W} - V_\infty \left[\frac{1}{2}\rho_\infty V_\infty^2 \left(\frac{W}{S}\right)^{-1} C_{D,0} + \frac{W}{S}\frac{2K}{\rho_\infty V_\infty^2}\right]$$

[5.119]

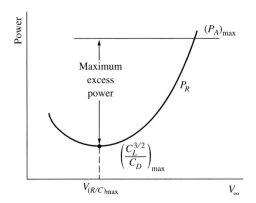

Figure 5.36 Conditions for maximum rate of climb for a propeller-driven airplane with power available constant with flight velocity.

Insert Eq. (5.118) into (5.119) to obtain $(R/C)_{max}$. At the moment, we will make this insertion only for the terms inside the square brackets in Eq. (5.119)—a convenience that will soon become apparent. Equation (5.119) becomes

$$(R/C)_{max} = \frac{\eta_{pr} P}{W}$$

$$-V_{(R/C)max} \left[\frac{1}{2}\rho_\infty \left(\frac{W}{S}\right)^{-1} C_{D,0} \frac{2}{\rho_\infty} \sqrt{\frac{K}{3C_{D,0}}\frac{W}{S}} + \frac{(W/S)2K}{\rho_\infty(2/\rho_\infty)\sqrt{K/(3C_{D,0})}(W/S)} \right]$$

which simplifies to

$$(R/C)_{max} = \frac{\eta_{pr} P}{W} - V_{(R/C)max} \left[\sqrt{\frac{KC_{D,0}}{3}} + \sqrt{3KC_{D,0}} \right] \qquad \textbf{[5.120]}$$

The last two terms in Eq. (5.120) combine as follows

$$\sqrt{\frac{KC_{D,0}}{3}} + \sqrt{3KC_{D,0}} = \left(\frac{1}{\sqrt{3}} + \sqrt{3}\right)\sqrt{KC_{D,0}} = \left(\frac{1}{\sqrt{3}} + \sqrt{3}\right)\frac{\sqrt{4KC_{D,0}}}{2}$$

$$= \frac{1/\sqrt{3} + \sqrt{3}}{2}\frac{1}{(L/D)_{max}} = \frac{1.155}{(L/D)_{max}}$$

Hence, Eq. (5.120) can be written as

$$(R/C)_{max} = \frac{\eta_{pr} P}{W} - V_{(R/C)max} \frac{1.155}{(L/D)_{max}} \qquad \textbf{[5.121]}$$

Finally, replacing $V_{(R/C)\text{max}}$ in Eq. (5.121) with Eq. (5.118), we obtain *for a propeller-driven airplane*,

$$(R/C)_{\text{max}} = \frac{\eta_{\text{pr}} P}{W} - \left[\frac{2}{\rho_\infty} \sqrt{\frac{K}{3C_{D,0}}} \left(\frac{W}{S} \right) \right]^{1/2} \frac{1.155}{(L/D)_{\text{max}}} \qquad \textbf{[5.122]}$$

Note from Eq. (5.122) that the dominant influence on $(R/C)_{\text{max}}$ is the power-to-weight ratio $\eta_{\text{pr}} P/W$. More power means a higher rate of climb—intuitively obvious. The effect of wing loading is secondary, but interesting. From Eq. (5.118), $V_{(R/C)\text{max}}$ increases with an increase in W/S. However, from Eq. (5.122), $(R/C)_{\text{max}}$ *decreases* with an increase in W/S. This is in contrast to the case of a jet-propelled airplane, where from Eq. (5.116) an increased wing loading increases $(R/C)_{\text{max}}$. Hence, propeller-driven airplanes are penalized in terms of $(R/C)_{\text{max}}$ if they have a high wing loading. Finally, the effect of increasing altitude is to increase $V_{(R/C)\text{max}}$ and decrease $(R/C)_{\text{max}}$. Even for a supercharged reciprocating engine assuming constant $\eta_{\text{pr}} P/W$ with increasing altitude, Eq. (5.122) shows that $(R/C)_{\text{max}}$ decreases with increasing altitude.

Example 5.13

For the Gulfstream IV considered in the previous examples, do the following: (*a*) Calculate and plot the rate of climb versus velocity at sea level. Also plot the hodograph diagram. From these plots, graphically obtain θ_{max}, $V_{\theta_{\text{max}}}$, $(R/C)_{\text{max}}$, and $V_{(R/C)\text{max}}$ at sea level. (*b*) Using the appropriate analytical expressions, calculate directly the values of θ_{max}, $V_{\theta_{\text{max}}}$, $(R/C)_{\text{max}}$, and $V_{(R/C)\text{max}}$ at sea level. Compare the results obtained from the graphical and analytical solutions.

Solution

(*a*) *Graphical solution.* Rate of climb is calculated from Eq. (5.80), where the excess power is the difference between the maximum power available and the power required, or

$$R/C = \frac{\text{excess power}}{W} = \frac{(P_A)_{\text{max}} - P_R}{W} = \frac{(T_A)_{\text{max}} V_\infty - D V_\infty}{W}$$

Here $(T_A)_{\text{max}} = 27{,}700$ lb and is constant; $W = 73{,}000$ lb. The power required P_R is calculated as shown in Example 5.7. In this example, we are at sea level, where $\rho_0 = 0.002377$ slug/ft^3. See Table 5.4, page 279, for $(P_A)_{\text{max}}$, P_R, and R/C at different values of V_∞.

Maximum power available, power required, and rate of climb are plotted versus velocity in Fig. 5.37. The hodograph diagram is shown in Fig. 5.38, where the same velocity scale is used on both axes. Using the same velocity scale produces a shallow hodograph curve, but this allows the measurement on Fig. 5.38 of the true angle for θ_{max}.

The graphical solutions given in Figs. 5.37 and 5.38 show that

$$\theta_{\text{max}} = 18°$$
$$V_{\theta_{\text{max}}} = 375 \text{ ft/s}$$
$$(R/C)_{\text{max}} = 180 \text{ ft/s}$$
$$V_{(R/C)\text{max}} = 750 \text{ ft/s}$$

Table 5.4

V_∞ (ft/s)	$(P_A)_{max}$ (ft·lb/s)	P_R (ft·lb/s)	R/C (ft/s)
150	4.155×10^6	2.574×10^6	21.7
200	5.540×10^6	2.023×10^6	48.2
300	8.310×10^6	1.716×10^6	90.3
400	1.108×10^7	2.028×10^6	124.0
500	1.385×10^7	2.872×10^6	150.4
600	1.662×10^7	4.288×10^6	168.9
700	1.939×10^7	6.348×10^6	178.7
750	2.078×10^7	7.648×10^6	179.8
			(essentially R/C_{max})
800	2.216×10^7	9.143×10^6	178.3
900	2.493×10^7	1.277×10^7	166.6
1,000	2.770×10^7	1.731×10^7	142.3
1,100	3.047×10^7	2.289×10^7	103.9
1,200	3.324×10^7	2.958×10^7	50.1
1,300	3.601×10^7	3.750×10^7	−20.4

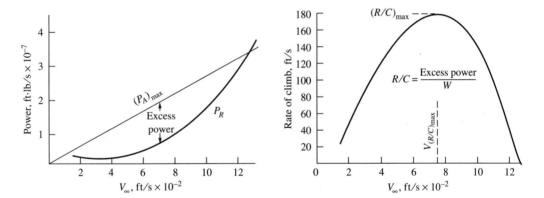

Figure 5.37 Power available, power required, and rate of climb versus flight velocity for the Gulfstream IV at sea level from Example 5.13.

(b) *Analytical solution.* Data necessary for the analytical solution are

$$\frac{T}{W} = \frac{27,700}{73,000} = 0.3795$$

$$\frac{W}{S} = \frac{73,000}{950} = 76.84$$

$$C_{D,0} = 0.015 \qquad K = 0.08$$

$$\left(\frac{L}{D}\right)_{max} = 14.43 \qquad \text{(from Example 5.4)}$$

$$\rho_\infty = 0.002377 \text{slug/ft}^3 \qquad \text{(sea level)}$$

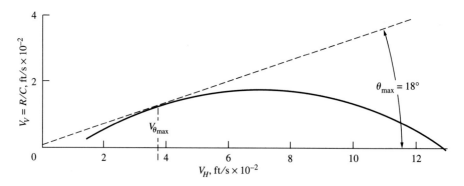

Figure 5.38   Hodograph diagram for the Gulfstream IV at sea level from Example 5.13.

From Eq. (5.94),

$$\sin\theta_{\max} = \frac{T}{W} - \frac{1}{(L/D)_{\max}} = 0.3795 - \frac{1}{14.43} = 0.3102$$

Hence,

$$\boxed{\theta_{\max} = 18.07°}$$

This result compares almost exactly with that obtained from the graphical solution, which is obtained by drawing in the tangent line shown in Fig. 5.38.

From Eq. (5.98),

$$V_{\theta_{\max}} = \sqrt{\frac{2}{\rho_\infty}\left(\frac{K}{C_{D,0}}\right)^{1/2}\frac{W}{S}\cos\theta_{\max}}$$

$$= \sqrt{\frac{2}{0.002377}\left(\frac{0.08}{0.015}\right)^{1/2}(76.84\cos 18.07°)} = \boxed{376.8 \text{ ft/s}}$$

Again, this result agrees remarkably well with the graphical solution.

From Eq. (5.116), where

$$Z = 1 + \sqrt{1 + \frac{3}{(L/D)^2_{\max}(T/W)^2}} = 1 + \sqrt{1 + \frac{3}{(14.43)^2(0.3795)^2}} = 2.0488$$

we have

$$(R/C)_{\max} = \left[\frac{(W/S)Z}{3\rho_\infty C_{D,0}}\right]^{1/2}\left(\frac{T}{W}\right)^{3/2}\left[1 - \frac{Z}{6} - \frac{3}{2(T/W)^2(L/D)^2_{\max}Z}\right]$$

$$= \left[\frac{(76.84)(2.0488)}{3(0.002377)(0.015)}\right]^{1/2}(0.3795)^{3/2}$$

$$\times\left[1 - \frac{2.0488}{6} - \frac{3}{2(0.3795)^2(14.43)^2(2.0488)}\right]$$

$$= (1,213.17)(0.23379)(1 - 0.34147 - 0.0244) = \boxed{179.86 \text{ ft/s}}$$

This agrees very well with the graphical solution.

From Eq. (5.112),

$$V_{(R/C)max} = \left\{ \frac{(T/W)(W/S)}{3\rho_\infty C_{D,0}} \left[1 + \sqrt{1 + \frac{3}{(L/D)^2_{max}(T/W)^2}} \right] \right\}^{1/2}$$

$$= \left[\frac{(T/W)(W/S)Z}{3\rho_\infty C_{D,0}} \right]^{1/2} = \left[\frac{(0.379)(76.84)(2.0488)}{3(0.002377)(0.015)} \right]^{1/2} = \boxed{747.36 \text{ ft/s}}$$

This result and that from the graphical solution agree very well.

Comment: It is conventional in aeronautics to quote rate of climb in units of feet per minute. In the above calculations, we have quoted rate of climb in the consistent units of feet per second, because those are the units that follow directly from the physical equations. Of course, the conversion is trivial, and we note that from the above results at sea level,

$$\boxed{(R/C)_{max} = 179.86 \text{ ft/s} = 10,792 \text{ ft/min}}$$

We should note that a rate of climb on the order of 10,000 ft/min is quite high for a conventional subsonic executive jet transport. The calculated value of 10,792 ft/min is due mainly to the relatively high thrust-to-weight ratio for our sample airplane. For our sample airplane, the Gulfstream IV ($T/W = 0.3795$), whereas for more typical subsonic jet transports, the values of T/W are on the order of 0.25. Clearly, for the worked examples in this chapter, we are dealing with a "hot" airplane. Also, note that the above calculation shows that this maximum rate of climb is achieved at a flight velocity of 747.36 ft/s = 510 mi/h. This means that the airplane must already be flying at high speed at sea level to achieve the calculated $(R/C)_{max}$. In actual practice, the airplane is at sea level at takeoff, and it enters its climb path at a much lower velocity than 510 mi/h. For example, at sea level the stalling velocity (from Example 5.12) is $V_{stall} = 164.5$ ft/s. This value was calculated for flaps fully deflected for landing. For takeoff, the flaps are only partially deployed to reduce the drag due to the flaps, and hence $(C_L)_{max}$ is smaller. From Table 5.3, we choose $(C_L)_{max}/\cos \Lambda = 2.1$ for takeoff, rather than the value of 2.7 used in Example 5.12 for landing. This increases V_{stall} to $(164.5)(2.7/2.1)^{1/2} = 187$ ft/s. If we assume that a safe takeoff velocity is $1.2V_{stall}$, then the airplane is flying at a velocity of 224 ft/s, or about 153 mi/h at takeoff. From Fig. 5.37, the unaccelerated rate of climb achievable by an airplane at this velocity is only 60 ft/s, or 3,600 ft/min. This value is a much more realistic estimate of the climb performance at sea level than the calculated value of $(R/C)_{max}$. It is also more consistent with the data in Ref. 36 which quotes for the Gulfstream IV a sea-level rate of climb of 4,000 ft/min.

DESIGN CAMEO

Once again we call attention to the effect of the velocity variation of T_A on the performance of the airplane. If we take into account this variation, the answer from Problem 5.20a gives the following results at sea level:

$$(R/C)_{max} = 5,028 \text{ ft/min}$$

$$V_{(R/C)max} = 440 \text{ ft/s}$$

These are to be compared with $(R/C)_{max} = 10,792$ ft/min and $V_{(R/C)max} = 747.36$ ft/s obtained in Example 5.13, which assumes that T_A is constant with velocity. Clearly, in the preliminary design process for a turbofan-powered airplane, it is essential to take into account the velocity variation of T_A.

5.10.3   Gliding (Unpowered) Flight

Whenever an airplane is flying such that the power required is *larger* than the power available, it will descend rather than climb. In the ultimate situation, there is no power at all; in this case, the airplane will be in gliding, or unpowered, flight. This will occur for a conventional airplane when the engine quits during flight (e.g., engine failure or running out of fuel). Also, this is the case for unpowered gliders and sailplanes. (Raymer in Ref. 25 adds a "cultural note" that distinguishes between sailplanes and gliders. He stated the following on p. 471: "In sailplane terminology, a 'sailplane' is an expensive, high-performance unpowered aircraft. A 'glider' is a crude, low-performance unpowered aircraft!") Gliding flight is a special (and opposite) case of our previous considerations dealing with climb; it is the subject of this subsection.

The force diagram for an unpowered aircraft in descending flight is shown in Fig. 5.39. For steady, unaccelerated descent, where θ is the equilibrium glide angle,

$$L = W \cos \theta \qquad\qquad\qquad \textbf{[5.123]}$$

$$D = W \sin \theta \qquad\qquad\qquad \textbf{[5.124]}$$

The equilibrium glide angle is obtained by dividing Eq. (5.124) by Eq. (5.123).

$$\frac{\sin \theta}{\cos \theta} = \frac{D}{L}$$

or

$$\boxed{\text{Tan } \theta = \frac{1}{L/D}} \qquad\qquad \textbf{[5.125]}$$

Clearly, the glide angle is strictly a function of the lift-to-drag ratio; the higher the L/D, the shallower the glide angle. From Eq. (5.125), the smallest equilibrium glide angle occurs at $(L/D)_{max}$.

$$\boxed{\text{Tan } \theta_{min} = \frac{1}{(L/D)_{max}}} \qquad\qquad \textbf{[5.126]}$$

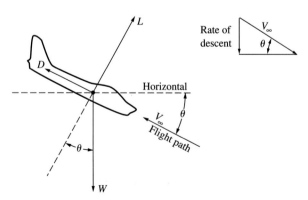

Figure 5.39   Force and velocity diagram for gliding flight.

For an aircraft at a given altitude h, this is the case for maximum horizontal distance covered over the ground (maximum range). This distance, denoted by R, is illustrated in Fig. 5.40 for a constant θ.

The simplicity reflected in Eqs. (5.125) and (5.126) is beautiful. The equilibrium glide angle θ does not depend on altitude or wing loading, or the like; it simply depends on the lift-to-drag ratio. However, to achieve a given L/D at a given altitude, the aircraft must fly at a specified velocity V_∞, called the *equilibrium glide velocity*, and this value of V_∞ *does* depend on the altitude and wing loading, as follows. Since

$$L = \tfrac{1}{2}\rho_\infty V_\infty^2 S C_L$$

Eq. (5.123) becomes

$$\tfrac{1}{2}\rho_\infty V_\infty^2 S C_L = W \cos\theta$$

or

$$V_\infty = \sqrt{\frac{2\cos\theta}{\rho_\infty C_L}\frac{W}{S}} \qquad [5.127]$$

In Eq. (5.127), V_∞ is the equilibrium glide velocity. Clearly, it depends on altitude (through ρ_∞) and wing loading. The value of C_L in Eq. (5.127) is that particular value which corresponds to the specific value of L/D used in Eq. (5.125). Recall that both C_L and L/D are aerodynamic characteristics of the aircraft that vary with angle of attack, as sketched in Fig. 5.41. Note from Fig. 5.41 that a specific value of L/D, say $(L/D)_1$, corresponds to a specific angle of attack α_1, which in turn dictates the lift coefficient $(C_L)_1$. If L/D is held constant throughout the glide path, then C_L is constant along the glide path. However, the equilibrium velocity along this glide path will change with altitude, decreasing with decreasing altitude.

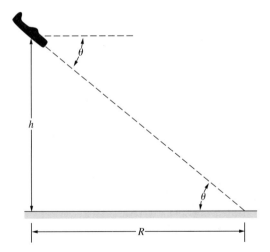

Figure 5.40   Range covered in an equilibrium guide.

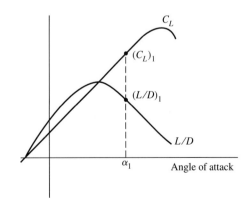

Figure 5.41   Sketch of the variation of C_L and L/D versus angle of attack for a given airplane.

Consider again the case for a minimum glide angle as treated by Eq. (5.126). For a typical modern airplane, $(L/D)_{max} = 15$, and for this case from Eq. (5.126), $\theta_{min} = 3.8°$—a small angle. Hence, we can reasonably assume $\cos \theta = 1$ for such cases. Recall from Eq. (5.30) that

$$\left(\frac{L}{D}\right)_{max} = \sqrt{\frac{1}{4C_{D,0}K}} \qquad \text{[5.30]}$$

and for $L = W$ (consistent with the assumption of $\cos \theta = 1$), the velocity at which L/D is maximum is given by Eq. (5.34)

$$V_{(L/D)_{max}} = \left(\frac{2}{\rho_\infty}\sqrt{\frac{K}{C_{D,0}}\frac{W}{S}}\right)^{1/2} \qquad \text{[5.34]}$$

Hence, for θ_{min}, the equilibrium velocity along the glide path is given by Eq. (5.34).

Example 5.14 | Consider the Gulfstream IV flying at 30,000 ft. Assume a total loss of engine thrust. Calculate (a) the minimum glide path angle, (b) the maximum range covered over the ground, and (c) the corresponding equilibrium glide velocity at 30,000 ft and at sea level.

Solution

(a) From Eq. (5.126)

$$\text{Tan } \theta_{min} = \frac{1}{(L/D)_{max}}$$

From Example (5.4), $(L/D)_{max} = 14.43$. Hence

$$\text{Tan } \theta_{min} = \frac{1}{14.43} = 0.0693$$

$$\boxed{\theta_{min} = 3.964°}$$

(b) From Fig. 5.40,

$$\frac{h}{R} = \text{Tan } \theta$$

Hence,

$$R_{max} = \frac{h}{\text{Tan } \theta_{min}} = \frac{30,000}{\text{Tan } \theta_{min}} = \frac{30,000}{0.0693} = \boxed{432,900 \text{ ft} = 82 \text{ mi}}$$

(c) At 30,000 ft, $\rho_\infty = 8.9068 \times 10^{-4}$ slug/ft³. From Eq. (5.34),

$$V_{(L/D)_{max}} = \left(\frac{2}{\rho_\infty}\sqrt{\frac{K}{C_{D,0}}\frac{W}{S}}\right)^{1/2} = \left[\frac{2}{8.9068 \times 10^{-4}}\sqrt{\frac{0.08}{0.015}}(76.84)\right]^{1/2} = \boxed{631.2 \text{ ft/s}}$$

At sea level, $\rho_\infty = 0.002377$ slug/ft³. Note that in Eq. (5.34), the only quantity that changes is ρ_∞. Hence, we can write from Eq. (5.34),

$$\left[V_{(L/D)\text{max}}\right]_{\text{sea level}} = \left[\frac{(\rho_\infty)_{30,000 \text{ ft}}}{(\rho_\infty)_{\text{sea level}}}\right]^{1/2} \left[V_{(L/D)\text{max}}\right]_{30,000 \text{ ft}}$$

$$= \left(\frac{8.9068 \times 10^{-4}}{0.002377}\right)^{1/2} (631.2) = \boxed{386.4 \text{ ft/s}}$$

The rate of descent, sometimes called the *sink rate*, is the downward vertical velocity of the airplane V_V. It is, for unpowered flight, the analog of rate of climb for powered flight. As seen in the insert in Fig. 5.39,

$$\text{Rate of descent} = V_V = V_\infty \sin\theta \qquad \textbf{[5.128]}$$

Rate of descent is a positive number in the downward direction. Multiplying Eq. (5.124) by V_∞ and inserting Eq. (5.128), we have

$$DV_\infty = WV_\infty \sin\theta = WV_V$$

or

$$V_V = \frac{DV_\infty}{W} \qquad \textbf{[5.129]}$$

By making the assumption of $\cos\theta = 1$, in Eq. (5.129), DV_∞ is simply the power required for steady, level flight. Hence, the variation of V_V with velocity is the same as the power required curve, divided by the weight. This variation is sketched in Fig. 5.42, with positive values of V_V increasing along the downward vertical axis (just to emphasize that the sink rate V_V is in the downward direction). Clearly, minimum sink rate occurs at the flight velocity for minimum power required. Hence the conditions for minimum sink rate are the same as those for $(P_R)_{\text{min}}$, which from Eqs. (5.41) and (5.57) are

1. $\dfrac{C_L^{3/2}}{C_D}$ is maximum

2. $(V_\infty)_{\text{min sink rate}} = \left(\dfrac{2}{\rho_\infty}\sqrt{\dfrac{K}{3C_{D,0}}\dfrac{W}{S}}\right)^{1/2}$

The hodograph diagram is sketched in Fig. 5.43 where a line from the origin tangent to the hodograph curve defines θ_{min}. This sketch is shown just to emphasize that the minimum sink rate does *not* correspond to the minimum glide angle. The flight velocity for the minimum sink rate (corresponding to maximum $C_L^{3/2}/C_D$) is less than that for minimum glide angle (corresponding to maximum C_L/C_D).

An analytical expression for the sink rate V_V can be obtained as follows. From Eq. (5.123)

$$W\cos\theta = L = \tfrac{1}{2}\rho_\infty V_\infty^2 S C_L$$

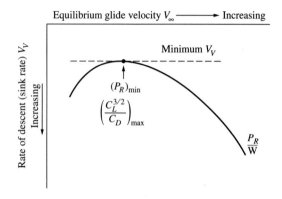

Figure 5.42   Rate of descent versus equilibrium guide velocity.

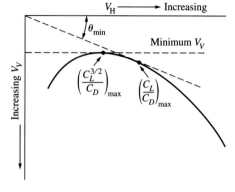

Figure 5.43   Hodograph for unpowered flight.

or

$$V_\infty = \sqrt{\frac{2W \cos\theta}{\rho_\infty S C_L}}$$ [5.130]

Substituting Eq. (5.130) into Eq. (5.128), we have

$$V_V = V_\infty \sin\theta = (\sin\theta)\sqrt{\frac{2\cos\theta}{\rho_\infty C_L} \frac{W}{S}}$$ [5.131]

Dividing Eq. (5.124) by Eq. (5.123), we obtain

$$\sin\theta = \frac{D}{L}\cos\theta = \frac{C_D}{C_L}\cos\theta$$ [5.132]

Inserting Eq. (5.132) into Eq. (5.131), we have

$$V_V = \sqrt{\frac{2\cos^3\theta}{\rho_\infty(C_L^3/C_D^2)} \frac{W}{S}}$$ [5.133]

By making the assumption that $\cos\theta = 1$, Eq. (5.133) is written as

$$V_V = \sqrt{\frac{2}{\rho_\infty(C_L^3/C_D^2)} \frac{W}{S}}$$ [5.134]

Equation (5.134) explicitly shows that $(V_V)_{\min}$ occurs at $(C_L^{3/2}/C_D)_{\max}$. It also shows that the sink rate decreases with decreasing altitude and increases as the square root of the wing loading.

Example 5.15   For the unpowered Gulfstream IV at 30,000 ft, calculate (*a*) the sink rate for the case of minimum glide angle and (*b*) the minimum sink rate.

Solution

(a) From Example 5.14, $\theta_{\min} = 3.964°$ and $(V_\infty)_{\theta_{\min}} = 631.2$ ft/s. Hence,

$$V_V = (V_\infty)_{\theta_{\min}} \sin \theta_{\min} = 631.2 \sin 3.964° = \boxed{43.6 \text{ ft/s}}$$

(b) From Example 5.4, $(C_L^{3/2}/C_D)_{\max} = 10.83$. Hence from Eq. (5.134),

$$(V_V)_{\min} = \sqrt{\frac{2}{\rho_\infty (C_L^3/C_D^2)_{\max}} \frac{W}{S}} = \sqrt{\frac{2(76.84)}{(8.9068 \times 10^{-4})(10.83)^2}}$$

$$= \sqrt{\frac{2(76.84)}{(8.9068 \times 10^{-4})(10.83)^2}} = \boxed{38.6 \text{ ft/s}}$$

Note that, as expected, the minimum sink rate of 38.6 ft/s is smaller than the sink rate of 43.6 ft/s for minimum glide angle.

Glider pilots take advantage of the different sink rates discussed above. When flying through an upward-lifting thermal, they fly at the velocity for minimum sink rate, so as to gain the greatest altitude. Out of the thermal, they accelerate to the flight velocity for minimum glide angle in order to cover the greatest distance before encountering the next thermal.

5.11 SERVICE AND ABSOLUTE CEILINGS

How high can an airplane fly in steady, level flight? The answer is straightforward—that altitude where the maximum rate of climb is zero is the highest altitude achievable in steady, level flight. This altitude is defined as the *absolute ceiling*, that is, that altitude where $(R/C)_{\max} = 0$. A more useful quantity is the *service ceiling*, conventionally defined as that altitude where $(R/C)_{\max} = 100$ ft/min. The service ceiling represents the *practical* upper limit for steady, level flight.

The absolute and service ceilings are denoted in Fig. 5.44, which also illustrates a simple graphical technique for finding these ceilings. In Fig. 5.44, the maximum rate of climb (on the abscissa) is plotted versus altitude (on the ordinate); for many conventional airplanes, this variation is almost (but not precisely) linear. The graphical solution for service and absolute ceilings is straightforward. For a given airplane:

1. Calculate $(R/C)_{\max}$ at a number of different altitudes. This calculation can be made by either the graphical or analytical solution discussed in Section 5.10.
2. Plot the results in the form shown in Fig. 5.44.
3. Extrapolate the curve to a value of $(R/C)_{\max} = 100$ ft/min, denoted by point 1 in Fig. 5.44. The corresponding value of h at point 1 is the service ceiling.
4. Extrapolate the curve to a value of $(R/C)_{\max} = 0$, denoted by point 2 in Fig. 5.44. The corresponding value of h at point 2 is the service ceiling.

An analytical solution is also straightforward. For a jet-propelled airplane, $(R/C)_{\max}$ is given by Eq. (5.116). The free-stream density ρ_∞ appears explicitly in the first term and implicitly through the altitude variation of T. By inserting $(R/C)_{\max} = 0$ in the left-hand side of Eq. (5.116), ρ_∞ can be obtained by solving

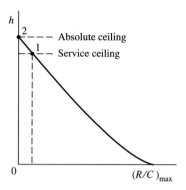

Figure 5.44   Sketch of variation of maximum rate of climb with altitude, illustrating absolute and service ceilings.

Eq. (5.116) by a trial-and-error, iterative process. In turn, this value of ρ_∞ will define the standard altitude at which $(R/C)_{max} = 0$, that is, the absolute ceiling. A similar process, wherein $(R/C)_{max} = 100$ ft/min $= 1.67$ ft/s is inserted in the left-hand side of Eq. (5.116), leads to a solution for the service ceiling. For a propeller-driven airplane, the absolute and service ceilings can be obtained from Eq. (5.122) by means of a similar trial-and-error solution.

Example 5.16

For the Gulfstream IV discussed in previous examples, plot the variation of $(R/C)_{max}$ versus altitude, and use this curve to graphically obtain the absolute ceiling. Also plot the variation of $V_{(R/C)_{max}}$ versus altitude.

Solution

Equation (5.116) is used to calculate $(R/C)_{max}$ at different altitudes from sea level to 60,000 ft, in increments of 2,000 ft. Similarly, Eq. (5.112) is used to calculate $V_{(R/C)_{max}}$ at the same altitudes. The data that are inserted in Eqs. (5.112) and (5.116) are the same as those used in Example 5.3, part (b), except that the appropriate values of ρ_∞ are used for the different altitudes, as obtained from Appendix B. The following is an abridged tabulation of the results.

h (ft)	$(R/C)_{max}$ (ft/s)	$V_{(R/C)_{max}}$ (ft/s)
0	179.9	747.4
10,000	156.6	798.0
20,000	133.8	858.3
30,000	111.0	931.9
40,000	85.9	1,033.4
50,000	58.2	1,176.6
60,000	30.1	1,358.7

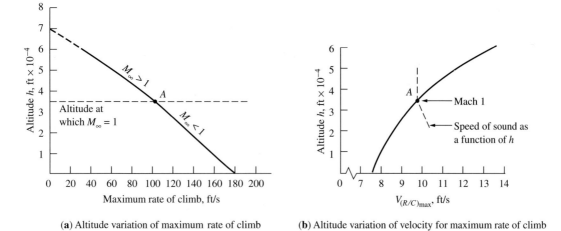

(a) Altitude variation of maximum rate of climb

(b) Altitude variation of velocity for maximum rate of climb

Figure 5.45 Variations of maximum rate of climb and the corresponding velocity with altitude, from Example 5.16.

All the calculated results are plotted in Fig. 5.45. In Fig. 5.45**a**, $(R/C)_{\text{max}}$ is plotted versus altitude, and in Fig. 5.45**b** the corresponding $V_{(R/C)_{\text{max}}}$ is shown versus altitude. The results in Fig. 5.45**a** are extrapolated to $(R/C)_{\text{max}} = 0$, yielding

$$\text{Absolute ceiling} = 70,000 \text{ ft}$$

This is the theoretical result based on the conventional drag polar for our given airplane. This, in combination with the high T/W for the Gulfstream IV, yields an inordinately high value for the absolute ceiling. In reality, an absolute ceiling of 70,000 ft will never be achieved, because compressibility effects become important at the higher altitudes. How does this happen?

For an answer, recall that $V_{(R/C)_{\text{max}}}$ increases with altitude; indeed, this is clearly seen in Fig. 5.45**b**, where the solid curve is the plot of $V_{(R/C)_{\text{max}}}$ versus altitude. Also shown in Fig. 5.45**b** is the variation of the free-stream speed of sound versus altitude, given for a limited range by the dashed line in Fig. 5.45**b**. At the intersection of these two curves, labeled point A, the flight velocity equals the speed of sound; that is, the airplane goes through Mach 1. For the present conditions, this occurs at an altitude of about 35,000 ft. For higher altitudes, where the present calculations predict an even higher velocity, drag-divergence effects will prevail. Hence, in reality the flight velocity will remain subsonic. Our present calculations do not include the drag-divergence effect, and hence the calculations for altitudes above 35,000 ft do not reflect the real situation. That is, in Fig 5.45**a** and **b**, the results at altitudes above 35,000 ft (above point A) are ramifications of our conventional drag polar without drag divergence, and hence are of academic interest only. Their purpose here is only to help illustrate the conventional technique. In reality, above 35,000 ft, drag divergence will prevent the airplane from flying at the theoretical velocity required to obtain maximum rate of climb. Hence, it will climb at a lower R/C, and the absolute ceiling will be less, indeed in the present calculation much less, than the absolute ceiling of 70,000 ft/s predicted in Fig. 5.45**a**. The practical absolute ceiling will not be much higher than 35,000 ft. In Ref. 36, the maximum operating altitude of the Gulfstream IV is listed as 45,000 ft.

5.12   TIME TO CLIMB

The rate of climb, by definition, is the vertical component of the airplane's velocity, which is simply the time rate of change of altitude dh/dt. Hence,

$$\frac{dh}{dt} = R/C$$

or

$$dt = \frac{dh}{R/C} \qquad [5.135]$$

In Eq. (5.135), R/C is a function of altitude, and dt is the small increment in time required to climb the small height dh at a given instantaneous altitude. The time to climb from one altitude h_1 to another h_2 is obtained by integrating Eq. (5.135) between the two altitudes:

$$t = \int_{h_1}^{h_2} \frac{dh}{R/C} \qquad [5.136]$$

Normally, the performance characteristic labeled *time to climb* is considered from sea level, where $h_1 = 0$. Hence, the time to climb from sea level to any given altitude h_2 is, from Eq. (5.136),

$$t = \int_0^{h_2} \frac{dh}{R/C} \qquad [5.137]$$

If in Eq. (5.137) the maximum rate of climb is used at each altitude, then t becomes the minimum time to climb to altitude h_2.

$$t_{\min} = \int_0^{h_2} \frac{dh}{(R/C)_{\max}} \qquad [5.138]$$

5.12.1   Graphical Approach

Consider a plot of $(R/C)^{-1}$ versus altitude, as shown in Fig. 5.46. The time to climb to altitude h_2 is simply the area under the curve, shown by the shaded area in Fig. 5.46.

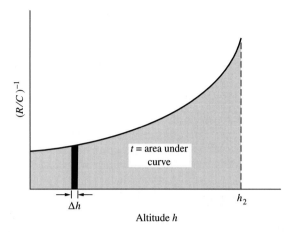

Figure 5.46 Graphical representation of the time to climb to altitude h_2.

Example 5.17

Using a graphical approach, calculate the minimum time to climb to 30,000 ft for the Gulfstream IV based on the data from previous examples.

Solution

The integral in Eq. (5.138) can be numerically evaluated by dividing the area shown in Fig. 5.46 into a large number of small vertical segments of width Δh and local height equal to $(R/C)^{-1}$. The area of each segment is then $\Delta h/(R/C)$. In turn,

$$t = \int_0^{h_2} \frac{dh}{R/C} \approx \sum_{i=1}^{n} \left(\frac{\Delta h}{R/C} \right)_i \qquad \text{[5.139]}$$

where n is the number of segments chosen. Since in Example 5.16 we calculated $(R/C)_{\text{max}}$ at every 2,000-ft interval, for convenience we choose here $\Delta h = 2,000$ ft. Also, for $(R/C)_i$ in Eq. (5.139), we use an average value of the rate of climb for each segment. For example, for the first segment from $h = 0$ to $h = 2,000$ ft,

$$\left(\frac{\Delta h}{R/C} \right)_1 = \frac{2,000}{\frac{1}{2} \left[(R/C)_0 + (R/C)_{2,000} \right]}$$

For the second segment,

$$\left(\frac{\Delta h}{R/C} \right)_2 = \frac{2,000}{\frac{1}{2} \left[(R/C)_{2,000} + (R/C)_{4,000} \right]}$$

and so forth. When this is carried out over the 15 segments from $h = 0$ to $h_2 = 30,000$ ft, using the values of $(R/C)_{\text{max}}$ calculated in Example 5.16, we have

$$t_{\min} = \int_0^{30,000} \frac{dh}{(R/C)_{\text{max}}} = \sum_{i=1}^{15} \left[\frac{2,000}{(R/C)_{\text{max}}} \right]_i = 210.5 \text{ s} = \boxed{3.51 \text{ min}}$$

Note: The minimum time to climb to 30,000 ft of 3.51 min calculated here is small because $(R/C)_{max}$ has been used at each altitude, and we already noted in the discussion following Example 5.13 that the actual airplane would have values of rate of climb lower than $(R/C)_{max}$.

5.12.2   Analytical Approach

The governing relation for time to climb is given by Eq. (5.136), with the minimum time to climb given by Eq. (5.138). There is no exact analytical formula for t that can be obtained from these equations because of the nonlinear variation of rate of climb with altitude. However, examination of Fig. 5.45a shows that, in an *approximate fashion*, the variation of $(R/C)_{max}$ is nearly linear with altitude. If we make the approximation, we can write

$$(R/C)_{max} = a + bh \qquad \textbf{[5.140]}$$

where h is altitude, a is the $h = 0$ intercept on the abscissa in Fig. 5.45a, and b is the slope of the approximate linear curve. Substituting Eq. (5.140) into Eq. (5.138), we have

$$t_{min} = \int_0^{h_2} \frac{dh}{a + bh} = \frac{1}{b}[\ln(a + bh_2) - \ln a] \qquad \textbf{[5.141]}$$

Example 5.18

Using the analytical approach described above, calculate the minimum time to climb to 30,000 ft for the Gulfstream IV, and compare your answer to the graphical result from Example 5.17.

Solution

To find the values of a and b in Eq. (5.140), refer to Fig. 5.45a; or better yet, see the table in Example 5.16, where

$$a = 179.9 \text{ ft/s}$$

Let us approximate the slope b by using the values of $(R/C)_{max}$ at 20,000 and 50,000 ft (a rather arbitrary choice), which are 133.8 and 58.2 ft/s, respectively.

$$b = \frac{58.2 - 133.8}{50,000 - 20,000} = -2.52 \times 10^{-3} \text{ s}^{-1}$$

Hence, the form of Eq. (5.140) used for this example is

$$(R/C)_{max} = 179.9 - 2.52 \times 10^{-3}h$$

[Note that, just as a check, the above relation evaluated at $h = 10,000$ ft gives $(R/C)_{max} = 154.7$ ft/s. The exact value from the table in Example 5.16 is $(R/C)_{max} = 156.6$ ft/s. Our approximate linear curve is accurate to within 1.2% at this altitude.] From Eq. (5.141), we have

$$t_{min} = \frac{1}{b}[\ln(a + bh_2) - \ln a]$$

$$= \frac{1}{-2.52 \times 10^{-3}}\left\{\ln[179.9 - (2.52 \times 10^{-3})(30{,}000)] - \ln 179.9\right\}$$

$$= 216.3 \text{ s} = \boxed{3.61 \text{ min}}$$

This is to be compared to the graphical result of $t_{min} = 3.51$ min obtained from Example 5.17. Our approximate analytical calculation agrees to within 2.8%.

5.13 RANGE

Imagine that you are getting ready to fly across the Atlantic Ocean, say, leaving Dulles airport in Washington, D.C., and flying to Heathrow airport near London. You may be going for business or pleasure, but in either case, when you step on the airplane and it takes off, you will not touch land again until you have covered the 3,665 mi between Dulles and Heathrow. You take for granted that the airplane can cover this distance without running out of fuel. Indeed, the aeronautical engineers who designed the airplane made certain that you will cover this distance on one load of fuel. They designed the airplane to have enough range to cross the Atlantic and get you safely to London.

How did they do this? What airplane design features and operating parameters ensure that you will cover enough distance to arrive safely at Heathrow? The general answer to this question is the subject of this section.

By definition, *range* is the total distance (measured with respect to the ground) traversed by an airplane on one load of fuel. We denote the range by R. We also consider the following weights:

W_0—gross weight of the airplane including *everything*; full fuel load, payload, crew, structure, etc.

W_f—weight of fuel; this is an instantaneous value, and it changes as fuel is consumed during flight.

W_1—weight of the airplane when the fuel tanks are empty.

At any instant during the flight, the weight of the airplane W is

$$W = W_1 + W_f \qquad \textbf{[5.142]}$$

Since W_f is decreasing during the flight, W is also decreasing. Indeed, the time rate of change of weight is, from Eq. (5.142),

$$\frac{dW}{dt} = \frac{dW_f}{dt} = \dot{W}_f \qquad \textbf{[5.143]}$$

where both dW/dt and $\dot{W}_f$ are negative numbers because fuel is being consumed, and hence both W and W_f are decreasing.

Range is intimately connected with engine performance through the specific fuel consumption, defined in Chapter 3. For a propeller-driven/reciprocating engine combination, the specific fuel consumption is defined by Eq. (3.37), repeated in a slightly different form here:

$$c \equiv -\frac{\dot{W}_f}{P} \qquad\qquad \text{[5.144]}$$

where P is the shaft power and the minus sign is necessary because $\dot{W}_f$ is negative and c is always treated as a positive quantity. For a jet-propelled airplane, the thrust specific fuel consumption is defined by Eq. (3.38), repeated in a similarly modified form here:

$$c_t \equiv -\frac{\dot{W}_f}{T} \qquad\qquad \text{[5.145]}$$

where T is the thrust available. However, as shown in Section 3.7, c can be expressed in terms of c_t and vice versa, via Eq. (3.43), repeated here:

$$c_t = \frac{cV_\infty}{\eta_{\text{pr}}} \qquad\qquad \text{[3.43]}$$

where η_{pr} is the propeller efficiency. Equation (3.43) is particularly useful for generating an equivalent "thrust" specific fuel consumption for propeller-driven airplanes. It is useful to review these matters from Section 3.7 before going further.

A general relation for the calculation of range can be obtained as follows. Consider an airplane in steady, level flight. Let s denote horizontal distance covered over the ground. Assuming a stationary atmosphere (no wind), the airplane's velocity V_∞ is

$$V_\infty = \frac{ds}{dt}$$

or

$$ds = V_\infty \, dt \qquad\qquad \text{[5.146]}$$

Return to Eq. (5.145), from which

$$c_t = -\frac{dW_f/dt}{T}$$

or

$$dt = -\frac{dW_f}{c_t T} \qquad\qquad \text{[5.147]}$$

Substitute Eq. (5.147) into Eq. (5.146).

$$ds = -\frac{V_\infty}{c_t T} \, dW_f \qquad\qquad \text{[5.148]}$$

From Eq. (5.142), $dW_f = dW$. Equation (5.148) then becomes

$$ds = -\frac{V_\infty}{c_t T} dW = -\frac{V_\infty}{c_t} \frac{W}{T} \frac{dW}{W} \qquad \textbf{[5.149]}$$

In steady, level flight, $L = W$ and $T = D$. Hence Eq. (5.149) can be written as

$$ds = -\frac{V_\infty}{c_t} \frac{L}{D} \frac{dW}{W} \qquad \textbf{[5.150]}$$

The range of the airplane is obtained by integrating Eq. (5.150) between $s = 0$, where the fuel tanks are full and hence $W = W_0$, and $s = R$, where the fuel tanks are empty and hence $W = W_1$.

$$R = \int_0^R ds = -\int_{W_0}^{W_1} \frac{V_\infty}{c_t} \frac{L}{D} \frac{dW}{W}$$

or

$$R = \int_{W_1}^{W_0} \frac{V_\infty}{c_t} \frac{L}{D} \frac{dW}{W} \qquad \textbf{[5.151]}$$

where W_0 is the gross weight (with full fuel tanks) and W_1 is the weight with the fuel tanks empty. Equation (5.151) is a general equation for range; the only restriction is for steady, level flight with no headwinds or tailwinds. Equation (5.151) holds for a jet-propelled airplane with c_t given directly by the engine performance, and for a propeller-driven airplane with a reciprocating engine, where an effective c_t can be obtained from c via Eq. (3.43).

Some parameters that influence range are clearly evident from Eq. (5.151). Not surprisingly, range is influenced by the lift-to-drag ratio, specific fuel consumption, velocity, and the initial amount of fuel (the difference between W_0 and W_1). However, these parameters are not all independent of one another. For example, L/D depends on angle of attack, which depends on V_∞, W, and altitude. For a given flight, if the variation of L/D, V_∞, c_t, and W are known throughout the duration of the flight, then Eq. (5.151) can be numerically integrated to *exactly* calculate the range.

For a preliminary performance analysis, Eq. (5.151) is usually simplified. If we assume flight at constant V_∞, c_t, and L/D, Eq. (5.151) becomes

$$R = \frac{V_\infty}{c_t} \frac{L}{D} \int_{W_1}^{W_0} \frac{dW}{W}$$

or

$$R = \frac{V_\infty}{c_t} \frac{L}{D} \ln \frac{W_0}{W_1} \qquad \textbf{[5.152]}$$

Equation (5.152) is frequently called the *Breguet range equation*, although the earliest form of the Breguet equation appeared at the end of World War I and was written in a slightly different form pertaining directly to propeller-driven airplanes with reciprocating engines. We will address such matters in Section 5.13.1.

At first glance, it would appear from Eq. (5.152) that to obtain the largest possible range, we would want to fly simultaneously at the highest possible velocity and at the largest possible value of L/D. Indeed, Eq. (5.152) clearly states the importance of high aerodynamic efficiency (high L/D) in obtaining large range. However, V_∞ and L/D are not independent. Keep in mind that for a given airplane L/D varies with angle of attack, which in turn changes as V_∞ changes in level flight. Hence, L/D is a function of V_∞ in this case. From Eq. (5.152), to obtain maximum range, we need to fly at a condition where the product $V_\infty(L/D)$ is maximized. This condition is different for propeller-driven and jet-propelled airplanes, and therefore we must consider each category of aircraft in turn. This is the subject of the next two subsections.

5.13.1 Range for Propeller-Driven Airplanes

As discussed in Chapter 3, the specific fuel consumption for propeller/reciprocating engine power plants is fundamentally expressed in terms of power and is given by Eq. (5.144). Hence, it is convenient to express the range equation for propeller-driven airplanes in terms of the specific fuel consumption c, rather than the thrust specific fuel consumption c_t. The relation between c and c_t is given by Eq. (3.43). Hence, Eq. (5.152) can be expressed as

$$R = \frac{V_\infty}{c_t}\frac{L}{D}\ln\frac{W_0}{W_1} = \frac{\eta_{\mathrm{pr}}}{cV_\infty}V_\infty\frac{L}{D}\ln\frac{W_0}{W_1}$$

or

$$\boxed{R = \frac{\eta_{\mathrm{pr}}}{c}\frac{L}{D}\ln\frac{W_0}{W_1}} \qquad\qquad \textbf{[5.153]}$$

Equation (5.153) is the *historical* Breguet range equation, and it dates back to before 1920.

Important: Note that V_∞ does not appear in Eq. (5.153); it is canceled when c is used instead of c_t. Therefore, the clear impact of the lift-to-drag ratio on range for a propeller-driven airplane is blatantly seen in Eq. (5.153).

For an airplane with a propeller/reciprocating engine power plant, how do you obtain the largest possible range? Equation (5.153) tells the story, namely, for maximum range:

1. Fly at maximum L/D.
2. Have the highest possible propeller efficiency.
3. Have the lowest possible specific fuel consumption.
4. Have the highest possible ratio of gross weight to empty weight (i.e., carry a lot of fuel).

The flight conditions associated with $(L/D)_{\mathrm{max}}$ have been discussed in Section 5.4.1. It follows that the theoretical maximum range for a propeller-driven airplane is obtained by flying at the velocity where zero-lift drag equals the drag due to lift, that is, where, from Eq. (5.28)

$$C_{D,0} = K C_L^2 \qquad \text{[5.28]}$$

This velocity is given by Eq. (5.34).

$$V_{(L/D)_{\max}} = \left(\frac{2}{\rho_\infty} \sqrt{\frac{K}{C_{D,0}} \frac{W}{S}} \right)^{1/2} \qquad \text{[5.34]}$$

The value of maximum lift-to-drag ratio is given by Eq. (5.30).

$$\left(\frac{L}{D} \right)_{\max} = \left(\frac{C_L}{C_D} \right)_{\max} = \sqrt{\frac{1}{4 C_{D,0} K}} \qquad \text{[5.30]}$$

5.13.2 Range for Jet-Propelled Airplanes

The simplified range equation for a jet-propelled airplane is given by Eq. (5.152), which is written directly in terms of the *thrust* specific fuel consumption c_t. Clearly, maximum range for a jet is *not* dictated by maximum L/D, but rather the maximum value of the product $V_\infty (L/D)$. Let us examine this product. For steady, level flight,

$$L = W = \tfrac{1}{2} \rho_\infty V_\infty^2 S C_L$$

or

$$V_\infty = \sqrt{\frac{2W}{\rho_\infty S C_L}}$$

Thus,

$$V_\infty \frac{L}{D} = \sqrt{\frac{2W}{\rho_\infty S C_L}} \frac{C_L}{C_D} = \sqrt{\frac{2W}{\rho_\infty S}} \frac{C_L^{1/2}}{C_D} \qquad \text{[5.154]}$$

Thus the product $V_\infty (L/D)$ is maximum when the airplane is flying at a maximum value of $C_L^{1/2}/C_D$.

Using Eq. (5.154), we obtain a more explicitly useful expression for the range of a jet-propelled airplane. Since Eq. (5.154) involves W, and Eq. (5.152) has already been integrated with respect to W, we have to return to the general range equation given by Eq. (5.151). Substituting Eq. (5.154) into Eq. (5.151) gives

$$R = \int_{W_1}^{W_0} \frac{1}{c_t} \sqrt{\frac{2W}{\rho_\infty S}} \frac{C_L^{1/2}}{C_D} \frac{dW}{W} \qquad \text{[5.155]}$$

Assuming c_t, ρ_∞, S, and $C_L^{1/2}/C_D$ are constant, Eq. (5.155) can be written as

$$R = \frac{1}{c_t} \sqrt{\frac{2}{\rho_\infty S}} \frac{C_L^{1/2}}{C_D} \int_{W_1}^{W_0} \frac{dW}{W^{1/2}}$$

or

$$R = \frac{2}{c_t} \sqrt{\frac{2}{\rho_\infty S} \frac{C_L^{1/2}}{C_D}} \left(W_0^{1/2} - W_1^{1/2} \right) \qquad \textbf{[5.156]}$$

Equation (5.156) is a simplified range equation for a jet-propelled airplane. From this equation, the flight conditions for maximum range for a jet-propelled airplane are

1. Fly at maximum $C_L^{1/2}/C_D$.
2. Have the lowest possible thrust specific fuel consumption.
3. Fly at high altitude, where ρ_∞ is small.
4. Carry a lot of fuel.

Note that Eq. (5.153) for the range of a propeller-driven airplane does not explicitly depend on ρ_∞, and hence the influence of altitude appears only implicitly via the altitude effects on η_{pr} and c. However, ρ_∞ appears directly in Eq. (5.156) for the range of a jet-propelled airplane, and hence the altitude has a first-order effect on range. This explains why, in part, when you fly in your jumbo jet across the Atlantic Ocean to London, you cruise at altitudes above 30,000 ft instead of skimming across the tops of the waves. Of course, Eq. (5.156), when taken in the limit of ρ_∞ going to zero, shows the range going to infinity. As you might expect, this is nonsense. The highest altitude that a given airplane can reach is limited by its absolute ceiling, and flight near the absolute ceiling does not yield maximum range.

The flight conditions associated with $(C_L^{1/2}/C_D)_{\max}$ have been discussed in Section 5.4.1. It follows that the theoretical maximum range for a jet-propelled airplane is obtained by flying at the velocity where the zero-lift drag is 3 times the drag due to lift, that is, where

$$C_{D,0} = 3K C_L^2 \qquad \textbf{[5.43]}$$

The velocity is given by Eq. (5.45).

$$V_{(C_L^{1/2}/C_D)_{\max}} = \left(\frac{2}{\rho_\infty} \sqrt{\frac{3K}{C_{D,0}} \frac{W}{S}} \right)^{1/2} \qquad \textbf{[5.45]}$$

The value of $(C_L^{1/2}/C_D)_{\max}$ is given by Eq. (5.44).

$$\left(\frac{C_L^{1/2}}{C_D} \right)_{\max} = \frac{3}{4} \left(\frac{1}{3K C_{D,0}^3} \right)^{1/4} \qquad \textbf{[5.44]}$$

Recall from Section 5.4.1 that the velocity for $(C_L^{1/2}/C_D)_{\max}$ is 1.32 times that for $(L/D)_{\max}$. Reflecting on the product $V_\infty(L/D)$ in Eq. (5.152), we see that for maximum range for a jet, although the airplane is flying such that L/D is less than its maximum value, the higher V_∞ is a compensating factor.

Estimate the maximum range at 30,000 ft for the Gulfstream IV. Also calculate the flight velocity required to obtain this range. The maximum usable fuel weight is 29,500 lb. The thrust specific fuel consumption of the Rolls-Royce Tay turbofan at 30,000 ft is 0.69 lb of fuel consumed per hour per pound of thrust.

Example 5.19

Solution

From Example 5.4,

$$\left(\frac{C_L^{1/2}}{C_D}\right)_{max} = 25$$

and

$$V_{(C_L^{1/2}/C_D)_{max}} = 830.8 \text{ ft/s} \qquad \text{(at 30,000 ft)}$$

Also, at 30,000 ft, $\rho_\infty = 8.9068 \times 10^{-4}$ slug/ft^3. From the given fuel weight, we have $W_1 = W_0 - W_f = 73,000 - 29,500 = 43,500$ lb. The thrust specific fuel consumption in consistent units (seconds, not hours) is

$$c_t = \frac{0.69}{3,600} = 1.917 \times 10^{-4} \text{ s}^{-1}$$

From Eq. (5.156),

$$R = \frac{2}{c_t}\sqrt{\frac{2}{\rho_\infty S}}\frac{C_L^{1/2}}{C_D}(W_0^{1/2} - W_1^{1/2}) = \frac{2}{1.917 \times 10^{-4}}\sqrt{\frac{2}{(8.9068 \times 10^{-4})(950)}}$$

$$\times 25[(73,000)^{1/2} - (43,500)^{1/2}] = 2.471 \times 10^7 \text{ ft}$$

In terms of miles,

$$R = \frac{2.471 \times 10^7}{5,280} = \boxed{4,680 \text{ mi}}$$

The use of Eq. (5.156) generally leads to an overestimation of the actual range, for reasons to be given in the next subsection. According to Ref. 36, the maximum range of the Gulfstream IV is 4,254 mi; in this case the above calculation gives a reasonable estimate of the actual range.

The velocity for maximum range has already been quoted at the beginning of this example, as obtained from Example 5.4. It is the velocity at 30,000 ft at which the airplane is flying at $(C_L^{1/2}/C_D)_{max}$.

$$V_\infty(\text{max. range}) = 830.8 \text{ ft/s} = \boxed{566 \text{ mi/h}}$$

This velocity is close to the cruising speed at 31,000 ft of 586 mi/h as listed in Ref. 36 for the real Gulfstream IV.

5.13.3 Other Considerations

There is a contingency in the assumption that led to Eqs. (5.152), (5.153), and (5.156), that is, the assumption that V_∞, L/D, and $C_L^{1/2}/C_D$ are constant throughout the flight. During the flight, fuel is being consumed, and therefore W is decreasing. Since

$L = W$ throughout the flight and

$$L = W = \tfrac{1}{2}\rho_\infty V_\infty^2 S C_L \qquad\qquad [5.157]$$

then the right-hand side of Eq. (5.157) must decrease during the flight. Because of the assumption that L/D or $C_L^{1/2}/C_D$ is constant, the angle of attack remains constant, and hence C_L is constant. Since V_∞ is also assumed constant, the only quantity on the right-hand side of Eq. (5.157) that can change is ρ_∞. Therefore, the contingency in our assumptions is that as the flight progresses and fuel is consumed, the altitude must be continuously increased in just the right manner so that C_L remains constant as W decreases. To take the conditions of Example 5.19 as a case in point, at the start of the flight, C_L is given by

$$C_L = \frac{W}{\tfrac{1}{2}\rho_\infty V_\infty^2 S} = \frac{73{,}000}{\tfrac{1}{2}(8.9068 \times 10^{-4})(830.8)^2(950)} = 0.25$$

At the end of the flight, when $W = 43{,}500$ lb, the value of ρ_∞ necessary to keep $C_L = 0.25$ is

$$\rho_\infty = \frac{2W}{V_\infty^2 S C_L} = \frac{2(43{,}500)}{(830.8)^2(950)(0.25)} = 5.307 \times 10^{-4} \text{ slug/ft}^3$$

This density corresponds to a standard altitude of about 42,000 ft. Hence, for the conditions of Example 5.19, the airplane starts out at an altitude of 30,000 ft, but must continually climb and will end up at an altitude of 42,000 ft in order to keep V_∞ and C_L (hence $C_L^{1/2}/C_D$) constant. Of course, this changing of altitude compromises the use of a fixed value of ρ_∞ in the range equation for a jet airplane, Eq. (5.156). The range equation for a propeller-driven airplane, Eq. (5.153), does not contain ρ_∞ and hence is *not* compromised in the same manner.

Air traffic control constraints do not usually allow an airplane to constantly increase its altitude during the flight, and hence at constant velocity the airplane is generally flying off its maximum value of L/D or $C_L^{1/2}/C_D$, as the case may be. However, on long flights, such as across the Atlantic Ocean, you may note that from time to time the pilot will put the airplane into a short climb to higher altitude. This "stairstepping" flight profile helps to increase the range.

Equations (5.152), (5.153), and (5.156) are useful for preliminary performance estimates for range. However, it is important to keep the above comments in mind when you interpret the results. Also, these equations do not account for takeoff, ascent to altitude, descent, and landing.

There are other scenarios for the calculation of range, such as constant-altitude constant-velocity flight (where the value of C_L changes, hence L/D and $C_L^{1/2}/C_D$ change), and constant-altitude constant-C_L flight (where the value of V_∞ changes). These scenerios lead to predictions of maximum range that are less than the constant-velocity constant-C_L scenario (the cruise-climb scenario) we have treated here. For a more in-depth discussion of various range scenarios, see the books of Hale (Ref. 49) and Mair and Birdsall (Ref. 41).

Another consideration has to do with the flight velocity necessary for maximum range. In Example 5.19, this velocity turned out to be a reasonably high value; the calculated velocity of 830.8 ft/s is equivalent to Mach 0.84. However, what happens when the velocity for maximum range turns out to be a fairly low value, considerably below the maximum velocity of the airplane? This would correspond to a low power setting for the engine—it would be throttled back considerably. To fly at such low power settings, hence low velocities, would result in an inordinately long time to arrive at the destination. Instead, the cruise velocity is set at some higher value in order to realize the full performance capability of the airplane, even though the range is reduced. This is analogous to driving your automobile on the highway. Your best fuel economy, that is, miles per gallon, usually occurs at a speed of between 40 and 50 mi/h. However, you will drive at the posted speed limit, say, 65 mi/h, in order to shorten your trip time, even though you will burn more gas to get to your destination. In the case of an airplane, to fly at higher velocity than that for maximum range is not as inefficient as one might think. For example, return to Fig. 5.11 where the aerodynamic ratios, including C_L/C_D and $C_L^{1/2}/C_D$, are plotted versus velocity. Note that the maximum values are relatively flat peaks, and the values of C_L/C_D and $C_L^{1/2}/C_D$ at speeds of at least 200 ft/s faster are still fairly close to their maximum values. Although Fig. 5.11 is for a specific case, it is representative of the general situation. Even though a penalty in range is paid by flying faster than the best-range speed, the penalty is usually small and does not outweigh the advantage of a shorter flight time.

Related to the above considerations, Bernard Carson, a professor of aerospace engineering at the U.S. Naval Academy, suggested another figure of merit that combines the concept of long range and higher velocity (Ref. 51). Maximum range occurs when the number of pounds of fuel consumed *per mile* is minimized. Recognizing that the flight velocity at this condition could be too small for practical situations, Carson reasoned that a more appropriate combination of both speed and economy would be flight in which the number of pounds of fuel consumed *per unit of velocity* were minimized, that is, when

$$\frac{|dW_f|}{V_\infty} \text{ is a minimum}$$

Let us consider a propeller-driven airplane, which was the focus of Carson's study. From Eq. (5.144),

$$\dot{W}_f = \frac{dW_f}{dt} = -cP$$

or

$$dW_f = -cP\,dt \qquad \textbf{[5.158]}$$

Since $V_\infty = ds/dt$ and $P = TV_\infty$, Eq. (5.158) can be written as

$$dW_f = -\frac{cP\,ds}{V_\infty} = -\frac{cTV_\infty\,ds}{V_\infty} = -cT\,ds \qquad \textbf{[5.159]}$$

By using Eq. (5.159), Carson's figure of merit becomes

$$\frac{|dW_f|}{V_\infty} = \frac{T}{V_\infty} c \, ds \qquad\qquad \textbf{[5.160]}$$

Clearly, this figure of merit is minimized when T/V_∞ is a minimum. We examine the aerodynamic condition that holds when T/V_∞ is a minimum, keeping in mind that $T = D$ and $L = W$.

$$\frac{T}{V_\infty} = \frac{D}{V_\infty} = \frac{D}{L}\frac{L}{V_\infty} = \frac{C_D}{C_L}\frac{W}{V_\infty} \qquad\qquad \textbf{[5.161]}$$

From the expression for lift $L = W = \frac{1}{2}\rho_\infty V_\infty^2 S C_L$, we have

$$V_\infty = \sqrt{\frac{2W}{\rho_\infty S C_L}} \qquad\qquad \textbf{[5.162]}$$

Substituting Eq. (5.162) into Eq. (5.161), we obtain

$$\frac{T}{V_\infty} = \frac{C_D}{C_L} W \sqrt{\frac{\rho_\infty S C_L}{2W}} = \frac{C_D}{C_L^{1/2}} \sqrt{\frac{\rho_\infty S W}{2}} \qquad\qquad \textbf{[5.163]}$$

From Eq. (5.163), minimum T/V_∞ occurs when the airplane is flying such that $C_D/C_L^{1/2}$ is a minimum, hence when $C_L^{1/2}/C_D$ is a *maximum*. We have already seen in Section 5.4.1 that the velocity for $(C_L^{1/2}/C_D)_{\max}$ is given by Eq. (5.45), and that this velocity is 1.32 times the velocity for $(L/D)_{\max}$.

   In short, to fly at the minimum number of pounds of fuel consumed per unit of velocity, the propeller-driven airplane must fly at $(C_L^{1/2}/C_D)_{\max}$. The corresponding velocity is faster than that for $(L/D)_{\max}$. This velocity has come to be called *Carson's speed* in parts of the aeronautical community:

$$\text{Carson's speed} = 1.32 V_{(L/D)_{\max}}$$

For the reasons mentioned earlier, Carson's speed is certainly a more practical cruise speed for propeller-driven airplanes than the lower speed for maximum L/D, although the resulting range will be less than the maximum possible range. Carson himself has put it quite succinctly: flight at this speed is "the least wasteful way of wasting fuel."

5.14   ENDURANCE

Imagine that you are on an air surveillance mission, on the watch for ground or sea activity of various sorts, or monitoring the path and characteristics of a hurricane. Your main concern is staying in the air for the longest possible time. You want the airplane to have long *endurance*. By definition, endurance is the amount of time that an airplane can stay in the air on one load of fuel.

   The flight conditions for maximum endurance are different from those for maximum range, discussed in the previous section. Also, the parameters for endurance are different for propeller-driven and jet-propelled airplanes. Let us consider these matters in more detail.

From Eq. (5.145),

$$\frac{dW_f}{dt} = -c_t T$$

or

$$dt = -\frac{dW_f}{c_t T} \qquad \textbf{[5.164]}$$

Since $T = D$ and $L = W$ in steady, level flight, Eq. (5.164) can be written as

$$dt = -\frac{dW_f}{c_t D} = -\frac{L}{D} \frac{1}{c_t} \frac{dW_f}{W} \qquad \textbf{[5.165]}$$

Integrating Eq. (5.165) from $t = 0$, where $W = W_0$, to $t = E$, where $W = W_1$, we have

$$E = -\int_{W_0}^{W_1} \frac{1}{c_t} \frac{L}{D} \frac{dW_f}{W} = \int_{W_1}^{W_0} \frac{1}{c_t} \frac{L}{D} \frac{dW_f}{W} \qquad \textbf{[5.166]}$$

Equation (5.166) is the general equation for the endurance E of an airplane. If the detailed variations of c_t, L/D, and W are known throughout the flight, Eq. (5.166) can be numerically integrated to obtain an *exact* result for the endurance.

For preliminary performance analysis, Eq. (5.166) is usually simplified. If we assume flight at constant c_t and L/D, Eq. (5.166) becomes

$$E = \frac{1}{c_t} \frac{L}{D} \int_{W_1}^{W_0} \frac{dW_f}{W}$$

or

$$\boxed{E = \frac{1}{c_t} \frac{L}{D} \ln \frac{W_0}{W_1}} \qquad \textbf{[5.167]}$$

Let us consider the individual cases of propeller-driven and jet-propelled aircraft.

5.14.1 Endurance for Propeller-Driven Airplanes

The specific fuel consumption for propeller-driven airplanes is given in terms of power rather than thrust. From Eq. (3.43), the relation between c and c_t is

$$c_t = \frac{cV_\infty}{\eta_{pr}}$$

Substituting this relation into Eq. (5.166), we have

$$E = \int_{W_1}^{W_0} \frac{\eta_{pr}}{cV_\infty} \frac{C_L}{C_D} \frac{dW_f}{W}$$

Substituting Eq. (5.162) into Eq. (5.167), we have

$$E = \int_{W_1}^{W_0} \frac{\eta_{pr}}{c} \sqrt{\frac{\rho_\infty S C_L}{2W}} \frac{C_L}{C_D} \frac{dW_f}{W}$$

or

$$E = \int_{W_1}^{W_0} \frac{\eta_{pr}}{c} \sqrt{\frac{\rho_\infty S}{2}} \frac{C_L^{3/2}}{C_D} \frac{dW_f}{W^{3/2}}$$ **[5.168]**

By making the assumptions of constant η_{pr}, c, ρ_∞, and $C_L^{3/2}/C_D$, Eq. (5.168) becomes

$$E = \frac{\eta_{pr}}{c} \sqrt{2\rho_\infty S} \frac{C_L^{3/2}}{C_D} \left(W_1^{-1/2} - W_0^{-1/2} \right)$$ **[5.169]**

The contingencies associated with the assumptions leading to Eq. (5.169) are the same as those discussed in Section 5.13.3 in regard to the range equation.

We note from Eq. (5.169) that maximum endurance for a propeller-driven airplane corresponds to the following conditions:

1. Fly at maximum $C_L^{3/2}/C_D$.
2. Have the highest possible propeller efficiency.
3. Have the lowest possible specific fuel consumption.
4. Have the highest possible difference between W_0 and W_1 (i.e., carry a lot of fuel).
5. Fly at sea level, where ρ_∞ is the largest value.

The flight conditions associated with $(C_L^{3/2}/C_D)_{max}$ have been discussed in Section 5.4.1. It follows that the theoretical maximum endurance for a propeller-driven airplane is obtained by flying at the velocity where zero-lift drag equals one-third of the drag due to lift

$$C_{D,0} = \tfrac{1}{3} K C_L^2$$ **[5.36]**

This velocity is given by Eq. (5.41).

$$V_{(C_L^{3/2}/C_D)_{max}} = \left(\frac{2}{\rho_\infty} \sqrt{\frac{K}{3C_{D,0}}} \frac{W}{S} \right)^{1/2}$$ **[5.41]**

Note that this velocity is *smaller* than that for maximum L/D by the factor 0.76, as given in Eq. (5.42). The value of $(C_L^{3/2}/C_D)_{max}$ is given by Eq. (5.38):

$$\left(\frac{C_L^{3/2}}{C_D} \right)_{max} = \frac{1}{4} \left(\frac{3}{K C_{D,0}^{1/3}} \right)^{3/4}$$ **[5.38]**

5.14.2 Endurance for Jet-Propelled Airplanes

Equation (5.167) is already expressed in terms of thrust specific fuel consumption, and it gives the endurance for a jet-propelled airplane directly. We repeat Eq. (5.167) for convenience:

$$E = \frac{1}{c_t} \frac{L}{D} \ln \frac{W_0}{W_1}$$ **[5.167]**

Note from Eq. (5.167) that maximum endurance for a jet-propelled airplane corresponds to the following conditions:

1. Fly at maximum L/D.

2. Have the lowest possible thrust specific fuel consumption.

3. Have the highest possible ratio of W_0 to W_1 (i.e., carry a lot of fuel).

The flight conditions associated with maximum L/D have already been discussed at length in Section 5.4.1, and repeated in Section 5.13.1. Hence, they will not be repeated below.

Estimate the maximum endurance for the Gulfstream IV, using the pertinent data from previous examples. | **Example 5.20**

Solution

From the data given in Example 5.19, the fuel weight is 29,500 lb and the specific fuel consumption is 0.69 lb of fuel consumed per hour per pound of thrust, which in consistent units gives $c_t = 1.917 \times 10^{-4}$ s^{-1}. From Example 5.4, the maximum value of L/D is 14.43. From Eq. (5.167),

$$E = \frac{1}{c_t} \frac{L}{D} \ln \frac{W_0}{W_1} = \frac{1}{1.917 \times 10^{-4}} 14.42 \ln \frac{73,000}{43,500} = 38,969 \text{ s}$$

In units of hours,

$$E = \frac{38,969}{3,600} = \boxed{10.8 \text{ h}}$$

5.15 RANGE AND ENDURANCE: A SUMMARY AND SOME GENERAL THOUGHTS

A rather detailed discussion of range and endurance has been given in Sections 5.13 and 5.14, respectively. It will be helpful to now step back from these details for a moment and to look at the more general picture. This is the purpose of this section.

5.15.1 More on Endurance

The simplest way to think about endurance is in terms of pounds of fuel consumed per hour. The smaller the number of pounds of fuel consumed per hour, the longer the airplane will be able to stay in the air, that is, the longer the endurance. Let us examine what dictates this parameter, first for a propeller-driven airplane and then for a jet airplane.

Propeller-Driven Airplane   The specific fuel consumption for a propeller-driven airplane is based on power. The conventional expression for specific fuel consumption (SFC) is given in terms of the inconsistent units of horsepower and hours.

$$\text{SFC} \equiv \frac{\text{lb of fuel consumed}}{(\text{shaft bhp}) (\text{h})} \qquad \textbf{[5.170]}$$

where the shaft brake horsepower is provided by the engine directly to the shaft. In turn, the horsepower available for the airplane is given by

$$HP_A = \eta_{\text{pr}} (\text{shaft bhp})$$

In steady, level flight, recall that power available equals power required: $HP_A = HP_R$. Hence, from Eq. (5.170), we can write the relation

$$\frac{\text{lb of fuel consumed}}{\text{hour}} \propto (\text{SFC}) (HP_R) \qquad \textbf{[5.171]}$$

Therefore, *minimum* pounds of fuel consumed per hour are obtained with *minimum* HP_R. This minimum point on the power required curve is labeled point 1 in Fig. 5.47. This point defines the conditions for maximum endurance for a propeller-driven airplane. Moreover, from Section 5.6.2, this point also corresponds to the aerodynamic condition of flying at $(C_L^{3/2}/C_D)_{\text{max}}$. The velocity at which this occurs is the flight velocity for best endurance for a propeller-driven airplane. All this information is labeled in association with point 1 in Fig. 5.47.

Jet-Propelled Airplane   The specific fuel consumption for a jet-propelled airplane is based on thrust. The conventional expression for thrust specific fuel consumption (TSFC) is given in terms of the inconsistent unit of hours.

$$\text{TSFC} = \frac{\text{lb of fuel connsumed}}{(\text{thrust}) (\text{h})} \qquad \textbf{[5.172]}$$

Hence, from Eq. (5.172), and noting that in steady, level flight $T_A = T_R$, we can write

$$\frac{\text{lb of fuel consumed}}{\text{h}} = T_R (\text{TSFC}) \qquad \textbf{[5.173]}$$

Therefore, *minimum* pounds of fuel consumed per hour are obtained with *minimum* T_R. This minimum point on the thrust required curve is labeled point 2 in Fig. 5.47. This point defines the conditions for maximum endurance for a jet-propelled airplane. Moreover, from Section 5.3.2, this point also corresponds to the aerodynamic

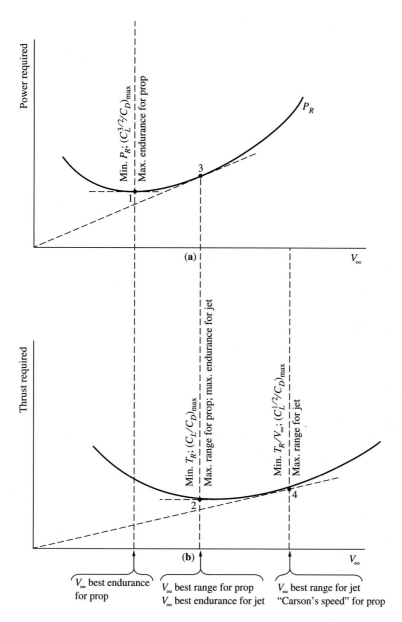

Figure 5.47 Graphical summary of conditions for maximum range and endurance.

condition of flying at $(L/D)_{max}$. The velocity at which this occurs is the flight velocity for best endurance for a jet-propelled airplane. All this information is labeled in association with point 2 in Fig. 5.47.

5.15.2   More on Range

The simplest way to think about range is in terms of pounds of fuel consumed per mile. The smaller the number of pounds of fuel consumed per mile, the larger the distance the airplane can fly, that is, the larger the range. Let us examine what dictates this parameter, first for a propeller-driven airplane and then for a jet airplane.

Propeller-Driven Airplane   The pounds of fuel consumed per mile for a propeller-driven airplane are given by

$$\frac{\text{lb of fuel consumed}}{\text{mi}} = \frac{(\text{SFC})HP_R}{\eta_{\text{pr}}V_\infty} \qquad \textbf{[5.174]}$$

where V_∞ is in miles per hour. Clearly, from Eq. (5.174) the *minimum* pounds of fuel consumed per mile are obtained with *minimum* HP_R/V_∞. Return to Fig. 5.47**a**. Imagine a straight line drawn from the origin to any arbitrary point on the power required curve (and consider the units of power to be horsepower). The slope of such a line is HP_R/V_∞. The minimum value of this slope occurs when the straight line is tangent to the HP_R curve; this tangent point is denoted by point 3 in Fig. 5.47**a**. Therefore, point 3 corresponds to the conditions for maximum range for a propeller-driven airplane. Since $P_R = T_R V_\infty$, then

$$\frac{HP_R}{V_\infty} \propto T_R$$

and therefore minimum HP_R/V_∞ corresponds to minimum T_R. From Section 5.3.2, this corresponds to flight at maximum L/D. This is also the flight condition for point 2 in Fig. 5.47**b**. Therefore, point 3 in Fig. 5.47**a** corresponds to the same flight velocity as point 2 in Fig. 5.47**b**. As itemized on Fig. 5.47, the flight conditions for maximum range for a propeller-driven airplane are the same as those for maximum endurance for a jet-propelled airplane.

Jet-Propelled Airplane   The pounds of fuel consumed per mile for a jet airplane are given by

$$\frac{\text{lb of fuel consumed}}{\text{mi}} = \frac{(\text{TSFC})T_R}{V_\infty} \qquad \textbf{[5.175]}$$

where V_∞ is in miles per hour. From Eq. (5.175), the *minimum* pounds of fuel consumed per mile are obtained with *minimum* T_R/V_∞. Return to Fig. 5.47**b**. Imagine a straight line drawn from the origin to any arbitrary point on the thrust required curve. The slope of such a line is T_R/V_∞. The minimum value of this slope occurs when the straight line is tangent to T_R; this tangent point is denoted by point 4 in Fig. 5.47**b**. Therefore, point 4 corresponds to the conditions for maximum range for a jet-propelled airplane. Furthermore, the aerodynamic condition that holds at point 4 is found as follows.

$$\frac{T_R}{V_\infty} = \frac{1}{2}\rho_\infty V_\infty S C_D = \frac{1}{2}\rho_\infty \sqrt{\frac{2W}{\rho_\infty S C_L}} S C_D = \left(\frac{\rho_\infty W S}{2}\right)^{1/2} \frac{C_D}{C_L^{1/2}} \qquad \textbf{[5.176]}$$

From Eq. (5.176), T_R/V_∞ is a minimum when $C_D/C_L^{1/2}$ is a minimum, or when $C_L^{1/2}/C_D$ is a maximum. Thus, at point 4, the flight conditions correspond to flight at $(C_L^{1/2}/C_D)_{\text{max}}$.

In addition, recall from Section 5.13.3 that Carson's speed for a propeller-driven airplane is given as the flight velocity that corresponds to a minimum value of T/V_∞. Hence, point 4 in Fig. 5.47**b** also corresponds to Carson's speed.

5.15.3 Graphical Summary

Study Fig. 5.47 carefully. It is an all-inclusive graphical construction that illustrates the various conditions for maximum range and endurance for propeller-driven and jet-propelled aircraft. In particular, note the flight velocities for these conditions, that is, the three velocities corresponding to points 1, 2 and 3, and 4. Maximum endurance for a propeller-driven airplane occurs at the lowest of these velocities (point 1). The velocity for maximum range for a propeller-driven airplane, and for maximum endurance for a jet airplane, is higher (points 2 and 3). The velocity for maximum range for a jet airplane (point 4) is the highest of the three. Denoting the three velocities by V_1, V_2, and V_3, the results of Section 5.4.1 show that, from Eq. (5.42)

$$V_1 = 0.76V_2 = 0.76V_3$$

and from Eq. (5.46)

$$V_4 = 1.32V_2$$

Also, note that the construction of a line through the origin tangent to either the P_R curve or the T_R curve yields useful information. This construction allows a simple method for dealing with the effect of wind, as discussed below.

5.15.4 The Effect of Wind

Most preliminary performance analyses assume that the airplane is flying through a stationary atmosphere, that is, there are no prevailing winds in the atmosphere. This has been the assumption underlying all our performance analyses in this chapter. Although not important for such a preliminary analysis, it is worthwhile to at least ask the question: How is endurance affected by a headwind or a tailwind? Similarly, how is range affected? Let us examine the answers to these questions.

First, we emphasize that the aerodynamic properties of the airplane depend on the velocity of the air *relative to the airplane* V_∞. The aerodynamics does not "care" whether there is a headwind or a tailwind. For example, in all our previous discussions, V_∞ is the velocity of the free stream *relative to the airplane*. It is the *true airspeed* of the airplane. In a stationary atmosphere, V_∞ is also the velocity of the airplane relative to the ground. However, when there is a headwind or tailwind, the velocity of the airplane relative to the air is different from that of the airplane relative to the ground. We denote the velocity of the airplane relative to the ground as simply the *ground speed* V_g. When there is a headwind or a tailwind, V_g is different from V_∞. Again, keep in mind that the aerodynamics of the airplane depends on V_∞, not V_g.

The relationship between V_∞ and V_g is illustrated in Fig. 5.48. In Fig. 5.48**a**, the airplane is flying into a headwind of velocity V_{HW}. The airplane's relative velocity through the air is V_∞, and its ground velocity is $V_g = V_\infty - V_{HW}$, as shown in Fig. 5.48**a**. Simiarly, in Fig. 5.48**b** the airplane is flying with a tailwind of velocity V_{TW}. Here, the airplane's ground speed is $V_g = V_\infty + V_{TW}$, as shown in Fig. 5.48**b**.

To return to the two questions asked at the beginning of this subsection, endurance is not influenced by the wind. The airplane's relative velocity V_∞ is simply that for maximum endurance, as explained in previous sections. The distance covered over the ground is irrelevant to the consideration of endurance.

The same *cannot* be said about range. Range is directly affected by wind. An extreme example occurs when the relative velocity of an airplane through the air is 100 mi/h, and there is a headwind of 100 mi/h. The ground speed is zero—the airplane just hovers over the same location, and the range is zero. Clearly, range depends on the wind.

Indeed, range is a function of ground speed V_g; the ground speed is what enters into the consideration of distance covered over the ground. For example, letting s denote the horizontal distance covered over the gound, we have

$$V_g = \frac{ds}{dt}$$

or

$$ds = V_g \, dt \qquad\qquad \textbf{[5.177]}$$

Compare Eqs. (5.177) and (5.146). They are the same relationship, because Eq. (5.146)

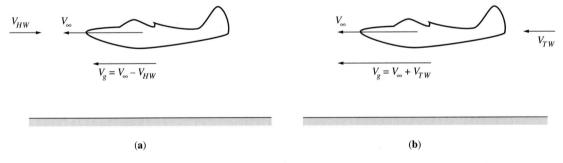

(**a**)   (**b**)

Figure 5.48   Relationship between flight velocity V_∞ and ground speed V_g for (**a**) headwind and (**b**) tailwind.

assumes a stationary atmosphere, that is, no wind. Hence, in Eq. (5.146), V_∞ represents the ground speed as well as the airspeed. However, with a wind, we have to remember that the fundamental relationship is Eq. (5.177), not Eq. (5.146). Following a derivation analogous to that for Eq. (5.152), Eq. (5.177) leads to the expression for range for a jet-propelled airplane

$$R = \frac{V_g}{c_t} \frac{L}{D} \ln \frac{W_0}{W_1} \qquad [5.178]$$

The power available for the airplane is, as before, the product of the thrust and the true airspeed of the airplane $T V_\infty$, no matter what the wind velocity may be. Hence, Eq. (3.43) still holds, namely, $c_t = c V_\infty / \eta_{\text{pr}}$. In turn, Eq. (5.178) can be written as

$$R = \frac{\eta_{\text{pr}}}{c} \frac{V_g}{V_\infty} \frac{L}{D} \ln \frac{W_0}{W_1} \qquad [5.179]$$

which is in a form convenient for calculating the range for a propeller-driven airplane.

The values of V_∞ that correspond to maximum range for a jet airplane and a propeller-driven airplane including the effect of wind can be found by differentiating Eqs. (5.178) and (5.179), respectively, with respect to V_∞ and setting the derivatives equal to zero. The details can be found in Refs. 41 and 49. Because of the appearance of V_g in Eqs. (5.178) and (5.179), the values of V_∞ that result in maximum range with wind effects are different from those we obtained earlier for the case of no wind. Indeed for both the jet airplane and the propeller-driven airplane, the best-range value of V_∞ with a headwind is higher than that for no wind, and the best-range value of V_∞ with a tailwind is lower than that for no wind. See Refs. 41 and 49 for analytical expressions for these best-range airspeeds with wind.

A graphical approach provides a more direct way of obtaining the best-range airspeeds with wind. First, consider a propeller-driven airplane. Range is determined by the pounds of fuel consumed *per mile covered over the ground*. Hence, analogous to Eq. (5.174), we write

$$\frac{\text{lb of fuel consumed}}{\text{mi}} = \frac{(\text{SFC}) HP_R}{\eta_{\text{pr}} V_g} \qquad [5.180]$$

Clearly, from Eq. (5.180) minimum number of pounds of fuel consumed per mile, which corresponds to maximum range, is obtained with *minimum* HP_R / V_g. Consider the power required curve sketched in Fig. 5.49. This is a plot of HP_R versus airspeed V_∞; it is our familiar power required curve as discussed throughout this chapter. It depends on the aerodynamics of the airplane, which depends on V_∞. Also, as discussed in Section 5.15.2, a line drawn from the origin tangent to the HP_R curve at point 1 defines the airspeed for maximum range without wind. This is shown by point 1 in Fig. 5.49. Now assume that a headwind with velocity V_{HW} exists. Hence, $V_g = V_\infty - V_{HW}$. If we want to use V_g as the abscissa rather than V_∞ in Fig. 5.49, we have to shift the origin to the right, to the tick mark labeled V_{HW}, and place the origin of the new abscissa at that point, as indicated by the new abscissa labeled V_g in Fig. 5.49. The power required curve stays where it is—it does not move because it depends on the airspeed V_∞. However, the condition for best range with wind is

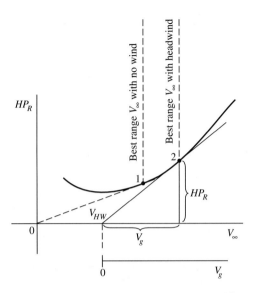

Figure 5.49   Effect of headwind on best-range airspeed for a propeller-driven airplane.

given by Eq. (5.180), and the minimum HP_R/V_g corresponds to the solid straight line through the point labeled V_{HW} tangent to the power required curve. The tangent point is point 2 as sketched in Fig. 5.49. The slope of this line is HP_R/V_g, as shown in Fig. 5.49, and it is the *minimum* value of the slope because it is tangent to the HP_R curve. Hence, from Eq. (5.180), point 2 corresponds to the flight conditions for maximum range with a headwind of strength V_{HW}. Point 2 identifies the value of the airspeed V_∞ for best range with a headwind. Note that this value is larger than that for best range with no wind, confirming our previous statement in the analytical discussion. The case for a tailwind is treated in a similar fashion, except the point for V_{TW} on the original abscissa is to the left of the origin, as shown in Fig. 5.50. Point 3 in Fig. 5.50 is the tangent point on the HP_R curve of a straight line drawn through the tick mark for V_{TW}. Point 3 identifies the value of the airspeed V_∞ for best range with a tailwind. Note that this value is smaller than that for best range with no wind, consistent with our earlier discussion.

Consider a jet-propelled airplane. Range is again determined by the pounds of fuel consumed per mile covered over the ground. Hence, analogous to Eq. (5.175), we write

$$\frac{\text{lb of fuel consumed}}{\text{mi}} = \frac{(\text{TSFC})T_R}{V_g} \qquad \textbf{[5.181]}$$

From Eq. (5.181), minimum number of pounds of fuel consumed per mile, which corresponds to maximum range, is obtained with minimum T_R/T_g. Consider the thrust required curve sketched in Fig. 5.51. This is a plot of T_R versus V_∞; it is the familiar thrust required curve discussed throughout this chapter. It depends on the

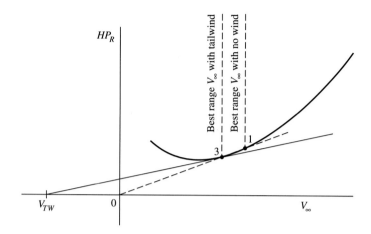

Figure 5.50 Effect of tailwind on best-range airspeed for a propeller-driven airplane.

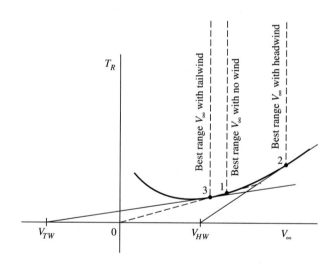

Figure 5.51 Effect of tailwind and headwind on best-range airspeed for a jet.

aerodynamics of the airplane, which depends on V_∞. Note that the solid lines drawn tangent to the curve, one through the tick mark V_{HW} and the other through the tick mark V_{TW}, identify the tangent points 2 and 3, respectively. Since the slopes of these lines are the minimum values of T_R/V_g, points 2 and 3 correspond to the values of V_∞ for best range for a headwind and a tailwind, respectively. The interpretation of Fig. 5.51 for the jet airplane is the same as that of Figs. 5.49 and 5.50 for the propeller-driven airplane. Hence, no further discussion is needed.

5.16   SUMMARY

By definition, the static performance analysis of an airplane assumes rectilinear motion with no acceleration. The material in this chapter provides the basis for a preliminary static performance analysis.

A few of the important aspects of this chapter are listed below. As you read through this list, if any items are unclear or uncertain to you, return to the pertinent section in the chapter and review the material until you are comfortable.

1. For steady, level flight, the equations of motion are the simple equilibrium relations

$$T = D \qquad \text{[5.3]}$$

$$L = W \qquad \text{[5.4]}$$

2. The basic aerodynamics needed for a performance analysis is the drag polar.

$$C_D = C_{D,0} + K C_L^2 \qquad \text{[5.5]}$$

3. A *thrust required curve* is a plot of $T_R = D$ versus velocity for a given airplane at a given altitude. A *thrust available curve* is a plot of T_A versus velocity for a given airplane at a given altitude. The high-speed intersection of the maximum thrust available and thrust required curves determines the maximum velocity of the airplane. Thrust required is inversely proportional to the lift-to-drag ratio

$$T_R = \frac{W}{L/D} \qquad \text{[5.7]}$$

4. The design parameters T_R/W and W/S play a strong role in airplane performance. An analytical expression for the resulting airplane velocity for a given T_R/W and W/S is

$$V_\infty = \left[\frac{(T_R/W)(W/S) \pm (W/S)\sqrt{(T_R/W)^2 - 4C_{D,0}K}}{\rho_\infty C_{D,0}} \right]^{1/2} \qquad \text{[5.18]}$$

5. A *power required curve* is a plot of P_R versus velocity for a given airplane at a given altitude. A *power available curve* is a plot of P_A versus velocity for a given airplane at a given altitude. The high-speed intersection of the maximum power available and the power required curves determines the maximum velocity of the airplane. The power required is inversely proportional to $C_L^{3/2}/C_D$

$$P_R = \sqrt{\frac{2W^3 C_D^2}{\rho_\infty S C_L^3}} \propto \frac{1}{C_L^{3/2}/C_D} \qquad \text{[5.56]}$$

6. The following aerodynamic relations are important for a static performance analysis.

a. Maximum L/D occurs when the zero-lift drag equals the drag due to lift:

$$C_{D,0} = K C_L^2 \qquad \text{[5.28]}$$

The value of $(L/D)_{\max}$ depends only on $C_{D,0}$ and K.

$$\left(\frac{L}{D}\right)_{\max} = \left(\frac{C_L}{C_D}\right)_{\max} = \sqrt{\frac{1}{4 C_{D,0} K}} \qquad \text{[5.30]}$$

The flight velocity at which $(L/D)_{\max}$ is achieved for a given airplane depends on the altitude and wing loading:

$$V_{(L/D)_{\max}} = \left(\frac{2}{\rho_\infty} \sqrt{\frac{K}{C_{D,0}}} \frac{W}{S}\right)^{1/2} \qquad \text{[5.34]}$$

Minimum T_R occurs when L/D is maximum.

b. Maximum $C_L^{3/2}/C_D$ occurs when the zero-lift drag is one-third of the drag due to lift:

$$C_{D,0} = \tfrac{1}{3} K C_L^2 \qquad \text{[5.36]}$$

The value of $(C_L^{3/2}/C_D)_{\max}$ depends only on $C_{D,0}$ and K:

$$\left(\frac{C_L^{3/2}}{C_D}\right)_{\max} = \frac{1}{4}\left(\frac{3}{K C_{D,0}^{1/3}}\right)^{3/4} \qquad \text{[5.38]}$$

The flight velocity at which $(C_L^{3/2}/C_D)_{\max}$ is achieved for a given airplane depends on the altitude and wing loading:

$$V_{(C_L^{3/2}/C_D)_{\max}} = \left(\frac{2}{\rho_\infty} \sqrt{\frac{K}{3 C_{D,0}}} \frac{W}{S}\right)^{1/2} \qquad \text{[5.41]}$$

Minimum P_R occurs when $C_L^{3/2}/C_D$ is maximum.

c. Maximum $C_L^{1/2}/C_D$ occurs when the zero-lift drag is 3 times the drag due to lift:

$$C_{D,0} = 3 K C_L^2 \qquad \text{[5.43]}$$

The value of $(C_L^{1/2}/C_D)_{\max}$ depends only on $C_{D,0}$ and K:

$$\left(\frac{C_L^{1/2}}{C_D}\right)_{\max} = \frac{3}{4}\left(\frac{1}{3 K C_{D,0}^3}\right)^{1/4} \qquad \text{[5.44]}$$

The flight velocity at which $(C_L^{1/2}/C_D)_{\max}$ is achieved for a given airplane depends on the altitude and wing loading:

$$V_{(C_L^{1/2}/C_D)_{\max}} = \left(\frac{2}{\rho_\infty} \sqrt{\frac{3K}{C_{D,0}}} \frac{W}{S}\right)^{1/2} \qquad \text{[5.45]}$$

    d. The flight velocities for maximum values of the above aerodynamic ratios are related in magnitude as follows:

$$V_{(C_L^{3/2}/C_D)_{\max}} : V_{(C_L/C_D)_{\max}} : V_{(C_L^{1/2}/C_D)_{\max}} = 0.76 : 1 : 1.32$$

7. The stall speed of a given airplane at a given altitude is dictated by $(C_L)_{\max}$ and the wing loading:

$$V_{\text{stall}} = \sqrt{\frac{2}{\rho_\infty} \frac{W}{S} \frac{1}{(C_L)_{\max}}} \tag{5.67}$$

The values of $(C_L)_{\max}$ can be increased by a variety of high-lift devices, such as trailing- and leading-edge flaps, slats, etc.

8. Rate of climb is given by

$$R/C = \frac{TV_\infty - DV_\infty}{W} = \frac{\text{excess power}}{W} \tag{5.78}$$

The various analytical expressions obtained for a rate of climb analysis show that R/C for a given airplane at a given altitude depends on wing loading and thrust-to-weight ratio.

9. For unpowered gliding flight, the glide angle θ is determined by

$$\text{Tan } \theta = \frac{1}{L/D} \tag{5.125}$$

10. Absolute ceiling is that altitude where $(R/C)_{\max} = 0$. Service ceiling is that altitude where $(R/C)_{\max} = 100$ ft/min.

11. The conditions for maximum range and maximum endurance are different. Moreover, they also depend on whether the airplane is propeller-driven or jet-propelled:

$$R = \frac{\eta_{\text{pr}}}{c} \frac{L}{D} \ln \frac{W_0}{W_1} \qquad \text{propeller-driven} \tag{5.153}$$

$$R = \frac{2}{c_t} \sqrt{\frac{2}{\rho_\infty S} \frac{C_L^{1/2}}{C_D}} \left(W_0^{1/2} - W_1^{1/2} \right) \qquad \text{jet} \tag{5.156}$$

$$E = \frac{\eta_{\text{pr}}}{c} \sqrt{2\rho_\infty S} \frac{C_L^{3/2}}{C_D} \left(W_1^{-1/2} - W_0^{-1/2} \right) \qquad \text{propeller-driven} \tag{5.169}$$

$$E = \frac{1}{c_t} \frac{L}{D} \ln \frac{W_0}{W_1} \qquad \text{jet} \tag{5.167}$$

Note that maximum endurance for a propeller-driven airplane occurs when the airplane is flying at $(C_L^{3/2}/C_D)_{\max}$. Maximum range for a propeller-driven airplane and maximum endurance for a jet occur when the airplane is flying at $(L/D)_{\max}$. Maximum range for a jet occurs when the airplane is flying at $(C_L^{1/2}/C_D)_{\max}$.

PROBLEMS

The Bede BD-5J is a very small single-seat home-built jet airplane which became **5.1** available in the early 1970s. The data for the BD-5J are as follows:

- Wing span: 17 ft
- Wing planform area: 37.8 ft^2
- Gross weight at takeoff: 960 lb
- Fuel capacity: 55 gal
- Power plant: one French-built Microturbo TRS 18 turbojet engine with maximum thrust at sea level of 202 lb and a specific fuel consumption of 1.3 lb/(lb·h)

We will approximate the drag polar for this airplane by

$$C_D = 0.02 + 0.062C_L^2$$

(a) Plot the thrust required and thrust available curves at sea level, and from these curves obtain the maximum velocity at sea level.
(b) Plot the thrust required and thrust available curves at 10,000 ft, and from these curves obtain the maximum velocity at 10,000 ft.

For the BD-5J (the airplane in Problem 5.1), calculate *analytically* (directly) (a) the **5.2** maximum velocity at sea level and (b) the maximum velocity at 10,000 ft. Compare these results with those from Problem 5.1.

Derive Eqs. (5.43), (5.44), and (5.45). **5.3**

Using the results of Section 5.4.1, repeat the task in Problem 2.11: Find an expression **5.4** for the maximum lift-to-drag ratio for a supersonic two-dimensional flat plate, and the angle of attack at which it occurs. Check your results with those from Problem 2.11. They should be identical.

For the BD-5J, calculate **5.5**
(a) The maximum value of C_L/C_D
(b) The maximum value of $C_L^{1/2}/C_D$
(c) The velocities at which they occur at sea level
(d) The velocities at which they occur at 10,000 ft

The BD-5J is equipped with plain flaps. The airfoil section at the wing root is an **5.6** NACA 64-212, and interestingly enough, it has a thicker section at the tip, an NACA 64-218 (Reference: *Jane's All the World's Aircraft*, 1975–76). Estimate the stalling speed of the BD-5J at sea level.

For the BD-5J, plot the power required and power available curves at sea level. From **5.7** these curves, estimate the maximum rate of climb at sea level.

Derive Eq. (5.85) for the rate of climb as a function of velocity, thrust-to-weight ratio, **5.8** wing loading, and the drag polar.

5.9   For the BD-5J use the analytical results to calculate directly
(a) Maximum rate of climb at sea level and the velocity at which it occurs. Compare with your graphical result from Problem 5.7.
(b) Maximum climb angle at sea level and the velocity at which it occurs.

5.10   For a turbojet-powered airplane with the altitude variation of thrust given by Eq. (3.19), show that as the altitude increases, the maximum velocity decreases.

5.11   Consider the BD-5J flying at 10,000 ft. Assume a sudden and total loss of engine thrust. Calculate (a) the minimum glide path angle, (b) the maximum range covered over the ground during the glide, and (c) the corresponding equilibrium glide velocities at 10,000 ft and at sea level.

5.12   For the BD-5J, plot the maximum rate of climb versus altitude. From this graph, estimate the service ceiling.

5.13   For the BD-5J, analytically calculate the service ceiling, and compare this result with the graphical solution obtained in Problem 5.12.

5.14   Using the analytical approach described in Section 5.12.2, calculate the minimum time to climb to 10,000 ft for the BD-5J.

5.15   For the BD-5J, estimate the maximum range at an altitude of 10,000 ft. Also, calculate the flight velocity required to obtain this range. (*Recall:* All the pertinent airplane data, including the thrust specific fuel consumption, are given in Problem 5.1.)

5.16   For the BD-5J, estimate the maximum endurance.

5.17   Calculate the maximum range at 10,000 ft for the BD-5J in a tailwind of 40 mi/h.

5.18   In the worked examples in this chapter, the thrust available is assumed to be constant with velocity for the reasons explained at the end of Section 5.1. However, in reality, the thrust from a typical turbofan engine decreases with an increase in velocity. The purpose of this and the following problems is to revisit some of the worked examples, this time including a velocity variation for the thrust available. In this fashion we will be able to examine the effect of such a velocity variation on the performance of the airplane. The airplane is the same Gulfstream IV examined in the worked examples, with the same wing loading, drag polar, etc. However, now we consider the variation of thrust available given by

At sea level:

$$\frac{T_A}{(T_A)_{V=0}} = 0.4M_\infty^{-0.6} \tag{1}$$

At 30,000 ft:

$$\frac{T_A}{(T_A)_{V=0}} = 0.222M_\infty^{-0.6} \tag{2}$$

Recall that $(T_A)_{V=0}$ is the thrust at sea level at zero velocity.
(a) At sea level, plot the thrust available curve using Eq. (1) above, and the thrust required curve, both on the same graph. From this, obtain V_{max} at sea level.

(b) At an altitude of 30,000 ft, plot the thrust available curve, using Eq. (2) above, and the thrust required curve, both on the same graph. From this obtain V_{max} at 30,000 ft.
(c) Compare the results obtained from (a) and (b) with the analytical results from Example 5.6.

5.19 For the Gulfstream IV with the thrust available variations given by Eqs. (1) and (2) in Problem 5.18, analytically (directly) calculate V_{max} at sea level and at 30,000 ft. Compare with the graphical results obtained in Problem 5.18. Comment on the increased level of difficulty of this calculation compared to that performed in Example 5.6 where the thrust was assumed constant with velocity.

5.20 For the Gulfstream IV with the thrust available variations given in Problem 5.18, do the following:
(a) Plot the power available and power required curves at sea level. From this graphical construction, obtain the maximum rate of climb at sea level and the velocity at which it is obtained. Compare with the results obtained in Example 5.13.
(b) Plot the power available and power required curves at 30,000 ft. From this graphical construction, obtain the maximum rate of climb at 30,000 ft and the velocity at which it is obtained.

5.21 When the thrust available variation is given by $T_A/(T_A)_{V=0} = AM_\infty^{-n}$, develop an analytical solution for the calculation of maximum rate of climb. Compare this with the simpler analytical approach discussed in Section 5.10.2 for the case of constant thrust available.

5.22 Use the development in Problem 5.21 to calculate analytically the maximum rate of climb at sea level and at 30,000 ft for the Gulfstream IV. Compare these analytical results with the graphical results from Problem 5.20.

5.23 Use the two data points for maximum rate of climb obtained in Problem 5.20 (or Problem 5.22) to make an approximate estimate of the absolute ceiling for the Gulfstream IV. Compare this result with that obtained in Example 5.16.

chapter

6

Airplane Performance: Acce

With its unique requirement for blending together such a wide range of the sciences, aviation has been one of the most stimulating, challenging, and prolific fields of technology in the history of mankind.

Morgan M. (Mac) Blair
Rockwell International, 1980

The success or otherwise of a design therefore depends to a large extent on the designer's knowledge of the physics of the flow, and no improvements in numerical and experimental design tools are ever likely to dispose of *the need for physical insight*.

Dietrich Kuchemann
Royal Aircraft Establishment,
England, 1978

6.1 INTRODUCTION

Our study of static performance (no acceleration) in Chapter 5 answered a number of questions about the capabilities of a given airplane—how fast it can fly, how far it can go, etc. However, there are more questions to be asked: How fast can it turn? How high can it "zoom"? What ground distances are covered during takeoff and landing? The answers to these questions involve *accelerated* flight, the subject of this chapter. To this end, we return to the general equations of motion derived in Chapter 4, which you should review before going further.

6.2   LEVEL TURN

The flight path and forces for an airplane in a level turn are sketched in Fig. 6.1. Here, the flight path is *curved*, in contrast to the rectilinear motion studied in Chapter 5. By definition, a *level* turn is one in which the curved flight path is in a horizontal plane parallel to the plane of the ground; that is, in a level turn the altitude remains constant. The relationship between forces required for a level turn is illustrated in

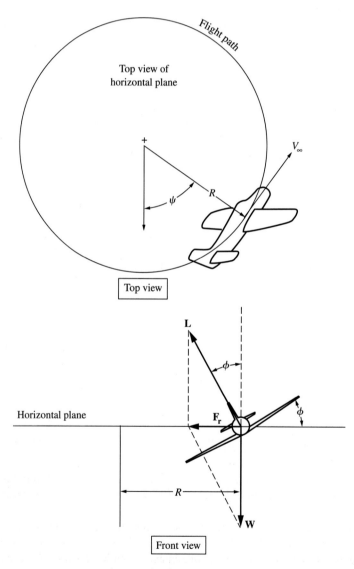

Figure 6.1   An airplane in a level turn.

Fig. 6.1. Here, the airplane is banked through the roll angle ϕ. The magnitude of the lift L and the value of ϕ are adjusted such that the vertical component of lift, denoted by $L \cos \phi$, exactly equals the weight, or

$$\boxed{L \cos \phi = W} \qquad \textbf{[6.1]}$$

Under this condition, the altitude of the airplane will remain constant. Hence, Eq. (6.1) applies only to the case of a level turn; indeed, it is the *necessary condition for a level turn*.

Another way of stating this necessary condition is to consider the resultant force $\mathbf{F_r}$, which is the vector sum of vectors $\mathbf{L}$ and $\mathbf{W}$. As shown in Fig. 6.1, for the case of the level turn, the magnitude and direction of $\mathbf{L}$ are adjusted to be just right so that the vector sum of $\mathbf{L}$ and $\mathbf{W}$ results in $\mathbf{F_r}$ always being in the horizontal plane. In this fashion the altitude remains constant.

The generalized force diagram for an airplane in climbing and banking flight is given in Fig. 4.3. When this figure is specialized for level flight, that is, $\theta = 0$, and assuming the thrust vector is parallel to the free-stream direction, that is, $\epsilon = 0$, then the force diagram for a level turn is obtained as sketched in Fig. 6.1. The governing equation of motion is given by Eq. (4.7), specialized for the case of $\theta = 0$ and $\epsilon = 0$, namely,

$$m \frac{V_\infty^2}{r_2} = L \sin \phi \qquad \textbf{[6.2]}$$

Recalling Fig. 4.5, we see that r_2 is the local radius of curvature of the flight path in the horizontal plane. This is the same as the radius R shown in Fig. 6.1. Hence, for a level turn, the governing equation of motion is, from Eq. (6.2),

$$\boxed{m \frac{V_\infty^2}{R} = L \sin \phi} \qquad \textbf{[6.3]}$$

Equation (6.3) is simply a physical statement that the centrifugal force $m V_\infty^2 / R$ is balanced by the radial force $L \sin \phi$.

The two performance characteristics of greatest importance in turning flight are

1. The turn radius R.
2. The turn rate $\omega \equiv d\psi/dt$, where ψ is defined in Fig. 6.1. The turn rate is simply the local *angular* velocity of the airplane along the curved flight path.

These characteristics are particularly germane to combat aircraft. For superior dog-fighting capability, the airplane should have the smallest possible turn radius R and the fastest possible turn rate ω. What aspects of the airplane determine R and ω? Let us examine this question.

First, take another look at Fig. 6.1. The airplane is turning due to the radial force $\mathbf{F_r}$. The larger the magnitude of this force F_r, the tighter and faster will be the turn. The magnitude F_r is the horizontal component of the lift $L \sin \phi$. As L increases, F_r increases for two reasons: (1) The length of the lift vector increases, and (2) ϕ

increases because for a level turn, $L \cos \phi$ must remain constant, namely, equal to W, as seen from Eq. (6.1). Hence, the lift vector **L** controls the turn; when a pilot goes to turn the airplane, she or he rolls the airplane in order to point the lift vector in the general direction of the turn. Keep in mind that L and ϕ are not independent; they are related by the condition for a level turn given by Eq. (6.1), which can be written as

$$\cos \phi = \frac{W}{L} = \frac{1}{L/W} \qquad [6.4]$$

In Eq. (6.4), the ratio L/W is an important parameter in turning performance; it is defined as the *load factor n*, where

$$n \equiv \frac{L}{W} \qquad [6.5]$$

Hence, Eq. (6.4) can be written as

$$\phi = \text{Arccos} \, \frac{1}{n} \qquad [6.6]$$

The roll angle ϕ depends *only* on the load factor; if you know the load factor, then you know ϕ, and vice versa. The turn performance of an airplane strongly depends on the load factor, as we will next demonstrate.

To obtain an expression for the turn radius, insert $m = W/g$ in Eq. (6.3) and solve for R.

$$R = \frac{m V_\infty^2}{L \sin \phi} = \frac{W}{L} \frac{V_\infty^2}{g \sin \phi} = \frac{V_\infty^2}{gn \sin \phi} \qquad [6.7]$$

From Eq. (6.4),

$$\cos \phi = \frac{1}{n}$$

and from the trigonometric identity

$$\cos^2 \phi + \sin^2 \phi = 1$$

we have

$$\left(\frac{1}{n}\right)^2 + \sin^2 \phi = 1$$

or

$$\sin \phi = \sqrt{1 - \frac{1}{n^2}} = \frac{1}{n}\sqrt{n^2 - 1} \qquad [6.8]$$

By substituting Eq. (6.8) into Eq. (6.7), the turn radius is expressed as

$$R = \frac{V_\infty^2}{g\sqrt{n^2 - 1}}$$

[6.9]

From Eq. (6.9), the turn radius depends only on V_∞ and n. To obtain the smallest possible R, we want

1. The highest possible load factor (i.e., the highest possible L/W).
2. The lowest possible velocity.

To obtain an expression for the turn rate ω, return to Fig. 6.1 and recall from physics that angular velocity is related to R and V_∞ as

$$\omega = \frac{d\psi}{dt} = \frac{V_\infty}{R}$$

[6.10]

Replacing R in Eq. (6.10) with Eq. (6.9), we have

$$\omega = \frac{g\sqrt{n^2 - 1}}{V_\infty}$$

[6.11]

From Eq. (6.11), to obtain the largest possible turn rate, we want

1. The highest possible load factor.
2. The lowest possible velocity.

These are exactly the same criteria for the smallest possible R.

This leads to the following questions. For a given airplane in a level turn, what is the highest possible load factor? Equations (6.9) and (6.11) show that R and ω depend only on V_∞ and n—design characteristics such as W/S, T/W, and the drag polar, as well as altitude, do not appear explicitly. The fact is that even though the expression for R and ω in general contains only V_∞ and n, there are specific *constraints* on the values of V_∞ and n for a given airplane, and these constants *do* depend on the design characteristics and altitude. Let us examine these constraints.

Constraints on Load Factor Return to Fig. 6.1, and note that as the airplane's bank angle ϕ is increased, the magnitude of the lift must increase. As L increases, *the drag due to lift increases*. Hence, to maintain a sustained level turn at a given velocity and a given bank angle ϕ, the thrust must be increased from its straight and level flight value to compensate for the increase in drag. If this increase in thrust pushes the required thrust beyond the maximum thrust available from the power plant, then the level turn cannot be sustained at the given velocity and bank angle. In this case, to maintain a turn at the given V_∞, ϕ will have to be decreased in order to decrease the drag sufficiently that the thrust required does not exceed the thrust available. Since the load factor is a function of ϕ via Eq. (6.6), written as

$$n = \frac{1}{\cos \phi} \qquad \textbf{[6.12]}$$

at any given velocity, the maximum possible load factor for a sustained level turn is constrained by the maximum thrust available.

This maximum possible load factor n_{max} can be calculated as follows. From the drag polar, the drag is

$$D = \tfrac{1}{2}\rho_\infty V_\infty^2 S \left(C_{D,0} + K C_L^2\right) \qquad \textbf{[6.13]}$$

For a level turn, the thrust equals the drag.

$$T = D \qquad \textbf{[6.14]}$$

Also,

$$L = nW = \tfrac{1}{2}\rho_\infty V_\infty^2 S C_L$$

or

$$C_L = \frac{2nW}{\rho_\infty V_\infty^2 S} \qquad \textbf{[6.15]}$$

Substituting Eqs. (6.14) and (6.15) into Eq. (6.13), we have

$$T = \frac{1}{2}\rho_\infty V_\infty^2 S \left[C_{D,0} + K \left(\frac{2nW}{\rho_\infty V_\infty^2 S} \right)^2 \right] \qquad \textbf{[6.16]}$$

Solving Eq. (6.16) for n (the details are left for a homework problem), we have

$$n = \left[\frac{\tfrac{1}{2}\rho_\infty V_\infty^2}{K(W/S)} \left(\frac{T}{W} - \frac{1}{2}\rho_\infty V_\infty^2 \frac{C_{D,0}}{W/S} \right) \right]^{1/2} \qquad \textbf{[6.17]}$$

Equation (6.17) gives the load factor (hence ϕ) for a given velocity and thrust-to-weight ratio. The maximum value of n is obtained by inserting $T = T_{max}$, or $(T/W)_{max}$, into Eq. (6.17).

$$n_{max} = \left\{ \frac{\tfrac{1}{2}\rho_\infty V_\infty^2}{K(W/S)} \left[\left(\frac{T}{W} \right)_{max} - \frac{1}{2}\rho_\infty V_\infty^2 \frac{C_{D,0}}{W/S} \right] \right\}^{1/2} \qquad \textbf{[6.18]}$$

Hence, although Eqs. (6.9) and (6.11) show that R and ω depend only on V_∞ and n, the load factor cannot be any arbitrary value. Rather, for a given V_∞, n can only range between

$$1 \le n \le n_{max}$$

where n_{max} is given by Eq. (6.18). Hence, there is a constraint on n imposed by the maximum available thrust. Moreover, from Eq. (6.18), n_{max} is dictated by the design parameters W/S, T/W, $C_{D,0}$, and K as well as the altitude (via ρ_∞).

The variation of n_{max} versus velocity for a given airplane, as calculated from Eq. (6.18), is shown in Fig. 6.2. The airplane considered here is the Gulfstream-like airplane treated in the examples in Chapter 5. The altitude is sea level; the results will be difficult for different altitudes. At the maximum velocity of the airplane, there is no excess power, hence no level turn is possible and $n = 1$. As V_∞ decreases, n_{max} increases, reaches a local maximum value at point B, and then decreases. For velocities higher than that at point B, the zero-lift drag (which increases with V_∞) dominates; and for velocities lower than that at point B, the drag due to lift (which decreases with V_∞) dominates. This is why the n_{max} curve first increases, then reaches a local maximum, and finally decreases with velocity.

At point B in Fig. 6.2, the airplane is flying at its maximum L/D. This is easily seen from the relation (recalling that $D = T$)

$$n \equiv \frac{L}{W} = \frac{L}{D}\frac{D}{W} = \frac{L}{D}\frac{T}{W} \qquad [6.19]$$

When T_{max} is inserted in Eq. (6.19), then n becomes n_{max}—the same quantity as calculated from Eq. (6.18):

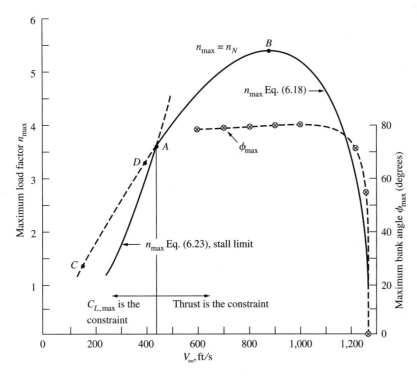

Figure 6.2 Thrust and $(C_L)_{max}$ constraints on maximum load factor and maximum bank angle versus flight velocity. $T/W = 0.3795$, $W/S = 76.84$ lb/ft², $C_{D,0} = 0.015$, and $K = 0.08$, and $(C_L)_{max} = 1.2$. Ambient conditions are standard sea level.

$$n_{max} = \frac{L}{D}\left(\frac{T}{W}\right)_{max} \qquad \text{[6.20]}$$

For each point along the n_{max} curve in Fig. 6.2, there is a different value of L/D, consistent with Eq. (6.20). When n_{max} reaches its local maximum at point B—which is the *maximum of the maximum* values of n, denoted by n_M—the value of L/D in Eq. (6.20) is at its maximum, that is,

$$n_M = \left(\frac{L}{D}\right)_{max}\left(\frac{T}{W}\right)_{max} \qquad \text{[6.21]}$$

For the Gulfstream treated in Chapter 5, we found in Example 5.4 that $(L/D)_{max} = 14.43$. Inserting this value along with $(T/W)_{max} = 0.3795$ into Eq. (6.21) yields $n_M = (14.43)(0.3795) = 5.47$, which is consistent with the value of n_{max} at point B in Fig. 6.2.

There is another, completely different constraint on the load factor having to do with the maximum lift coefficient $(C_L)_{max}$. In Fig. 6.2, for each point on the n_{max} curve obtained from Eq. (6.18), there is a different value of C_L; indeed, as V_∞ decreases, C_L increases. It is easy to see why. As V_∞ (and hence the dynamic pressure) decreases, the magnitude of L is maintained (or increased) by increasing C_L, that is, by increasing the angle of attack of the airplane. Obviously, C_L cannot increase indefinitely—it is limited by its maximum value at stall $(C_L)_{max}$. The velocity at which $(C_L)_{max}$ is reached is denoted by point A in Fig. 6.2. At lower velocities, less than at point A, the maximum load factor is constrained by $(C_L)_{max}$, not by available thrust.

When n_{max} is constrained by $(C_L)_{max}$, the value of n_{max} can be obtained as follows.

$$n \equiv \frac{L}{W} = \frac{1}{2}\rho_\infty V_\infty^2 C_L \frac{S}{W} \qquad \text{[6.22]}$$

In Eq. (6.22), when $C_L = (C_L)_{max}$, then $n = n_{max}$.

$$\boxed{n_{max} = \frac{1}{2}\rho_\infty V_\infty^2 \frac{(C_L)_{max}}{W/S}} \qquad \text{[6.23]}$$

The solid curve to the left of point A in Fig. 6.2 is obtained from Eq. (6.23). It represents the value of n_{max} at low velocities where $(C_L)_{max}$ is the constraint. In Fig. 6.2, a value of $(C_L)_{max} = 1.2$ is assumed. This is representative of a jet transport with moderate wing sweep and no high-lift devices employed.

Values of ϕ_{max} corresponding to values of n_{max} can be obtained from Eq. (6.12), written as

$$\cos\phi_{max} = \frac{1}{n_{max}}$$

These values of ϕ_{max} are also plotted in Fig. 6.2 as a function of V_∞. It is interesting to note that the variation of ϕ_{max} is relatively flat over a wide range of velocity.

Finally, we note that the structural design limits of a given airplane represent a practical, mechanical constraint on the load factor. This constraint will be discussed in Section 6.5.

Constraints on V_∞   Returning to Eqs. (6.9) and (6.11), which show that R and ω depend only on V_∞ and n, we have already stated that n cannot be any arbitrary value. Although for high performance these equations dictate that n should be as large as possible, there are definite limits on the value of n that are associated with the design aspects of the airplane.

Equations (6.9) and (6.11) also show that for high performance V_∞ should be as small as possible. However, V_∞ cannot be reduced indefinitely without encountering stall. Hence, the stall limit is a constraint on V_∞. Indeed, when the airplane is at a bank angle ϕ, the stalling velocity is increased above that for straight and level flight. The stalling velocity is a function of the load factor. To show this, recall that

$$L = nW = \tfrac{1}{2}\rho_\infty V_\infty^2 S C_L$$

Hence,

$$V_\infty = \sqrt{\frac{2nW}{\rho_\infty S C_L}} \qquad\qquad \textbf{[6.24]}$$

When $C_L = (C_L)_{max}$ is inserted into Eq. (6.24), then $V_\infty = V_{stall}$.

$$V_{stall} = \sqrt{\frac{2}{\rho_\infty}\frac{W}{S}\frac{n}{(C_L)_{max}}} \qquad\qquad \textbf{[6.25]}$$

Equation (6.25) is a more general result for V_{stall} than that for straight and level flight given by Eq. (5.67). When $n = 1$ is inserted into Eq. (6.25), Eq. (5.67) is obtained. Hence, when the airplane is in a level turn with a load factor $n > 1$, the stalling velocity increases proportionally to $n^{1/2}$. This stalling velocity is a constraint on the minimum value of V_∞ that can be inserted in Eqs. (6.9) and (6.11) for R and ω.

6.2.1 Minimum Turn Radius

With all the above discussion on the level turn in mind, we now return to Eq. (6.9) and ask: What is the smallest possible value of R for a given airplane? The minimum R does not necessarily correspond to $n_{max} = n_M$ given by point B in Fig. 6.2, because R also depends on V_∞, and the minimum R may occur at a set of values (n_{max}, V_∞) different from those at point B. Let us investigate this matter.

The conditions for minimum R are found by setting $dR/dV_\infty = 0$. The algebra will be simpler if we deal with dynamic pressure, $q = \tfrac{1}{2}\rho_\infty V_\infty^2$, rather than V_∞. Hence, from Eq. (6.9), written in terms of q_∞,

$$R = \frac{2q_\infty}{g\rho_\infty\sqrt{n^2 - 1}} \qquad\qquad \textbf{[6.26]}$$

Differentiating Eq. (6.26) with respect to q_∞, remembering that n is a function of V_∞ hence q_∞, and setting the derivative equal to zero, we have

$$\frac{dR}{dq_\infty} = \frac{2g\rho_\infty\sqrt{n^2-1} - 2g\rho_\infty q_\infty n(n^2-1)^{-1/2}\, dn/dq_\infty}{g^2\rho_\infty^2(n^2-1)} = 0$$

or

$$2g\rho_\infty\sqrt{n^2-1} - 2g\rho_\infty q_\infty \frac{n}{\sqrt{n^2-1}}\frac{dn}{dq_\infty} = 0$$

or

$$n^2 - 1 - q_\infty n \frac{dn}{dq_\infty} = 0 \qquad\qquad \textbf{[6.27]}$$

The load factor n as a function of q_∞ is given by Eq. (6.17), written in terms of q_∞.

$$n^2 = \frac{q_\infty}{K(W/S)}\left(\frac{T}{W} - q_\infty\frac{C_{D,0}}{W/S}\right) \qquad\qquad \textbf{[6.28]}$$

Differentiating Eq. (6.28) with respect to q_∞ gives

$$n\frac{dn}{dq_\infty} = \frac{T/W}{2K(W/S)} - \frac{q_\infty C_{D,0}}{K(W/S)^2} \qquad\qquad \textbf{[6.29]}$$

Substituting Eqs. (6.28) and (6.29) into Eq. (6.27), we have

$$\frac{q_\infty}{K(W/S)}\frac{T}{W} - \frac{q_\infty^2 C_{D,0}}{K(W/S)^2} - 1 - \frac{q_\infty(T/W)}{2K(W/S)} + \frac{q_\infty^2 C_{D,0}}{K(W/S)^2} = 0$$

Combining and canceling terms, we get

$$\frac{q_\infty(T/W)}{2K(W/S)} = 1$$

or

$$q_\infty = \frac{2K(W/S)}{T/W} \qquad\qquad \textbf{[6.30]}$$

Since $q_\infty = \frac{1}{2}\rho_\infty V_\infty^2$, Eq. (6.30) becomes

$$\boxed{(V_\infty)_{R_{\min}} = \sqrt{\frac{4K(W/S)}{\rho_\infty(T/W)}}} \qquad\qquad \textbf{[6.31]}$$

Equation (6.31) gives the value of V_∞ which corresponds to the minimum turning radius; this velocity is denoted by $(V_\infty)_{R_{\min}}$ in Eq. (6.31). In turn, the load factor corresponding to this velocity is found by substituting Eq. (6.30) into Eq. (6.28)

$$n^2 = \frac{q_\infty(T/W)}{K(W/S)} - \frac{q_\infty^2 C_{D,0}}{K(W/S)^2}$$

$$= \frac{2K(W/S)(T/W)}{(T/W)K(W/S)} - \frac{4K^2(W/S)^2 C_{D,0}}{(T/W)^2 K(W/S)^2} = 2 - \frac{4KC_{D,0}}{(T/W)^2}$$

or

$$n_{R_{min}} = \sqrt{2 - \frac{4KC_{D,0}}{(T/W)^2}}$$ [6.32]

Equation (6.32) gives the load factor corresponding to the minimum turning radius, denoted by $n_{R_{min}}$. Finally, the expression for minimum turning radius is obtained by substituting Eqs. (6.31) and (6.32) into Eq. (6.9), written as

$$R_{min} = \frac{(V_\infty)^2_{R_{min}}}{g\sqrt{n^2_{R_{min}} - 1}} = \frac{4K(W/S)}{\rho_\infty(T/W)} \frac{1}{g\sqrt{2 - 4KC_{D,0}/(T/W)^2 - 1}}$$

$$R_{min} = \frac{4K(W/S)}{g\rho_\infty(T/W)\sqrt{1 - 4KC_{D,0}/(T/W)^2}}$$ [6.33]

Example 6.1

Calculate the minimum turning radius at sea level for the Gulfstream-like airplane treated in the worked examples in Chapter 5, and locate the corresponding conditions on Fig. 6.2.

Solution

From Eq. (6.33),

$$R_{min} = \frac{4(0.08)(76.84)}{(32.2)(0.002377)(0.3795)\sqrt{1 - 4(0.08)(0.015)/(0.3795)^2}} = 861 \text{ ft}$$

The corresponding load factor and velocity are obtained from Eqs. (6.32) and (6.31), respectively.

$$n_{R_{min}} = \sqrt{2 - \frac{4KC_{D,0}}{(T/W)^2}} = \sqrt{2 - \frac{4(0.08)(0.015)}{(0.3795)^2}} = 1.4$$

and

$$(V_\infty)_{R_{min}} = \sqrt{\frac{4K(W/S)}{\rho_\infty(T/W)}} = \sqrt{\frac{4(0.08)(76.84)}{(0.002377)(0.3795)}} = 165 \text{ ft/s}$$

These values of n and V_∞ locate point C on the n_{max} curve shown in Fig. 6.2. Right away, we see the value of $R_{min} = 861$ ft is *unobtainable*; it corresponds to a velocity below the stalling velocity. Point C is beyond the $(C_L)_{max}$ constraint in Fig. 6.2. Indeed, the lift coefficient corresponding to point C is

$$C_L = \frac{2n}{\rho_\infty V_\infty^2} \frac{W}{S} = \frac{2(1.4)(76.84)}{(0.002377)(165)^2} = 3.32$$

This value is well beyond the assumed value of $(C_L)_{max} = 1.2$ used in Eq. (6.23) for the generation of the $(C_L)_{max}$ constraint curve in Fig. 6.2. Therefore, for the airplane considered here, the minimum turning radius is constrained by stall, and it is not predicted by Eq. (6.33).

Rather, the smallest turning radius will actually be that corresponding to point A in Fig. 6.2, where, from the graph

$$V_A = 445 \text{ ft/s} \qquad n_A = 3.6$$

Hence,

$$R_{\min} = \frac{V_A^2}{g\sqrt{n_A^2 - 1}} = \frac{(445)^2}{32.2\sqrt{(3.6)^2 - 1}} = \boxed{1{,}778 \text{ ft}}$$

Note that the conditions for minimum turn radius are far different from those for the maximum value of $n_{\max}$; points C and A in Fig. 6.2 are far removed from point B.

6.2.2   Maximum Turn Rate

The thought process given for maximum turn rate parallels that given above for minimum turn radius—only the details are different. The conditions for maximum turn rate $\omega_{\max}$ are obtained by differentiating Eq. (6.11) and setting the derivative equal to zero. The details are left for a homework problem. The results are

$$(V_\infty)_{\omega_{\max}} = \left[\frac{2(W/S)}{\rho_\infty}\right]^{1/2} \left(\frac{K}{C_{D,0}}\right)^{1/4} \tag{6.34}$$

$$n_{\omega_{\max}} = \left(\frac{T/W}{\sqrt{KC_{D,0}}} - 1\right)^{1/2} \tag{6.35}$$

$$\omega_{\max} = q\sqrt{\frac{\rho_\infty}{W/S}\left[\frac{T/W}{2K} - \left(\frac{C_{D,0}}{K}\right)^{1/2}\right]} \tag{6.36}$$

Example 6.2   Calculate the maximum turning rate and the corresponding values of load factor and velocity for our Gulfstream-like airplane at sea level.

Solution
   From Eq. (6.36),

$$\omega_{\max} = 32.2\sqrt{\frac{0.002377}{76.84}\left[\frac{0.3795}{2(0.08)} - \left(\frac{0.015}{0.08}\right)^{1/2}\right]} = 0.25 \text{ rad/s}$$

Recalling that 1 rad $= 57.3°$, we get

$$\omega_{\max} = (0.25)(57.3) = 14.3 \text{ deg/s}$$

The corresponding value of n is obtained from Eq. (6.35).

$$n_{\omega_{\max}} = \left[\frac{0.3795}{\sqrt{(0.08)(0.015)}} - 1\right]^{1/2} = 3.16$$

The corresponding value of V_∞ is obtained from Eq. (6.34)

$$(V_\infty)_{\omega_{max}} = \left[\frac{2(76.84)}{0.002377}\right]^{1/2}\left(\frac{0.08}{0.015}\right)^{1/4} = 386 \text{ ft/s}$$

These values of n and V_∞ locate point D in Fig. 6.2. Once again, this is beyond the stall limit, but only slightly. The value of ω_{max} will be different from 14.27 deg/s calculated above, because ω_{max} is constrained by $(C_L)_{max}$. Indeed, ω_{max} will correspond to point A in Fig. 6.2, where $V_A = 445$ ft/s and $n_A = 3.6$. For this case, from Eq. (6.11)

$$\omega_{max} = \frac{g\sqrt{n_A^2 - 1}}{V_A} = \frac{32.2\sqrt{(3.6)^2 - 1}}{445} = 0.25 \text{ rad/s} = \boxed{14.3 \text{ deg/s}}$$

Note that, within roundoff error, this is the same value as calculated earlier from Eq. (6.36). This is because, for this case, points D and A in Fig. 6.2 are so close, and because ω has a rather flat variation with V_∞ in the vicinity of ω_{max}.

DESIGN CAMEO

Minimum turn radius and maximum turning rate are important performance characteristics for a fighter airplane; they are much less so for a commercial transport or heavy bomber. For the design of a high-performance fighter, the results of this section reveal some of the design features desirable for good turning performance.

For example, an examination of Eqs. (6.33) and (6.36) shows that wing loading and thrust-to-weight ratio dominate the values of R_{min} and ω_{max}. For good turn performance (low R_{min} and high ω_{max}), W/S should be *low* and T/W should be *high*. For the design of a modern high-performance fighter, T/W is usually dictated by other requirements than turn performance, such as the need to have a high supersonic maximum velocity, or a constraint on takeoff length. Wing loading is usually dictated by landing velocity (i.e., stall velocity) considerations. However, airplane design is always a compromise, and both T/W and W/S can be "adjusted" within some margins to enhance turning performance for a design where such performance is important. The designer can choose to make W/S slightly smaller and T/W slightly larger than would otherwise be the case, just to give the new airplane "an edge" in turning performance over the competition. For example, there has been some discussion of designing a new wing for the F-15 supersonic fighter, a mainline

aircraft for the U.S. Air Force since 1974. The new wing would be *larger*, hence reducing W/S, for the purpose of enhancing subsonic combat maneuverability, at the cost of some decrease in maximum velocity.

In regard to combat maneuverability (today, the word *agility* is used to describe the overall concept of high maneuverability), an examination of Eqs. (6.33) and (6.36) shows that minimum turn radius and maximum turn rate depend on ρ_∞, that is, altitude. Turning performance increases with ρ_∞. Hence, the best turning performance is achieved at sea level. Moreover, we have noted from Fig. 6.2 that R_{min} and ω_{max} occur at relatively low velocities (e.g., denoted by points C and A, respectively, in Fig. 6.2). When modern fighters with supersonic capability engage in dogfights, their altitude generally decreases and the flight velocities are rapidly lowered, generally below Mach 1. Hence, the "combat arena," even for a Mach 3 airplane, is usually in the subsonic range.

From Eqs. (6.33) and (6.36), good turning performance is also enhanced by good aerodynamics, that is, by low values of $C_{D,0}$ and K. In airplane design, good streamlining will usually result in lower $C_{D,0}$, with a weaker but still beneficial reduction in K. However, from the discussion surrounding Eq. (2.46), the drag-due-to-lift factor K is of the form

(*continued*)

$$K = a + \frac{b}{AR} \qquad [6.37]$$

Hence, for subsonic flight, the most direct way of reducing K is to increase the aspect ratio AR. An airplane designed for good turn performance will benefit aerodynamically by having a high-aspect-ratio wing. Indeed, from Eq. (6.33) we see that R_{min} varies slightly more strongly than K to the first power (due to the added enhancement of K in the denominator).

However, structural design limitations place a major constraint on the allowable design aspect ratio. This is particularly true for airplanes designed for high turning performance; here the large load factors result in large bending moments at the wing root. The *wingspan* is really more germane than the aspect ratio in this consideration. Airplanes with high maneuver performance simply do not have large wingspans, in order to keep the wing bending moments within reasonable design limits. Some typical design features of subsonic high-performance fighters are tabulated below.

	Wingspan (ft)	Aspect Ratio
North American P-51 Mustang (World War II)	37	5.86
Grumman F6F Hellcat (World War II)	42.8	5.34
North American F-86 Saberjet	37.1	4.78

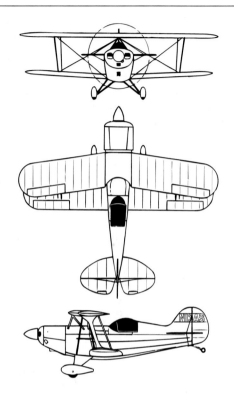

Figure 6.3   Pitts S-2A Special. Span of upper wing = 20 ft. Span of lower wing = 19 ft. Overall (total of both wings) planform wing area = 125 ft². This yields an approximate aspect ratio of 6.4.

These airplanes are all *monoplanes*, that is, a single-wing design. A way to have a short wingspan and a reasonably high aspect ratio at the same time is to go to a *biplane configuration*; here, the necessary lift is generated by two smaller wings rather than one larger wing. A perfect example is the famous aerobatic airplane, the Pitts Special, shown in Fig. 6.3. For this airplane, the wingspan is only 20 ft, and yet the aspect ratio of each wing is (approximately) a respectable 6.4. The biplane configuration has good structural advantages, which is one reason why it is appealing for aerobatic airplanes which routinely are subjected to high stresses. Also, a shorter wingspan leads to a smaller rolling moment of inertia, and hence higher roll rates. These advantages were

among the reasons why the biplane configuration was favored during the early part of the twentieth century (see Section 1.2.2). However, the biplane configuration suffers from increased zero-lift drag due to the interwing struts and bracing wires, and there is usually an unfavorable aerodynamic interaction between the two wings which results in lower lift coefficients and higher induced drag coefficients. Hence biplanes are not usually efficient for high-speed flight. This, in concert with the development of the cantilevered, stressed-skin wing in the late 1920s, eventually led to the demise of the biplane (except for special applications) in favor of the monoplane.

An important design feature which has a direct impact on turning performance is $(C_L)_{max}$. We have

(continued)

already discussed how $(C_L)_{max}$ can constrain R_{min} and ω_{max}. In Fig. 6.2, the curve at the left, generated from Eq. (6.23), reflects the constraint on turning performance due to stall. This constraint dictates that R_{min} and ω_{max} correspond to point A rather than the more favorable thrust-limited values given by points C and D, respectively. However, the turning performance associated with points C and D could be achieved by shifting the stall limit curve sufficiently to the left. In turn, this can be achieved by a sufficient increase in $(C_L)_{max}$. Hence, in the design of an airplane, turning performance can be enhanced by choosing a high-lift airfoil shape and/or incorporating high-lift devices that can be deployed during a turn. However, the primary factor in the design choice for $(C_L)_{max}$ is usually the landing speed, not turning performance. Nevertheless, for those airplane designs where turning performance is particularly important, some extra emphasis on achieving a high $(C_L)_{max}$ is important and appropriate.

Finally, in a similar vein, turning performance can be greatly enhanced by orienting the engine thrust vector in the direction of the turn. We have not considered this case in the present discussion. However, return to Fig. 4.5, and note that in general the thrust has a component $T \sin \epsilon \cos \phi$ in the direction of the turn. For a jet-powered airplane, by designing the engine nozzles to rotate relative to the axis of the rest of the engine, the value of ϵ can be greatly increased, markedly increasing the magnitude of $T \sin \epsilon \cos \phi$ in Fig. 4.5, and hence greatly increasing turning performance. Such vectoring nozzles are used on some vertical takeoff and landing (VTOL) airplanes for a different purpose, namely to provide a vertical thrust force; the Harrier fighter (Fig. 1.36) is an example. However, Harrier pilots in combat have used the vectored thrust feature to also obtain enhanced turn performance. The consideration of using vectored thrust to improve agility (which includes turning performance) is part of the new design philosophy for high-performance jet fighters. The new Lockheed-Martin F-22 (Fig. 6.4) incorporates two-dimensional exhaust nozzles (convergent-divergent nozzles of rectangular

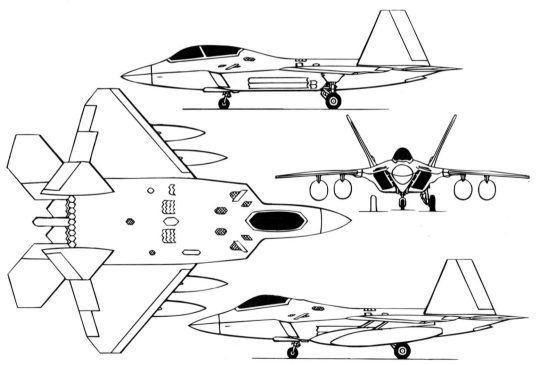

Figure 6.4 The Lockheed-Martin F-22.

(*continued*)

cross section, in contrast to the conventional axisymmetric nozzle with a circular cross section) which can be rotated up or down for changing the direction of the thrust vector in the symmetry plane of the aircraft. An added advantage of two-dimensional exhaust nozzles is that they are easier to "hide" in the fuselage, reducing the radar cross section, hence improving the stealth characteristic of the airplane.

We end this design cameo on the following note. To this author's knowledge, turning performance has never been *the* sole driver in the design of any airplane. Indeed, the 1903 *Wright Flyer* with its wing loading of 1.2 lb/ft^2 can outturn and outmaneuver any modern high-performance fighter of today, such as the Lockheed-Martin F-16 with a wing loading of 74 lb/ft^2. However, the *Wright Flyer* cannot begin to carry out the supersonic, high-altitude missions for which the F-16 is primarily designed. So this is an "apples and oranges" comparison. On the other hand, a certain level of turning performance is frequently included in the specifications for a new fighter design, and the designer must be familiar with the design factors which optimize turn performance in order to meet the specifications. Those factors have been highlighted in this design cameo.

6.3   THE PULL-UP AND PULLDOWN MANEUVERS

Consider an airplane initially in straight and level flight, where $L = W$. The pilot suddenly pitches the airplane to a higher angle of attack such that the lift suddenly increases. Because $L > W$, the airplane will arch upward, as sketched in Fig. 6.5. The flight path becomes curved in the vertical plane, with a turn radius R and turn rate $d\theta/dt$. This is called the *pull-up maneuver*.

The general picture of the flight path in the vertical plane and the components of force which act in the vertical plane, are sketched in Fig. 4.4. For the pull-up maneuver, the roll angle is zero, that is, $\phi = 0$. The picture shown in Fig. 6.5 is a

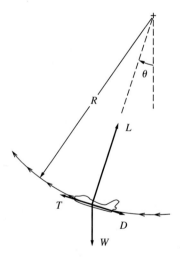

Figure 6.5   The pull-up maneuver.

specialized case of that shown in Fig. 4.4, where $\phi = 0$ and $\epsilon = 0$. The appropriate equations of motion associated with the flight path in Fig. 4.4 are Eqs. (4.5) and (4.6). In particular, Eq. (4.6) specialized for $\phi = 0$ and $\epsilon = 0$ becomes

$$m\frac{V_\infty^2}{R} = L - W\cos\theta \qquad \textbf{[6.38]}$$

where r_1 in Eq. (4.6) is replaced by R, as shown in Fig. 6.5. Equation (6.38) is a governing equation of motion for the flight path shown in Fig. 6.5.

Unlike the level turn discussed in Section 6.2, where we considered a *sustained* turn (constant flight properties during the level turn), in the pull-up maneuver we will focus on an *instantaneous* turn, where we are interested in the turn radius and turn rate at the instant that the maneuver is initiated. Airplanes frequently execute sustained level turns, but rarely a sustained pull-up maneuver with constant flight properties. The instantaneous pull-up is of much greater interest, and we will focus on it. Moreover, we assume the instantaneous pull-up is initiated from straight and level horizontal flight; this corresponds to $\theta = 0$ in Fig. 6.5. For this case, Eq. (6.38) becomes

$$m\frac{V_\infty^2}{R} = L - W \qquad \textbf{[6.39]}$$

As in the case of the level turn, the pull-up performance characteristics of greatest interest are the turn radius R and turn rate $\omega = d\theta/dt$. The instantaneous turn radius is obtained from Eq. (6.39) as follows.

$$R = \frac{mV_\infty^2}{L - W} = \frac{W}{g}\frac{V_\infty^2}{L - W} = \frac{V_\infty^2}{g(L/W - 1)} \qquad \textbf{[6.40]}$$

Noting that L/W is the load factor n, we see that Eq. (6.40) can be written as

$$R = \frac{V_\infty^2}{g(n - 1)} \qquad \textbf{[6.41]}$$

The instantaneous turn rate (angular velocity) is given by $\omega = V_\infty/R$. Hence, from Eq. (6.41) we have

$$\omega = \frac{g(n - 1)}{V_\infty} \qquad \textbf{[6.42]}$$

A related case is the pulldown maneuver, sketched in Fig. 6.6. Here, an airplane initially in straight and level flight is suddenly rolled to an inverted position, such that both L and W are pointing downward. The airplane will begin to turn downward in a flight path with instantaneous turn radius R and turn rate $\omega = d\theta/dt$. For this case, the equation of motion is still Eq. (6.38) with θ taken as $180°$ (see Fig. 6.6). For this

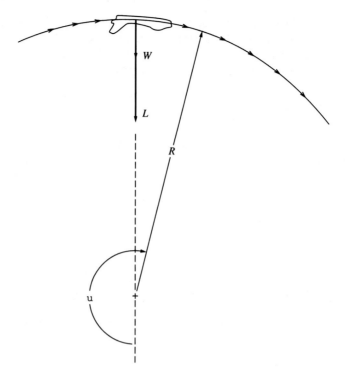

Figure 6.6   The pulldown maneuver.

case, Eq. (6.38) is written as

$$m\frac{V_\infty^2}{R} = L + W \qquad [6.43]$$

Hence,

$$R = \frac{mV_\infty^2}{L + W} = \frac{W}{g}\frac{V_\infty^2}{L + W} = \frac{V_\infty^2}{g(L/W + 1)} \qquad [6.44]$$

Since $n = L/W$, Eq. (6.44) becomes

$$\boxed{R = \frac{V_\infty^2}{g(n + 1)}} \qquad [6.45]$$

and $\omega = V_\infty/R$ becomes

$$\boxed{\omega = \frac{g(n + 1)}{V_\infty}} \qquad [6.46]$$

Note the similarity between Eqs. (6.41) and (6.42) for pull-up and between Eqs. (6.45) and (6.46) for pulldown; the difference is only a minus and plus sign in the parentheses. Also note that, as in the case for the level turn, for the pull-up and pulldown, R and ω depend only on the flight characteristics V_∞ and n.

DESIGN CAMEO

The airplane design features for good pull-up and pull-down performance are the same as those for good turning performance, as discussed at the end of Section 6.2.

This is because the roles of V_∞ and n in Eqs. (6.41), (6.42), (6.45), and (6.46) are qualitatively the same as those in Eqs. (6.9) and (6.11) for the level turn.

6.4 LIMITING CASE FOR LARGE LOAD FACTOR

Consider the turn radius equations for the level turn, pull-up, and pulldown maneuvers, as repeated here:

$$\text{Level turn} \qquad R = \frac{V_\infty^2}{g\sqrt{n^2 - 1}} \qquad\qquad \textbf{[6.9]}$$

$$\text{Pull-up} \qquad R = \frac{V_\infty^2}{g(n - 1)} \qquad\qquad \textbf{[6.41]}$$

$$\text{Pulldown} \qquad R = \frac{V_\infty^2}{g(n + 1)} \qquad\qquad \textbf{[6.45]}$$

In the limit of large load factor $n \gg 1$, these three equations reduce to the same form, namely,

$$R = \frac{V_\infty^2}{gn} \qquad\qquad \textbf{[6.47]}$$

Similarly, consider the expressions for turn rate for the level turn, pull-up, and pulldown maneuvers, as repeated here:

$$\text{Level turn} \qquad \omega = \frac{g\sqrt{n^2 - 1}}{V_\infty} \qquad\qquad \textbf{[6.11]}$$

$$\text{Pull-up} \qquad \omega = \frac{g(n - 1)}{V_\infty} \qquad\qquad \textbf{[6.42]}$$

$$\text{Pulldown} \qquad \omega = \frac{g(n+1)}{V_\infty} \qquad\qquad \textbf{[6.46]}$$

In the limit of large load factor, these equations reduce to the same form, namely,

$$\boxed{\omega = \frac{gn}{V_\infty}} \qquad\qquad \textbf{[6.48]}$$

The physical reason why the same form for R is obtained in the limiting case of large n for all three maneuvers is that the magnitude of the lift is so large that the weight is unimportant by comparison. In all three cases, the lift vector dominates the dynamics. The same is true for ω.

The pull-up and pulldown maneuvers considered in Section 6.3 are treated as *instantaneous*, in contrast to the sustained maneuver (sustained turn) discussed in Section 6.2. For instantaneous maneuvers, the thrust limitations discussed in Section 6.2 are not relevant. Why? An instantaneous maneuver is initiated by a sudden change in lift, achieved by a sudden increase in angle of attack. The drag is suddenly increased as well, causing the airplane to experience a deceleration. However, at the instant the maneuver is initiated, the instantaneous velocity is V_∞ as it appears in Eqs. (6.41), (6.42), (6.45), and (6.46), and as it appears in the limit impressions given by Eqs. (6.47) and (6.48). So even though the airplane will feel a sudden increase in drag and therefore a sudden deceleration, the velocity decreases only after the instant of initiation of the maneuver. Any increase in thrust to counteract the increase in drag comes "after the fact." So, by definition of an instantaneous maneuver, the type of thrust limitation discussed in regard to the sustained turn in Section 6.2 is not relevant to the instantaneous maneuver.

Let us use the limiting equations, Eqs. (6.47) and (6.48), to examine those characteristics of the airplane which are important to an instantaneous maneuver. In this category we will include the instantaneous turn as well as the pull-up or pulldown—Eqs. (6.47) and (6.48) govern all three types of instantaneous maneuvers in the limit of large n. In these equations, V_∞ can be replaced as follows.

Since

$$L = \tfrac{1}{2}\rho_\infty V_\infty^2 SC_L$$

then

$$V_\infty^2 = \frac{2L}{\rho_\infty SC_L} \qquad\qquad \textbf{[6.49]}$$

Substituting Eq. (6.49) into Eqs. (6.47) and (6.48), we have

$$R = \frac{V_\infty^2}{gn} = \frac{2L}{\rho_\infty SC_L gn} = \frac{2L}{\rho_\infty SC_L g(L/W)} = \frac{2}{\rho_\infty C_L g}\frac{W}{S} \qquad\qquad \textbf{[6.50]}$$

and

$$\omega = \frac{gn}{V_\infty} = \frac{gn}{\sqrt{2L/(\rho_\infty SC_L)}} = \frac{gn}{\sqrt{[2n/(\rho_\infty C_L)](W/S)}} = g\sqrt{\frac{n\rho_\infty C_L}{2(W/S)}} \qquad\qquad \textbf{[6.51]}$$

Examining Eqs. (6.50) and (6.51), we see clearly R will be a minimum and ω will be a maximum when both C_L and n are maximum. That is,

$$R_{min} = \frac{2}{\rho g (C_L)_{max}} \frac{W}{S}$$ **[6.52]**

and

$$\omega_{max} = g \sqrt{\frac{\rho_\infty (C_L)_{max} n_{max}}{2(W/S)}}$$ **[6.53]**

However, keep in mind that n_{max} is itself limited by $(C_L)_{max}$ via Eq. (6.23), repeated here:

$$n_{max} = \frac{1}{2} \rho_\infty V_\infty^2 \frac{(C_L)_{max}}{W/S}$$ **[6.23]**

DESIGN CAMEO

For an instantaneous maneuver, the two design characteristics that are important are the maximum lift coefficient $(C_L)_{max}$ and wing loading W/S. The minimum turn radius can be made smaller, and the maximum turn rate can be made larger, by designing the airplane with a higher $(C_L)_{max}$ and a smaller W/S. For an instantaneous maneuver, the thrust-to-weight ratio does not play a role.

6.5 THE V–n DIAGRAM

There are structural limitations on the maximum load factor allowed for a given airplane. These structural limitations were not considered in the previous sections; let us examine them now.

There are two categories of structural limitations in airplane design:

1. *Limit load factor.* This is the boundary associated with *permanent* structural deformation of one or more parts of the airplane. If n is less than the limit load factor, the structure may deflect during a maneuver, but it will return to its original state when $n = 1$. If n is greater than the limit load factor, then the airplane structure will experience a permanent deformation, that is, it will incur *structural damage.*

2. *Ultimate load factor.* This is the boundary associated with outright *structural failure.* If n is greater than the ultimate load factor, parts of the airplane will break.

Both the aerodynamic and structural limitations for a given airplane are illustrated in the *V–n diagram*, a plot of load factor versus flight velocity, as given in Fig. 6.7. A V–n diagram is a type of "flight envelope" for a given airplane; it establishes the maneuver boundaries. Let us examine Fig. 6.7 in greater detail.

The curve between points A and B in Fig. 6.7 represents the aerodynamic limit on load factor imposed by $(C_L)_{max}$. This curve is literally a plot of Eq. (6.23). The region above curve AB in the V–n diagram is the stall region. To understand the significance of curve AB better, consider an airplane flying at velocity V_1, where V_1 is shown in Fig. 6.7. Assume the airplane is at an angle of attack such that $C_L < (C_L)_{max}$. This flight condition is represented by point 1 in Fig. 6.7. Now assume the angle of attack is increased to that for $(C_L)_{max}$, keeping the velocity constant at V_1. The lift increases to its maximum value for the given V_1, and hence the local factor $n = L/W$ reaches its maximum value for the given V_1. This value of n_{max} is given by Eq. (6.23), and the corresponding flight condition is given by point 2 in Fig. 6.7. If the angle of attack is increased further, the wing stalls and the load factor decreases. Therefore, point 3 in Fig. 6.7 is unobtainable in flight. Point 3 is in the stall region of the V–n diagram. Consequently, point 2 represents the highest possible load factor that can be obtained

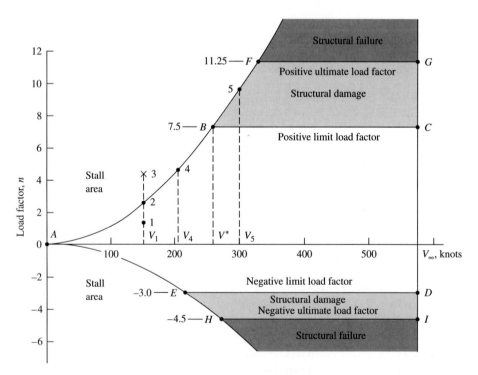

Figure 6.7   The V–n diagram for a typical jet trainer aircraft. Free-stream velocity V_∞ is given in knots. 1 knot (kn) = 1.15 mi/h.

at the given velocity V_1. As V_∞ is increased, say, to a value of V_4, then the maximum possible load factor n_{max} also increases, as given by point 4 in Fig. 6.7. However, n_{max} cannot be allowed to increase indefinitely. It is constrained by the structural limit load factor, given by point B in Fig. 6.7.

The horizontal line BC denotes the *positive limit load factor* in the V–n diagram. The flight velocity corresponding to B is designated as V^*. At velocities higher than V^*, say, V_5, the airplane must fly at values of C_L less than $(C_L)_{max}$ so that the positive limit load factor is not exceeded. If flight at $(C_L)_{max}$ is obtained at velocity V_5, corresponding to point 5 in Fig. 6.7, then structural damage or possibly structural failure will occur. The right-hand side of the V–n diagram, line CD, is a high-speed limit. At flight velocities higher than this limit (to the right of line CD), the dynamic pressure is higher than the design range for the airplane. This will exacerbate the consequences of other undesirable phenomena that may occur in high-speed flight, such as encountering a critical gust and experiencing destructive flutter, aileron reversal, wing or surface divergence, and severe compressibility buffeting. Any one of these phenomena in combination with the high dynamic pressure could cause structural damage or failure. The high-speed limit velocity is a *red-line speed* for the airplane; it should never be exceeded. By design, it is higher than the level flight maximum cruise velocity V_{max}, determined in Chapter 5, by at least a factor of 1.2. It may be as high as the terminal dive velocity of the aircraft. The bottom part of the V–n diagram, given by curve AE and the horizontal line ED in Fig. 6.7, corresponds to negative absolute angles of attack, that is, negative lift, and hence the load factors are negative quantities. Curve AE defines the stall limit. (If the wing is pitched downward to a large enough negative angle of attack, the flow will separate from the bottom surface of the wing and the negative lift will decrease in magnitude; that is, the wing "stalls.") Line ED gives the negative limit load factor, beyond which structural damage will occur. Line HI gives the negative ultimate load factor beyond which structural failure will occur.

For instantaneous maneuver performance, point B on the V–n diagram in Fig. 6.7 is very important. This point is called the *maneuver point*. At this point, both C_L and n are simultaneously at their highest possible values that can be obtained anywhere throughout the allowable flight envelope of the airplane. In turn, from Eqs. (6.52) and (6.53), this point simultaneously corresponds to the smallest possible instantaneous turn radius and the largest possible instantaneous turn rate for the airplane. The velocity corresponding to point B is called the *corner velocity* and is designated by V^* in Fig. 6.7. The corner velocity can be obtained by solving Eq. (6.23) for velocity, yielding

$$V^* = \sqrt{\frac{2n_{max}}{\rho_\infty (C_L)_{max}} \frac{W}{S}} \qquad \textbf{[6.54]}$$

In Eq. (6.54), the value of n_{max} corresponds to that at point B in Fig. 6.7. The corner velocity is an interesting dividing line. At flight velocities less than V^*, it is not possible to structurally damage the airplane due to the generation of too much lift. In contrast, at velocities greater than V^*, lift can be obtained that can structurally damage the aircraft (e.g., point 5 in Fig. 6.7), and the pilot must make certain to avoid such a case.

Example 6.3

For our Gulfstream-like aircraft, assume the positive limit load factor is 4.5. Calculate the airplane's corner velocity at sea level.

Solution

From previous examples, for this airplane $W/S = 76.84$ lb/ft^2 and $(C_L)_{max} = 1.2$. Since $\rho_\infty = 0.002377$ slug/ft^3 at sea level, Eq. (6.54) yields

$$V^* = \sqrt{\frac{2n_{max}}{\rho_\infty (C_L)_{max}} \frac{W}{S}} = \sqrt{\frac{2(4.5)(76.84)}{(0.002377)(1.2)}} = \boxed{492.4 \text{ ft/s}}$$

DESIGN CAMEO

For airplane design, the limit load factor depends on the type of aircraft. Some typical values for limit load factors are given below (Ref. 25).

Aircraft Type	n_{pos}	n_{neg}
Normal general aviation	2.5–3.8	−1−−1.5
Aerobatic general aviation	6	−3
Civil transport	3–4	−1−−2
Fighter	6.5–9	−3−−6

The values shown in Fig. 6.7, namely, $n_{pos} = 7.5$ and $n_{neg} = -3.0$, are for a typical military trainer aircraft.

Note from the above table and Fig. 6.7 that the magnitudes of n_{neg} are smaller than those for n_{pos}. This is a design decision which reflects that airplanes rarely fly under conditions of negative lift.

Because most airplanes are constructed primarily from aluminum alloys, for which the ultimate allowable stress is about 50% greater than the yield stress, a factor of safety of 1.5 is generally used between the ultimate load factor and the limit load factor. Note in Fig. 6.7 that the positive ultimate load factor is $7.5 \times 1.5 = 11.25$, and the negative ultimate load factor is $-3.0 \times 1.5 = -4.5$.

6.6   ENERGY CONCEPTS: ACCELERATED RATE OF CLIMB

The discussion of rate of climb in Section 5.10 was limited to the equilibrium case, that is, no acceleration. Indeed, some of the analysis in Section 5.10 included the assumption that the steady-state climb angle θ was small enough that $\cos \theta \approx 1$. In this section we remove those constraints and deal with the general case of accelerated rate of climb at any climb angle. Unlike the approach taken in all our performance analyses to this point, where we dealt with forces and invoked Newton's second law for our fundamental dynamic equation, in this present section we take a different approach where we will deal with energy concepts. Energy methods have been used since the 1970s for the analysis of airplane performance with acceleration. The subject of this section is an example of such energy methods.

Energy Height Consider an airplane of mass m in flight at some altitude h and with some velocity V_∞. Due to its altitude, the airplane has *potential energy* equal to mgh. Due to its velocity, the airplane has kinetic energy equal to $\frac{1}{2}mV_\infty^2$. The total energy of the airplane is the sum of these energies, that is,

$$\text{Total aircraft energy} = mgh + \tfrac{1}{2}mV_\infty^2 \qquad \textbf{[6.55]}$$

The *specific energy*, denoted by H_e, is defined as total energy per unit weight and is obtained by dividing Eq. (6.55) by $W = mg$. This yields

$$H_e = \frac{mgh + \frac{1}{2}mV_\infty^2}{W} = \frac{mgh + \frac{1}{2}mV_\infty^2}{mg}$$

or

$$\boxed{H_e = h + \frac{V_\infty^2}{2g}} \qquad \textbf{[6.56]}$$

The specific energy H_e has units of height and is therefore also called the *energy height* of the aircraft. Thus, let us become accustomed to quoting the energy of an airplane in terms of its energy height H_e, always remembering that it is simply the sum of the potential and kinetic energies of the airplane per unit weight. Contours of constant H_e are given in Fig. 6.8, which is an "altitude–Mach number map." Here the ordinate and abscissa are altitude h and Mach number M, respectively, and the dashed curves are lines of constant energy height.

We can draw an analogy between energy height and money in the bank. Say that you have a sum of money in the bank split between a checking account and a savings account. Say that you transfer part of your money in the savings account into your checking account. You still have the same total; the distribution of funds between the two accounts is just different. Energy height is analogous to the total of money in the bank; the distribution between kinetic energy and potential energy can change, but the total will be the same. For example, consider two airplanes, one flying at an altitude of 30,000 ft at Mach 0.81 (point A in Fig. 6.8) and the other flying at an altitude of 10,000 ft at Mach 1.3 (point B in Fig. 6.8). Both airplanes have the same energy height of 40,000 ft (check this yourself). However, airplane A has more potential energy and less kinetic energy (per unit weight) than airplane B. If both airplanes maintain their same states of total energy, then both are capable of "zooming" to an altitude of 40,000 ft at zero velocity (point C in Fig. 6.8) simply by trading all their kinetic energy for potential energy.

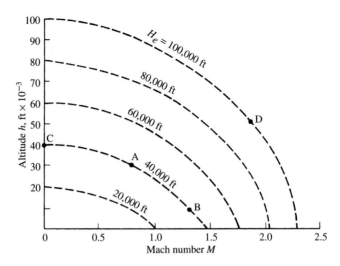

Figure 6.8   Altitude–Mach number map showing curves of constant energy height. These are universal curves that represent the variation of kinetic and potential energies per unit mass. They do not depend on the specific design factors of a given airplane.

Consider another airplane, flying at an altitude of 50,000 ft at Mach 1.85, denoted by point D in Fig. 6.8. This airplane will have an energy height of 100,000 ft and is indeed capable of zooming to an actual altitude of 100,000 ft by trading all its kinetic energy for potential energy. Airplane D is in a much higher energy state ($H_e = 100{,}000$ ft) than airplanes A and B (which have $H_e = 40{,}000$ ft). Therefore airplane D has a much greater capability for speed and altitude performance than airplanes A and B. In air combat, everything else being equal, it is advantageous to be in a higher energy state (have a higher H_e) than your adversary.

Example 6.4

Consider an airplane flying at an altitude of 30,000 ft at a velocity of 540 mi/h. Calculate its energy height.

Solution

$$V_\infty = 540 \times \frac{88}{60} = 792 \text{ ft/s}$$

From Eq. (6.56),

$$H_e = h + \frac{V_\infty^2}{2g} = 30{,}000 + \frac{(792)^2}{2(32.2)} = \boxed{39{,}740 \text{ ft}}$$

Specific Excess Power How does an airplane change its energy state; for example, in Fig. 6.8, how could airplanes A and B increase their energy heights to equal that of D? The answer to this question has to do with specific excess power, defined below.

Examine again Fig. 4.2, which illustrates an airplane in motion in the vertical plane, and Fig. 4.4, which gives the forces in the vertical plane. The equation of motion along the flight path is given by Eq. (4.5) which, assuming zero bank angle ($\phi = 0$) and the thrust aligned in the direction of V_∞ ($\epsilon = 0$), reduces to

$$T - D - W \sin\theta = m\frac{dV_\infty}{dt} \qquad \textbf{[6.57]}$$

Since $m = W/g$, Eq. (6.57) can be written as

$$T - D = W\left(\sin\theta + \frac{1}{g}\frac{dV_\infty}{dt}\right)$$

Multiplying by V_∞/W, we obtain

$$\frac{TV_\infty - DV_\infty}{W} = V_\infty \sin\theta + \frac{V_\infty}{g}\frac{dV_\infty}{dt} \qquad \textbf{[6.58]}$$

Recall from Eq. (5.79) that

$$TV_\infty - DV_\infty = \text{excess power} \qquad \textbf{[5.79]}$$

We define *specific excess power*, denoted by P_s, as the excess power per unit weight. From Eq. (5.79),

$$P_s \equiv \frac{\text{excess power}}{W} = \frac{TV_\infty - DV_\infty}{W} \qquad \textbf{[6.59]}$$

Also, recall from Eq. (5.77) that the rate of climb R/C is expressed by

$$R/C = V_\infty \sin\theta \qquad \textbf{[5.77]}$$

Since rate of climb is simply the time rate of change of altitude $R/C = dh/dt$, Eq. (5.77) can be written as

$$V_\infty \sin\theta = \frac{dh}{dt} \qquad \textbf{[6.60]}$$

Substituting Eqs. (6.59) and (6.60) into Eq. (6.58), we have

$$\boxed{P_s = \frac{dh}{dt} + \frac{V_\infty}{g}\frac{dV_\infty}{dt}} \qquad \textbf{[6.61]}$$

Equation (6.61) shows that an airplane with excess power can use this excess for rate of climb (dh/dt) or to accelerate along its flight path (dV/dt) or for a combination of both. Equation (6.61) helps to put our discussion in Section 5.10 in perspective. In Section 5.10 we assumed no acceleration, that is, $dV_\infty/dt = 0$. For this case, Eq. (6.61) becomes

$$P_s = \frac{dh}{dt} \qquad \textbf{[6.62]}$$

In Section 5.10, our governing relation for steady climb was Eq. (5.80), rewritten here:

$$\frac{\text{Excess power}}{W} = R/C \qquad \textbf{[6.63]}$$

Equations (6.62) and (6.63) are the same equation. So our discussion in Section 5.10 was based on a special form of Eq. (6.61), namely, Eq. (6.62).

Specific excess power allows an increase in the energy height of an airplane, as follows. Return to the definition of energy height given by Eq. (6.56). Differentiating this expression with respect to time, we have

$$\frac{dH_e}{dt} = \frac{dh}{dt} + \frac{V_\infty}{g}\frac{dV_\infty}{dt} \qquad \textbf{[6.64]}$$

The right-hand sides of Eqs. (6.61) and (6.64) are identical. Hence

$$\boxed{P_s = \frac{dH_e}{dt}} \qquad \textbf{[6.65]}$$

That is, the *time rate of change of energy height is equal to the specific excess power.* An airplane can increase its energy height simply by the application of excess power. In Fig. 6.8, airplanes A and B can reach the energy height of airplane D *if* they have enough specific excess power to do so.

Question: How can we ascertain whether a given airplane has enough P_s to reach a certain energy height? The answer has to do with contours of constant P_s on an altitude–Mach number map. Let us see how such contours can be constructed.

P_s Contours   Return to Fig. 5.33, and recall that excess power is the difference between power available and power required. For a given altitude, say, h, the excess power (hence P_s) can be plotted versus velocity (or Mach number); P_s first increases with velocity, then reaches a maximum, and finally decreases to zero as the velocity approaches V_{max} for the airplane. This variation is sketched in Fig. 6.9a, which is a graph of P_s versus Mach number. Three curves are shown, each one corresponding to a given altitude. These results can be cross-plotted on an altitude-Mach number map using P_s as a parameter, as illustrated in Fig. 6.9b. For example, consider all the points on Fig. 6.9a where $P_s = 0$; these correspond to points along a horizontal axis through $P_s = 0$, that is, points along the abscissa in Fig. 6.9a. Such points are labeled a, b, c, d, e, and f in Fig. 6.9a. Now replot these points on the altitude–Mach number map in Fig. 6.9b. Here, points a, b, c, d, e, and f form a bell-shaped curve, along which $P_s = 0$. This curve is called the P_s *contour* for $P_s = 0$. Similarly, all points with $P_s = 200$ ft/s are on the horizontal line AB in Fig. 6.9a, and these points can be cross-plotted to generate the $P_s = 200$ ft/s contour in Fig. 6.9b. In this fashion, an entire series of P_s contours can be generated in the altitude–Mach number map.

The shapes of the curves shown in Fig. 6.9a and b are typical of a subsonic airplane. They look somewhat different for a supersonic airplane because of the effect of the drag-divergence phenomenon on drag, hence excess power. For a supersonic airplane, the P_s–Mach number curves at different altitudes will appear as sketched in

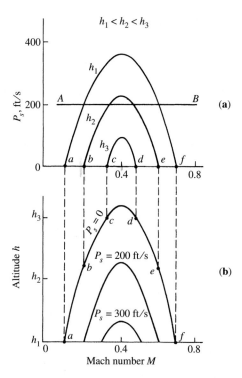

Figure 6.9 Construction of the specific excess power contours in the altitude–Mach number map for a subsonic airplane below the drag-divergence Mach number. These contours are constructed for a fixed load factor; if the load factor is changed, the P_s contours will shift.

Fig. 6.10**a**. The "dent" in the curves around Mach 1 is due to the large drag increase in the transonic flight regime. For modern fighters, such as the Lockheed-Martin F-16, the maximum thrust available from the engine is so large compared to the thrust required near Mach 1 that the dent in the transonic region is much less pronounced. In turn, the curves in Fig. 6.10**a** can be cross-plotted on the altitude–Mach number map, producing the P_s contours as illustrated in Fig. 6.10**b**. Owing to the double-hump shape of the P_s curves in Fig. 6.10**a**, the P_s contours in Fig. 6.10**b** have different shapes in the subsonic and supersonic regions.

We can now answer the question of how to ascertain whether a given airplane has enough P_s to reach a certain energy height. Let us overlay the P_s contours, say, from Fig. 6.10**b**, and the curves for constant energy height shown in Fig. 6.8—all on an altitude–Mach number map. This overlay is given in Fig. 6.11. The P_s contours pertain to a specific airplane at a given load factor; the curves for constant energy height are universal curves that have nothing to do with any given airplane. The usefulness of Fig. 6.11 is that it clearly establishes what energy heights are obtainable by a given airplane. The regime of sustained flight for the airplane lies *inside* the

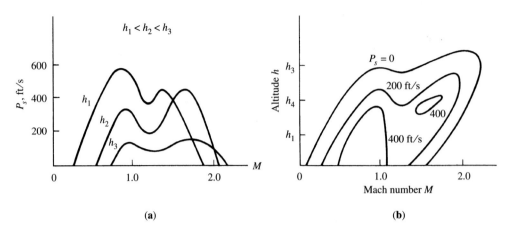

(a)   **(b)**

Figure 6.10   Specific excess power contours for a supersonic airplane.

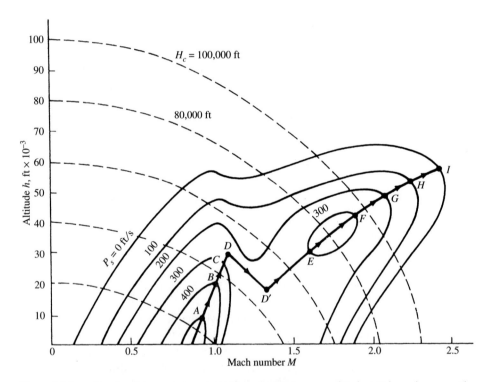

Figure 6.11   Overlay of P_s contours and specific energy states on an altitude–Mach number map. The P_s values shown here approximately correspond to a Lockheed F-104G supersonic fighter. Load factor $n = 1$ and $W = 18,000$ lb. Airplane is at maximum thrust. The path given by points A through I is the flight path for minimum time to climb.

envelope formed by the $P_s = 0$ contour. Hence, all values of H_e inside this envelope are obtainable by the airplane. A comparison of figures like Fig. 6.11 for different airplanes will clearly show in what regions of altitude and Mach number an airplane has maneuver advantages over another.

Rate of Climb and Time to Climb (Accelerated Performance) Accelerated rate of climb and time to climb can be treated by energy considerations. The rate of climb with acceleration is easily found from Eq. (6.61), repeated here:

$$P_s = \frac{dh}{dt} + \frac{V_\infty}{g}\frac{dV_\infty}{dt}$$

[6.61]

Rate of climb   Acceleration

Consider an airplane at a given altitude and Mach number. This flight condition is represented by a specific point in the altitude–Mach number map, such as Fig. 6.9b. At this point, the airplane will have a certain value of P_s. Assume the airplane is accelerating, with a specified value of $dV_\infty/dt \equiv A$. The rate of climb for this specified accelerated condition is, from Eq. (6.61),

$$\boxed{\frac{dh}{dt} = P_s - \frac{V_\infty}{g}A}$$

[6.66]

In Eq. (6.66), all quantities on the right-hand side are known or specified; the equation gives the instantaneous maximum rate of climb that can be achieved at the instantaneous velocity V_∞ and the instantaneous acceleration A.

The time required for an airplane to change from one energy height $H_{e,1}$ to a larger energy height $H_{e,2}$ can be obtained as follows. From Eq. (6.65),

$$dt = \frac{dH_e}{P_s}$$

[6.67]

Integrating Eq. (6.67) between time t_1 where $H_e = H_{e,1}$ and time t_2 where $H_e = H_{e,2}$, we have

$$t_2 - t_1 = \int_{H_{e,1}}^{H_{e,2}} \frac{dH_e}{P_s}$$

[6.68]

Since $H_e = h + V_\infty^2/(2g)$, then

$$H_{e,2} - H_{e,1} = h_2 - h_1 + \frac{1}{2g}\left(V_{\infty,2}^2 - V_{\infty,1}^2\right)$$

or

$$h_2 - h_1 = H_{e,2} - H_{e,1} - \frac{1}{2g}\left(V_{\infty,2}^2 - V_{\infty,1}^2\right)$$

[6.69]

For a given $V_{\infty,2}$ at energy height $H_{e,2}$, and a given $V_{\infty,1}$ at energy height $H_{e,1}$, Eq. (6.69) gives the change in altitude $h_2 - h_1$ between these two conditions. Equation (6.68) gives the time required to achieve this change in altitude; that is, it gives the time to climb from altitude h_1 to altitude h_2 when the airplane has accelerated (or decelerated) from velocity $V_{\infty,1}$ at altitude h_1 to velocity $V_{\infty,2}$ at altitude h_2.

The time to climb $t_2 - t_1$ between $H_{e,1}$ and $H_{e,2}$ is *not* a unique value—*it depends on the flight path taken in the altitude–Mach number map.* Examine again Fig. 6.11. In changing from $H_{e,1}$ to $H_{e,2}$, there are an infinite number of variations of altitude and Mach number that will get you there. In terms of Eq. (6.68), there are an infinite number of different values of the integral because there are an infinite number of different possible variations of dH_e/P_s between $H_{e,1}$ and $H_{e,2}$. However, once a *specific path* in Fig. 6.11 is chosen between $H_{e,1}$ and $H_{e,2}$ then dH_e/P_s has a definite variation along this path, and a specific value of $t_2 - t_1$ is obtained.

This discussion has particular significance to the calculation of *minimum* time to climb to a given altitude, which is a unique value. There is a unique path in the altitude–Mach number map that corresponds to minimum time to climb. We can see how to construct this path by examining Eq. (6.68). The time to climb will be a minimum when P_s is a maximum value. Looking at Fig. 6.11, for each H_e curve, we see there is a point where P_s is a maximum. Indeed, at this point the P_s curve is *tangent* to the H_e curve. Such points are illustrated by points A to I in Fig. 6.11. The heavy curve through these points illustrates the variation of altitude and Mach number along the flight path for minimum time to climb. Along this path (the heavy curve), dH_e/P_s varies in a definite way, and when these values of dH_e/P_s are used in calculating the integral in Eq. (6.68), the resulting value of $t_2 - t_1$ *is* the minimum time to climb between $H_{e,1}$ and $H_{e,2}$. In general, there is no analytical form of the integral in Eq. (6.68); it is usually evaluated numerically.

We note in Fig. 6.11 that the segment of the flight path between D and D' represents a constant energy dive to accelerate through the drag-divergence region near Mach 1. We also note that Eq. (6.68) gives the time to climb between two energy heights, not necessarily that between two different altitudes. However, at any given constant energy height, kinetic energy can be traded for potential energy, and the airplane can "zoom" to higher altitudes until all the kinetic energy is spent. For example, in Fig. 6.11 point I corresponds to $P_s = 0$. The airplane cannot achieve any further increase in energy height. However, after arriving at point I, the airplane can zoom to a minimum altitude equal to the value of H_e at point I—in Fig. 6.11, a maximum altitude well above 100,000 ft. After the end of the zoom, $V_\infty = 0$ (by definition) and the corresponding value of h is the maximum obtainable altitude for accelerated flight conditions, achieved in a minimum amount of time.

Example 6.5 | Consider an airplane with an instantaneous acceleration of 8 ft/s^2 at an instantaneous velocity of 800 ft/s. At the existing flight conditions, the specific excess power is 300 ft/s. Calculate the instantaneous maximum rate of climb that can be obtained at these accelerated flight conditions.

Solution

From Eq. (6.66)

$$\frac{dh}{dt} = P_s - \frac{V_\infty}{g} A = 300 - \frac{800}{32.2}(8) = \boxed{101 \text{ ft/s}}$$

6.7 TAKEOFF PERFORMANCE

For the performance characteristics discussed so far in this book, we have considered the airplane in full flight in the air. However, for the next two sections, we come back to earth, and we explore the characteristics of takeoff and landing, many of which are concerned with the airplane rolling along the ground. These are accelerated performance problems of a special nature.

Consider an airplane standing motionless at the end of a runway. This is denoted by location 0 in Fig. 6.12. The pilot releases the brakes and pushes the throttle to maximum takeoff power, and the airplane accelerates down the runway. At some distance from its starting point, the airplane lifts into the air. How much distance does the airplane cover along the runway before it lifts into the air? This is the central question in the analysis of takeoff performance. Called the *ground roll* (or sometimes the *ground run*) and denoted by s_g in Fig. 6.12, it is a major focus of this section. However, this is not the whole consideration. The total takeoff distance also includes the extra distance covered over the ground after the airplane is airborne but before it clears an obstacle of a specified height. This is denoted by s_a in Fig. 6.12. The height of the obstacle is generally specified to be 50 ft for military aircraft and 35 ft for commercial aircraft. The sum of s_g and s_a is the total takeoff distance for the airplane.

The ground roll s_g is further divided into intermediate segments, as shown in Fig. 6.13. These segments are defined by various velocities, as follows:

1. As the airplane accelerates from zero velocity, at some point it will reach the stalling velocity V_{stall}, as noted in Fig. 6.13.

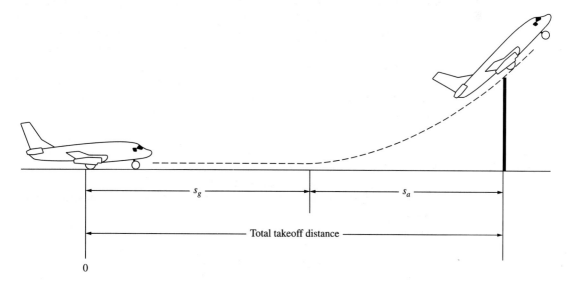

Figure 6.12 Illustration of ground roll s_g, airborne distance s_a, and total takeoff distance.

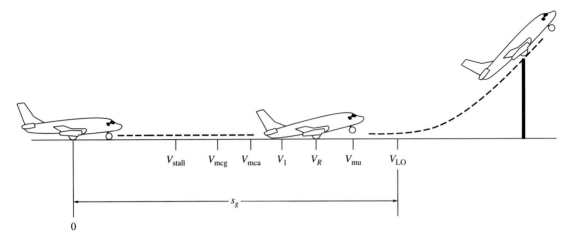

Figure 6.13   Intermediate segments of the ground roll.

2.   The airplane continues to accelerate until it reaches the *minimum control speed on the ground*, denoted by V_{mcg} in Fig. 6.13. This is the minimum velocity at which enough aerodynamic force can be generated on the vertical fin with rudder deflection while the airplane is still rolling along the ground to produce a yawing moment sufficient to counteract that produced when there is an engine failure for a multiengine aircraft.

3.   If the airplane were in the air (without the landing gear in contact with the ground), the minimum speed required for yaw control in case of engine failure is slightly greater than V_{mcg}. This velocity is called the *minimum control speed in the air*, denoted by V_{mca} in Fig. 6.13. For the ground roll shown in Fig. 6.13, V_{mca} is essentially a reference speed—the airplane is still on the ground when this speed is reached.

4.   The airplane continues to accelerate until it reaches the *decision speed*, denoted by V_1 in Fig. 6.13. This is the speed at which the pilot can successfully continue the takeoff even though an engine failure (in a multiengine aircraft) would occur at that point. This speed must be equal to or larger than V_{mcg} in order to maintain control of the airplane. A more descriptive name for V_1 is the *critical engine failure speed*. If an engine fails before V_1 is achieved, the takeoff must be stopped. If an engine fails after V_1 is reached, the takeoff can still be achieved.

5.   The airplane continues to accelerate until the takeoff rotational speed, denoted by V_R in Fig. 6.13, is achieved. At this velocity, the pilot initiates by elevator deflection a rotation of the airplane in order to increase the angle of attack, hence to increase C_L. Clearly, the maximum angle of attack achieved during rotation should not exceed the stalling angle of attack. Actually, all that is needed is an angle of attack high enough to produce a lift at the given velocity larger than the weight, so that the airplane will lift off the ground. However, even this angle of attack may not

be achievable because the tail may drag the ground. (Ground clearance for the tail after rotation is an important design feature for the airplane, imposed by takeoff considerations.)

6. If the rotation of the airplane is limited by ground clearance for the tail, the airplane must continue to accelerate while rolling along the ground after rotation is achieved, until a higher speed is reached where indeed the lift becomes larger than the weight. This speed is called the *minimum unstick speed*, denoted by V_{mu} in Fig. 6.13. For the definition of V_{mu}, it is assumed that the angle of attack achieved during rotation is the maximum allowable by the tail clearance.

7. However, for increased safety, the angle of attack after rotation is slightly less than the maximum allowable by tail clearance, and the airplane continues to accelerate to a slightly higher velocity, called the *liftoff speed*, denoted by V_{LO} in Fig. 6.13. This is the point at which the airplane actually lifts off the ground. The total distance covered along the ground to this point is the ground roll s_g.

The relative values of the various velocities discussed above, and noted on Fig. 6.13, are all sandwiched between the value of V_{stall} and that for V_{LO}, where usually $V_{LO} \approx 1.1 V_{stall}$. A nice discussion of the relative values of the velocities noted in Fig. 6.13 is contained in Ref. 41, which should be consulted for more details.

Related to the above discussion is the concept of *balanced field length*, defined as follows. The decision speed V_1 was defined earlier as the minimum velocity at which the pilot can successfully continue the takeoff even though an engine failure would occur at that point. What does it mean that the pilot "can successfully continue the takeoff" in such an event? The answer is that when the airplane reaches V_1, if an engine fails at that point, then the additional distance required to clear the obstacle at the end of takeoff is exactly the same distance as required to bring the airplane to a stop on the ground. If we let A be the distance traveled by the airplane along the ground from the original starting point (point 0 in Fig. 6.13) to the point where V_1 is reached, and we let B be the additional distance traveled with an engine failure (the same distance to clear an obstacle or to brake to a stop), then the balanced field length is by definition the *total* distance $A + B$.

6.7.1 Calculation of Ground Roll

The forces acting on the airplane during takeoff are shown in Fig. 6.14. In addition to the familiar forces of thrust, weight, lift, and drag, there is a rolling resistance R, caused by friction between the tires and the ground. This resistance force is given by

$$R = \mu_r (W - L) \qquad \textbf{[6.70]}$$

where μ_r is the coefficient of rolling friction and $W - L$ is the net normal force exerted between the tires and the ground. Summming forces parallel to the ground and employing Newton's second law, we have from Fig. 6.14

$$m \frac{dV_\infty}{dt} = T - D - R$$

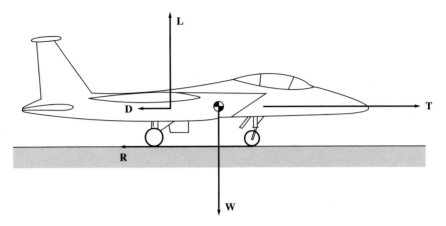

Figure 6.14   Forces acting on an airplane during takeoff and landing.

or

$$m \frac{dV_\infty}{dt} = T - D - \mu_r(W - L) \qquad \text{[6.71]}$$

Equation (6.71) is the equation of motion for the airplane during takeoff.

Let us examine the terms on the right-hand side of Eq. (6.71). The engine thrust T in general varies with velocity during the ground roll. For a reciprocating engine/propeller combination, the power available is reasonably constant with V_∞ (see Section 3.3.1). Since $P = T V_\infty$, during the ground roll,

$$\text{Reciprocating engine/propeller} \qquad T = \frac{\text{const}}{V_\infty} \qquad \text{[6.72]}$$

For a turbojet engine, T is reasonably constant with V_∞ for the ground roll (see Section 3.4.1).

$$\text{Turbojet} \qquad T = \text{const} \qquad \text{[6.73]}$$

For a turbofan engine, T deceases slightly with V_∞ during the ground roll. An example for the Rolls-Royce RB211-535E4 turbofan was given by Eq. (3.22). Following this example, we can write for a turbofan engine during the ground roll

$$\text{Turbofan} \qquad T = k_1^* - k_2^* V_\infty + k_3^* V_\infty^2 \qquad \text{[6.74]}$$

where the values of k_1^*, k_2^*, and k_3^* are constants obtained from the performance characteristics for a given engine.

The drag D in Eq. (6.71) varies with velocity according to

$$D = \tfrac{1}{2}\rho_\infty V_\infty^2 S C_D \qquad \text{[6.75]}$$

However, during the ground roll, C_D in Eq. (6.75) is *not* the same value as given by the conventional drag polar used for full flight in the atmosphere; the conventional drag polar is given by Eq. (2.47), repeated here:

$$C_D = C_{D,0} + K C_L^2 \qquad \textbf{[2.47]}$$

This is for two primary reasons: (1) With the landing gear fully extended, $C_{D,0}$ is larger than when the landing gear is retracted; and (2) there is a reduction in the induced drag due to the close proximity of the wings to the ground—part of the "ground effect." An approximate expression for the increase in $C_{D,0}$ due to the extended landing gear is given in Ref. 41 as

$$\Delta C_{D,0} = \frac{W}{S} K_{uc} m^{-0.215} \qquad \textbf{[6.76]}$$

where W/S is the wing loading, m is the maximum mass of the airplane, and the factor K_{uc} depends on the amount of flap deflection. With flap deflection, the average airflow velocity over the bottom of the wing is lower than it would be with no flap deflection; that is, the deflected flap partially blocks the airflow over the bottom surface. Hence, the landing gear drag is less with flap deflection than its value with no flap deflection. In Eq. (6.76), when W/S is in units of newtons per square meter and m is in units of kilograms, $K_{uc} = 5.81 \times 10^{-5}$ for a zero flap deflection and 3.16×10^{-5} for maximum flap deflection. These values are based on correlations for a number of civil transports, and are approximate only. In regard to the induced drag during the ground roll, the downwash is somewhat inhibited by the proximity of the ground, and hence the induced drag contribution is less than that included in Eq. (2.47); that is, for the ground roll, K in Eq. (2.47) must be reduced below that for the airplane in flight. The reduction in the induced drag coefficient can be approximated by the relation from Ref. 50 given here

$$\frac{C_{D_i} \text{ (in-ground effect)}}{C_{D_i} \text{ (out-of-ground effect)}} \equiv G = \frac{(16h/b)^2}{1 + (16h/b)^2} \qquad \textbf{[6.77]}$$

where h is the height of the wing above the ground and b is the wingspan. To return to our discussion of the drag polar in Section 2.9.2, the value of k_3 in Eq. (2.44) is reduced by the factor G due to ground effect. Also, because wave drag does not occur at the low speeds for takeoff, $k_2 = 0$ in Eq. (2.44). With all the above in mind, the conventional drag polar given by Eq. (2.47) should be modified to account for the effects during ground roll. Returning to Eqs. (2.44) and (2.45), we can approximate the drag polar during ground roll as

$$C_D = C_{D,0} + \Delta C_{D,0} + (k_1 + G k_3) C_L^2 \qquad \textbf{[6.78]}$$

where $C_{D,0}$ and k_1 in Eq. (6.78) are the same values given by the conventional flight drag polar—$\Delta C_{D,0}$ is given by Eq. (6.76), and G is given by Eq. (6.77).

The value of the coefficient of rolling friction in Eq. (6.71) depends on the type of ground surface that the airplane is rolling on. It also depends on whether the wheel brakes are off or on. Obviously, during takeoff the brakes are off, and during landing the brakes are usually on. Some representative values of μ_r are listed in Table 6.1, as obtained from Ref. 25.

Table 6.1

	μ_r(Typical Values)	
Surface	Brakes off	Brakes on
Dry concrete/asphalt	0.03–0.05	0.3–0.5
Wet concrete/asphalt	0.05	0.15–0.3
Icy concrete/asphalt	0.02	0.06–0.10
Hard turf	0.05	0.4
Firm dirt	0.04	0.3
Soft turf	0.07	0.2
Wet grass	0.08	0.2

To return to the right-hand side of Eq. (6.71), the weight W is usually considered to be constant, although it is slightly decreasing due to the fuel's being consumed during takeoff. The lift L is given by

$$L = \tfrac{1}{2}\rho_\infty V_\infty^2 S C_L \qquad\qquad \textbf{[6.79]}$$

In Eq. (6.79), the lift coefficient is that for the angle of attack of the airplane rolling along the ground. In turn, the angle of attack during ground roll is essentially a design feature of the airplane, determined by the built-in incidence angle of the wing chord line relative to the fuselage, and by the built-in orientation of the centerline of the airplane relative to the ground due to the different height of the main landing gear relative to that of the nose (or tail) wheel. The value of C_L in Eq. (6.79) also depends on the extent to which the wing high-lift devices are employed during takeoff. Raymer (Ref. 25) states that C_L for the ground roll is typically less than 0.1. Of course, during the rotation phase near the end of the ground roll (see Fig. 6.13), the value of C_L will increase, and it is frequently limited by the amount of tail clearance. Hence, C_L in Eq. (6.79) is primarily determined (and limited) by features of the geometric design configuration of the airplane rolling along the ground.

A detailed calculation of the ground roll can be made by numerically solving Eq. (6.71) for $V_\infty = V(t)$, where in this equation T, D, and L are variables which take on their appropriate instantaneous values at each instant during the ground roll. For the numerical solution of Eq. (6.71), T can be expressed by Eq. (6.72) for a reciprocating engine/propeller combination, Eq. (6.73) for a turbojet, or Eq. (6.74) for a turbofan. The drag is expressed by Eq. (6.75) where C_D is given by Eq. (6.78). The lift is given by Eq. (6.79). From the numerical solution of Eq. (6.71), we obtain a tabulation of V_∞ versus t, starting with $V_\infty = 0$ at $t = 0$ and ending when $V_\infty = V_{LO}$. The value of V_{LO} is prescribed in advance; it is usually set equal to $1.1 V_{mu}$, where V_{mu} is the minimum unstick speed described earlier. Because of the limited tail clearance of many airplanes, the minimum unstick speed corresponds to a value of $C_L < (C_L)_{max}$, and hence $V_{mu} > V_{stall}$, as clearly shown in Fig. 6.13. As a result, the specified value of V_{LO} is close to $1.1 V_{stall}$. The value of t that exists when V_∞ reaches V_{LO} is denoted by t_{LO}, the liftoff time. The ground roll s_g, can then be obtained from

$$ds = \frac{ds}{dt} dt = V_\infty \, dt$$

or

$$\int_0^{s_g} ds = \int_0^{t_{LO}} V_\infty \, dt$$

or

$$s_g = \int_0^{t_{LO}} V_\infty \, dt \qquad \qquad \textbf{[6.80]}$$

The integral in Eq. (6.80) is evaluated numerically, using the tabulated values of V_∞ versus t obtained from the numerical solution of Eq. (6.71).

Approximate Analysis of Ground Roll   The numerical solution of the governing equation of motion described above does not readily identify the governing design parameters that determine takeoff performance. Let us extract these parameters from an approximate analysis of ground roll as follows.

Recalling that s is the distance along the ground, we can write

$$ds = \frac{ds}{dt} dt = V_\infty \, dt = V_\infty \frac{dt}{dV_\infty} dV_\infty$$

or

$$ds = \frac{V_\infty \, dV_\infty}{dV_\infty/dt} = \frac{d(V_\infty^2)}{2(dV_\infty/dt)} \qquad \qquad \textbf{[6.81]}$$

Let us now construct an appropriate expression for dV_∞/dt to be inserted into Eq. (6.81). Returning to Eq. (6.71), we have

$$\frac{dV_\infty}{dt} = \frac{1}{m}[T - D - \mu_r(W - L)] \qquad \qquad \textbf{[6.82]}$$

Substituting Eqs. (6.75) and (6.79) into Eq. (6.82), and noting that $m = W/g$, we have

$$\frac{dV_\infty}{dt} = \frac{g}{W}\left[T - \frac{1}{2}\rho_\infty V_\infty^2 S C_D - \mu_r\left(W - \frac{1}{2}\rho_\infty V_\infty^2 S C_L\right)\right]$$

or

$$\frac{dV_\infty}{dt} = g\left[\frac{T}{W} - \mu_r - \frac{\rho_\infty}{2(W/S)}(C_D - \mu_r C_L)V_\infty^2\right] \qquad \qquad \textbf{[6.83]}$$

In Eq. (6.83), C_D is given by Eq. (6.78). Hence, recalling that $k_3 \equiv 1/(\pi e AR)$, we have

$$\frac{dV_\infty}{dt} = g\left\{\frac{T}{W} - \mu_r - \frac{\rho_\infty}{2(W/S)}\left[C_{D,0} + \Delta C_{D,0} + \left(k_1 + \frac{G}{\pi e AR}\right)C_L^2 - \mu_r C_L\right]V_\infty^2\right\}$$

$$\textbf{[6.84]}$$

This is the expression for dV_∞/dt that is inserted into Eq. (6.81). To simplify the appearance of the following equations, we define the symbols K_T and K_A as

$$K_T = \frac{T}{W} - \mu_r \qquad [6.85]$$

$$K_A = -\frac{\rho_\infty}{2(W/S)}\left[C_{D,0} + \Delta C_{D,0} + \left(k_1 + \frac{G}{\pi e AR}\right)C_L^2 - \mu_r C_L\right] \qquad [6.86]$$

Then Eq. (6.84) can be written as

$$\frac{dV_\infty}{dt} = g\left(K_T + K_A V_\infty^2\right) \qquad [6.87]$$

Substituting Eq. (6.87) into Eq. (6.81), we have

$$ds = \frac{d(V_\infty^2)}{2g\left(K_T + K_A V_\infty^2\right)} \qquad [6.88]$$

Integrating Eq. (6.88) between $s = 0$ where $V_\infty = 0$ and $s = s_g$ where $V_\infty = V_{LO}$, we have

$$s_g = \int_0^{V_{LO}} \frac{d(V_\infty^2)}{2g(K_T + K_A V_\infty^2)} \qquad [6.89]$$

Up to this point, no simplifications have been made. The values of K_T and K_A vary with V_∞ during the ground roll; and if this variation is properly taken into account, a numerical evaluation of the integral in Eq. (6.89) will yield an accurate value of s_g.

The integral in Eq. (6.89) can be evaluated analytically by assuming that K_T and K_A are constant during the ground roll. Examining the definitions of K_T and K_A given by Eqs. (6.85) and (6.86), we see that this implies that

1. T/W is constant. This is a good approximation for a turbojet, but it is not good for a reciprocating engine/propeller aircraft or for a high-bypass-ratio turbofan. In the latter two cases where T varies with velocity, a frequently made assumption is to consider T in Eq. (6.85) to be a constant equal to its value at $V_\infty = 0.7V_{LO}$.

2. C_L is constant. This is a reasonable assumption during the ground roll to the point of rotation, because the angle of attack is fixed by the landing gear features and the design incidence angle of the wing chord with respect to the fuselage. During rotation C_L increases to a value that is equal to or less than $(C_L)_{max}$, depending on the degree of tail clearance with the ground. However, the distance covered over the ground during the rotation phase is relatively small compared to the total ground roll. Raymer (Ref. 25) suggests that the time between the initiation of rotation and actual liftoff is typically 3 s for large aircraft and about 1 s for small aircraft, and that the velocity of the airplane changes very little during this phase. Hence the ground distance covered during the rotation period can be approximated by $3V_{LO}$ for large airplanes and V_{LO} for small airplanes.

Thus, by assuming K_T and K_A are constant in Eq. (6.89), performing the integration, and allowing a distance equal to NV_{LO} for the rotation phase (where $N = 3$ for large aircraft and $N = 1$ for small aircraft), the ground roll can be approximated by

$$s_g = \frac{1}{2gK_A} \ln \left(1 + \frac{K_A}{K_T} V_{LO}^2 \right) + NV_{LO} \qquad \textbf{[6.90]}$$

With Eq. (6.90), a quick analytical evaluation of the ground roll can be made.

An analytic form for s_g that more clearly illustrates the design parameters that govern takeoff performance can be obtained by substituting Eq. (6.82) directly into Eq. (6.81), obtaining

$$ds = \frac{m}{2} \frac{d(V_\infty^2)}{T - D - \mu_r(W - L)} \qquad \textbf{[6.91]}$$

Integrating Eq. (6.91) from point 0 to liftoff, and again noting that $m = W/g$, we have

$$s_g = \frac{W}{2g} \int_0^{V_{LO}} \frac{d(V_\infty^2)}{T - D - \mu_r(W - L)} \qquad \textbf{[6.92]}$$

In Eq. (6.92), $T - D - \mu_r(W - L)$ is the net force acting in the horizontal direction on the airplane during takeoff. In Fig. 6.15, a schematic is shown of the variation of the forces acting during takeoff as a function of distance along the ground. Note that the net force $T - D - \mu_r(W - L)$, specifically identified in Fig. 6.15, does not vary greatly. This gives some justification to assuming that the expression $T - D - \mu_r(W - L)$ is constant up to the point of rotation. If we take this net force to be constant at a value

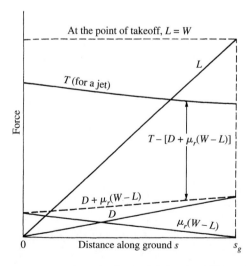

Figure 6.15 Schematic of a typical variation of forces acting on an airplane during takeoff.

equal to its value at $V_\infty = 0.7V_{LO}$, then Eq. (6.92) is easily integrated, giving

$$s_g = \frac{WV_{LO}^2}{2g}\left[\frac{1}{T - D - \mu_r(W - L)}\right]_{0.7V_{LO}} + NV_{LO} \qquad \textbf{[6.93]}$$

where the term NV_{LO} has been added to account for that part of the ground roll during rotation, as noted earlier.

The velocity at liftoff V_{LO} should be no less than $1.1V_{stall}$, where from Eq. (5.67)

$$V_{stall} = \sqrt{\frac{2}{\rho_\infty}\frac{W}{S}\frac{1}{(C_L)_{max}}} \qquad \textbf{[5.67]}$$

In Eq. (5.67), $(C_L)_{max}$ is that value with the flaps extended for takeoff; also keep in mind that $(C_L)_{max}$ may be a smaller value if the angle of attack is limited by tail clearance with the ground. Setting $V_{LO} = 1.1V_{stall}$ and inserting Eq. (5.67) into Eq. (6.93), we have

$$s_g = \frac{1.21(W/S)}{g\rho_\infty(C_L)_{max}\,[T/W - D/W - \mu_r\,(1 - L/W)]_{0.7V_{LO}}} + 1.1N\sqrt{\frac{2}{\rho_\infty}\frac{W}{S}\frac{1}{(C_L)_{max}}}$$

$$\textbf{[6.94]}$$

The design parameters that have an important effect on takeoff ground roll are clearly seen from Eq. (6.94). Specifically, s_g depends on wing loading, thrust-to-weight ratio, and maximum lift coefficient. From Eq. (6.94) we note that

1. s_g increases with an increase in W/S.
2. s_g decreases with an increase in $(C_L)_{max}$.
3. s_g decreases with an increase in T/W.

Equation (6.94) can be simplified by assuming that T is much larger than $D + \mu_r(W - L)$; as seen from Fig. 6.15, this is a reasonable assumption. Also, neglect the contribution to s_g due to the rotation segment. With this, Eq. (6.94) can be approximated by

$$s_g \approx \frac{1.21(W/S)}{g\rho_\infty(C_L)_{max}(T/W)} \qquad \textbf{[6.95]}$$

Equation (6.95) clearly illustrates some important physical trends:

1.   The ground roll is very sensitive to the weight of the airplane via both W/S and T/W. For example, if the weight is doubled, everything else being the same, then W/S is doubled and T/W is halved, leading to a factor-of-4 increase in s_g. Essentially, s_g varies as W^2.

2.   The ground roll is dependent on the ambient density, through both the explicit appearance of ρ_∞ in Eq. (6.95) and the effect of ρ_∞ on T. If we assume that $T \propto \rho_\infty$, then Eq. (6.95) shows that

$$s_g \propto \frac{1}{\rho_\infty^2}$$

This is why on hot, summer days, when the air density is less than that on cooler days, a given airplane requires a longer ground roll to get off the ground. Also, longer ground rolls are required at airports located at higher altitudes (such as Denver, Colorado, a mile above sea level).

3. The ground roll can be decreased by increasing the wing area (decreasing W/S), increasing the thrust (increasing T/W), and increasing $(C_L)_{max}$, all of which simply make common sense.

6.7.2 Calculation of Distance While Airborne to Clear an Obstacle

Return to Fig. 6.12 and recall that the total takeoff distance is equal to the ground roll s_g and the extra distance required to clear an obstacle after becoming airborne s_a. In this section, we consider the calculation of s_a.

The flight path after liftoff is sketched in Fig. 6.16. This is essentially the pull-up maneuver discussed in Section 6.3. In Fig. 6.16, R is the turn radius given by Eq. (6.41), repeated here:

$$R = \frac{V_\infty^2}{g(n-1)}$$ [6.41]

During the airborne phase, Federal Air Regulations (FAR) require that V_∞ increase from $1.1V_{stall}$ at liftoff to $1.2V_{stall}$ as it clears the obstacle of height h_{OB}. Therefore, we assume that V_∞ in Eq. (6.41) is an average value equal to $1.15\,V_{stall}$. The load factor n in Eq. (6.41) is obtained as follows. The average lift coefficient during this airborne phase is kept slightly less than $(C_L)_{max}$ for a margin of safety; we assume $C_L = 0.9(C_L)_{max}$. Hence,

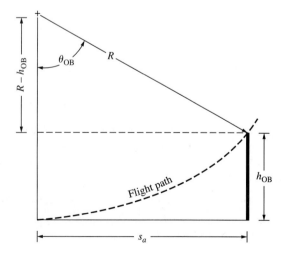

Figure 6.16   Sketch for the calculation of distance while airborne.

$$n = \frac{L}{W} = \frac{\frac{1}{2}\rho_\infty(1.15V_{stall})^2 S(0.9)(C_L)_{max}}{W} \tag{6.96}$$

From Eq. (5.67), the weight in Eq. (6.96) can be expressed in terms of $(C_L)_{max}$ and V_{stall} as

$$W = \frac{1}{2}\rho_\infty(V_{stall})^2 S(C_L)_{max} \tag{6.97}$$

Substituting Eq. (6.97) into Eq. (6.96), we have

$$n = \frac{\frac{1}{2}\rho_\infty(1.15V_{stall})^2 S(0.9)(C_L)_{max}}{\frac{1}{2}\rho_\infty(V_{stall})^2 S(C_L)_{max}}$$

or

$$n = 1.19$$

Returning to Eq. (6.41), with $V_\infty = 1.15V_{stall}$ and $n = 1.19$, we have

$$R = \frac{(1.15V_{stall})^2}{g(1.19 - 1)}$$

or

$$R = \frac{6.96(V_{stall}^2)}{g} \tag{6.98}$$

In Fig. 6.16, θ_{OB} is the included angle of the flight path between the point of takeoff and that for clearing the obstacle of height h_{OB}. From this figure, we see that

$$\text{Cos } \theta_{OB} = \frac{R - h_{OB}}{R} = 1 - \frac{h_{OB}}{R}$$

or

$$\theta_{OB} = \cos^{-1}\left(1 - \frac{h_{OB}}{R}\right) \tag{6.99}$$

Also from the geometry of Fig. 6.16, we have

$$s_a = R \sin \theta_{OB} \tag{6.100}$$

In summary, to calculate the distance along the ground covered by the airborne segment:

1. Calculate R from Eq. (6.98).
2. For the given obstacle height h_{OB}, calculate θ_{OB} from Eq. (6.99).
3. Calculate s_a from Eq. (6.100).

Example 6.6 | Calculate the total takeoff distance for our Gulfstream-like airplane at standard sea level, assuming a takeoff gross weight of 73,000 lb. The design features of the airplane are the same as those given in Example 5.1, with the added information that the wingspan is 75 ft. Assume that the variation of engine thrust with velocity during takeoff is given by Eq. (6.74), where

$k_1^* = 27,700$ lb, $k_2^* = 21.28$ lb·s/ft, and $k_3^* = 1.117 \times 10^{-2}$ lb·s/ft)^2. The height of the wing above the ground during the ground roll is 5.6 ft. Assume the runway is dry concrete, with $\mu_r = 0.04$.

Solution

The total takeoff distance, as shown in Fig. 6.12, is the sum of s_g and s_a. Let us first calculate the ground roll s_g, using Eq. (6.90). The information needed in Eq. (6.90) is obtained as follows.

The liftoff velocity V_{LO} is chosen to be equal to $1.1 V_{stall}$. Our Gulfstream-like airplane is equipped with single-slotted Fowler flaps; based on the data in Table 5.3 for flaps deflected in the takeoff position, $(C_L)_{max} / \cos \Lambda = 2.1$. As mentioned in Example 5.12, the wing sweep angle $\Lambda = 27°40'$. Hence, $(C_L)_{max} = 2.1 \cos 27°40' = 1.86$. From Eq. (5.7),

$$V_{stall} = \sqrt{\frac{2}{\rho_\infty} \frac{W}{S} \frac{1}{(C_L)_{max}}} = \sqrt{\frac{2(76.84)}{(0.002377)(1.86)}} = 186.4 \text{ ft/s}$$

Hence,

$$V_{LO} = 1.1 V_{stall} = 1.1(186.4) = 205.1 \text{ ft/s}$$

From Eq. (6.85), evaluated at $V_\infty = 0.7 V_{LO}$,

$$K_T = \left(\frac{T}{W} - \mu_r\right)_{0.7 V_{LO}}$$

To evaluate T at $0.7 V_{LO}$, use Eq. (6.74) as follows.

$$T = k_1^* - k_2^* V_\infty + k_3^* V_\infty^2 = 27,700 - 21.28 V_\infty + 1.117 \times 10^{-2} V_\infty^2$$

Since $0.7 V_{LO} = 0.7(205.1) = 143.6$ ft/s,

$$T = 27,700 - 21.28(143.6) + 1.117 \times 10^{-2}(143.6)^2$$
$$= 27,700 - 3,055.8 + 230.3 = 24,875 \text{ lb}$$

Thus,

$$K_T = \left(\frac{T}{W} - \mu_r\right)_{0.7 V_{LO}} = \frac{24,875}{73,000} - 0.04 = 0.301$$

For the evaluation of K_A as given by Eq. (6.86), the following information is needed. From Example 5.1, $C_{D,0} = 0.015$. The increase in the zero-lift drag coefficient due to the extended landing gear is estimated in Eq. (6.76), where we will assume that K_{uc} is approximately 4.5×10^{-5} for the case of moderate flap deflection (see previous discussion in Section 6.7.1). Equation (6.76) is repeated here:

$$\Delta C_{D,0} = \frac{W}{S} K_{uc} m^{-0.215}$$

where W/S is in units of newtons per square meter and m is in units of kilograms. Since 1 lb $= 4.448$ N, 1 ft $= 0.3048$ m, and 1 lb$_m = 0.4536$ kg, we have

$$\frac{W}{S} = \frac{73,000 \text{ lb}}{950 \text{ ft}^2} \frac{4.448 \text{ N}}{1 \text{ lb}} \left(\frac{1 \text{ ft}}{0.3048 \text{ m}}\right)^2 = 3,679 \text{ N/m}^2$$

$$m = 73,000 \text{ lb}_m \frac{0.4536 \text{ kg}}{1 \text{ lb}_m} = 33,113 \text{ kg}$$

Therefore,

$$\Delta C_{D,0} = \frac{W}{S} K_{uc} m^{-0.215} = (3,679)(4.5 \times 10^{-5})(33,113)^{-0.215} = 0.0177$$

It is interesting to note that the zero-lift drag coefficient is more than doubled by the extended landing gear. The value of k_1 in Eq. (6.86) was given in Example 5.1; $k_1 = 0.02$. Also given in Example 5.1 is $e = 0.9$. In Eq. (6.86), G is obtained from Eq. (6.77).

$$G = \frac{(16h/b)^2}{1 + (16h/b)^2} = \frac{[16(5.6/75)]^2}{1 + [16(5.6/75)]^2} = \frac{1.427}{2.427} = 0.588$$

Finally, as discussed in Section 6.7.1, we will assume that $C_L = 0.1$ during the ground roll. Therefore, from Eq. (6.86),

$$K_A = -\frac{\rho_\infty}{2(W/S)}\left[C_{D,0} + \Delta C_{D,0} + \left(k_1 + \frac{G}{\pi e AR}\right)C_L^2 - \mu_r C_L\right]$$

$$= \frac{-0.002377}{2(76.48)}\left\{0.015 + 0.0177 + \left[0.02 + \frac{0.588}{\pi(0.9)(5.92)}\right](0.1)^2 - (0.04)(0.1)\right\}$$

$$= -(1.547 \times 10^{-5})(0.0327 + 0.00055 - 0.004)$$

$$= -(1.547) \times 10^{-5})(0.02925) = -4.525 \times 10^{-7}$$

In the above calculation for K_A, note that the contribution due to zero-lift drag

$$C_{D,0} + \Delta C_{D,0} = 0.0327$$

is much larger than that for drag due to lift

$$\left(k_1 + \frac{G}{\pi e AR}\right)C_L^2 = 0.00055$$

From Eq. (6.90), letting $N = 3$, we have

$$S_g = \frac{1}{2g K_A} \ln\left(1 + \frac{K_A}{K_T} V_{LO}^2\right) + N V_{LO}$$

$$= \frac{1}{2(32.2)(-4.525 \times 10^{-7})} \ln\left[1 + \frac{-4.525 \times 10^{-7}}{0.301}(205.1)^2\right] + 3(205.1)$$

$$= 2,242 + 615 = 2,857 \text{ ft}$$

To calculate the airborne segment of the total takeoff distance s_a, that is, that distance covered over the ground while airborne necessary to clear a 35-ft obstacle, we use Eqs. (6.98) to (6.100). From Eq. (6.98)

$$R = \frac{6.96(V_{stall})^2}{g} = \frac{6.96(186.4)^2}{32.2} = 7,510 \text{ ft}$$

From Eq. (6.99)

$$\theta_{OB} = \cos^{-1}\left(1 - \frac{h_{OB}}{R}\right) = \cos^{-1}\left(1 - \frac{35}{7,510}\right) = 5.534°$$

From Eq. (6.100)

$$s_a = R \sin \theta_{OB} = 7,510 \sin 5.534° = 724 \text{ ft}$$

Hence,

$$\text{Total takeoff distance} = s_g + s_a = 2{,}857 + 724 = \boxed{3{,}581 \text{ ft}}$$

It is interesting to compare the above calculation for s_g with the more approximate relation given by Eq. (6.95)

$$s_g \approx \frac{1.21(W/S)}{g\rho_\infty (C_L)_{\max}(T/W)}$$

Evaluating T/W at $V_\infty = 0.7V_{LO}$, as carried out earlier in this example, we have $T/W = 24{,}875/73{,}000 = 0.341$. Hence, Eq. (6.95) yields

$$s_g \approx \frac{(1.21)(76.84)}{(32.2)(0.002377)(1.86)(0.341)} = 1{,}915 \text{ ft}$$

If we add the 615 ft covered during the rotation phase, which is neglected in Eq. (6.95), we have

$$s_g = 1{,}915 + 615 = 2{,}530 \text{ ft}$$

This is to be compared with the value $s_g = 2{,}857$ ft obtained earlier. Hence, the greatly simplified expression given by Eq. (6.95) leads to a value for s_g that is only 11% lower than that obtained by using our more precise analysis carried out above.

6.8 LANDING PERFORMANCE

The analysis of the landing performance of an airplane is somewhat analogous to that for takeoff, only in reverse. Consider an airplane on a landing approach. The landing distance, as sketched in Fig. 6.17, begins when the airplane clears an obstacle, which is taken to be 50 ft in height. At that instant the airplane is following a straight approach path with angle θ_a, as noted in Fig. 6.17. The velocity of the airplane at the instant it clears the obstacle, denoted by V_a, is required to be equal to $1.3V_{\text{stall}}$ for commercial airplanes and $1.2V_{\text{stall}}$ for military airplanes. At a distance h_f above the ground, the airplane begins the flare, which is the transition from the straight approach path to the horizontal ground roll. The flight path for the flare can be considered a circular arc with radius R, as shown in Fig. 6.17. The distance measured along the ground from the obstacle to the point of initiation of the flare is the approach distance s_a. Touchdown occurs when the wheels touch the ground. The distance over the ground covered during the flare is the flare distance s_f. The velocity at the touchdown V_{TD} is $1.15V_{\text{stall}}$ for commercial airplanes and $1.1V_{\text{stall}}$ for military airplanes. After touchdown, the airplane is in free roll for a few seconds before the pilot applies the brakes and/or thrust reverser. The free-roll distance is short enough that the velocity over this length is assumed constant, equal to V_{TD}. The distance that the airplane rolls on the ground from touchdown to the point where the velocity goes to zero is called the *ground roll* s_g.

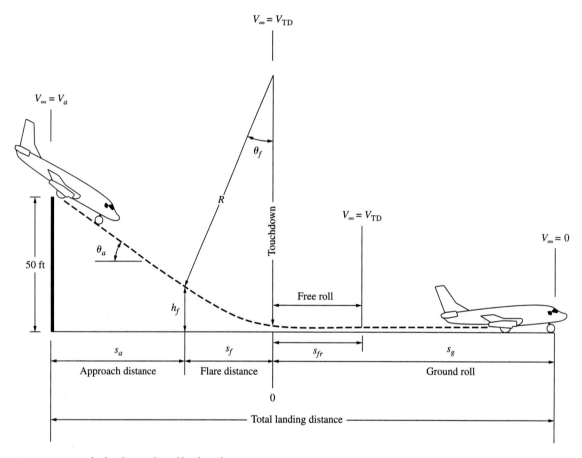

Figure 6.17   The landing path and landing distance.

6.8.1   Calculation of Approach Distance

Examining Fig. 6.17, we see that the approach distance s_a depends on the approach angle θ_a and the flare height h_f. In turn, θ_a depends on T/W and L/D. This can be seen from Fig. 6.18, which shows the force diagram for an aircraft on the approach flight path. Assuming equilibrium flight conditions, from Fig. 6.18,

$$L = W \cos \theta_a \qquad\qquad \textbf{[6.101]}$$

$$D = T + W \sin \theta_a \qquad\qquad \textbf{[6.102]}$$

From Eq. (6.102),

$$\sin \theta_a = \frac{D - T}{W} = \frac{D}{W} - \frac{T}{W} \qquad\qquad \textbf{[6.103]}$$

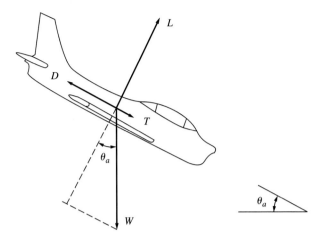

Figure 6.18 Force diagram for an airplane on the landing approach flight path.

The approach angle is usually small for most cases. For example, Raymer (Ref. 25) states that for transport aircraft $\theta_a \leq 3°$. Hence, $\cos \theta_a \approx 1$ and from Eq. (6.101), $L \approx W$. In this case, Eq. (6.103) can be written as

$$\sin \theta_a = \frac{1}{L/D} - \frac{T}{W} \qquad \textbf{[6.104]}$$

The flare height h_f, shown in Fig. 6.17, can be calculated from the construction shown in Fig. 6.19 as follows.

$$h_f = R - R \cos \theta_f \qquad \textbf{[6.105]}$$

However, because the circular arc flight path of the flare is tangent to both the approach path and the ground, as shown in Fig. 6.19, $\theta_f = \theta_a$. Hence, Eq. (6.105) becomes

$$h_f = R(1 - \cos \theta_a) \qquad \textbf{[6.106]}$$

In Eq. (6.106), R is obtained from Eq. (6.41) by assuming that V_∞ varies from $V_a = 1.3V_{\text{stall}}$ for commercial aircraft and $1.2V_{\text{stall}}$ for military aircraft to $V_{\text{TD}} = 1.15V_{\text{stall}}$ for commercial aircraft and $1.1V_{\text{stall}}$ for military aircraft, yielding an average velocity during the flare of $V_f = 1.23V_{\text{stall}}$ for commercial airplanes and $1.15V_{\text{stall}}$ for military airplanes. With the load factor n stipulated as $n = 1.2$, Eq. (6.41) yields

$$R = \frac{V_f^2}{0.2g} \qquad \textbf{[6.107]}$$

Finally, with R given by Eq. (6.107) and θ_a from Eq. (6.104), h_f can be calculated from Eq. (6.106). In turn, s_a is obtained from Fig. 6.17 as

$$s_a = \frac{50 - h_f}{\text{Tan}\,\theta_a} \qquad \textbf{[6.108]}$$

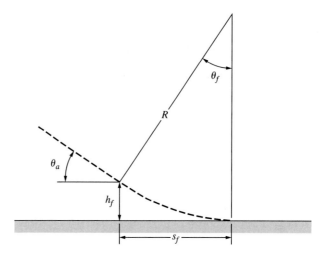

Figure 6.19   Geometry of the landing flare.

6.8.2   Calculation of Flare Distance

The flare distance s_f, shown in Figs. 6.17 and 6.19, is given by

$$s_f = R \sin \theta_f$$

Since $\theta_f = \theta_a$, this becomes

$$s_f = R \sin \theta_a \qquad \textbf{[6.109]}$$

6.8.3   Calculation of Ground Roll

The force diagram for the airplane during the landing ground roll is the same as that shown in Fig. 6.14. Hence, the equation of motion is the same as Eq. (6.71). However, normal landing practice assumes that upon touchdown, the engine thrust is reduced to idle (essentially zero). In this case, with $T = 0$, Eq. (6.71) becomes

$$m \frac{dV_\infty}{dt} = -D - \mu_r(W - L) \qquad \textbf{[6.110]}$$

Many jet aircraft are equipped with thrust reversers which typically produce a negative thrust equal in magnitude to 40% or 50% of the maximum forward thrust. Some reciprocating engine/propeller-driven airplanes are equipped with reversible propellers that can produce a negative thrust equal in magnitude to about 40% of the static forward thrust. For turboprops, this increases to about 60%. In such cases, if T_{rev} denotes the absolute magnitude of the reverse thrust, then Eq. (6.71) becomes

$$m \frac{dV_\infty}{dt} = -T_{rev} - D - \mu_r(W - L) \qquad \textbf{[6.111]}$$

Also, the value of D in Eqs. (6.110) and (6.111) can be increased by deploying spoilers, speed brakes, or drogue chutes. Note that in both Eqs. (6.110) and (6.111), dV_∞/dt will be a negative quantity; that is, the airplane will *decelerate* (obviously) during the landing ground run.

An expression for s_g can be obtained in the same fashion as in Section 6.7.1. From Eq. (6.111),

$$
\begin{aligned}
\frac{dV_\infty}{dt} &= -\frac{g}{W}\left[T_{\text{rev}} + \frac{1}{2}\rho_\infty V_\infty^2 S C_D + \mu_r\left(W - \frac{1}{2}\rho_\infty V_\infty^2 S C_L\right)\right] \\
&= -g\left[\frac{T_{\text{rev}}}{W} + \mu_r + \frac{\rho_\infty}{2(W/S)}\,(C_D - \mu_r C_L)\,V_\infty^2\right] \\
&= -g\left\{\frac{T_{\text{rev}}}{W} + \mu_r + \frac{\rho_\infty}{2(W/S)}\left[C_{D,0} + \Delta C_{D,0} + \left(k_1 + \frac{G}{\pi e AR}\right)C_L^2 - \mu_r C_L\right]V_\infty^2\right\}
\end{aligned}
$$

[6.112]

Defining the symbols

$$
J_T \equiv \frac{T_{\text{rev}}}{W} + \mu_r \qquad\qquad\text{[6.113]}
$$

$$
J_A \equiv \frac{\rho_\infty}{2(W/S)}\left[C_{D,0} + \Delta C_{D,0} + \left(k_1 + \frac{G}{\pi e AR}\right)C_L^2 - \mu_r C_L\right] \qquad\text{[6.114]}
$$

We can write Eq. (6.112) as

$$
\frac{dV_\infty}{dt} = -g\left(J_T + J_A V_\infty^2\right) \qquad\qquad\text{[6.115]}
$$

Substituting Eq. (6.115) into Eq. (6.81), we have

$$
ds = \frac{d(V_\infty^2)}{2(dV_\infty/dt)} = -\frac{d(V_\infty^2)}{2g\left(J_T + J_A V_\infty^2\right)} \qquad\text{[6.116]}
$$

Let us apply Eq. (6.116) beginning at the end of the free-roll segment shown in Fig. 6.17. Integrating Eq. (6.116) between the end of the roll, where $s = s_{\text{fr}}$ and $V_\infty = V_{\text{TD}}$, and the complete stop, where $s = s_g$ and $V_\infty = 0$, we have

$$
\int_{s_{\text{fr}}}^{s_g} ds = -\int_{V_{\text{TD}}}^{0} \frac{d(V_\infty^2)}{2g(J_T + J_A V_\infty^2)}
$$

or

$$
s_g - s_{\text{fr}} = \int_{0}^{V_{\text{TD}}} \frac{d(V_\infty^2)}{2g(J_T + J_A V_\infty^2)} \qquad\text{[6.117]}
$$

Equation (6.117) for landing is directly analogous to Eq. (6.89) for takeoff. Note that no simplifications have been made in obtaining Eq. (6.117); the values of J_T and J_A vary with V_∞ during the ground roll.

However, if we assume that J_T and J_A in Eq. (6.117) can be assumed constant, Eq. (6.117) becomes

$$s_g - s_{fr} = \frac{1}{2g J_A} \ln \left(1 + \frac{J_A}{J_T} V_{TD}^2 \right) \qquad \textbf{[6.118]}$$

According to Raymer (Ref. 25), the free roll depends partly on pilot technique and usually lasts for 1 to 3 s. Letting N be the time increment for the free roll, we have $s_{fr} = N V_{TD}$. Then Eq. (6.118) yields for the total ground roll s_g

$$\boxed{s_g = N V_{TD} + \frac{1}{2g J_A} \ln \left(1 + \frac{J_A}{J_T} V_{TD}^2 \right)} \qquad \textbf{[6.119]}$$

Equation (6.119) for the landing ground roll is analogous to Eq. (6.90) for the takeoff ground roll. With Eq. (6.119), a quick analytical solution of the ground roll for landing can be made.

An analytic form for s_g that more closely illustrates the design parameters that govern landing performance can be obtained by substituting Eq. (6.111) directly into Eq. (6.81), obtaining

$$ds = \frac{m}{2} \frac{d(V_\infty^2)}{-T_{rev} - D - \mu_r(W - L)} \qquad \textbf{[6.120]}$$

Integrating Eq. (6.120) from s_{fr} to s_g and noting that $m = W/g$, we have

$$s_g - s_{fr} = \frac{W}{2g} \int_{V_{TD}}^{0} \frac{d(V_\infty^2)}{-T_{rev} - D - \mu_r(W - L)}$$

or

$$s_g = N V_{TD} + \frac{W}{2g} \int_{0}^{V_{TD}} \frac{d(V_\infty^2)}{T_{rev} + D + \mu_r(W - L)} \qquad \textbf{[6.121]}$$

In Eq. (6.121), $T_{rev} + D + \mu_r(W - L)$ is the net force acting in the horizontal direction on the airplane during the landing ground roll. In Fig. 6.20, a schematic is shown of the variation of forces acting during the landing ground roll, with the exception of T_{rev}. The application of thrust reversal is a matter of pilot technique, and it may be applied only for a certain segment of the ground roll. If this is the case, Eq. (6.121) must be integrated in segments, with and without T_{rev}. In any event, the force $D + \mu_r(W - L)$ in Fig. 6.20 is reasonably constant with s. If we assume that T_{rev} is also constant, then it is reasonable to assume that the expression $T_{rev} + D + \mu_r(W - L)$ is a constant, evaluated at a value equal to its value at $V_\infty = 0.7 V_{TD}$. Then Eq. (6.121) is easily integrated, giving

$$s_g = N V_{TD} + \frac{W V_{TD}^2}{2g} \left[\frac{1}{T_{rev} + D + \mu_r(W - L)} \right]_{0.7 V_{TD}} \qquad \textbf{[6.122]}$$

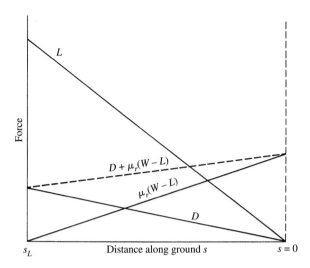

Figure 6.20   Schematic of a typical variation of forces acting on an airplane during landing.

The touchdown velocity V_{TD} should be no less than $j V_{\text{stall}}$, where $j = 1.15$ for commercial airplanes and $j = 1.1$ for military airplanes. Since from Eq. (5.67)

$$V_{\text{stall}} = \sqrt{\frac{2}{\rho_\infty} \frac{W}{S} \frac{1}{(C_L)_{\text{max}}}}$$ **[5.67]**

then Eq. (6.122) can be written as

$$s_g = jN \sqrt{\frac{2}{\rho_\infty} \frac{W}{S} \frac{1}{(C_L)_{\text{max}}}} + \frac{j^2(W/S)}{g\rho_\infty(C_L)_{\text{max}} \left[T_{\text{rev}}/W + D/W + \mu_r\left(1 - L/W\right)\right]_{0.7V_{\text{TD}}}}$$

[6.123]

Equation (6.123) for the landing ground roll is analogous to Eq. (6.94) for the takeoff ground roll. The design parameters that have an important effect on landing ground roll are clearly seen from Eq. (6.123). Specifically, s_g depends on wing loading, maximum lift coefficient, and (if used) the reverse thrust-to-weight ratio. From Eq. (6.123), we note that

1. s_g increases with an increase in W/S.
2. s_g decreases with an increase in $(C_L)_{\text{max}}$.
3. s_g decreases with an increase in T_{rev}/W.
4. s_g increases with a decrease in ρ_∞.

Clearly, by comparing Eqs. (6.123) and (6.94), we see that W/S and $(C_L)_{\text{max}}$ play identical roles in determining both landing and takeoff ground rolls. However, the forward engine thrust-to-weight ratio is a major player during takeoff, whereas the engine thrust is in idle (or in reverse) for the landing ground roll.

Example 6.7

Calculate the total landing distance for our Gulfstream-like airplane at standard sea level, assuming that (for conservatism) the landing weight is the same as the takeoff gross weight of 73,000 lb. Assume that no thrust reversal is used and that the runway is dry concrete with a brakes-on value of $\mu_r = 0.4$. The approach angle is $3°$.

Solution

Let us first calculate the stalling velocity for landing. From Table 5.3 for single-slotted Fowler flaps deflected in the landing position, we take $(C_L)_{max}/\cos \Lambda = 2.7$. With the wing sweep angle of $\Lambda = 27°40'$, we have $(C_L)_{max} = 2.7\cos(27°40') = 2.39$. From Eq. (5.67),

$$V_{stall} = \sqrt{\frac{2}{\rho_\infty}\frac{W}{S}\frac{1}{(C_L)_{max}}} = \sqrt{\frac{2(76.84)}{(0.002377)(2.39)}} = 164.5 \text{ ft/s}$$

For a commercial airplane, the average flight velocity during the flare is

$$V_f = 1.23V_{stall} = 1.23(164.5) = 202.3 \text{ ft/s}$$

and the touchdown velocity is

$$V_{TD} = 1.15V_{stall} = 1.15(164.5) = 189.2 \text{ ft/s}$$

From Eq. (6.107)

$$R = \frac{V_f^2}{0.2g} = \frac{(202.3)^2}{(0.2)(32.2)} = 6,354.9 \text{ ft}$$

From Eq. (6.106)

$$h_f = R(1 - \cos\theta_a) = 6,354.9(1 - \cos 3°) = 8.71 \text{ ft}$$

The approach distance is obtained from Eq. (6.108):

$$S_a = \frac{50 - h_f}{\text{Tan } \theta_a} = \frac{50 - 8.71}{\text{Tan } 3°} = 788 \text{ ft}$$

The flare distance is given by Eq. (6.109).

$$s_f = R \sin\theta_a = 6,354.9 \sin 3° = 333 \text{ ft}$$

The ground roll is obtained from Eq. (6.119). In this equation, the values of J_T and J_A are given by Eqs. (6.113) and (6.114), respectively. For J_T, we have

$$J_T = \frac{T_{rev}}{W} + \mu_r = 0 + 0.4 = 0.4$$

To calculate J_A, we note from Example 6.6 that $G = 0.588$, $k_1 = 0.02$, and $C_L = 0.1$ for the ground roll. The value of $\Delta C_{D,0} = 0.0177$ calculated in Example 6.6 was for moderate flap deflection for takeoff. In contrast, for landing we assume full flap deflection. From the discussion in Section 6.7.1, K_{uc} in Eq. (6.76) can be taken as 3.16×10^{-5} for maximum flap deflection. In Example 6.6, the value used for K_{uc} was 4.5×10^{-5}. Hence, from Eq. (6.76), the value of $\Delta C_{D,0}$ calculated in Example 6.6 for takeoff should be reduced for landing by the ratio $3.16 \times 10^{-5}/4.5 \times 10^{-5} = 0.702$. Thus, recalling that $\Delta C_{D,0} = 0.0177$ as calculated in Example 6.6, we have for the present case

$$\Delta C_{D,0} = (0.702)(0.0177) = 0.0124$$

Therefore, from Eq. (6.114),

$$J_A = \frac{\rho_\infty}{2(W/S)} \left[C_{D,0} + \Delta C_{D,0} + \left(k_1 + \frac{G}{\pi eAR} \right) C_L^2 - \mu_r C_L \right]$$

$$= \frac{0.002377}{2(76.84)} \left\{ 0.015 + 0.0124 + \left[0.02 + \frac{0.588}{\pi(0.9)(5.92)} \right] (0.1)^2 - (0.4)(0.1) \right\}$$

$$= 1.547 \times 10^{-5} (0.0274 + 0.00055 - 0.04) = 1.864 \times 10^{-7}$$

From Eq. (6.119), assuming that $N = 3$ s,

$$s_g = N V_{\text{TD}} + \frac{1}{2g J_A} \ln \left(1 + \frac{J_A}{J_T} V_{\text{TD}}^2 \right)$$

$$= 3(189.2) + \frac{1}{(2)(32.2)(-1.864 \times 10^{-7})} \ln \left[1 + \frac{-1.864 \times 10^{-7}}{0.4} (189.2) \right]$$

$$= 568 + 1{,}401 = 1{,}969 \text{ ft}$$

Finally,

$$\text{Total landing distance} = s_a + s_f + s_g = 788 + 333 + 1{,}969 = \boxed{3{,}090 \text{ ft}}$$

6.9 SUMMARY

This chapter has dealt with some special cases of *accelerated* airplane performance, that is, cases where the acceleration of the airplane is not zero. In some of these cases, the load factor n, defined as

$$n = \frac{L}{W} \qquad \textbf{[6.5]}$$

plays an important role.

The turn radius and turn rate in a level turn are given, respectively, by

$$R = \frac{V_\infty^2}{g\sqrt{n^2 - 1}} \qquad \textbf{[6.9]}$$

and

$$\omega = \frac{g\sqrt{n^2 - 1}}{V_\infty} \qquad \textbf{[6.11]}$$

For a sustained turn (one with constant flight characteristics), the load factor in Eqs. (6.9) and (6.11) is limited by the maximum available thrust-to-weight ratio. In turn, for such a case, the minimum turn radius and maximum turning rate are functions of both wing loading and the thrust-to-weight ratio.

For the pull-up and pulldown maneuvers, the turn radius and turn rate are given by, respectively,

$$R = \frac{V_\infty^2}{g(n \pm 1)} \qquad \text{[(6.41)/(6.45)]}$$

and

$$\omega = \frac{g(n \mp 1)}{V_\infty} \qquad \text{[(6.42)/(6.46)]}$$

where the minus sign pertains to the pull-up maneuver and the plus sign pertains to the pulldown maneuver.

In the limiting case for large load factor, these equations become the same for instantaneous level turn, pull-up, and pulldown maneuvers, namely,

$$R = \frac{V_\infty^2}{gn} \qquad \text{[6.47]}$$

and

$$\omega = \frac{gn}{V_\infty} \qquad \text{[6.48]}$$

There are various practical limitations on the maximum load factor that can be experienced or allowed; these are embodied in the V–n diagram for a given airplane.

Accelerated climb performance can be analyzed by energy methods, where the total aircraft energy (potential plus kinetic) is given by the energy height

$$H_e = h + \frac{V_\infty^2}{2g} \qquad \text{[6.56]}$$

A change in energy height can be accomplished by the application of specific excess power P_e;

$$\frac{dH_e}{dt} = P_s \qquad \text{[6.65]}$$

where P_s is given by

$$P_s = \frac{\text{excess power}}{W} \qquad \text{[6.59]}$$

The takeoff distance is strongly dependent on W/S, T/W, and $(C_L)_{\max}$; it increases with an increase in W/S and decreases with increases in T/W and $(C_L)_{\max}$. The landing distance is dictated mainly by $(C_L)_{\max}$ and W/S; it decreases with an increase in $(C_L)_{\max}$ and increases with an increase in W/S. The effect of thrust on landing distance appears in two ways: (1) Increased T/W decreases the approach angle, hence lengthening the landing distance; and (2) the use of reversed thrust after touchdown decreases the landing distance.

PROBLEMS

For the BD-5J small, kit-built jet airplane described in Problem 5.1, calculate the **6.1** minimum turning radius and maximum turn rate at sea level.

For the conditions of maximum turn rate, derive the expressions for (*a*) the load factor, **6.2** Eq. (6.35); (*b*) the velocity, Eq. (6.34); (*c*) the maximum turn rate, Eq. (6.36).

Assume that the positive limit load factor for the BD-5J is 5. Calculate the corner **6.3** velocity at sea level for the airplane.

Derive Eq. (6.17). **6.4**

Consider an airplane flying at 620 mi/h at 35,000 ft. Calculate its energy height. **6.5**

Consider an airplane in an accelerated climb. At a given instant in this climb, the **6.6** specific excess power is 120 ft/s, the instantaneous velocity is 500 ft/s, and the instantaneous rate of climb is 3,000 ft/min. Calculate the instantaneous acceleration.

For the BD-5J (see Problem 5.1), calculate the total takeoff distance at sea level, **6.7** assuming clearing a 35-ft obstacle. The height of the wing above the ground during the ground roll is 1.5 ft. Assume the runway is firm dirt with a coefficient of rolling friction of 0.04.

For the BD-5J (see Problem 5.1), calculate the total landing distance, starting with the **6.8** clearance of a 50-ft obstacle, assuming the landing weight is the same as the takeoff gross weight. The runway is firm dirt with a brakes-on coefficient of rolling friction of 0.3. The approach angle is 4°.

3

AIRPLANE DESIGN

The capstone of most aeronautical research and development is a flying machine—an airplane, missile, space shuttle, etc. How does the existing technology in aerodynamics, propulsion, and flight mechanics, as highlighted in Parts 1 and 2 of this book, lead to the design of a flying machine? This is the central question addressed in Part 3. Here we will focus on the philosophy and general methodology of airplane design. After a general introduction to the design process, we will illustrate this process in separate chapters dealing with the design of a propeller-driven airplane, a subsonic high-speed jet-propelled airplane, and a supersonic airplane. We will further illustrate the design process with case histories of the design of several historically significant aircraft that revolutionized flight in the twentieth century.

7

The Philosophy of Airplane Design

The . . . line of argument draws, or attempts to draw, a clear line between pure science and technology (which is tending to become a pejorative word). This is a line that once I tried to draw myself: but, though I can still see the reasons, I shouldn't now. The more I have seen of technologists at work, the more untenable the distinction has come to look. If you actually see someone design an aircraft, you can find him going through the same experience—aesthetic, intellectual, moral—as though he was setting up an experiment in particle physics."

> C. P. Snow, *The Two Cultures:*
> *and a Second Look*, 1963,
> Cambridge University Press

A beautiful aircraft is the expression of the genius of a great engineer who is also a great artist."

> Neville Shute, British aeronautical
> engineer and novelist. From
> *No Highway*, 1947.

7.1 INTRODUCTION

Airplane design is both an art and a science. In that respect it is difficult to learn by reading a book; rather, it must be experienced and practiced. However, we can offer the following definition and then attempt in this book to explain it. Airplane design is the intellectual engineering process of creating on paper (or on a computer screen) a flying machine to (1) meet certain specifications and requirements established by potential users (or as perceived by the manufacturer) and/or (2) pioneer innovative, new ideas and technology. An example of the former is the design of most commerical transports, starting at least with the Douglas DC-1 in 1932, which was designed to meet or exceed various specifications by an airline company. (The airline was TWA, named Transcontinental and Western Air at that time.) An example of the latter is the design of the rocket-powered Bell X-1, the first airplane to exceed the speed of sound in level or climbing flight (October 14, 1947). The design process is indeed an intellectual activity, but a rather special one that is tempered by good intuition developed via experience, by attention paid to successful airplane designs that have been used in the past, and by (generally proprietary) design procedures and databases (handbooks, etc.) that are a part of every airplane manufacturer.

The remainder of this book focuses on the philosophy and general methodology of airplane design, that is, the intellectual activity. It is not intended to be a handbook, nor does it directly impart intuition, which is something that grows with experience. Rather, our intent is to provide some feeling and appreciation for the design experience. In this respect, this book is intended to serve as an intellectural steppingstone and natural companion to the several mainline airplane design texts presently available, such as Refs. 25, 35, and 52 to 54. These design texts are replete with detailed design procedures and data—all necessary for the successful design of an airplane. This book takes a more philosophical approach which is intended to provide an intellectual framework on which the reader can hang all those details presented elsewhere and then stand back and see the broader picture of the airplane design process. This author hopes that by studying two (or more) books—this book and one (or more) of the detailed mainline design texts—the reader will enjoy a greatly enhanced learning process. To paraphrase a currently popular television commercial in the United States, the present book is not intended to make the course in airplane design; it is intended to make the course in airplane design *better*.

7.2   PHASES OF AIRPLANE DESIGN

From the time that an airplane first materializes as a new thought in the mind of one or more persons to the time that the finished product rolls out of the manufacturer's door, the complete design process has gone through three distinct phases that are carried out in sequence. These phases are, in chronological order, conceptual design, preliminary design, and detail design. They are characterized as follows.

7.2.1   Conceptual Design

The design process starts with a set of specifications (requirements) for a new airplane, or much less frequently as the response to the desire to implement some pioneering, innovative new ideas and technology. In either case, there is a rather concrete goal toward which the designers are aiming. The first steps toward achieving that goal constitute the *conceptual design phase*. Here, within a certain somewhat fuzzy latitude, the overall shape, size, weight, and performance of the new design are determined. The product of the conceptual design phase is a layout (on paper or on a computer screen) of the airplane configuration. But we have to visualize this drawing as one with flexible lines, capable of being *slightly* changed during the second design phase, the preliminary design phase. However, the conceptual design phase determines such fundamental aspects as the shape of the wings (swept back, swept forward, or straight), the location of the wings relative to the fuselage, the shape and location of the horizontal and vertical tail, the use of a canard surface or not, engine size and placement, etc. Figure 7.1 is an example of the level of detail in a configuration layout at the end of the conceptual design phase. (Shown in Fig. 7.1 is the World War II vintage Bell P-39 Airacobra, chosen for its historical significance and aesthetic beauty.)

The major drivers during the conceptual design process are aerodynamics, propulsion, and flight performance. The first-order question is: Can the design meet the specifications? If the answer is yes, then the next question is: Is the design optimized, that is, is it the *best* design that meets the specifications? These questions are answered during the conceptual design phase by using tools primarily from aerodynamics, propulsion, and flight performance (e.g., material from Chapters 2, 3, 5, and 6). Structural and control system considerations are not dealt with in any detail. However, they are not totally absent. For example, during the conceptual design phase, the designer is influenced by such qualitative aspects as the increased structural loads imposed by a high horizontal T-tail versus a more conventional horizontal tail location through the fuselage, and the difficulties associated with cutouts in the wing structure if the landing gear are to retract into the wing rather than the fuselage or engine nacelle. No part of the design process is ever carried out in a total vacuum unrelated to the other parts.

7.2.2 Preliminary Design

In the preliminary design phase, only minor changes are made to the configuration layout (indeed, if major changes were demanded during this phase, the conceptual design process would have been seriously flawed to begin with). It is in the preliminary design phase that serious structural and control system analysis and design take place. During this phase also, substantial wind tunnel testing will be carried out, and major computational fluid dynamic (CFD) calculations of the complete flow field over the airplane configuration will be made. It is possible that the wind tunnel tests and/or the CFD calculations will uncover some undesirable aerodynamic interference, or some unexpected stability problems, which will promote changes to the configuration layout. At the end of the preliminary design phase, the airplane configuration is frozen and precisely defined. The drawing process called *lofting* is carried out which mathematically models the precise shape of the outside skin of the airplane, making certain that all sections of the aircraft properly fit together. (Lofting is a term carried over to airplane design from ship design. Historically, shipbuilders designed the shape of the hull in the loft, an area located above the shipyard floor.)

The end of the preliminary design phase brings a major decision—to commit to the manufacture of the airplane or not. The importance of this decision point for modern aircraft manufacturers cannot be understated, considering the tremendous costs involved in the design and manufacture of a new airplane. This is no better

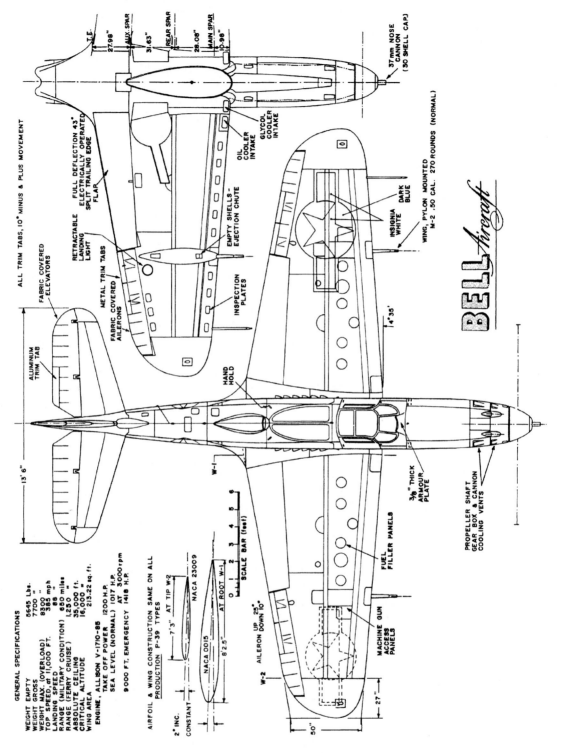

Figure 7.1 Detailed layout for the Bell P-39 Airacobra of World War II fame. (*Courtesy of Aviation Heritage, Destin, Florida.*) (*continued*)

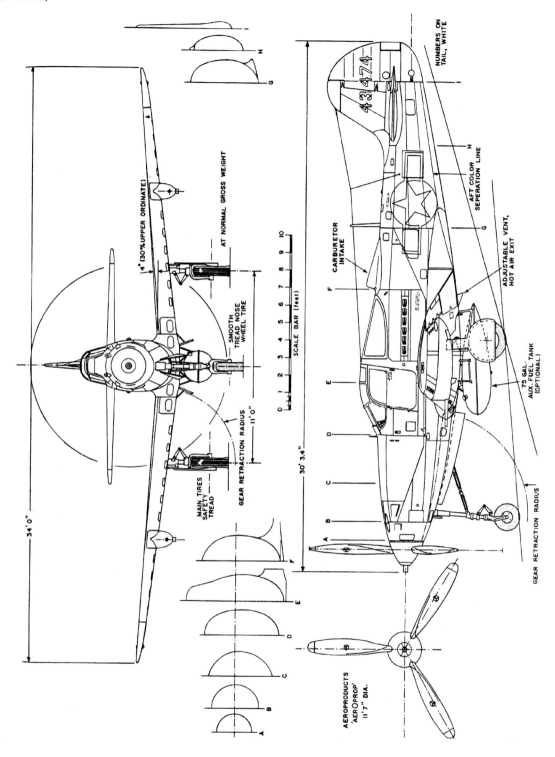

4° (30% UPPER ORDINATE)

AT NORMAL GROSS WEIGHT

SMOOTH TREAD NOSE WHEEL TIRE

MAIN TIRES SAFETY TREAD

GEAR RETRACTION RADIUS 11' 0"

34' 0"

SCALE BAR (feet)

0 1 2 3 4 5 6 7 8 9 10

30' 3.4"

NUMBERS ON TAIL, WHITE

43474

AFT COLOR SEPERATION LINE

ADJUSTABLE VENT, HOT AIR EXIT

CARBURETOR INTAKE

75 GAL. AUX. FUEL TANK (OPTIONAL)

GEAR RETRACTION RADIUS

AEROPRODUCTS 'AEROPROP' 11' 7" DIA.

385

illustrated than with the Boeing Airplane Company's decision in 1966 to proceed with the manufacture of the 747 wide-body transport (Fig. 1.34) after the preliminary design was finished. As noted in Section 1.2.4, the failure of the 747 would have financially ruined Boeing. It is no longer unusual for such decisions in the aircraft industry to be one of "you bet your company" on the full-scale development of a new airplane.

7.2.3   Detail Design

The detail design phase is literally the "nuts and bolts" phase of airplane design. The aerodynamic, propulsion, structures, performance, and flight control analyses have all been finished with the preliminary design phase. For detail design, the airplane is now simply a machine to be fabricated. The precise design of each individual rib, spar, and section of skin now takes place. The size, number, and location of fasteners (rivets, welded joints, etc.) are determined. Manufacturing tools and jigs are designed. At this stage, flight simulators for the airplane are developed. And these are just a few of the many detailed requirements during the detail design phase. At the end of this phase, the aircraft is ready to be fabricated.

7.2.4   Interim Summary

Figure 7.2 is a schematic intended, in a very simple manner, to visually illustrate the distinction between the products of the three phases of airplane design. The product of conceptual design is represented in Fig. 7.2**a**. Here, the basic configuration of the airplane is determined, but only within a certain (hopefully small) fuzzy latitude. The product of preliminary design is represented in Fig. 7.2**b**. Here, the precise configuration (precise dimensions) is determined. Finally, the product of detail design is represented in Fig. 7.2**c**. Here, the precise fabrication details are determined, represented by the precise rivet sizes and locations.

When students first study the subject of airplane design, it is the conceptual design phase that is treated. For example, the mainline design texts (Refs. 25, 35, and 52 to 54) are essentially conceptual design texts. The subjects of preliminary and detail

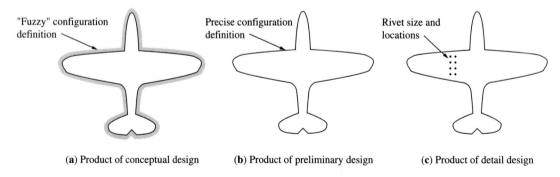

(**a**) Product of conceptual design   (**b**) Product of preliminary design   (**c**) Product of detail design

Figure 7.2   Schematic illustrating the difference between conceptual, preliminary, and detail designs.

design are much too extensive and specialized for a first study of airplane design. This book is no different in that respect; we will limit our discussions to aspects of conceptual design as defined in Section 7.2.1.

7.3 THE SEVEN INTELLECTUAL PIVOT POINTS FOR CONCEPTUAL DESIGN

The design process is an act of creativity, and like all creative endeavors, there is no one correct and absolute method to carry it out. Different people, different companies, different books all approach the subject from different angles and with a different sequence of events. However, this author suggests that, on a philosophical basis, the overall conceptual design process is anchored by seven intellectual (let us say) "pivot points"—seven aspects that anchor the conceptual design thought process, but which allow different, more detailed thinking to reach out in all directions from each (hypothetical) pivot point. Hence, conceptual design can be imagined as an array of the seven pivot points anchored at strategic locations in some kind of intellectual space, and these pivot points are connected by a vast web of detailed approaches. The webs constructed by different people would be different, although the pivot points should be the same, due to their fundamental significance. These seven pivot points are listed in the block array shown in Fig. 7.3 and are described and discussed below.

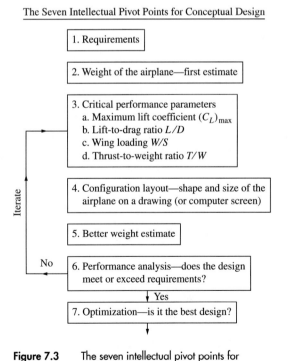

Figure 7.3   The seven intellectual pivot points for conceptual design.

Fixing these pivot points in your mind will serve to create an intellectual framework on which you can hang all the details of conceptual design, no matter how different these details may be from one design group to another.

Let us now consider in turn each of the seven intellectual pivot points listed in Fig. 7.3.

7.3.1   Requirements

Imagine that you are now ready to begin the design of a new airplane. Where and how do you start? With a clear statement of the requirements to be satisfied by the new airplane. The requirements may be written by the people who are going to buy the new airplane—the customer. For military aircraft, the customer is the government. For civilian transports, the customer is the airlines. On the other hand, for general aviation aircraft—from executive jet transports owned by private businesses (and some wealthy individuals) to small, single-engine recreation airplanes owned by individual private pilots—the requirements are usually set by the manufacturer in full appreciation of the needs of the private airplane owner. [An excellent historical example was the design of the famous Ercoupe by Engineering and Research Corporation (ERCO) in the late 1930s, where in the words of Fred Weick, its chief designer, the company set as its overall goal the design of an airplane "that would be unusually simple and easy to fly and free from the difficulties associated with stalling and spinning." The Ercoupe is shown in Fig. 7.4.] If the general aviation aircraft manufacturer has done its homework correctly, the product will be bought by the private airplane owner.

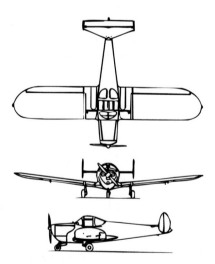

Figure 7.4   The ERCO Ercoupe, circa 1940.

Requirements for a new airplane design are as unique and different from one airplane to another as fingerprints are from one human being to another. Hence, we cannot stipulate in this section a specific, standard form to use to write requirements—there is none. All we can say is that for any new airplane design, there must be *some* established requirements which serve as the jumping-off point for the design process, and which serve as the focused goal for the completed design. Typical aspects that are frequently stipulated in the requirements are some combination of the following:

1. Range.
2. Takeoff distance.
3. Stalling velocity.
4. Endurance [usually important for reconnaissance airplanes; an overall dominating factor for the new group of very high-altitude uninhabited air vehicles (UAVs) that are of great interest at present].
5. Maximum velocity.
6. Rate of climb.
7. For dogfighting combat aircraft, maximum turn rate and sometimes minimum turn radius.
8. Maximum load factor.
9. Service ceiling.
10. Cost.
11. Reliability and maintainability.
12. Maximum size (so that the airplane will fit inside standard hangers and/or be able to fit in a standard gate at airline terminals).

These are just a few examples, to give you an idea as to what constitutes "requirements." Today, the design requirements also include a host of details associated with both the interior and exterior mechanical aspects of the airplane. An interesting comparison is between the one page of U.S. Army Signal Corps requirements (reproduced in Fig. 7.5) set forth on January 20, 1908 for the first army airplane, and the thick, detailed general design document that the government usually produces today for establishing the requirements for new military aircraft. (The requirements shown in Fig. 7.5 were satisfied by the Wright brothers' type A airplane. This airplane was purchased by the Army, and became known as the Wright *Military Flyer*.)

7.3.2 Weight of the Airplane—First Estimate

No airplane can get off the ground unless it can produce a lift greater than its weight. And no airplane design process can "get off the ground" without a first estimate of the gross takeoff weight. The fact that a weight estimate, albeit crude, is the next pivot point after the requirements is also satisfying from an historical point of view. Starting with George Cayley in 1799, the efforts to design a successful

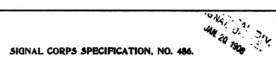

SIGNAL CORPS SPECIFICATION, NO. 486.

ADVERTISEMENT AND SPECIFICATION FOR A HEAVIER-THAN-AIR FLYING MACHINE.

To the Public:

Sealed proposals, in duplicate, will be received at this office until 12 o'clock noon on February 1, 1908, on behalf of the Board of Ordnance and Fortification for furnishing the Signal Corps with a heavier-than-air flying machine. All proposals received will be turned over to the Board of Ordnance and Fortification at its first meeting after February 1 for its official action.

Persons wishing to submit proposals under this specification can obtain the necessary forms and envelopes by application to the Chief Signal Officer, United States Army, War Department, Washington, D. C. The United States reserves the right to reject any and all proposals.

Unless the bidders are also the manufacturers of the flying machine they must state the name and place of the maker.

Preliminary.—This specification covers the construction of a flying machine supported entirely by the dynamic reaction of the atmosphere and having no gas bag.

Acceptance.—The flying machine will be accepted only after a successful trial flight, during which it will comply with all requirements of this specification. No payments on account will be made until after the trial flight and acceptance.

Inspection.—The Government reserves the right to inspect any and all processes of manufacture.

GENERAL REQUIREMENTS.

The general dimensions of the flying machine will be determined by the manufacturer, subject to the following conditions:

1. Bidders must submit with their proposals the following:
   (a) Drawings to scale showing the general dimensions and shape of the flying machine which they propose to build under this specification.
   (b) Statement of the speed for which it is designed.
   (c) Statement of the total surface area of the supporting planes.
   (d) Statement of the total weight.
   (e) Description of the engine which will be used for motive power.
   (f) The material of which the frame, planes, and propellers will be constructed. Plans received will not be shown to other bidders.

2. It is desirable that the flying machine should be designed so that it may be quickly and easily assembled and taken apart and packed for transportation in army wagons. It should be capable of being assembled and put in operating condition in about one hour.

3. The flying machine must be designed to carry two persons having a combined weight of about 350 pounds, also sufficient fuel for a flight of 125 miles.

4. The flying machine should be designed to have a speed of at least forty miles per hour in still air, but bidders must submit quotations in their proposals for cost depending upon the speed attained during the trial flight, according to the following scale:

    40 miles per hour, 100 per cent.
    39 miles per hour, 90 per cent.
    38 miles per hour, 80 per cent.
    37 miles per hour, 70 per cent.
    36 miles per hour, 60 per cent.
    Less than 36 miles per hour rejected.
    41 miles per hour, 110 per cent.
    42 miles per hour, 120 per cent.
    43 miles per hour, 130 per cent.
    44 miles per hour, 140 per cent.

5. The speed accomplished during the trial flight will be determined by taking an average of the time over a measured course of more than five miles, against and with the wind. The time will be taken by a flying start, passing the starting point at full speed at both ends of the course. This test subject to such additional details as the Chief Signal Officer of the Army may prescribe at the time.

6. Before acceptance a trial endurance flight will be required of at least one hour during which time the flying machine must remain continuously in the air without landing. It shall return to the starting point and land without any damage that would prevent it immediately starting upon another flight. During this trial flight of one hour it must be steered in all directions without difficulty and at all times under perfect control and equilibrium.

7. Three trials will be allowed for speed as provided for in paragraphs 4 and 5. Three trials for endurance as provided for in paragraph 6, and both tests must be completed within a period of thirty days from the date of delivery. The expense of the tests to be borne by the manufacturer. The place of delivery to the Government and trial flights will be at Fort Myer, Virginia.

8. It should be so designed as to ascend in any country which may be encountered in field service. The starting device must be simple and transportable. It should also land in a field without requiring a specially prepared spot and without damaging its structure.

9. It should be provided with some device to permit a safe descent in case of an accident to the propelling machinery.

10. It should be sufficiently simple in its construction and operation to permit an intelligent man to become proficient in its use within a reasonable length of time.

11. Bidders must furnish evidence that the Government of the United States has the lawful right to use all patented devices or appurtenances which may be a part of the flying machine, and that the manufacturers of the flying machine are authorized to convey the same to the Government. This refers to the unrestricted right to use the flying machine sold to the Government, but does not contemplate the exclusive purchase of patent rights for duplicating the flying machine.

12. Bidders will be required to furnish with their proposal a certified check amounting to ten per cent of the price stated for the 40-mile speed. Upon making the award for this flying machine these certified checks will be returned to the bidders, and the successful bidder will be required to furnish a bond, according to Army Regulations, of the amount equal to the price stated for the 40-mile speed.

13. The price quoted in proposals must be understood to include the instruction of two men in the handling and operation of this flying machine. No extra charge for this service will be allowed.

14. Bidders must state the time which will be required for delivery after receipt of order.

JAMES ALLEN,
Brigadier General, Chief Signal Officer of the Army.

Signal Office,
Washington, D. C., *December 23, 1907.*

Figure 7.5 Facsimile of the first published mission requirement for a military aircraft, January 20, 1908. This request led to the purchase of the Wright brothers' military type A *Flyer* later that year for the U.S. Army.

heavier-than-air flying machine in the nineteenth century were dominated by two questions: (1) Can enough aerodynamic lift be produced in a practical manner to exceed the weight? (2) If so, can it be done without producing so much drag that the power plant required to produce the opposing thrust would be impractically large and heavy? In particular, Lilienthal, Langley, and the Wright brothers were acutely aware of the importance of weight; they knew that more weight meant more drag, which dictated an engine with more power, which meant even more weight. In the conceptual design of an airplane, we cannot go any further until we have a first estimate of the takeoff gross weight.

7.3.3 Critical Performance Parameters

The design requirements stipulate the required performance of the new airplane. In Chapters 5 and 6, we found out that airplane performance is critically dependent on several parameters, especially (1) maximum lift coefficient $(C_L)_{\text{max}}$; (2) lift-to-drag ratio L/D, usually at cruise; (3) wing loading W/S; and (4) thrust-to-weight ratio T/W. We saw in particular how W/S and T/W appeared in many governing equations for airplane performance. Therefore, the next pivot point is the calculation of first estimates for W/S and T/W that are necessary to achieve the performance as stipulated by the requirements. In the subsequent chapters, we will see how these first estimates can be made.

7.3.4 Configuration Layout

The configuration layout is a drawing of the shape and size (dimensions) of the airplane as it has evolved to this stage. The critical performance parameters (Section 7.3.3) in combination with the initial weight estimate (Section 7.3.2) give enough information to approximately size the airplane and to draw the configuration.

7.3.5 Better Weight Estimate

By this stage, the overall size and shape of the airplane are coming more into focus. Because of the dominant role played by weight, the pivot point at this stage is an improved estimate of weight, based upon the performance parameters determined in Section 7.3.3, a detailed component weight breakdown based on the configuration layout in Section 7.3.4, and a more detailed estimate of the fuel weight necessary to meet the requirements.

7.3.6 Performance Analysis

At this pivot point, the airplane as drawn in Section 7.3.4 is put through a preliminary performance analysis using the techniques (or the equivalent) discussed in Chapters 5 and 6. This pivot point is where "the rubber meets the road"—where the configuration drawn in Section 7.3.4 is judged as to whether it can meet all the original specifications set forth in Section 7.3.1. This is obviously a critical point in the conceptual design process. It is unlikely that the configuration, as first obtained, will indeed meet *all* the specifications; it may exceed some, but not measure up to others. At this stage, the creative judgment of the designer is particularly important. An iterative process is initiated wherein the configuration is modified, with the expectation of coming closer to meeting the requirements. The design process returns to step 3 in Fig. 7.3 and readjusts the critical performance parameters in directions that will improve performance. These readjustments in turn readjust the configuration in step 4 and the better weight estimate in step 5. The new (hopefully improved) performance is assessed in step 6. The iteration is repeated until the resulting airplane design meets the requirements.

At this stage, some mature judgment on the part of the design team is critical, because the iterative process might not lead to a design that meets all the requirements. This may be because some of the specifications are unrealistic, or that the existing technology is not sufficiently advanced, or that costs are estimated to be prohibitive, or for a host of other reasons. As a result, in collaboration with the customer, some specifications may be relaxed in order to achieve other requirements that take higher priority. For example, if high speed is critical, but the high wing loading that allows this high speed increases the takeoff and landing distances beyond the original specifications, then the takeoff and landing requirements might be relaxed.

7.3.7   Optimization

When the design team is satisfied that the iterative process between steps 3 and 6 in Fig. 7.3 has produced a viable airplane, the next question is: Is it the *best* design? This leads to an optimization analysis, which is the seventh and final pivot point listed in Fig. 7.3. The optimization may be carried out by a systematic variation of different parameters, such as T/W and W/S, producing a large number of different airplanes via steps 3 to 6, and plotting the performance of all these airplanes on graphs which provide a sizing matrix or a carpet plot from which the optimum design can be found. In recent years, the general field of optimization has grown into a discipline of its own. Research in optimization theory had led to more mathematical sophistication which is finding its way into the design process. It is likely that airplane designers in the early twenty-first century will have available to them optimized design programs which may revolutionize the overall design process.

7.3.8   Constraint Diagram

Some of the intellectual activity described in Sections 7.3.6 and 7.3.7 can be aided by constructing a constraint diagram, which identifies the allowable *solution space* for the airplane design, subject to various constraints imposed by the initial requirements and the laws of physics. We have seen that the thrust-to-weight ratio and wing loading are two of the most important design parameters. A constraint diagram consists of plots of the sea-level thrust-to-takeoff weight ratio T_0/W_0 versus the wing loading at takeoff W_0/S that are determined by various requirements set up in our intellectual pivot point 1. A schematic of a constraint diagram is shown in Fig. 7.6, where the curves labeled A, B, and C pertain to constraints imposed by different specific requirements. Let us examine each curve in turn.

Curve A: Takeoff Constraint If the requirements specify a maximum takeoff length, then curve A gives the allowed variation of T_0/W_0 versus W_0/S for which this requirement is exactly satisfied. For example, for simplicity, let us approximate the takeoff distance by the expression for the ground roll given by Eq. (6.95), repeated here:

$$s_g = \frac{1.21(W/S)}{g\rho_\infty (C_L)_{\max}(T/W)} \qquad \textbf{[6.95]}$$

In Eq. (6.95), s_g is a given number. Solving Eq. (6.95) for T/W, we have

$$\frac{T}{W} = \left[\frac{1.21}{g\rho_\infty (C_L)_{\max} s_g} \right] \frac{W}{S} \qquad \textbf{[7.1]}$$

Noting that the factor in brackets is a constant and applying Eq. (7.1) to takeoff conditions at sea level, we have

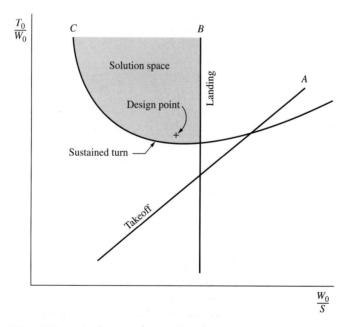

Figure 7.6   A schematic of a constraint diagram.

$$\frac{T_0}{W_0} = \text{constant} \times \frac{W_0}{S} \qquad \textbf{[7.2]}$$

For the takeoff constraint, T_0/W_0 is a linear function of W_0/S; this is given by curve A in Fig. 7.6. *Important:* Any value of T_0/W_0 above this curve will satisfy the takeoff constraint by resulting in a takeoff distance *smaller* than the required value. So the area above curve A is "allowable" from the point of view of the takeoff constraint.

Curve B: Landing Constraint   If the requirements specify a maximum landing length, then curve B represents this constraint. Equation (6.123) gives the landing ground roll. Let us represent the landing distance by Eq. (6.123), repeated here:

$$s_g = jN\sqrt{\frac{2}{\rho_\infty}\frac{W}{S}\frac{1}{(C_L)_{\max}}} + \frac{j^2(W/S)}{g\rho_\infty(C_L)_{\max}[T_{\mathrm{rev}}/D + D/W + \mu_r(1 - L/W)]}$$

$$\textbf{[6.123]}$$

For a given value of s_g, there is only one value of W/S that satisfies this equation. Hence, the landing constraint is represented by a vertical line through this particular value of W/S. This is shown by curve B in Fig. 7.6. Values of W_0/S to the left of this vertical line will satisfy the constraint by resulting in a landing distance *smaller* than the required value. So the area to the left of curve B is "allowable" from the point of view of the landing constraint.

Curve C: Sustained Level Turn   If the requirements specify a sustained level turn with a given load factor at a given altitude and speed, then curve C represents this constraint. Equation (6.18), repeated here, relates load factor, T/W, and W/S for a sustained level turn.

$$n_{\max} = \left\{ \frac{\frac{1}{2}\rho_\infty V_\infty^2}{K(W/S)}\left[\left(\frac{T}{W}\right)_{\max} - \frac{1}{2}\rho_\infty V_\infty^2 \frac{C_{D,0}}{W/S}\right]\right\}^{1/2} \qquad \textbf{[6.18]}$$

For the given constraint, all quantities in Eq. (6.18) are given except the two variables T/W and W/S. Solving Eq. (6.18) for T/W, we can write

$$\frac{T_0}{W_0} = C_1\frac{W}{S} + \frac{C_2}{W/S} \qquad \textbf{[7.3]}$$

where C_1 and C_2 are constants. Equation (7.3) is represented by curve C in Fig. 7.6. Values of T_0/W_0 above curve C will satisfy the sustained turn requirements. The area above curve C is "allowable" from the point of view of the sustained turning requirement

Assuming curves A, B, and C represent the only constraints, the area in Fig. 7.6 that is common to the three allowable areas is the shaded area identified as the *solution space*. An airplane with any combination of T_0/W_0 and W_0/S that falls within this solution space will satisfy the constraints imposed by the requirements.

By constructing the constraint diagram as shown in Fig. 7.6, the airplane designer can intelligently decide where to start the preliminary design, hence avoiding some trial designs that later prove not to satisfy one or more of the requirements. Looking at the constraint diagram, the designer can choose to start at a selected design point, indentified by the cross in Fig. 7.6. It makes sense to pick a design point with a relatively low T_0/W_0, but which is still in the solution space, so that the aircraft design is not unduly overpowered, hence costing more than necessary.

7.3.9 Interim Summary

Figure 7.3 illustrates the seven intellectual pivot points in the conceptual design of an airplane. To actually carry out the conceptual design process, we must visualize these seven pivot points interconnected by a web of detailed considerations. For example, we must

1. Make a selection of the airfoil section.
2. Determine the wing geometry (aspect ratio, sweep angle, taper ratio, twist, incidence angle relative to the fuselage, dihedral, vertical location on the fuselage, wing-tip shape, etc.)
3. Choose the geometry and arrangement of the tail. Would a canard be more useful?
4. Decide what specific power plants are to be used. What are the size, number, and placement of the engines?
5. Decide what high-lift devices will be necessary.

These are just a few elements of the web of details that surrounds and interconnects the seven pivot points listed in Fig. 7.3. Moreover, there is nothing unique about this web of details; each designer or design team spins this web as suits their purposes. The next two chapters spin some simple webs that are illustrative of the design process for a propeller-driven airplane, a jet-propelled subsonic airplane, and a supersonic airplane, respectively. They are intended to be illustrative only; the reader should not attempt to actually construct and fly a flying machine from the designs presented in subsequent chapters. Recall that our purpose in this book is to give insight into the design *philosophy*. It is intended to be studied as a precursor and as a companion to the more detailed design texts exemplified by Refs. 25 and 52 to 54.

So, let us get on with spinning these webs.

chapter

8

Design of a Propeller-Driven Airplane

There is nothing revolutionary in the airplane business. It is just a matter of development. What we've got today is the Wright brothers' airplane developed and refined. But the basic principles are just what they always were.

> Donald W. Douglas, July 1, 1936.
> Comment made at the presentation of
> the Collier Trophy to Douglas for the
> design of the DC-3. President
> Roosevelt presented the award to
> Douglas at the White House.

When you design it . . . think about how you would feel if you had to fly it! Safety first!

> Sign on the wall of the design office at
> Douglas Aircraft Company, 1932.

8.1 INTRODUCTION

The purpose of this chapter is to illustrate the process and philosophy of the design of a subsonic propeller-driven airplane. In a sense, this chapter (and the subsequent chapter) is just one large "worked example." We will use the seven pivot points described in Chapter 7 to anchor our thinking, and we will draw from Chapters 1 to 6 to construct our web of details around these pivot points.

8.2   REQUIREMENTS

We are given the job of designing a light, business transport aircraft which will carry five passengers plus the pilot in relative comfort in a pressurized cabin. The specified performance is to be as follows:

1. Maximum level speed at midcruise weight: 250 mi/h.
2. Range: 1,200 mi.
3. Ceiling: 25,000 ft.
4. Rate of climb at sea level: 1,000 ft/min.
5. Stalling speed: 70 mi/h.
6. Landing distance (to clear a 50-ft obstacle): 2,200 ft.
7. Takeoff distance (to clear a 50-ft obstacle): 2,500 ft.

In addition, the airplane should be powered by one (or more) conventional reciprocating engine.

The stipulation of these requirements constitutes an example of the first pivot point in Fig. 7.3.

8.3   THE WEIGHT OF AN AIRPLANE AND ITS FIRST ESTIMATE

As noted in Fig. 7.3, the second pivot point in our conceptual design analysis is the preliminary (almost crude) estimation of the gross weight of the airplane. Let us take this opportunity to discuss the nature of the weight of an airplane in detail.

There are various ways to subdivide and categorize the weight components of an airplane. The following is a common choice.

1. *Crew weight* W_{crew}. The crew comprises the people necessary to operate the airplane in flight. For our airplane, the crew is simply the pilot.
2. *Payload weight* $W_{payload}$. The payload is what the airplane is intended to transport—passengers, baggage, freight, etc. If the airplane is intended for military combat use, the payload includes bombs, rockets, and other disposable ordnance.
3. *Fuel weight* W_{fuel}. This is the weight of the fuel in the fuel tanks. Since fuel is consumed during the course of the flight, W_{fuel} is a variable, decreasing with time during the flight.
4. *Empty weight* W_{empty}. This is the weight of everything else—the structure, engines (with all accessory equipment), electronic equipment (including radar, computers, communication devices, etc.), landing gear, fixed equipment (seats, galleys, etc.), and anything else that is not crew, payload, or fuel.

The sum of these weights is the total weight of the airplane W. Again, W varies throughout the flight because fuel is being consumed, and for a military combat airplane, ordnance may be dropped or expended, leading to a decrease in the payload weight.

The design takeoff gross weight W_0 is the weight of the airplane at the instant it begins its mission. It includes the weight of all the fuel on board at the beginning of the flight. Hence,

$$W_0 = W_{\text{crew}} + W_{\text{payload}} + W_{\text{fuel}} + W_{\text{empty}} \qquad \textbf{[8.1]}$$

In Eq. (8.1), W_{fuel} is the weight of the full fuel load at the beginning of the flight.

In Eq. (8.1), W_0 is the important quantity for which we want a first estimate; W_0 is the desired result from pivot point 2 in Fig. 7.3. To help make this estimate, Eq. (8.1) can be rearranged as follows. If we denote W_{fuel} by W_f and W_{empty} by W_e (for notational simplicity), Eq. (8.1) can be written as

$$W_0 = W_{\text{crew}} + W_{\text{payload}} + W_f + W_e \qquad \textbf{[8.2]}$$

or

$$W_0 = W_{\text{crew}} + W_{\text{payload}} + \frac{W_f}{W_0} W_0 + \frac{W_e}{W_0} W_0 \qquad \textbf{[8.3]}$$

Solving Eq. (8.3) for W_0, we have

$$\boxed{W_0 = \frac{W_{\text{crew}} + W_{\text{payload}}}{1 - W_f / W_0 - W_e / W_0}} \qquad \textbf{[8.4]}$$

The form of Eq. (8.4) is particularly useful. Although at this stage we do not have a value of W_0, we can fairly readily obtain values of the *ratios* W_f / W_0 and W_e / W_0, as we will see next. Then Eq. (8.4) provides a relation from which W_0 can be obtained in an iterative fashion. [The iteration is required because in Eq. (8.4), W_f / W_0 and W_e / W_0 may themselves be functions of W_0.]

8.3.1 Estimation of W_e / W_0

Most airplane designs are evolutionary rather than revolutionary; that is, a new design is usually an evolutionary change from previously existing airplanes. For this reason, historical, statistical data on previous airplanes provide a starting point for the conceptual design of a new airplane. We will use such data here. In particular, Fig. 8.1 is a plot of W_e / W_0 versus W_0 for a number of reciprocating engine, propeller-driven airplanes. Data for 19 airplanes covering the time period from 1930 to the present are shown. The data show a remarkable consistency. The values of W_e / W_0 tend to cluster around a horizontal line at $W_e / W_0 = 0.62$. For gross weights above 10,000 lb, W_e / W_0 tends to be slightly higher for some of the aircraft. However, there is no technical reason for this; rather, the higher values for the heavier airplanes are most likely an historical phenomenon. The P-51, B-10, P-38, DC-3, and B-26 are all examples of 1930's technology. A later airplane, the Lockheed P2V Neptune, is based

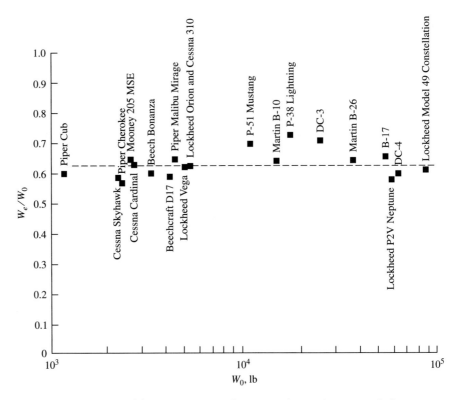

Figure 8.1   Variation of the empty–gross weight ratio W_e/W_0 with gross weight for reciprocating-engine, propeller-dirven airplanes.

on 1940s' technology, and it has a relatively low value of $W_e/W_0 = 0.57$. Eclipsed by jet-propelled airplanes, the design of heavy reciprocating engine/propeller-driven airplanes in the gross weight class above 10,000 lb has virtually ceased since the 1950s. The last major airplanes of this class were the Douglas DC-7 and the Lockheed Super Constellation, both large, relatively luxurious passenger transports. Hence, reflected in Fig. 8.1, no modern airplanes are represented on the right side of the graph. In contrast, the data shown at the left of the graph, for gross weight less than 10,000 lb, are a mixture, representing airplanes from 1930 to the present.

As a result of the data shown in Fig. 8.1, we choose for our first estimate

$$\frac{W_e}{W_0} = 0.62 \qquad\qquad \textbf{[8.5]}$$

8.3.2   Estimation of W_f/W_0

The amount of fuel required to carry out the mission depends critically on the efficiency of the propulsion device—the engine specific fuel consumption and the propeller efficiency. It also depends critically on the aerodynamic efficiency—the lift-to-drag ratio. These factors are principal players in the Brequet range equation given by Eq. (5.153), repeated here:

$$R = \frac{\eta_{pr}}{c} \frac{L}{D} \ln \frac{W_0}{W_1} \qquad \textbf{[5.153]}$$

Equation (5.153) is very important in our estimation of W_f / W_0, as defined below.

The total fuel consumed during the mission is that consumed from the moment the engines are turned on at the airport to the moment they are shut down at the end of the flight. Between these times, the flight of the airplane can be described by a *mission profile*, a conceptual sketch of altitude versus time such as shown in Fig. 8.2. As stated in the specifications, the mission of our airplane is that of a business light transport, and therefore its mission profile is that for a simple cruise from one location to another. This is the mission profile shown in Fig. 8.2. It starts at the point labeled 0, when the engines are first turned on. The takeoff segment is denoted by the line segment 0–1, which includes warm-up, taxiing, and takeoff. Segment 1–2 denotes the climb to cruise altitude (the use of a straight line here is only schematic and is *not* meant to imply a constant rate of climb to altitude). Segment 2–3 represents the cruise, which is by far the largest segment of the mission. Segment 2–3 shows an increase in altitude during cruise, consistent with an attempt to keep C_L (and hence L/D) constant as the airplane weight decreases because of the consumption of fuel. This is discussed at length in Section 5.13.3. Segment 3–4 denotes the descent, which generally includes loiter time to account for air traffic delays; for design purposes, a loiter time of 20 min is commonly used. Segment 4–5 represents landing.

The mission profile shown in Fig. 8.2 is particularly simple. For other types of missions, especially those associated with military combat aircraft, the mission profiles will include such aspects as combat dogfighting, weapons drop, in-flight refueling, etc. For a discussion of such combat mission profiles, see, for example,

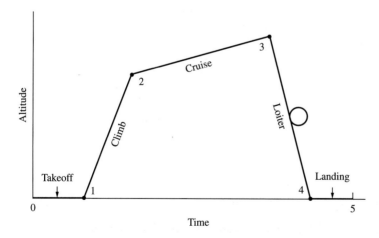

Figure 8.2 Mission profile for a simple cruise.

Ref. 25. For our purposes, we will deal only with the simple cruise mission profile sketched in Fig. 8.2.

The mission profile is a useful bookkeeping tool to help us estimate fuel weight. Each segment of the mission profile is associated with a *weight fraction*, defined as the airplane weight at the end of the segment divided by the weight at the beginning of the segment.

$$\text{Mission segment weight fraction} = \frac{W_i}{W_{i-1}}$$

For example, the cruise weight fraction is W_3/W_2, where W_3 is the airplane weight at the end of the cruise and W_2 is the weight at the beginning of cruise. The fuel weight ratio W_f/W_0, can be obtained from the product of the mission segment weight fractions as follows. Using the mission profile in Fig. 8.2, the ratio of the airplane weight at the end of the mission to the initial gross weight is W_5/W_0. In turn,

$$\frac{W_5}{W_0} = \frac{W_1}{W_0}\frac{W_2}{W_1}\frac{W_3}{W_2}\frac{W_4}{W_3}\frac{W_5}{W_4} \qquad \textbf{[8.6]}$$

The right side of Eq. (8.6) is simply the product of the individual mission segment weight fractions. Also, keep in mind that for the simple cruise mission shown in Fig. 8.2, the change in weight during each segment is due to the consumption of fuel. If, at the end of the flight, the fuel tanks were completely empty, then

$$W_f = W_0 - W_5$$

or

$$\frac{W_f}{W_0} = 1 - \frac{W_5}{W_0} \qquad \textbf{[8.7]}$$

However, at the end of the mission, the fuel tanks are not completely empty—by design. There should be some fuel left in reserve at the end of the mission in case weather conditions or traffic problems require that the pilot of the airplane divert to another airport, or spend a longer-than-normal time in a holding pattern. Also, the geometric design of the fuel tanks and the fuel system leads to some trapped fuel that is unavailable at the end of the flight. Typically, a 6% allowance is made for reserve and trapped fuel. Modifying Eq. (8.7) for this allowance, we have

$$\frac{W_f}{W_0} = 1.06\left(1 - \frac{W_5}{W_0}\right) \qquad \textbf{[8.8]}$$

Hence, the sequence for the calculation of W_f/W_0 that appears in the denominator of Eq. (8.4) is as follows:

1. Calculate each individual mission segment weight fraction W_1/W_0, W_2/W_1, etc., that appears in Eq. (8.6).

2. Calculate W_5/W_0 from Eq. (8.6).

3. Calculate W_f/W_0 from Eq. (8.8).

Let us proceed to make this calculaton for our business transport airplane.

For takeoff, segment 0–1, historical data show that W_1/W_0 is small, on the order of 0.97. Hence, we assume

$$\frac{W_1}{W_0} = 0.97 \qquad \textbf{[8.9]}$$

For climb, segment 1–2, we again rely on historical data for a first estimate, which indicate that W_2/W_1 is also small, on the order of 0.985. Hence, we assume

$$\frac{W_2}{W_1} = 0.985 \qquad \textbf{[8.10]}$$

For cruise, segment 2–3, we make use of the Brequet range equation, Eq. (5.153). This requires an estimate of L/D. At this stage of our design process (pivot point 2 in Fig. 7.3), we cannot carry out a detailed aerodynamic analysis to predict L/D—we have not even laid out the shape of the airplane yet (which comes later in the process, pivot point 4 in Fig. 7.3). Therefore, we can only make a crude approximation, again based on data from existing airplanes. Loftin (Ref. 13) has tabulated the values of $(L/D)_{max}$ for a number of famous aircraft over the past century. The values for some representative reciprocating engine/propeller-driven airplanes of the size likely to carry four to six people are tabulated below, obtained by Loftin.

Airplane	$(L/D)_{max}$
Cessna 310	13.0
Beach Bonanza	13.8
Cessna Cardinal	14.2

Hence, a reasonable first approximation for our airplane is

$$(L/D)_{max} = 14 \qquad \textbf{[8.11]}$$

Also needed in the range equation, Eq. (5.153), are the specific fuel consumption c and propeller efficiency η. As stated in Section 3.3.1, a typical value of specific fuel consumption for current aircraft reciprocating engines is 0.4 lb of fuel consumed per horsepower per hour. In consistent units, noting that 1 hp = 550 ft·lb/s, we have

$$c = 0.4 \frac{lb}{hp\cdot h} \frac{1\ hp}{550\ ft\cdot lb/s} \frac{1\ h}{3{,}600\ s}$$

$$c = 2.02 \times 10^{-7} \frac{lb}{(ft\cdot lb/s)}(s) \qquad \textbf{[8.12]}$$

A reasonable value for the propeller efficiency, assuming a variable-pitch propeller (we will make this choice now—for a business aircraft, the cost of a variable-pitch propeller over that for a cheaper, fixed-pitch propeller is justified) is, from Section 3.3.2,

$$\eta_{pr} = 0.85 \qquad \textbf{[8.13]}$$

Returning to Eq. (5.153), the ratio W_0/W_1 in that equation is replaced for the mission segment 2–3 by W_2/W_3. Hence, from Eq. (5.153),

$$R = \frac{\eta_{pr}}{c}\frac{L}{D}\ln\frac{W_2}{W_3} \qquad \text{[8.14]}$$

Solving Eq. (8.14) for W_2/W_3, we have

$$\ln\frac{W_2}{W_3} = \frac{c}{\eta_{pr}}\frac{R}{L/D} \qquad \text{[8.15]}$$

In Eq. (8.15), the range is stipulated in the requirements as $R = 1{,}200$ mi $= 6.64 \times 10^6$ ft. Also inserting the values given by Eqs. (8.11) to (8.13) into Eq. (8.15), we have

$$\ln\frac{W_2}{W_3} = \frac{2.02 \times 10^{-7}}{0.85}\frac{6.64 \times 10^6}{14} = 0.1127$$

Hence,

$$\frac{W_2}{W_3} = e^{0.1127} = 1.119$$

or

$$\frac{W_3}{W_2} = \frac{1}{1.119} = 0.893 \qquad \text{[8.16]}$$

The loiter segment 3–4 in Fig. 8.2 is essentially the descent from cruise altitude to the landing approach. For our approximate calculations here, we will ignore the details of fuel consumption during descent, and just assume that the horizontal distance covered during descent is part of the required 1,200-mi range. Hence, for this assumption

$$\frac{W_4}{W_3} = 1 \qquad \text{[8.17]}$$

Finally, the fuel consumed during the landing process, segment 4–5, is also based on historical data. The amount of fuel used for landing is small, and based on previous airplanes, the value of W_5/W_4 is approximately 0.995. Hence, we assume for our airplane that

$$\frac{W_5}{W_4} = 0.995 \qquad \text{[8.18]}$$

Collecting the various segment weight fractions from Eqs. (8.9), (8.10), (8.16), (8.17), and (8.18), we have from Eq. (8.6)

$$\frac{W_5}{W_0} = \frac{W_1}{W_0}\frac{W_2}{W_1}\frac{W_3}{W_2}\frac{W_4}{W_3}\frac{W_5}{W_4} = (0.97)(0.985)(0.893)(1)(0.995) = 0.85 \qquad \text{[8.19]}$$

Inserting the value of W_5/W_0 from Eq. (8.19) into Eq. (8.8), we have

$$\frac{W_f}{W_0} = 1.06\left(1 - \frac{W_5}{W_0}\right) = 1.06(1 - 0.85)$$

or

$$\frac{W_f}{W_0} = 0.159 \qquad\qquad \textbf{[8.20]}$$

8.3.3 Calculation of W_0

Return to Eq. (8.4) for the design takeoff gross weight W_0. We have obtained a value for W_e/W_0 given by Eq. (8.5). We have also obtained a value for W_f/W_0 given by Eq. (8.20). All we need to obtain W_0 from Eq. (8.4) are values for the crew and payload weights W_{crew} and $W_{payload}$, respectively.

Corning (Ref. 55) suggests the average passenger weight of 160 lb, plus 40 lb of baggage per passenger. A more recent source is Raymer (Ref. 25) who suggests an average passenger weight of 180 lb (dressed and with carry-on bags), plus 40 to 60 lb of baggage per person in the cargo hold. For our airplane, there are five passengers and one pilot, six people in total. Let us assume the average weight per person is 170 lb. Hence, since the only crew is the pilot, we assume

$$W_{crew} = 170 \text{ lb} \qquad\qquad \textbf{[8.21]}$$

The payload is the five passengers, plus the baggage for all six people. The type of short business trip for which this airplane will most likely be used would require less baggage than a longer, intercontinental trip. Hence, it is reasonable to assume 20 lb of baggage per person rather than the 40 lb mentioned above. Thus, including the pilot's baggage, we have

$$W_{payload} = 5(170) + 6(20) = 970 \text{ lb} \qquad\qquad \textbf{[8.22]}$$

Inserting the values from Eqs. (8.5) and (8.20) to (8.22) into Eq. (8.4), we have

$$W_0 = \frac{W_{crew} + W_{payload}}{1 - W_f/W_0 - W_e/W_0} = \frac{170 + 970}{1 - 0.159 - 0.62}$$
$$= \frac{1{,}140}{0.221} = 1{,}140(4.525) = \boxed{5{,}158 \text{ lb}} \qquad\qquad \textbf{[8.23]}$$

This is our first estimate of the gross weight of the airplane. We have now completed pivot point 2 in Fig. 7.3.

Important comment. The calculation in Eq. (8.23) clearly shows the amplified impact of crew and payload weight on the gross weight of the airplane. The amplification factor is 4.525; that is, for every increase of 1 lb of payload weight, the airplane's gross weight increases by 4.525 lb. For example, if we had allowed each person 40 lb of baggage rather than the 20 lb we chose, the design gross weight of the airplane would have increased by $(6)(20)(4.525) = 543$ lb, that is, more than a

10% increase in the gross weight. This is a clear demonstration of the importance of weight in airplane design. For our example, 1 lb saved in any manner—payload reduction, reduced structural weight, reduced fuel weight, etc.—results in a 4.525-lb reduction in overall gross weight. It is easy to see why aeronautical engineers are so weight-conscious.

We also note that in our calculation of W_0 we have assumed that W_e/W_0 is independent of W_0, that is, independent of the gross weight of the airplane. This assumption was based on previous piston-engine airplanes, as shown in Fig. 8.1, where we chose $W_e/W_0 = 0.62$, independent of W_0. This is not usually the case for most classes of aircraft; in general, W_e/W_0 is a function of W_0. Indeed, Raymer (Ref. 25) gives empirical equations for this function for 13 different classes of aircraft. When W_e/W_0 is treated as a function of W_0, then the calculation of W_0 from Eq. (8.4) becomes an iteration. First, W_0 has to be assumed. Then W_e/W_0 is obtained for this assumed W_0. Next, a new value of W_0 is calculated from Eq. (8.4). This new value of W_0 is then used to estimate a new value of W_e/W_0, and the calculation of W_0 from Eq. (8.4) is repeated. This iteration is continued until convergence is obtained. In our calculation above, an iterative process was not required because we assumed that W_e/W_0 was a fixed value.

Finally, let us calculate the fuel weight; this will become important later in sizing the fuel tanks. From Eq. (8.20), $W_f/W_0 = 0.159$. Hence,

$$W_f = \frac{W_f}{W_0} W_0 = (0.159)(5,158) = 820 \text{ lb}$$

The weight of aviation gasoline is 5.64 lb/gal. Hence, the capacity of the fuel tank (or tanks) should be

$$\text{Tank capacity} = \frac{820}{5.64} = 145.4 \text{ gal}$$

8.4 ESTIMATION OF THE CRITICAL PERFORMANCE PARAMETERS

We now move to pivot point 3 in Fig. 7.3, namely, an estimation of the critical performance parameters $(C_L)_{\max}$, L/D, W/S, and T/W. These parameters are dictated by the requirements given in Section 8.2; that is, they will be determined by such aspects as maximum speed, range, ceiling, rate of climb, stalling speed, landing distance, and takeoff distance.

8.4.1 Maximum Lift Coefficient

This is the stage in the design process where we make an initial choice for the airfoil shape for the wing. Historically, general aviation airplanes have employed the NACA four-digit, five-digit, and 6-series airfoil sections—the laminar-flow series.

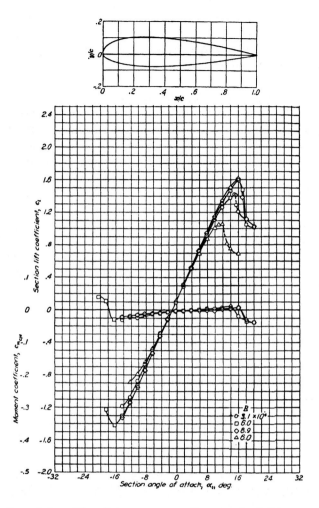

Figure 8.3 Lift coefficient, moment coefficient, and airfoil shape for the NACA 23018 airfoil.

The NACA five-digit airfoils have been particularly favored by the general aviation industry in the United States. These airfoils, such as the NACA 23018 and the NACA 23012 shown in Figs. 8.3 and 8.4, respectively, were designed in the middle 1930s. The maximum camber was placed closer to the leading edge (at $0.15c$ for the two airfoils shown) than was the case for the earlier NACA four-digit airfoils. A benefit of this design is a higher $(c_l)_{\max}$ compared to the earlier airfoils. A disadvantage is the very sharp stalling behavior, as seen in Figs. 8.3 and 8.4.

For many airplanes, including some general aviation aircraft, one airfoil section is used at the wing root, and another airfoil shape is used at the wing tip, with the airfoil

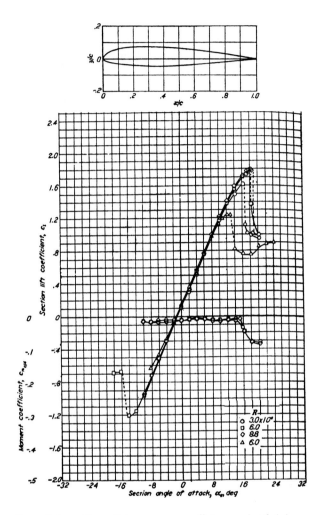

Figure 8.4   Lift coefficient, moment coefficient, and airfoil shape
for the NACA 23012 airfoil.

sections between the root and tip being a linear interpolation between the root and
tip sections. Several examples from existing general aviation airplanes are tabulated
below.

Airplane	Root Section	Tip Section
Beechcraft Bonanza	NACA 23016.5	NACA 23012
Beechcraft Baron	NACA 23015.5	NACA 23010.5
Cessna Caravan	NACA 23017.4	NACA 23012
Piper Cheyene	NACA 63A415	NACA 63A212

In these examples, the root airfoil section is relatively thick (about 15% to 17%), and the wing airfoil shape tapers to a thinner section at the tip (about 12%). There are good reasons for this. Structurally, the wing bending moment is greatest at the root; a thicker airfoil readily allows the design for greater structural strength at the root. Aerodynamically, an 18% airfoil will stall at a lower angle of attack than a 12% airfoil. Hence, a wing which has airfoil sections which taper from 18% thick at the root to 12% thick at the tip will tend to stall first at the wing root, with attached flow still at the tip. The resulting buffeting that occurs at stall at the root is a warning to the pilot, while at the same time the ailerons remain effective because flow is still attached at the tip—both distinct advantages. Finally, a thicker wing section at the root allows additional volume for the storage of fuel in the wing.

For all these reasons, we make an initial choice of the airfoil section for our airplane design as follows: at the root, an NACA 23018 section (Fig. 8.3); at the tip, an NACA 23012 section (Fig. 8.4). We will assume that a linear interpolation between the root and tip defines the local airfoil sections elsewhere along the wing. The resulting $(C_L)_{max}$ for the wing will be an average of the root and tip section values, depending on the planform taper ratio and the degree of geometric twist of the wing (if there is any). Also $(C_L)_{max}$ for the finite wing is less than that for the airfoil due to three-dimensional flow effects. Since we have not laid out the planform shape or twist distribution yet, we will assume that $(C_L)_{max}$ is a simple average of those for the airfoil sections at the root and tip, reduced by 10% for the effect of a finite aspect ratio. For the NACA 23018, from Fig. 8.3, $(c_l)_{max} = 1.6$; for the NACA 23012, from Fig. 8.4, $(c_l)_{max} = 1.8$. Taking the average, we have for the averge airfoil maximum lift coefficient for our wing

$$\text{Average } (c_l)_{max} = \frac{1.6 + 1.8}{2} = 1.7$$

To aid in the takeoff and landing performance, we will design the wing with trailing-edge flaps. For simplicity (and hence production cost savings), we choose a simple plain flap. From Fig. 5.28, such a flap deflected 45° will yield an increase in the airfoil maximum lift coefficient $\Delta(c_l)_{max} = 0.9$. Hence, for our average airfoil maximum lift coefficient, we have

$$\text{Average } (c_l)_{max} \text{ with } 45° \text{ flap deflection } = 1.7 + 0.9 = 2.6$$

Finally, to account for the three-dimensional effect of the finite aspect ratio, Raymer (Ref. 25) suggests that, for finite wings with aspect ratio greater than 5,

$$(C_L)_{max} = 0.9(c_l)_{max} \qquad \textbf{[8.24]}$$

Since we are designing a low-speed business, general aviation airplane, where efficient cruise is important, we most certainly will have a wing with an asepct ratio greater than 5. Hence, we use, as a preliminary estimate of maximum lift coefficient, from

Eq. (8.24)

$$(C_L)_{max} = 0.9(2.6)$$

$$\boxed{(C_L)_{max} = 2.34}$$ **[8.25]**

We will treat this as $(C_L)_{max}$ for the complete airplane, ignoring for the time being the effect of the fuselage, tail, and other parts of the configuration.

8.4.2   Wing Loading W/S

In most airplane designs, wing loading is determined by considerations of V_{stall} and landing distance. However, W/S also plays a role in the maximum velocity of the airplane [see Eq. (5.50)]; V_{max} increases as W/S increases. For our current airplane design, which is a low-speed aircraft, the primary constraints on W/S will be V_{stall} and landing, and we will take that approach. From Eq. (5.67), repeated here:

$$V_{stall} = \sqrt{\frac{2}{\rho_\infty}\frac{W}{S}\frac{1}{(C_L)_{max}}}$$ **[5.67]**

solving Eq. (5.67) for W/S, we have

$$\frac{W}{S} = \frac{1}{2}\rho_\infty V_{stall}^2 (C_L)_{max}$$ **[8.26]**

The requirements specify $V_{stall} \leq 70mi/h = 102.7$ ft/s. Using $(C_L)_{max}$ from Eq. (8.25) and making the calculation at sea level, where $\rho_\infty = 0.002377$ slug/ft^3, we have from Eq. (8.26)

$$\frac{W}{S} = \frac{1}{2}(0.002377)(102.7)^2(2.34) = 29.3 \text{ lb/ft}^2$$ **[8.27]**

Equation (8.27) gives us the value of W/S constrained by the stalling velocity.

Let us examine the constraint imposed by the specified landing distance. In Fig. 6.17, the landing distance is the sum of the approach distance s_a, the flare distance s_f, and the ground roll s_g. The approach angle θ_a is given by Eq. (6.104), which requires knowledge of L/D and T/W. Since we have not made estimates of either quantity yet, we assume, based on the rule of thumb that $\theta_a \leq 3°$ for transport aircraft, that $\theta_a = 3°$. Following the discussion of approach distance in Section 6.8.1, we have, from Eq. (6.107) for the flight path radius during flare,

$$R = \frac{V_f^2}{0.2g}$$ **[6.107]**

In Eq. (6.107), V_f is the average velocity during flare, given by $V_f = 1.23V_{stall}$. From our design, $V_f = 1.23(102.7) = 126.3$ ft/s. From Eq. (6.107)

$$R = \frac{(126.3)^2}{0.2(32.2)} = 2,477 \text{ ft}$$

From Eq. (6.106), the flare height h_f is given by

$$h_f = R(1 - \cos\theta_a) = 2{,}477(1 - \cos 3°) = 3.4 \text{ ft}$$

Finally, from Eq. (6.108), the approach distance required to clear a 50-ft obstacle is given by

$$s_a = \frac{50 - h_f}{\text{Tan } \theta_a} = \frac{50 - 3.4}{\text{Tan } 3°} = 889 \text{ ft}$$

The flare distance s_f is given by Eq. (6.109):

$$s_f = R \sin\theta_a = 2{,}477 \sin 3° = 130 \text{ ft}$$

An approximation for ground roll s_g is given by Eq. (6.123). In that equation, let us assume that lift has been intentionally made small by retracting the flaps combined with a small angle of attack due to the rather level orientation of the airplane relative to the ground. (We are assuming that we will use tricycle landing gear for the airplane.) Furthermore, assuming no provision for thrust reversal, and ignoring the drag compared to the friction force between the tires and the ground, Eq. (6.123) simplifies further to

$$s_g = jN\sqrt{\frac{2}{\rho_\infty}\frac{W}{S}\frac{1}{(C_L)_{\max}}} + \frac{j^2(W/S)}{g\rho_\infty(C_L)_{\max}\mu_r} \qquad \textbf{[8.28]}$$

As stated in Section 6.8.3, $j = 1.15$ for commercial airplanes. Also, N is the time increment for free roll immediately after touchdown, before the brakes are applied. By assuming $N = 3$ s and $\mu = 0.4$, Eq. (8.28) becomes

$$s_g = (1.15)(3)\sqrt{\frac{2}{0.002377}\frac{W}{S}\frac{1}{2.34}} + \frac{(1.15)^2(W/S)}{(32.2)(0.002377)(2.34)(0.4)}$$

or

$$s_g = 65.4\sqrt{\frac{W}{S}} + 18.46\frac{W}{S} \qquad \textbf{[8.29]}$$

Since the allowable landing distance is specified in the requirements as 2,200 ft, and we have previously estimated that $s_a = 889$ ft and $s_f = 130$ ft, the allowable value for s_g is

$$s_g = 2{,}200 - 889 - 130 = 1{,}181 \text{ ft}$$

Inserting this value for s_g into Eq. (8.29), we have

$$1{,}181 = 65.4\sqrt{\frac{W}{S}} + 18.46\frac{W}{S} \qquad \textbf{[8.30]}$$

Equation (8.30) is a quadratic equation for $\sqrt{W/S}$. Using the quadratic formula, we obtain

$$\frac{W}{S} = 41.5 \text{ lb/ft}^2 \qquad \textbf{[8.31]}$$

Compare the value of $W/S = 41.5$ lb/ft^2 obtained from the landing distance constraint, Eq. (8.31), with the value of $W/S = 29.3$ lb/ft^2 obtained from the stall constraint, Eq. (8.27). Clearly, if $W/S < 41.5$ lb/ft^2, the landing distance will be shorter than 2,200 ft, clearly satisfying the requirements. Hence, for our airplane design, W/S is determined from the specified stall velocity, namely,

$$\boxed{\frac{W}{S} = 29.3 \text{ lb/ft}^2}$$ [8.32]

The value of W/S from Eq. (8.32) along with that for W_0 from Eq. (8.23) allows us to obtain the wing area.

$$S = \frac{W_0}{W/S} = \frac{5,158}{29.3} = \boxed{176 \text{ ft}^2}$$ [8.33]

8.4.3 Thrust-to-Weight Ratio

The value of T/W determines in part the takeoff distance, rate of climb, and maximum velocity. To obtain the design value of T/W, we have to examine each of these three constraints.

First, let us consider the takeoff distance, which is specified as 2,500 ft to clear a 50-ft obstacle. Using Eq. (6.95) to estimate the ground roll, we have

$$s_g = \frac{1.21(W/S)}{g\rho_\infty (C_L)_{\max}(T/W)}$$ [6.95]

where $(C_L)_{\max}$ in Eq. (6.95) is that value with the flaps only partially extended, consistent with their takeoff setting. Hence, we need to recalculate $(C_L)_{\max}$ for this case. Following the guidance provided in Table 5.3, we assume a flap deflection of 20° for takeoff. To return to Fig. 5.28, the $\Delta(c_l)_{\max}$ for a 45° flap deflection is 0.9. Assuming a linear variation of $\Delta(c_l)_{\max}$ with flap deflection angle, we have for takeoff $\Delta(c_l)_{\max} = 0.9(25/45) = 0.5$. Hence, for the wing, the average $(c_l)_{\max}$ with a 20° flap deflection is $1.7 + 0.5 = 2.2$. Taking into account the finite aspect ratio, as discussed in Section 8.4.1, we have for the wing

$$(C_L)_{\max} = 0.9(c_l)_{\max} = 0.9(2.2) = 1.98$$

This is the takeoff value of $(C_L)_{\max}$ that will be used in Eq. (6.95). Returning to Eq. (6.95), we have

$$s_g = \frac{1.21(W/S)}{g\rho_\infty (C_L)_{\max}(T/W)} = \frac{(1.21)(29.3)}{(32.2)(0.002377)(1.98)(T/W)} = \frac{233.9}{T/W}$$ [8.34]

Recall from our discussion in Section 6.7.1 that when T varies with velocity, as it does for a propeller-driven airplane, the value of T/W in Eq. (6.95) is assumed to be that for a velocity $V_\infty = 0.7V_{LO}$, where V_{LO} is the liftoff velocity, taken as $V_{LO} = 1.1V_{\text{stall}}$.

To calculate the distance while airborne to clear an obstacle (Section 6.7.2), we need the value of V_{stall} corresponding to the $(C_L)_{\max}$ with flaps in the takeoff position, that is, corresponding for our case to $(C_L)_{max} = 1.98$. From Eq. (5.67)

$$V_{stall} = \sqrt{\frac{2}{\rho_\infty} \frac{W}{S} \frac{1}{(C_L)_{max}}} = \sqrt{\frac{2(29.3)}{(0.002377)(1.98)}} = 111.6 \text{ ft/s}$$

From Eq. (6.98), the flight path radius is

$$R = \frac{6.96(V_{stall})^2}{g} = \frac{6.96(111.6)^2}{32.2} = 2{,}692 \text{ ft}$$

From Eq. (6.99), the included flight path angle is

$$\theta_{OB} = \text{Cos}^{-1}\left(1 - \frac{h_{OB}}{R}\right) \qquad \textbf{[6.99]}$$

where h_{OB} is the obstacle height, $h_{OB} = 50$ ft, so

$$\theta_{OB} = \text{Cos}^{-1}\left(1 - \frac{50}{2{,}692}\right) = 11.06°$$

From Eq. (6.100), the airborne distance is

$$s_a = R\sin\theta_{OB} = 2{,}692 \sin 11.06° = 516.4 \text{ ft} \qquad \textbf{[8.35]}$$

Combining Eqs. (8.34) and (8.35), we have

$$s_g + s_a = 2{,}500 = \frac{233.9}{T/W} + 516.4$$

or

$$\left(\frac{T}{W}\right)_{0.7V_{LO}} = \frac{233.9}{2{,}500 - 516.4} = 0.118 \qquad \textbf{[8.36]}$$

This is the value of required T/W at a velocity

$$V_\infty = 0.7V_{LO} = 0.7(1.1V_{stall}) = 0.7(1.1)(111.6) = 85.9 \text{ ft/s}$$

At this velocity, the power required to take off at the gross weight $W_0 = 5{,}158$ lb [see Eq. (8.23)] is

$$P_R = TV_\infty = \frac{T}{W}W_0V_\infty = (0.118)(5{,}158)(85.9) = 5.228 \times 10^4 \text{ ft·lb/s} \qquad \textbf{[8.37]}$$

This power required must equal the power available P_A, obtained from Eq. (3.13).

$$P_A = \eta_{pr}P \qquad \textbf{[3.13]}$$

Solving Eq. (3.13) for the shaft brake power P, we have

$$P = \frac{P_A}{\eta_{pr}} \qquad \textbf{[8.38]}$$

Typical propeller efficiencies are shown in Fig. 3.7. In our design we choose to use a constant-speed propeller. Hence, from Fig. 3.7, it appears reasonable to assume

$\eta_{pr} = 0.8$. Hence, the shaft brake power from the engine should be at least [from Eq. (8.38)]

$$P = \frac{P_A}{\eta_{pr}} = \frac{5.228 \times 10^4}{0.8} = 6.535 \times 10^4 \text{ ft·lb/s}$$

Since 550 ft·lb/s = 1 hp, we have

$$P = \frac{6.535 \times 10^4}{550} = 118.8 \text{ hp}$$

As stated in Section 3.3.1, for a reciprocating engine P is reasonably constant with V_∞. Hence, to satisfy the takeoff constraint, the total power must be at least

$$P \geq 118.8 \text{ hp} \tag{8.39}$$

Next, let us consider the constraint due to the specified rate of climb of 1,000 ft/min at sea level. Here, we need to make an estimate of the zero-lift drag coefficient $C_{D,0}$. We will use the same approach as illustrated in Example 2.4. From Fig. 2.54, for single-engine general aviation airplanes, the ratio of wetted area to the wing reference area is approximately $S_{wet}/S_{ref} = 4$. The skin-friction coefficient C_{fe} is shown as a function of Reynolds number in Fig. 2.55, where some data points for various jet airplanes are also plotted. Our airplane design will probably be about the same size as that of some early jet fighters, but with about one-third the speed. Hence, based on mean length, a relevant Reynolds number for us is about 10^7. For this case, Fig. 2.55 yields

$$C_{fe} = 0.0043$$

Hence, from Eq. (2.37)

$$C_{D,0} = \frac{S_{wet}}{S} C_{fe} = (4)(0.0043)$$

or

$$\boxed{C_{D,0} = 0.017} \tag{8.40}$$

We also need an estimate for the coefficient K that appears in the drag polar, Eq. (2.47), repeated here:

$$C_D = C_{D,0} + K C_L^2 \tag{2.47}$$

where, from Eqs. (2.43) to (2.46),

$$K = k_1 + k_2 + k_3 = k_1 + k_2 + \frac{C_L^2}{\pi e \text{AR}} \tag{8.41}$$

In Eq. (8.41), e is the span efficiency factor to account for a nonelliptical lift distribution along the span of the wing, and $C_L^2/(\pi e \text{AR})$ is the induced drag coefficient.

Let us estimate the value of K to be consistent with the earlier assumed value of $(L/D)_{max} = 14$ [see Eq. (8.11)]. From Eq. (5.30),

$$\left(\frac{L}{D}\right)_{max} = \sqrt{\frac{1}{4C_{D,0}K}} \qquad \textbf{[5.30]}$$

we have

$$K = \frac{1}{4C_{D,0}(L/D)^2_{max}} = \frac{1}{4(0.017)(14)^2}$$

or

$$\boxed{K = 0.075} \qquad \textbf{[8.42]}$$

This estimate for K also allows an estimate of the aspect ratio, as follows. It is conventional to define another efficiency factor, the Oswald efficiency e_o, as

$$\frac{C_L^2}{\pi e_o AR} \equiv k_1 + k_2 + \frac{C_L^2}{\pi e AR} \equiv K C_L^2 \qquad \textbf{[8.43]}$$

A reasonable estimate for e_o for a low-wing general aviation airplane is 0.6 (see McCormick, Ref. 50). From Eq. (8.42),

$$AR = \frac{1}{\pi e_o K} = \frac{1}{\pi(0.6)(0.075)}$$

or

$$\boxed{AR = 7.07} \qquad \textbf{[8.44]}$$

Finally, to return to the consideration of rate of climb, Eq. (5.122) gives an expression for maximum rate of climb for a propeller-driven airplane as

$$(R/C)_{max} = \frac{\eta_{pr}P}{W} - \left(\frac{2}{\rho_\infty}\sqrt{\frac{K}{3C_{D,0}}}\frac{W}{S}\right)^{1/2}\frac{1.155}{(L/D)_{max}} \qquad \textbf{[5.122]}$$

Solving for the power term, we have

$$\frac{\eta_{pr}P}{W} = (R/C)_{max} + \left(\frac{2}{\rho_\infty}\sqrt{\frac{K}{3C_{D,0}}}\frac{W}{S}\right)^{1/2}\frac{1.155}{(L/D)_{max}} \qquad \textbf{[8.45]}$$

Everything on the right side of Eq. (8.45) is known, including $(R/C)_{max}$ which from the specifications is 1,000 ft/min = 16.67 ft/s at sea level. Hence, from Eq. (8.45),

$$\frac{\eta_{pr}P}{W} = 16.67 + \left[\frac{2}{0.002377}\sqrt{\frac{0.075}{3(0.017)}}(29.3)\right]^{1/2}\frac{1.155}{14} \qquad \textbf{[8.45a]}$$

$$\frac{\eta_{pr}P}{W} = 16.67 + 14.26 = 30.93 \text{ ft/s}$$

Assuming W is equal to the takeoff gross weight $W_0 = 5,158$ lb (ignoring the small amount of fuel burned during the takeoff run), and recalling our estimate of $\eta_{pr} = 0.8$, we have from Eq. (8.45)

$$P = \frac{30.93 W_0}{\eta_{pr}} = \frac{(30.93)(5,158)}{0.8} = 1.994 \times 10^5 \text{ ft·lb/s}$$

In terms of horsepower,

$$P = \frac{1.994 \times 10^5}{550} = 362.5 \text{ hp} \tag{8.46}$$

Thus, to satisfy the constraint on rate of climb, the power must be

$$P \geq 362.5 \text{ hp} \tag{8.47}$$

The third constraint on T/W (or P/W) is the maximum velocity V_{max}. The requirements stipulate $V_{max} = 250$ mi/h $= 366.7$ ft/s at midcruise weight. The altitude for the specified V_{max} is not stated. However, the requirements call for a pressurized cabin, and we can safely assume that an altitude of 20,000 ft would be comfortable for the pilot and passengers. Therefore, we assume that the specified V_{max} is associated with level flight at 20,000 ft. In level flight, $T = D$, and the drag D is given by Eq. (5.12)

$$T = D = \frac{1}{2}\rho_\infty V_\infty^2 S C_{D,0} + \frac{2KS}{\rho_\infty V_\infty^2}\left(\frac{W}{S}\right)^2 \tag{5.12}$$

Couching Eq. (5.12) in terms of the thrust-to-weight ratio, we have

$$\frac{T}{W} = \frac{1}{2}\rho_\infty V_\infty^2 \frac{C_{D,0}}{W/S} + \frac{2K}{\rho_\infty V_\infty^2}\frac{W}{S} \tag{8.48}$$

Since the requirements stipulate V_{max} *at midcruise weight*, the value of W that appears in Eq. (8.48) is less than the gross takeoff weight. To return to our weight estimates in Section 8.3, W_2 and W_3 are the weights at the beginning and end of cruise, respectively. We have, from Section 8.3.2,

$$\frac{W_2}{W_0} = \frac{W_1}{W_0}\frac{W_2}{W_1} = (0.97)(0.985) = 0.955$$

Hence,

$$W_2 = 0.955 W_0 = 0.955(5,158) = 4,926 \text{ lb}$$

At midcruise (defined here as when one-half of the fuel needed to cover the full cruise range is consumed), we have for the midcruise weight W_{MC}

$$W_{MC} = W_2 - \tfrac{1}{2}(W_2 - W_3)$$

or

$$\frac{W_{MC}}{W_2} = \frac{1}{2}\left(1 + \frac{W_3}{W_2}\right) \tag{8.49}$$

The weight fraction W_3/W_2 has been estimated in Eq. (8.16) as $W_3/W_2 = 0.893$. Hence, from Eq. (8.49)

$$\frac{W_{MC}}{W_2} = \frac{1}{2}(1 + 0.893) = 0.9465$$

Since $W_2 = 4{,}926$ as obtained above, we have

$$W_{MC} = (0.9465)(4{,}926) = 4{,}662 \text{ lb} \qquad \textbf{[8.50]}$$

This weight is used to define the new wing loading that goes into Eq. (8.48). This value is [recalling from Eq. (8.33) that $S = 176 \text{ ft}^2$]

$$\frac{W_{MC}}{S} = \frac{4{,}662}{176} = 26.5 \text{ lb/ft}^2$$

Returning to Eq. (8.48), written in terms of the midcruise weight, we have

$$\frac{T}{W_{MC}} = \frac{1}{2}\rho_\infty V_\infty^2 \frac{C_{D,0}}{W_{MC}/S} + \frac{2K}{\rho_\infty V_\infty^2} \frac{W_{MC}}{S} \qquad \textbf{[8.51]}$$

From Appendix B, at 20,000 ft, $\rho_\infty = 0.0012673 \text{ slug/ft}^3$. Also, inserting $V_{max} = 366.7$ ft/s for V_∞ in Eq. (8.51), we have

$$\frac{T}{W_{MC}} = \frac{1}{2}(0.0012673)(366.7)^2 \frac{0.017}{26.5} + \frac{2(0.075)(26.5)}{(0.0012673)(366.7)^2} \qquad \textbf{[8.52]}$$

$$= 0.0547 + 0.0233 = 0.0780$$

Comment: It is interesting to note that the two terms on the right side of Eq. (8.51) represent the effects of zero-lift drag and drag due to lift, respectively. In the above calculation, the zero-lift drag is about 2.3 times larger than the drag due to lift. This is consistent with the usual situation that as speed increases, the drag due to lift becomes a smaller percentage of the total drag. In this case, the drag due to lift is $0.0233/0.0780 = 0.3$ of the total drag, or less than one-third of the total drag.

The shaft power required P is given by

$$\eta_{pr} P = T V_\infty \qquad \textbf{[8.53]}$$

At V_{max} at midcruise weight, Eq. (8.53) is written as

$$P = \frac{1}{\eta_{pr}} \frac{T}{W_{MC}} W_{MC} V_{max} = \frac{1}{0.8}(0.0780)(4{,}662)(366.7) = 1.667 \times 10^5 \text{ ft·lb/s}$$

In terms of horsepower,

$$P = \frac{1.667 \times 10^5}{550} = 303.1 \text{ hp} \qquad \textbf{[8.54]}$$

To summarize the results from this section, the three constraints on power required for our airplane design have led to the following:

Takeoff	$P \geq 118.8$ hp
Rate of climb	$P \geq 362.5$ hp
Maximum velocity	$P \geq 303.1$ hp

Clearly, the specification of the maximum rate of climb at sea level of 1,000 ft/min is the determining factor of the required power from the engine. For our airplane design, the engine should be capable of producing a maximum power of 362.5 hp or greater.

We can couch this result in terms of the more relevant performance parameters T/W or P/W. When these parameters are quoted for a given airplane, the weight is usually taken as the gross takeoff weight W_0. Hence, for our design

$$\text{Power-to-weight ratio} = \frac{362.5 \text{ hp}}{5,158 \text{ lb}} = 0.07 \text{ hp/lb}$$

For a propeller-driven airplane, the power-to-weight ratio is more relevant than the thrust-to-weight ratio, which makes more sense to quote for jet airplanes. For a reciprocating engine/propeller-driven airplane, the shaft power is essentially constant with velocity, whereas the thrust decreases with velocity, as discussed in Chapter 3. Hence, for a reciprocating engine/propeller-driven airplane, to quote the power-to-weight ratio makes more sense. In the aeronautical literature, historically the *power loading*, which is the reciprocal of the power-to-weight ratio, is frequently given.

$$\text{Power loading} \equiv \frac{W}{P}$$

The definition of the power loading is semantically analogous to that for the wing loading W/S. The wing loading is the weight divided by wing area; the power loading is the weight divided by the power. For our airplane, we have estimated that

$$\text{Power loading} \quad \frac{W}{P} = \frac{1}{0.07} = 14.3 \text{ lb/hp}$$

[We note that Raymer (Ref. 25) quotes a typical value of 14 lb/hp for general aviation single-engine airplanes—so our estimation appears to be very reasonable.]

There is something important that is implicit in our discussion of the engine characteristics. Although the engine is sized at 362.5 hp to meet the rate-of-climb specification at sea level, it must also produce 303.1 hp at 20,000 ft to achieve the specified maximum velocity. Since the power of a conventional reciprocating engine is proportional to the air density [see Eq. (3.11)], such a conventional engine sized at 362.5 hp at sea level will produce only 193 hp at 20,000 ft—clearly unacceptable for meeting our specifications. Hence, the engine for our airplane must be *supercharged* to maintain sea-level power to an altitude of 20,000 ft.

8.5 SUMMARY OF THE CRITICAL PERFORMANCE PARAMETERS

We have now completed pivot point 3 in Fig. 7.3, namely, the first estimate of the critical performance parameters from airplane design. They are summarized as follows:

Maximum lift coefficient $\qquad (C_L)_{max} = 2.34$

Maximum lift-to-drag ratio $\qquad \left(\dfrac{L}{D}\right)_{max} = 14$

Wing loading $\qquad \dfrac{W}{S} = 29.3 \text{ lb/ft}^2$

Power loading $\qquad \dfrac{W}{P} = 14.3 \text{ lb/hp}$

In the process of estimating these performance parameters, we have found other characteristics of our airplane:

Takeoff gross weight $\qquad W_0 = 5,158 \text{ lb}$

Fuel weight $\qquad W_f = 820 \text{ lb}$

Fuel tank capacity $\qquad 145.4 \text{ gal}$

Wing area $\qquad S = 176 \text{ ft}^2$

High-lift device $\qquad$ Single-slotted trailing-edge flaps

Zero-lift drag coefficient $\qquad C_{D,0} = 0.017$

Drag-due-to-lift coefficient $\qquad K = 0.075$

Aspect ratio $\qquad AR = 7.07$

Propeller efficiency $\qquad 0.8$

Engine power, supercharged to 20,000 ft $\qquad 362.5 \text{ hp}$

We are now ready to draw a picture of our airplane design, that is, to construct a configuration layout. This is the subject of the next section.

8.6 CONFIGURATION LAYOUT

We now move to pivot point 4 in Fig. 7.3—the configuration layout. Based on the information we have calculated so far in this chapter, we are ready to draw a picture, with dimensions, of our airplane. Even though the data summarized in Section 8.5 clearly define a certain type of airplane, there are still an infinite number of different sizes and shapes that could satisfy these data. There are no specific laws or rules that tell us exactly what the precise dimensions and shape ought to be. Therefore, pivot point 4 in our intellectual process of airplane design is where intuition, experience, and the *art* of airplane design most strongly come into play. It is impossible to convey these assets in one particular section of one particular book. Rather, our purpose here is to simply illustrate the philosophy that goes into the configuration layout.

8.6.1   Overall Type of Configuration

There are some basic configuration decisions to make up front. Do we use one or two engines? Do we use a tractor (propeller in front) or a pusher (propeller in back) arrangement (or both)? Will the wing position be low-wing, mid-wing, or high-wing? (Indeed, do we have two wings, i.e., a biplane configuration? This is not very likely in modern airplane designs; the biplane configuration was essentially phased out in the 1930s, although today there are good reasons to consider the biplane for aerobatic and agricultural airplanes. We will not consider the biplane configuration here.)

First, let us consider the number of engines. The weight of 5,158 lb puts our airplane somewhat on the borderline of single- and twin-engine general aviation airplanes. We could have a rather heavy single-engine airplane, or a light twin-engine one. We need 362.5 hp—can we get that from a single, existing piston engine? (We have to deal with an existing engine; rarely is a new general aviation airplane design enough incentive for the small engine manufacturers to go to the time and expense of designing a new engine.) Examining the available piston engines at the time of writing, we find that the Textron Lycoming TIO/LTIO-540-V is rated at 360 hp supercharged to 18,000 ft. This appears to be the engine for us. It is only 2.5 hp less than our calculations show is required based on the rate-of-climb specification. We could tweak the airplane design, say, by slightly reducing the weight or slightly increasing the aspect ratio, both of which would reduce the power required for climb and would allow us to meet the performance specification with this engine. The fact that it is supercharged to 18,000 ft, not the 20,000 ft we assumed for our consideration of V_{max}, is not a problem. The free-stream density ratio between 20,000 and 18,000 ft is $1.2673/1.3553 = 0.935$. Hence, the engine power at 20,000 ft will be $(360 \text{ hp})(0.935) = 336.6$ hp. This is more than enough to meet the calculated requirement of 303.1 hp for V_{max} at 20,000 ft. *Therefore, we choose a single-engine configuration*, using the following engine with the following characteristics:

Textron Lycoming TIO/LTIO-540-V Piston Engine

Rated power output at sea level (turbosupercharged to 18,000 ft): 360 hp

Number of cylinders: 6

Compression ratio: 7.3

Dry weight: 547 lb

Length: 53.21 in

Width: 34.88 in

Height: 24.44 in

Question: Do we adopt a tractor or a pusher configuration? The tractor configuration—engine and propeller at the front—is illustrated in Fig. 8.5**a**; the pusher configuration—engine and propeller at the back—is illustrated in Fig. 8.5**b**. Some of the advantages and disadvantages of these configurations are itemized below.

Tractor Configuration   Advantages:

1. The heavy engine is at the front, which helps to move the center of gravity forward and therefore allows a smaller tail for stability considerations.
2. The propeller is working in an undisturbed free stream.
3. There is a more effective flow of cooling air for the engine.

Disadvantages:

1. The propeller slipstream disturbs the quality of the airflow over the fuselage and wing root.
2. The increased velocity and flow turbulence over the fuselage due to the propeller slipstream increase the local skin friction on the fuselage.

Pusher Configuration   Advantages:

1. Higher-quality (clean) airflow prevails over the wing and fuselage.
2. The inflow to the rear propeller induces a favorable pressure gradient at the rear of the fuselage, allowing the fuselage to close at a steeper angle without flow separation (see Fig. 8.5**b**). This in turn allows a shorter fuselage, hence smaller wetted surface area.
3. Engine noise in the cabin area is reduced.
4. The pilot's front field of view is improved.

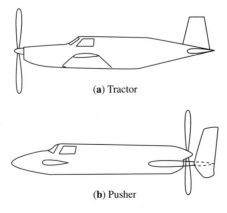

(a) Tractor

(b) Pusher

Figure 8.5   Comparison of a tractor and a pusher configuration.

Disadvantages:

1. The heavy engine is at the back, which shifts the center of gravity rearward, hence reducing longitudinal stability.
2. Propeller is more likely to be damaged by flying debris at landing.
3. Engine cooling problems are more severe.

The *Wright Flyer* was a pusher aircraft (Fig. 1.3). However, over the past century of airplane design, the tractor configuration has been the prevalent choice. Because we have a rather large, powerful reciprocating engine for our aiplane design, we wish to minimize any engine cooling problems. Therefore, we will be traditional and choose the tractor configuration.

8.6.2   Wing Configuration

Here, we have two considerations, the geometric shape of the wing and its location relative to the fuselage. First, we consider the shape.

Referring to Fig. 8.6, the wing geometry is described by (**a**) aspect ratio, (**b**) wing sweep, (**c**) taper ratio, (**d**) variation of airfoil shape and thickness along the span, and (**e**) geometric twist (change in airfoil chord incidence angle along the span). The aspect ratio is given by b^2/S, as shown in Fig. 8.6**a**. There are two sweep angles of importance, the leading-edge sweep angle Λ_{LE} and the sweep angle of the quarter-chord line $\Lambda_{c/4}$, as shown in Fig. 8.6**b**. The leading-edge sweep angle is most relevant to supersonic airplanes because to reduce wave drag, the leading edge should be swept behind the Mach cone (see Fig. 2.30 and the related discussion in Chapter 2). The sweep angle of the quarter-chord line $\Lambda_{c/4}$ is of relevance to high-speed subsonic airplanes near the speed of sound. The taper ratio is the ratio of the tip chord to the root chord c_t/c_r, illustrated in Fig. 8.6**c**. The possible variation of airfoil shape and thickness along the span is illustrated in Fig. 8.6**d**. Geometric twist is illustrated in Fig. 8.6**e**, where the root and tip chord lines are at different incidence angles. Shown in Fig. 8.6**e** is the case when the tip chord incidence angle is smaller than that of the root chord; this configuration is called *washout*. The opposite case, when the tip is at a higher incidence angle than the root, is called *wash-in*.

Let us proceed with the determination of the planform shape (top view) of the wing of our airplane. The decision in regard to a swept wing versus an unswept wing is easy. The maximum design velocity of our airplane is 250 mi/h—far below the transonic regime; hence, we have no aerodynamic requirement for a swept wing. We choose to use a conventional straight wing. For minimum induced drag, we noted in Section 2.8.3 that we want to have a spanwise elliptical lift distribution, which for an untwisted wing implies an elliptical planform shape. However, the higher production costs associated with a wing with curved leading and trailing edges in the planform view are usually not justified in view of the cheaper costs of manufacturing wings with straight leading and trailing edges. Moreover, by choosing the correct taper ratio, the elliptical lift distribution can be closely approximated. Recall from Eq. (2.31) that the span efficiency factor e is given by the ratio $1/(1 + \delta)$, where δ is plotted in Fig. 2.39 as a function of aspect ratio and taper ratio. Reflecting again on Fig. 2.39, we

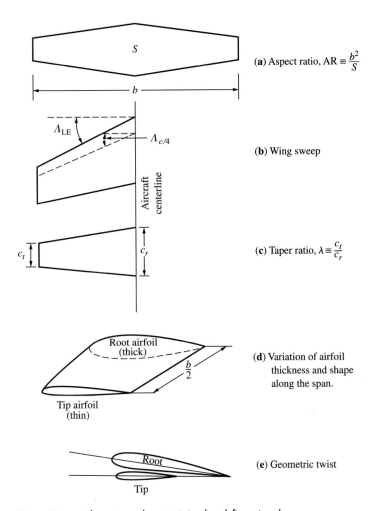

(a) Aspect ratio, $AR \equiv \dfrac{b^2}{S}$

(b) Wing sweep

(c) Taper ratio, $\lambda \equiv \dfrac{c_t}{c_r}$

(d) Variation of airfoil thickness and shape along the span.

(e) Geometric twist

Figure 8.6 The various characteristics that define wing shape.

see that, for our aspect ratio of 7.07, the minimum value of $\delta = 0.005$ occurs at a taper ratio $c_t/c_r = 0.3$. That is, a tapered wing with a taper ratio of 0.3 and an aspect ratio of 7 will have an induced drag that is only 0.5% higher than that for the elliptical wing. Clearly, the design choice of a straight, tapered wing with straight leading and trailing edges is justified.

The next question is: What taper ratio should be used? To expedite our consideration, we denote taper ratio by the symbol λ: $\lambda \equiv c_t/c_r$. At first glance, it would be tempting to choose $\lambda = 0.3$ on the basis of the theoretical results in Fig. 2.39. However, there are several competing considerations, as follows:

1.   The smaller the taper ratio, the lighter the wing structure. Why? The answer has to do with the spanwise lift distribution and the resulting bending moment it creates at the wing root. As λ decreases from 1.0 (a rectangular wing, as shown in

Fig. 8.7**a**) to 0 (a triangular wing with a pointed tip, as shown in Fig. 8.7**b**), the preponderance of the lifting force shifts inboard, closer to the wing root. This is clearly seen from the lift distributions (obtained from lifting-line theory) shown in Fig. 8.8; as the taper decreases, the centroid of the lift distribution (center of pressure) moves closer to the root of the wing. In turn, the moment arm from the root to the center of pressure deceases, and the bending moment at the root decreases, the lift staying the same. As a result, the wing structure can be made lighter. This trend is a *benefit* obtained from using a small taper ratio.

2.   On the other hand, wings with low taper ratios exhibit undesirable flow separation and stall behavior. This is illustrated schematically in Fig. 8.9, which shows the different regions of flow separation at the beginning of stall for wings at three different taper ratios. A rectangular wing, $\lambda = 1.0$, shown in Fig. 8.9**a**, will develop flow separation first in the root region. This location for flow separation has two advantages: (1) The separated, turbulent flow trails downstream from the root region and causes buffeting as it flows over the horizontal tail, thus serving as a dramatic stall warning to the pilot. (2) The wing-tip region still has attached flow, and because the ailerons (for lateral control) are located in this region, the pilot still has full aileron control. However, as the taper ratio decreases, the region where flow separation first develops moves out toward the tip, which is shown in Fig. 8.9**b** for a taper ratio on the order of 0.5. When λ is reduced to 0, as shown in Fig. 8.9**c**, the stall region first occurs at the tip region, with consequent total loss of aileron control. This characteristic is usually not tolerated in an airplane, and this is why we see virtually no airplanes designed with wings with zero (or very small) taper ratios. This trend is definitely a *detriment* associated with using small taper ratios.

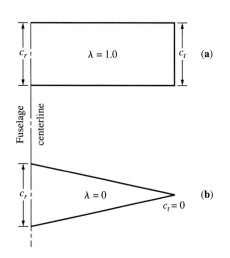

Figure 8.7   Illustrations of wings with taper ratio equal to 1 and 0.

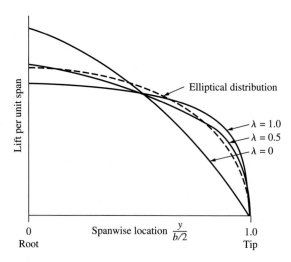

Figure 8.8   Effect of taper ratio on lift distribution.

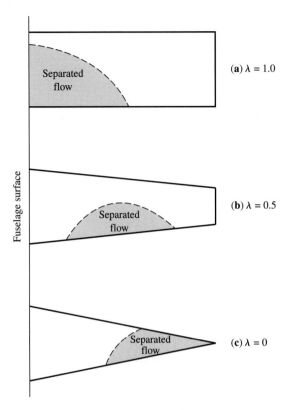

Figure 8.9 Effect of taper ratio on wing region of flow separation at near-stall conditions.

So, as usual, airplane design is a compromise—in this case a compromise between the structural benefit of small λ and the aerodynamic benefit of large λ. Historically, most straight-wing airplanes incorporate taper ratios on the order of 0.4 to 0.6. In fact, some general aviation airplanes, for the sake of minimizing the cost of wing manufacture, have rectangular wings ($\lambda = 1.0$). For our airplane design, we will choose a taper ratio of 0.5—a suitable compromise. Note from Fig. 2.39 that for $\lambda = 0.5$ and AR $= 7$, $\delta = 0.013$; with $\lambda = 0.5$ the induced drag is still only about 1.3% larger than the theoretical minimum. Hence, $\lambda = 0.5$ appears to be an acceptable choice. The plan view for our wing design, with AR $= 7.07$, $S = 176$ ft, and $\lambda = 0.5$, can now be drawn to scale; it is shown in Fig. 8.10. The linear dimensions shown in Fig. 8.10 are readily obtained as follows.

$$AR \equiv \frac{b^2}{S}$$

Hence,

$$b = \sqrt{(S)(AR)} = \sqrt{(176)(7.07)} = 35.27 \text{ ft}$$

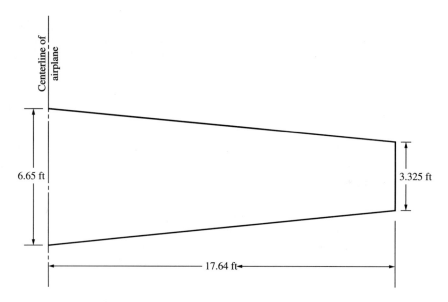

Figure 8.10   Scale plan view of our wing design.

The planform shape is a trapezoid. The area of a trapezoid is given by $\frac{1}{2}(a + b)h$, where a and b are the two parallel sides and h is the altitude. The area shown in Fig. 8.10 is one-half the total wing area, and the parallel sides a and b are c_t and c_r. The altitude is $b/2$. Hence, from the formula for the trapezoidal area, we have

$$\frac{S}{2} = \frac{1}{2}(c_t + c_r)\frac{b}{2}$$

or

$$2S = (c_t + c_r)b \qquad\qquad [8.55]$$

Dividing Eq. (8.55) by c_r, we have

$$\frac{2S}{c_r} = (\lambda + 1)b$$

or

$$c_r = \frac{2S}{(\lambda + 1)b} = \frac{2(176)}{(0.5 + 1)(35.27)} = 6.65 \text{ ft}$$

and

$$c_t = \lambda c_r = 0.5(6.65) = 3.325 \text{ ft}$$

The wing semispan is

$$\frac{b}{2} = \frac{35.27}{2} = 17.64 \text{ ft}$$

These are the dimensions shown in Fig. 8.10.

The mean aerodynamic chord $\bar{c}$ is defined as the chord length that, when multiplied by the wing area, the dynamic pressure, and the moment coefficient about the aerodynamic center, yields the value of the aerodynamic moment about the airplane's aerodynamic center. It can be calculated from

$$\bar{c} = \frac{1}{S} \int_{-b/2}^{b/2} c^2 \, dy \qquad \textbf{[8.56]}$$

where c is the local value of chord length at any spanwise location. Raymer (Ref. 25) gives a convenient geometric construction for finding the length and the spanwise location of $\bar{c}$, as illustrated in Fig. 8.11. Lay off the midchord line fg, as shown in Fig. 8.11. Lay off c_t from the root chord and c_r from the tip chord in the manner shown in Fig. 8.11 and connect the ends by line jk. The intersection of lines fg and jk defines the spanwise location of the mean aerodynamic chord, and the length $\bar{c}$ is simply measured from the drawing. Alternatively, the spanwise location of the mean aerodynamic chord $\bar{y}$ and its length $\bar{c}$ can be calculated as follows (Raymer, Ref. 25):

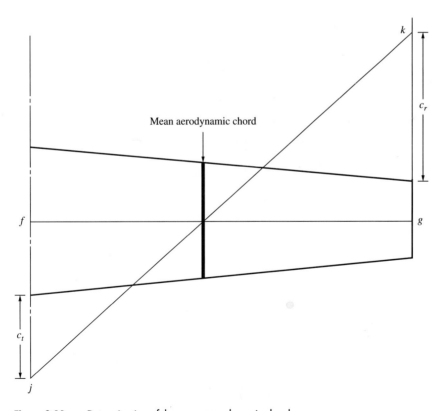

Figure 8.11   Determination of the mean aerodynamic chord.

$$\bar{y} = \frac{b}{6}\left(\frac{1+2\lambda}{1+\lambda}\right) = \frac{35.27}{6}\left(\frac{1+1}{1+0.5}\right) = \boxed{7.84 \text{ ft}} \qquad \textbf{[8.57]}$$

and

$$\bar{c} = \frac{2}{3}c_r\left(\frac{1+\lambda+\lambda^2}{1+\lambda}\right) = \frac{2}{3}(6.65)\left(\frac{1+0.5+2.5}{1+0.5}\right) = \boxed{5.17 \text{ ft}} \qquad \textbf{[8.58]}$$

Recall that the shape of the airfoil section changes along the span; we have chosen an NACA 23018 section at the wing root and an NACA 23012 section at the tip, for the reasons stated in Section 8.4.1. This, along with the taper ratio of 0.5, should give us a reasonable stall pattern, with separated flow occurring first near the root, thus maintaining reasonable aileron control in the attached flow region near the tip. Hence, we assume that geometric twist (washout) will not be required. Figure 8.12 illustrates the change in airfoil shape from the root to the tip. There are several geometric interpolation methods for generating the shapes of the airfoils between the root and tip, such as section AB in Fig. 8.12. See Raymer (Ref. 25) for the lofting details.

This finishes our discussion of the geometric shape of the wing. Let us now address its location relative to the fuselage. There are three basic vertical locations of the wings relative to the fuselage: (1) high wing; (2) mid-wing; (3) low wing. These are illustrated in Fig. 8.13. Some of the advantages and disadvantages of these different locations are as follows.

High Wing   Examining Fig. 8.13a, a high-wing configuration, with its low-slung fuselage, allows the fuselage to be placed lower to the ground. This is a distinct advantage for transport aircraft, because it simplifies the loading and unloading processes. The high-wing configuration is also more stable in terms of lateral, rolling motion. Dihedral (wings bent upward, as shown in Fig. 8.13c) is usually incorporated on an airplane to enhance lateral (rolling) stability. When an airplane rolls, the lift vector tilts away from the vertical, and the airplane sideslips in the direction of the

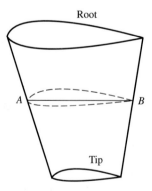

Figure 8.12   Change in airfoil shape along the span.

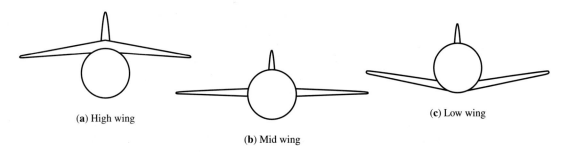

(a) High wing

(b) Mid wing

(c) Low wing

Figure 8.13   Comparison of high-wing, mid-wing, and low-wing configurations.

lowered wing. If there is dihedral, the extra component of flow velocity due to the sideslipping motion creates an increased lift on the lowered wing, hence tending to restore the wings to the level equilibrium position—this is the essense of lateral stability. For the high-wing configuration, even with *no* dihedral, the extra component of flow velocity due to the sideslipping motion in roll creates a region of higher pressure in the flow interaction region between the side of the fuselage and the undersurface of the lowered wing in the vicinity of the wing root. This increased pressure on the bottom surface of the lowered wing tends to roll the wings back to the level equilibrium. In fact, the high-wing position can be so strongly stable in roll that it becomes a disadvantage, and anhedral (wings bent downward as shown in Fig. 8.13**a**) is used on some high-wing airplanes to partially negate this overly stable behavior in roll.

Mid-wing   The mid-wing location (Fig. 8.13**b**) usually provides the lowest drag of any of the three locations because the wing-body interference is minimized. Both the high- and low-wing configurations require a fillet to help decrease this interference. (Fillets will be described below.) A major disadvantage is structural. The bending moment due to the wing lift must be carried through the fuselage in some manner. For a high-wing or low-wing configuration, this is most simply done by an extension of the wing box straight through the fuselage; such an extension does not get in the way of the internal cargo-carrying or people-carrying space of the fuselage. However, if this structural arrangement were to be used for a mid-wing configuration, there could be an unacceptable obstruction through the middle of the fuselage. To avoid this, the wing bending moments can be transmitted across the fuselage by a series of heavy ring frames in the fuselage shell, which inordinately increases the empty weight of the airplane. Note that, similar to the discussion on the high-wing configuration, the mid-wing arrangement requires little if any dihedral; hence, Fig. 8.13**b** shows no dihedral.

Low-wing   The major advantage of the low-wing configuration is in the design of the landing gear. Here, the landing gear can be retracted directly into the wing box, which is usually one of the strongest elements of the aircraft structure. For multiengine propeller-driven airplanes, the landing gear can most conveniently retract into the engine nacelles. However, because the fuselage requires some ground clearance for engine or propeller installation, the landing gear needs to be long enough to provide the proper height above the ground, hence adding weight. Also, for lateral stability, the low-wing configuration requires some dihedral, as shown in Fig. 8.13**c**.

To minimize the undesirable aerodynamic interference at the wing-body juncture for low- and high-wing configurations, a fillet is used. The source of this interference is sketched in Fig. 8.14. For simplicity, assume a constant-diameter cylindrical fuselage in the wing root region. Imagine a stream tube of the flow in the wing-body juncture region; such a stream tube is shown as the shaded region in Fig. 8.14. Examine section A-A near the maximum thickness of the wing (Fig. 8.14**b**). The cross-sectional area of the stream tube is small here, with a consequent higher flow velocity and lower pressure. Now examine section B-B near the trailing edge of the wing (Fig. 8.14**c**). The cross-sectional area of the stream tube is larger here, with a consequent lower flow velocity and higher pressure. That is, the stream tube between sections A-A and B-B has an *adverse pressure gradient*, which promotes flow separation with its attendant higher drag and unsteady buffeting. By filling in this region of the wing-body juncture with a suitably contoured surface, this adverse pressure gradient and consequent flow separation can be minimized. This contoured surface is called a *fillet*. Such a fillet is highlighted in the two-view shown in Fig. 8.15. For the mid-wing configuration, the wing root joins the fuselage at the 90° location around the cylindrical cross section, which geometrically minimizes the change in the stream tube area at the juncture. For this reason, mid-wing configurations are frequently designed without a fillet.

In light of all the above considerations, we choose a *low-wing configuration*, mainly due to the structural and landing gear considerations. We will employ a fillet with this configuration.

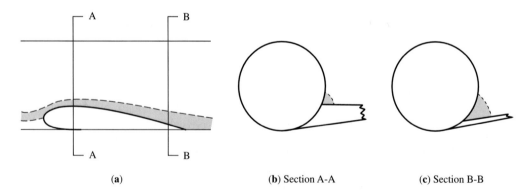

(a)   (**b**) Section A-A   (**c**) Section B-B

Figure 8.14   Expanding cross-sectional area of a stream tube at the wing-body juncture of a low-wing airplane.

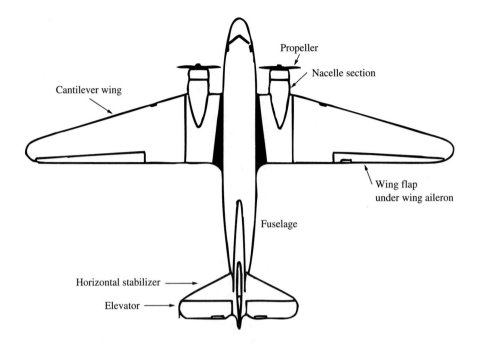

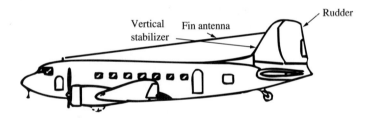

Figure 8.15 Two-view of the Douglas DC-3, showing the fillet (shaded region) at the wing-body juncture.

8.6.3 Fuselage Configuration

The fuselage must be large enough to contain the engine in the nose, the pilot and five passengers in the cabin, the baggage, and the fuel if it is decided to store it in the fuselage. Let us first examine the question of where to put the fuel. For enhanced safety to the occupants, it is extremely desirable to store the fuel in the wings rather than the fuselage. Also, with the fuel storage in the wings, the shift in the airplane's center of gravity as fuel is consumed is usually much less than if fuel were stored in the fuselage. Is our wing large enough, with enough internal volume, to hold the fuel? We need a fuel tank capacity of 145.4 gal. One gallon occupies 231 in^3, or 0.134 ft^3. Thus, our fuel tank must have a volume of $(0.134)(145.4) = 19.4$ ft^3. Let us assume that the internal wing structure includes a front spar located at 12% of the chord from the leading edge, and a rear spar located at 60% from the leading edge; this is shown on the wing planview in Fig. 8.16. A trapezoidal tank, 0.8 ft high, with a base of dimensions 3.27 ft, 2.9 ft, and 3.93 ft, as shown in the planview in Fig. 8.16, will have a volume of 9.7 ft^3; with a tank of equal volume in the other wing, the total capacity will be the required 19.4 ft^3. As shown in Fig. 8.16, this tank will fit nicely into the wing. Hence, we will not store the fuel within the interior of the fuselage; rather, we will place it in the wing, as shown in Fig. 8.16.

To size the fuselage, we recall that the engine size was given in Section 8.6.1; the length, width, and height of the engine are 4.43, 2.91, and 2.037 ft, respectively. The layout shown in Fig. 8.17 is a fuselage where the engine fits easily into the forward portion; the engine is shaded for emphasis. The passenger compartment is sized for six people. Using guidance from Raymer (Ref. 25), the seat size is chosen as follows: width, 1.67 ft; back height, 2.7 ft; pitch (distance between the backs of two seats, one directly ahead of the other), 3.0 ft. The resulting seat arrangement is shown in Fig. 8.17. Here, the side view and top view of a candidate fuselage, Fig. 8.17**a** and **b**, respectively, are shown that contain both the heavy engine and the people and payload. The engine, represented by the dark, shaded regions in Fig. 8.17, fits into the nose. The dual rows of three seats, a total of 9 ft long, are also shown in Fig. 8.17. The width and depth of the fuselage as shown in sections A-A and B-B are dictated by the engine size and passenger cabin, respectively. There is some art in the fairing in of the rest of the fuselage. We have drawn a fuselage that gently reduces to a zero

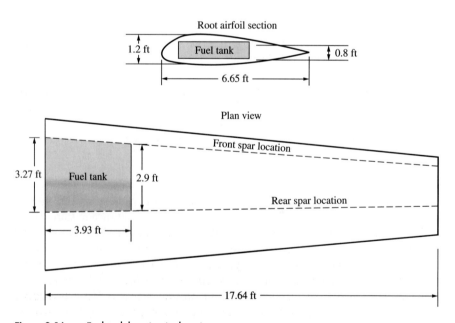

Figure 8.16   Fuel tank location in the wing.

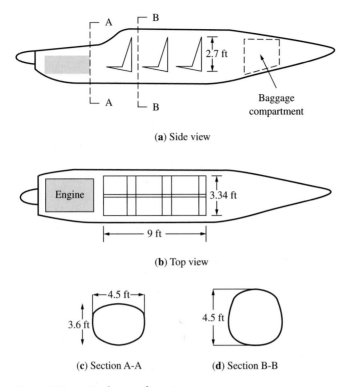

(a) Side view

(b) Top view

(c) Section A-A   (d) Section B-B

Figure 8.17   Fuselage configuration.

cross section at the back end, with enough room for a baggage compartment. Caution must be taken not to taper the back section of the fuselage at too large an angle, or else flow separation will occur. For subsonic airplanes, the taper angle should be no larger than about 15°. Also, the length of the fuselage behind the center of gravity should be long enough to provide a sufficient moment arm for both the horizontal and vertical tails. At this stage of our design, we have not yet determined the location of the center of gravity or the tail moment arm. These considerations are discussed in the next section.

8.6.4   Center-of-Gravity Location: First Estimate

The major weight components for which we have some idea of their location are the engine, the passengers and pilot, and the baggage. Using this information, we can make a very preliminary estimate of the location of the center of gravity, hereafter denoted by c.g. The tail, fuselage, and wing also contribute to the c.g. location; however, as yet we do not know the size and location of the vertical and horizontal tails. We can take into account the wing and fuselage, but again in only an approximate fashion, as we will see.

The weights of the engine, people, and baggage are shown in Fig. 8.18, along with the locations of their respective individual c.g. locations measured relative to the nose of the airplane, just behind the spinner. The effective c.g. of these three weights, located by $\bar{x}$ in Fig. 8.18, is calculated by summing moments about the nose and dividing by the sum of the weights. The result is

$$\bar{x} = \frac{(2.7)(765.8) + 10.1(1,020) + 19.6(120)}{765.8 + 1,020 + 120} = \frac{14,722}{1,905.8} = 7.72 \text{ ft}$$

In the above calculation, the weight of the *installed* engine is taken as 1.4 times the given dry weight of 547 lb, as suggested by Raymer (Ref. 25); hence the installed engine weight is taken as 765.8 lb.

Usual design procedure calls for locating the wing relative to the fuselage such that the mean aerodynamic center of the wing is close to the c.g. of the airplane. (Indeed, for static longitudinal stability, the aerodynamic center of the airplane, also called the *neutral point*, should be located behind the c.g. of the airplane.) To account for the weight of the wing at this stage of our calculation, we assume that the mean aerodynamic center of the wing is placed at the c.g. location calculated above; that is, place the mean aerodynamic center at $\bar{x} = 7.72$ ft. (Later in the design process, the wing will be relocated to ensure that the aerodynamic center of the airplane is behind the center of gravity.) The geometry of the mean aerodynamic chord, the mean aerodynamic center, and the wing c.g. location are shown in Fig. 8.19. Raymer (Ref. 25) suggests that we estimate the weight of the wing by multiplying the planform area by 2.5; hence $W_{\text{wing}} = 2.5(176) = 440$ lb. We also assume that the wing aerodynamic center is 25% of the mean aerodynamic chord from the leading edge, and that the wing center of gravity is at 40% of the mean aerodynamic chord. These points are

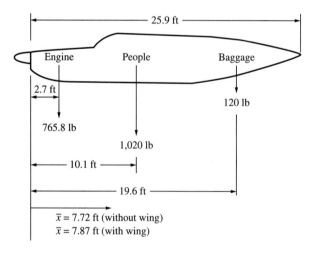

Figure 8.18   Sketch for the calculation of moments about the nose.

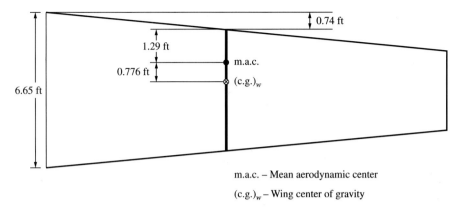

m.a.c. – Mean aerodynamic center

(c.g.)$_w$ – Wing center of gravity

Figure 8.19   Wing geometry, showing locations of the mean aerodynamic center and wing center of gravity.

located in Fig. 8.19, using the above assumptions. With this, a new center-of-gravity location for the airplane, including the weight of the wing, can be estimated by adding to our earlier calculation the weight of the wing, 440 lb, acting through the moment arm $(7.72 + 0.776)$. The result is

$$\bar{x} = \frac{14{,}722 + (440)(7.72 + 0.776)}{1{,}905.8 + 440} = \frac{18{,}460}{2{,}345.8} = 7.87 \text{ ft}$$

Under these assumptions, note that the addition of the wing has shifted the c.g. only a small amount rearward, from $\bar{x} = 7.72$ ft to $\bar{x} = 7.87$ ft. For the time being, measured from the nose, we will assume the airplane c.g. to be at

$$\text{Center-of-gravity location} \qquad \bar{x} = 7.87 \text{ ft} \qquad\qquad \textbf{[8.59]}$$

8.6.5   Horizontal and Vertical Tail Size

One of the most empirical and least precise aspects of the airplane design process is the sizing of the tail. The primary function of the horizontal tail is to provide longitudinal stability; the control surface on the horizontal tail—the elevator—provides longitudinal control and trim. The primary function of the vertical tail is to provide directional (yawing) stability; the control surface on the vertical tail—the rudder—provides directional control. The size of the horizontal and vertical tails must be sufficient to provide the necessary stability and control of the airplane. A detailed stability and control analysis can provide some information on tail sizing. However, we will not present such an analysis here; rather, we will size the tail based on historical, empirical data. The approach is consistent with our intent to present the overall philosophy of the airplane design process rather than get too involved with the design details.

We will make use of the tail volume ratios, defined as follows:

$$\text{Horizontal tail} \qquad V_{HT} = \frac{l_{HT} S_{HT}}{\bar{c} S} \qquad\qquad \textbf{[8.60]}$$

$$\text{Vertical tail} \qquad V_{VT} = \frac{l_{VT} S_{VT}}{b S} \qquad\qquad \textbf{[8.61]}$$

where V_{HT} and V_{VT} are the horizontal and vertical tail volume ratios, respectively, l_{HT} is the horizontal distance between the c.g. of the airplane and the aerodynamic center of the horizontal tail, l_{VT} is the horizontal distance between the c.g. of the airplane and the aerodynamic center of the vertical tail, S_{HT} is the planform area of the horizontal tail, S_{VT} is the sideview area of the vertical tail, $\bar{c}$ is the mean aerodynamic chord of the wing, b is the wingspan, and S is the wing planform area. Based on previous single-engine general aviation airplanes, suggested values of these volume ratios (from Raymer, Ref. 25) are

$$V_{HT} = 0.7 \qquad\qquad \textbf{[8.62]}$$

$$V_{VT} = 0.04 \qquad\qquad \textbf{[8.63]}$$

We will use these values for our design.

The conventional location for the horizontal tail is centered on the tail end of the fuselage, as shown in Fig. 8.20**a**. There are many other possible tail configurations, such as the T tail (the horizontal tail mounted at the top of the vertical tail) shown in Fig. 8.20**b**, and the cruciform tail shown in Fig. 8.20**c**. The configuration shown in Fig. 8.20**a** is called *conventional* simply because it is found on over 70% of airplanes. It is favored because of its low structural weight compared to the other configurations in Fig. 8.20 while at the same time providing reasonable stability and control. However, the horizontal tail should be sufficiently far back that at stall the wake of the horizontal tail does not mask the rudder on the vertical tail. For the T tail (Fig. 8.20**b**), the structure is heavier; the vertical tail must be strengthened to support the aerodynamic

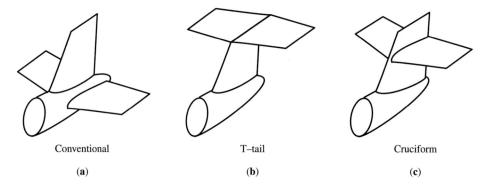

Conventional   T–tail   Cruciform

(**a**)   (**b**)   (**c**)

Figure 8.20   Some different tail configurations.

load and weight of the horizontal tail. On the other hand, the horizontal tail acts as an endplate on the vertical tail, allowing the vertical tail to experience a smaller induced drag and a higher lift slope; hence the aspect ratio of the vertical tail can be made smaller when the T tail configuration is used. Another advantage of the T tail is that the rudder is not blanketed at stall. Also, for a jet airplane with engines mounted in pods on the aft fuselage (such as the McDonnell-Douglas MD-80), the T tail is virtually necessary in order that the horizontal tail not be immersed in the jet exhaust. The cruciform tail (Fig. 8.20c) is basically a compromise between the conventional tail and T tail.

There are almost a dozen other possible tail configurations; these are nicely shown and discussed by Raymer (Ref. 25).

We choose to use a conventional tail configuration (Fig. 8.20a), primarily for its more light-weight structure. The length of the fuselage is 25.9 ft, as shown in Fig. 8.18. Our design logic will be to somewhat arbitrarily locate the aerodynamic center of the horizontal tail at a distance of 25 ft from the nose, as shown in Fig. 8.21, and then to calculate the horizontal tail area S_{HT} from Eqs. (8.60) and (8.62). Since the location of the c.g. is $\bar{x} = 7.87$ ft, then the moment arm from the center of gravity to the aerodynamic center of the horizontal tail is

$$l_{HT} = 25.0 - 7.87 = 17.13 \text{ ft} \qquad \textbf{[8.64]}$$

From Eqs. (8.60) and (8.62), we have

$$\frac{l_{HT} S_{HT}}{\bar{c} S} = 0.7$$

or

$$S_{HT} = \frac{0.7 \bar{c} S}{l_{HT}} \qquad \textbf{[8.65]}$$

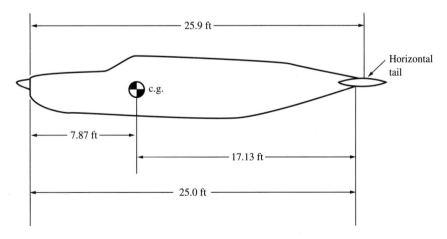

Figure 8.21   Moment arm of the horizontal tail.

From Eq. (8.58), $\bar{c} = 5.17$ ft. Also $S = 176$ ft^2. Thus, from Eq. (8.65), we have for the planview area of the horizontal tail

$$S_{\text{HT}} = \frac{0.7(5.17)(176)}{17.13} = \boxed{37.2 \text{ ft}^2} \qquad\qquad \textbf{[8.66]}$$

Similarly for the vertical tail, from Eqs. (8.61) and (8.63), and recalling that $b = 35.27$ ft, we have

$$S_{\text{VT}} = \frac{(0.04)bs}{l_{\text{VT}}} \qquad\qquad \textbf{[8.67]}$$

Again somewhat arbitrarily, let us place the mean aerodynamic center of the vertical tail 1.13 ft forward of that of the horizontal tail; that is, we assume $l_{\text{VT}} = 16$ ft. From Eq. (8.67), we have

$$S_{\text{VT}} = \frac{(0.04)(35.27)(176)}{16} = \boxed{15.5 \text{ ft}^2} \qquad\qquad \textbf{[8.68]}$$

To determine the *shape* of the tail surfaces, we quote from Raymer (Ref. 25) that "tails are little wings." Therefore, we use somewhat of the same logic that was employed in Section 8.6.2 in determining the shape of the wing.

Unlike the wing, whose function is to generate lift strong enough to sustain the airplane in the air, the aerodynamic forces generated on the tail surfaces are relatively small; they need only be large enough to maintain stability and control. Also, the aerodynamic forces on the tail readily change directions, depending on whether the airplane is yawing right or left, and/or pitching up or down, also depending on which direction the rudder and elevator are deflected. Hence, it makes no sense to use a cambered airfoil for the tail sections; rather, the horizontal tail and vertical tail on almost all airplanes use a symmetric airfoil section. A popular choice is the NACA 0012 airfoil. We will use the same for our design on both the horizontal and vertical tails.

First, let us lay out the horizontal tail. Wings of lower aspect ratio, although aerodynamically less efficient, stall at higher angles of attack than wings with higher aspect ratio. Hence, if the horizontal tail has a lower aspect ratio than the wing, when the wing stalls, the tail still has some control authority. To achieve this advantage, we choose an aspect ratio for the horizontal tail less than that for the wing; we choose a value AR = 4. Also, we choose a taper ratio the same as that of the wing, $\lambda = 0.5$. Thus, the span of the horizontal tail b_t is

$$b_t = \sqrt{(S_{\text{HT}})\text{AR}} = \sqrt{(37.2)(4)} = 12.2 \text{ ft}$$

The tail root chord c_{rt} is [see Eq. (8.55)]

$$c_{rt} = \frac{2S_{\text{HT}}}{(\lambda + 1)b_t} = \frac{2(37.2)}{(0.5 + 1)(12.2)} = 4.07 \text{ ft}$$

The tail tip chord c_{tt} is

$$c_{tt} = \lambda c_{rt} = (0.5)(4.07) = 2.035 \text{ ft}$$

The spanwise location of the mean aerodynamic chord for the horizontal tail is [see Eq. (8.57)]

$$\bar{y}_{HT} = \frac{b_t}{6}\frac{1+2\lambda}{1+\lambda} = \frac{12.2}{6}\frac{1+1}{1+0.5} = 2.71 \text{ ft}$$

and the mean aerodynamic chord for the horizontal tail is [see Eq. (8.58)]

$$\bar{c}_{HT} = \frac{2}{3}c_{rt}\frac{1+\lambda+\lambda^2}{1+\lambda} = \frac{2}{3}(4.07)\frac{1.75}{1.5} = 3.16 \text{ft}$$

This allows us to lay out the horizontal tail as shown in Fig. 8.22**a**.

We now lay out the vertical tail. Typical aspect ratios for vertical tails AR_{VT} range from 1.3 to 2.0, where the aspect ratio is based on the root-to-tip height h_{VT} (span from tip to tip does not have any meaning here).

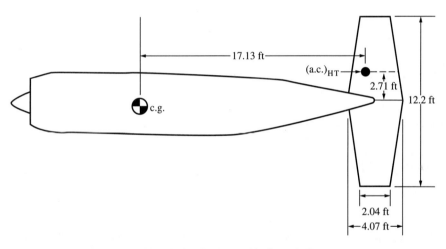

(a) Plan view — fuselage and horizontal tail

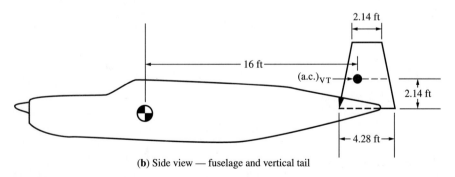

(b) Side view — fuselage and vertical tail

Figure 8.22   Layout of **(a)** the horizontal tail and **(b)** the vertical tail.

$$AR_{VT} = \frac{(h_{VT})^2}{S_{VT}} \qquad \textbf{[8.69]}$$

We will choose an aspect ratio of 1.5. Thus, from Eq. (8.69)

$$h_{VT} = \sqrt{(AR_{VT})S_{VT}} = \sqrt{(1.5)(15.5)} = 4.82 \text{ ft}$$

Consistent with our choice for the wing and horizontal tail, we choose a taper ratio of 0.5 for the vertical tail. Hence, the root chord is [see Eq. (8.55)]

$$c_{rvt} = \frac{2S_{VT}}{(\lambda + 1)(h_{VT})} = \frac{2(15.5)}{(0.5 + 1)(4.82)} = 4.28 \text{ ft}$$

The tip chord is

$$c_{tvt} = \lambda c_{rvt} = 0.5(4.28) = 2.14 \text{ ft}$$

The vertical location of the mean aerodynamic chord of the vertical tail, referenced to the root chord, is [see Eq. (8.57)].

$$\bar{z}_{VT} = \frac{2h_{VT}}{6}\frac{1 + 2\lambda}{1 + \lambda} = \frac{2(4.82)}{6}\frac{1 + 1}{1.5} = 2.14 \text{ ft}$$

The mean aerodynamic chord for the vertical tail is [see Eq. (8.58)]

$$\bar{c}_{VT} = \frac{2}{3}c_{rvt}\frac{1 + \lambda + \lambda^2}{1 + \lambda} = \frac{2}{3}(4.28)\frac{1.75}{1.5} = 3.32 \text{ ft}$$

This allows us to lay out the vertical tail, as shown in Fig. 8.22b.

8.6.6 Propeller Size

At this stage, we are not concerned with the details of the propeller design—the blade shape, twist, airfoil section, etc. Indeed, for a general aviation airplane of our design type, the propeller would be bought off the shelf from a propeller manufacturer. However, for the configuration layout, we need to establish the propeller diameter, because that will dictate the length (hence weight) of the landing gear.

The function of the propeller is to take the shaft power from the reciprocating engine and turn it into thrust power to propel the airplane forward. This is never accomplished without some losses, hence the propeller efficiency η_{pr} is always less than unity.

$$\eta_{pr} = \frac{\text{thrust power}}{\text{shaft power}} = \frac{TV_\infty}{P} < 1 \qquad \textbf{[8.70]}$$

Propeller efficiency is improved as the diameter of the propeller gets larger. The reason for this can be found in the discussion of propulsive efficiency in Section 3.2. Essentially, the larger the propeller diameter, the larger the mass flow of air processed by the propeller. Therefore, for the same thrust, the larger propeller requires a smaller flow velocity increase across the propeller disk. The smaller the increase in flow velocity across any propulsive device, the higher the propulsive efficiency.

There are two practical constraints on propeller diameter: (1) The propeller tips must clear the ground when the airplane is on the ground, and (2) the propeller tip speed should be less than the speed of sound, or else severe compressibility effects will occur that ruin the propeller performance. At the same time, the propeller must be large enough to absorb the engine power. (Imagine a small, carnival-variety pinwheel attached to our 360-hp engine—there would be virtually no thrust and no torque on the shaft, and the engine would simply "run away.") The power absorption by the propeller is enhanced by increasing the diamter and/or increasing the number of blades on the propeller. Two-blade propellers are common on general aviation aircraft. For the powerful combat airplanes of World War II, and the large propeller-driven commercial transports that immediately followed, three- and four-blade propellers were common.

For the purpose of initial sizing, Raymer (Ref. 25) gives an empirical relation for propeller diameter D as a function of engine horsepower, as follows:

$$\text{Two-blade} \qquad D = 22(HP)^{1/4} \qquad\qquad \textbf{[8.71]}$$

$$\text{Three-blade} \qquad D = 18(HP)^{1/4} \qquad\qquad \textbf{[8.72]}$$

where D is in inches. For our airplane design, we choose a two-blade, constant-speed, propeller. From Eq. (8.71), the propeller diameter is approximated as

$$D = 22(360)^{1/4} = 95.83 \text{ in } = 8 \text{ ft}$$

Question: Is this diameter too large to avoid adverse compressibility effects at the tip? Let us check the tip speed. The rated RPM (revolutions per minute) for our chosen Textron Lycoming TIO/LTIO-540-V engine is 2,600 (Ref. 36). The tip speed of the propeller when the airplane is standing still, denoted by $(V_{\text{tip}})_0$, is

$$(V_{\text{tip}})_0 = \pi n D \qquad\qquad \textbf{[8.73]}$$

where n is the shaft revolutions per second and D is in feet. Hence,

$$(V_{\text{tip}})_0 = \pi \frac{\text{RPM}}{60} D = \pi \left(\frac{2,600}{60}\right)(8) = 1,089 \text{ ft/s}$$

When the maximum forward velocity of the airplane is vectorally added to $(V_{\text{tip}})_0$, we have the actual tip velocity relative to the airflow V_{tip}.

$$V_{\text{tip}} = \sqrt{(V_{\text{tip}})_0^2 + V_\infty^2} \qquad\qquad \textbf{[8.74]}$$

The specified V_{max} is 250 mph $= 366.7$ ft/s. Hence,

$$V_{\text{tip}} = \sqrt{(1,089)^2 + (366.7)^2} = 1,149 \text{ ft/s} \qquad\qquad \textbf{[8.75]}$$

The speed of sound at standard sea level is 1,117 ft/s; our propeller tip speed exceeds the speed of sound, which is not desirable.

So we have to change our initial choice of a two-blade propeller to a three-blade propeller. From Eq. (8.72)

$$D = 18(\text{HP})^{1/4} = 18(360)^{1/4} = 78.4 \text{ in} = 6.53 \text{ ft}$$

The static tip speed is

$$(V_{tip})_0 = \pi \frac{RPM}{60} D = \pi \left(\frac{2,600}{60} \right) (6.53) = 889 \text{ ft/s}$$

Hence,

$$V_{tip} = \sqrt{(V_{tip})_0^2 + V_\infty^2} = \sqrt{(889)^2 + (366.7)^2} = 962 \text{ ft/s}$$

This is still a relatively high tip speed, but it is certainly more acceptable than our previous result. Therefore, for our propeller, we choose the following configuration:

Three-blade $D = 6.53$ ft

We make the assumption that we can find an off-the-shelf propeller that comes close to matching this size. Propeller design is an expensive process, and the nature of our airplane seems not to warrant the expense of designing a new propeller. Indeed, if we cannot find an existing propeller that satisfies our needs, then the propeller becomes a design constraint itself, and in subsequent iterations of our conceptual design process, the airplane will have to be sized to allow the use of an existing propeller. For example, if the takeoff gross weight of our airplane W_0 were reduced, the power requirement would be reduced, and hence a smaller engine would be needed. In turn, from Eqs. (8.71) and (8.72), the required propeller diameter would be reduced, possibly fitting more closely an existing, off-the-shelf item. Such a process as just described is an example of the type of constant compromising that is inherent in airplane design.

8.6.7 Landing Gear, and Wing Placement

In Section 8.4.2, relative to our discussion on landing distance and how it affects W/S, we made the decision to use a tricycle landing gear for our design. This configuration is illustrated in Fig. 8.23; here, the side view is shown, and the "footprint" of the wheels on the ground is sketched above the side view. An advantage of tricycle landing gear

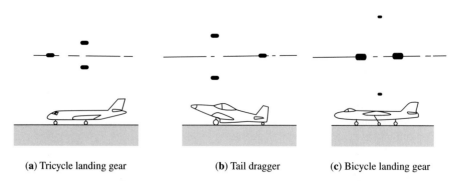

(**a**) Tricycle landing gear   (**b**) Tail dragger   (**c**) Bicycle landing gear

Figure 8.23   Some common landing gear configurations: side view and, above each side view, the landing gear footprint: (**a**) Tricycle; (**b**) tail dragger; (**c**) bicycle.

is that the cabin floor for passengers and cargo is horizontal when the airplane is on the ground. Also, forward visibility is improved on the ground for the pilot. The tricycle landing gear requires that the c.g. of the airplane be ahead of the main wheels, as shown in Fig. 8.23, and this enhances stability during the ground roll, allowing the airplane to "crab" into a cross-wing; that is, the fuselage does not have to be aligned parallel to the runway. There are numerous other possible gear configurations (see Raymer, Ref. 25); two of the more common arrangements are sketched in Fig. 8.23**b** and **c**. The tail dragger (Fig. 8.23**b**) was the most conventional configuration during the period from 1909 to 1945. Because the fuselage of the tail dragger is inclined on the ground, the propeller clearance is greater. Also, during the takeoff ground run, the wing can create more lift because it is already naturally at a higher angle of attack. For this configuration, the main wheels must be ahead of the c.g., as shown in Fig. 8.23**b**. This is an inherently unstable configuration during the ground roll; if the airplane (for whatever reason) starts to turn during the ground roll, the c.g. tends to swing around, causing the turn to get tighter. This can end up in a dangerous ground loop. To avoid such an event, the pilot must keep the airplane always aligned with the runway, constantly manipulating the rudder pedals. The bicycle landing gear, illustrated in Fig. 8.23**c**, is useful for high-wing airplanes. However, to prevent the airplane from tipping over on the ground, lightweight outrigger wheels are required near each wing tip.

We choose the tricycle configuration shown in Fig. 8.23**a**. Landing gear design is a specialty by itself—indeed, there exist complete books just on the design of landing gear, such as Refs. 56 and 57. Raymer (Ref. 25) devotes a complete chapter to the subject, oriented to the conceptual design phase. We will treat the subject here only to the extent of determining the length of the landing gear (struts plus wheels) and the wheel size for our airplane.

The landing gear should be long enough to give the propeller tip at least 9-in clearance above the ground. We choose a clearance of 1 ft. Since the propeller diameter is 6.53 ft, the radius is 3.265 ft. This places the spinner centerline 4.265 ft above the ground, as shown in Fig. 8.24. The landing gear needs to be designed to provide this height above the ground.

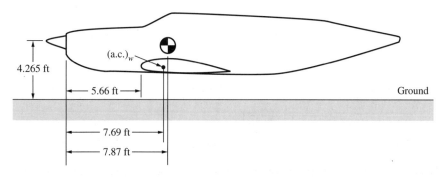

Figure 8.24   Placement of the wing.

At this stage we need to estimate the size of the tires. However, the tire size depends on the load carried by each tire. To calculate how the weight of the airplane is distributed over the two main wheels and the nosewheel, we need to locate the wheels relative to the airplane's center of gravity. Since the landing gear will retract into the wing, this means that we have to locate the wing relative to the fuselage. So let us redirect our attention for a moment to the question: Where should the wing be located?

This question was addressed to some extent in Section 8.6.4, where we estimated the location of the c.g. of the airplane. In that section, we arbitrarily placed the mean aerodynamic center of the wing at the location of our first estimate for the c.g., namely, at $\bar{x} = 7.72$ ft. Then, with the wing at this location, we recalculated the location of the c.g. including the weight of the wing; the result was $\bar{x} = 7.87$ ft, which is the c.g. location shown in Fig. 8.24. This location of the wing was preliminary; it was adopted only for the purpose of obtaining an approximation for the airplane's c.g. location. We are free to change the location of the wing at this stage in our iterative design process. We do so based on the following argument.

From considerations of longitudinal stability, the aerodynamic center of the airplane must lie behind the airplane's center of gravity. The aerodynamic center of the airplane is also called the *neutral point* for the airplane; the neutral point is, by definition, that location of the airplane's c.g. that would result in the pitching moment about the c.g. being independent of angle of attack. We have not discussed the subject of stability and control in this book simply because of a lack of space. However, all we need here is just the basic idea of longitudinal stability. Indeed, reference is made to the introductory discussion in chapter 7 of Ref. 3. There, the following relation was given between the location of the aerodynamic center of the wing body x_{acwb} and the location of the neutral point x_n as

$$x_{acwb} = x_n - V_{HT}\frac{a_t}{a} \qquad \text{[8.76]}$$

where V_{HT} is the horizontal tail volume ratio, defined by Eq. (8.60), and a_t and a are the lift slopes for the horizontal tail and the complete airplane, respectively. In Eq. (8.76), the influence of the downwash angle behind the wing and ahead of the tail is neglected. Furthermore, the *static margin* is defined as

$$\text{Static margin} \equiv \frac{x_n - \bar{x}}{\bar{c}} \qquad \text{[8.77]}$$

where $\bar{x}$ is the location of the airplane's c.g. and $\bar{c}$ is the wing mean aerodynamic chord. For conventional general aviation airplanes, the static margin should be on the order of 5% to 10%. Let us assume the 10% value for our airplane:

$$\text{Static margin} \equiv \frac{x_n - \bar{x}}{\bar{c}} = 0.1 \qquad \text{[8.78]}$$

Using $\bar{x} = 7.87$ ft and $\bar{c} = 5.17$ ft as obtained earlier, we find from Eq. (8.78) that

$$x_n = 0.1\bar{c} + \bar{x} = 0.1(5.17) + 7.87 = 8.387 \text{ ft}$$

In Eq. (8.76), we will assume for simplicity that the aerodynamic center of the wing-body (wing-fuselage) combination is the same as the aerodynamic center of the wing $x_{acwb} = (x_{ac})_{wing}$. Also, we assume for simplicity that the lift slope of the tail and that for the whole airplane are essentially the same, or $a_t = a$. Thus, from Eq. (8.76), we obtain for the longitudinal position of the wing aerodyamic center, recalling from Eq. (8.62) that $V_{HT} = 0.7$,

$$(x_{ac})_{wing} = x_n - V_{HT} = 8.387 - 0.7 = 7.69 \text{ ft} \qquad \textbf{[8.79]}$$

Hence, we will locate the wing such that its mean aerodynamic center is 7.69 ft behind the nose of the airplane. This location is shown in Fig. 8.24. Furthermore, from the wing layout in Fig. 8.19, this places the leading edge of the root chord at $x = 7.69 - 1.29 - 0.74 = 5.66$ ft, also shown in Fig. 8.24.

With the placement of the wing now established, we return to our consideration of the size and location of the landing gear. For structural and space reasons, we will locate the main gear at the center of the wing. As shown in Fig. 8.24, the root leading edge is at $x = 5.66$ ft. Since the root chord is 6.65 ft, then the location of the center of the wing is at $x_c = 5.66 + 6.65/2 = 8.99$ ft. This is shown in Fig. 8.25. The main landing gear is located 8.99 ft behind the nose of the airplane. Let us locate the nosewheel so that it can be conveniently folded rearward and upward into the fuselage. Setting a nosewheel location of 2.25 ft, as shown in Fig. 8.25, satisfies this criterion. Hence, Fig. 8.25 shows the location of the landing gear.

The size of the tires depends on the load distribution between the main wheels and the nosewheel. The loads on the tires can be calculated with the aid of Fig. 8.26. Here, points A and B are the contact points of the nosewheel and main wheels, respectively, with the ground. The load carried by each wheel is represented by the equal and opposite forces exerted by the ground on the wheel (the tire). And F_N and F_M are these forces on the nosewheel and main wheels, respectively. (Note that F_M is the combined load on the two main wheels; the load on each main wheel is $F_M/2$.)

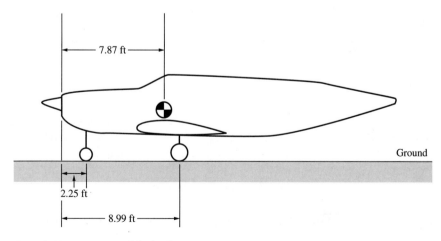

Figure 8.25   Location of the landing gear.

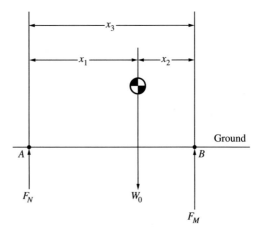

Figure 8.26   Force diagram for obtaining the load distribution among the tires.

The takeoff gross weight W_0 acts through the center of gravity. The distance between the line of action of F_N and the c.g. is x_1; the distance between the line of action of F_M and the c.g. is x_2. The distance between F_N and F_M is $x_3 = x_1 + x_2$. Taking moments about point A, we have

$$F_M x_3 = W_0 x_1$$

or

$$F_M = \frac{W_0 x_1}{x_3} \qquad\qquad \textbf{[8.80]}$$

Taking moments about point B, we have

$$F_N x_3 = W_0 x_2$$

or

$$F_N = \frac{W_0 x_2}{x_3} \qquad\qquad \textbf{[8.81]}$$

Equations (8.80) and (8.81) give the forces carried by the main wheels and the nose-wheel, respectively. Comparing Figs. 8.25 and 8.26, we find that, for our airplane,

$$x_3 = 8.99 - 2.25 = 6.74 \text{ ft}$$

$$x_1 = 7.87 - 2.25 = 5.62 \text{ ft}$$

$$x_2 = x_3 - x_1 = 6.74 - 5.62 = 1.12 \text{ ft}$$

Substituting these values into Eqs. (8.80) and (8.81), we obtain

$$F_M = \frac{W_0 x_1}{x_3} = \frac{(5,158)(5.62)}{6.74} = 4,301 \text{ lb}$$

and

$$F_N = \frac{W_0 x_2}{x_3} = \frac{(5,158)(1.12)}{6.74} = 857 \text{ lb}$$

Hence, the load on the nosewheel is 857 lb, and the load on *each* main wheel is $F_M/2 = 4,301/2 = 2,150.5$ lb.

With this information, the tire sizes can be estimated. Raymer (Ref. 25) gives empirically determined relations for wheel diameter and width in terms of the load on each tire.

$$\text{Wheel diameter or width (in)} = A W^B \qquad\qquad \textbf{[8.82]}$$

where, for general aviation airplanes, the values of A and B are as follows:

	A	B
Wheel diameter (in)	1.51	0.349
Wheel width (in)	0.715	0.312

For our design, from Eq. (8.82) we have

Main wheels:

$$\text{Diameter} = A \left(\frac{F_M}{2} \right)^B = 1.51 \left(\frac{4,301}{2} \right)^{0.349} = 21.98 \text{ in}$$

$$\text{Width} = A \left(\frac{F_M}{2} \right)^B = 0.715 \left(\frac{4,301}{2} \right)^{0.312} = 7.84 \text{ in}$$

Nosewheel:

$$\text{Diameter} = A F_N^B = 1.51(857)^{0.349} = 15.94 \text{ in}$$

$$\text{Width} = A F_N^B = 0.715(857)^{0.312} = 5.88 \text{ in}$$

As in the case of the propeller, we must use off-the-shelf tires from the manufacturers. From a tire catalog, we would choose the tires that most closely match the sizes calculated above.

Before we end this section, please note a detail that we did not take into account, namely, the shift in the position of the center of gravity. In all our previous calculations, we assumed a fixed c.g. location. However, due to changes in the distribution of payload and fuel during the flight, the c.g. shifts position. In a more detailed analysis, we would estimate the most forward and rearward positions to be expected for the center of gravity. Among other things, this would affect the calculation of the maximum static loads carried by the wheels. In Eq. (8.80) for the load on the main wheels, x_1 would correspond to the most rearward position of the c.g.; and in Eq. (8.81) for the load on the nosewheel, x_2 would correspond to the most forward position of the center of gravity. However, we will not account for this effect in our calculations here.

8.6.8   The Resulting Layout

All aspects of Section 8.6 have been aimed at achieving the configuration layout—a drawing of our first iteration for the shape and size of the airplane. Our various considerations—the wing's size, shape, and placement relative to the fuselage; the tail's size and placement; etc.—lead to the configuration shown in Fig. 8.27—our first configuration layout.

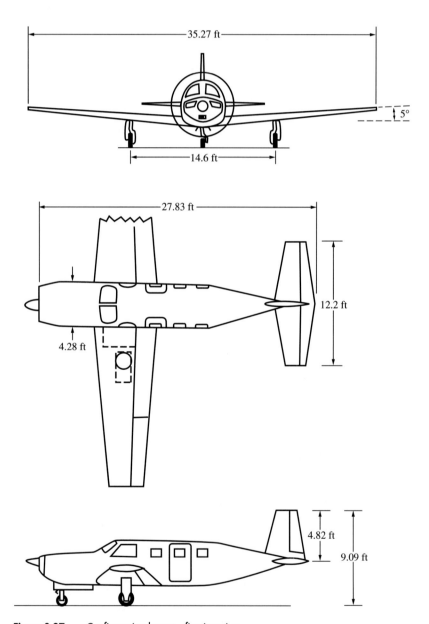

Figure 8.27   Configuration layout—first iteration.

In Fig. 8.27, a few additional features are shown. A wing dihedral of $5°$ is shown; this is based on previous general aviation airplane designs where the dihedral angle is on the order of $5°$ to $7°$ (Ref. 25). The ailerons, flaps, elevator, and rudder are shown, with a width equal to 30% of the local chord, a typical width. A more detailed design and sizing of the control surfaces are performed during later iterations of the configuration layout, and are based on a control analysis that has not been discussed in this book; see Ref. 3 for an introductory discussion of stability and control. (In our effort to present the *philosophy* of airplane design in this book, a detailed analysis of control is beyond our scope.) The tentative position of windows and doors is also shown in Fig. 8.27. The main landing gear is placed 14.6 ft apart so that it will retract into the wing without interfering with the space for the fuel tank; the fuel tank (see Fig. 8.16) and the retracted landing gear are shown in Fig. 8.27 as dashed lines in the plan view. One of the functions of the configuration layout is to see whether things fit internally in the airplane.

We have now completed pivot point 4 in Fig. 7.3. Let us move on to pivot point 5—a better weight estimate.

8.7   A BETTER WEIGHT ESTIMATE

In Section 8.3 we made a first estimate of the gross takeoff weight W_0 on the basis of historical data from previous airplanes. We had no other choice because at that stage we did not know the size and shape of our airplane design. However, with Fig. 8.27 we now have a configuration layout with which we can attempt a component weight buildup—estimating the weight of the various parts of the airplane and adding them to obtain the total empty weight.

Weight estimation in airplane design is critical. In most airplane companies, this job is carried out by specialized weight engineers, who draw from many disciplines such as structures, mechanical design, and statistics. Moreover, each company has its own established procedures and detailed formulas for estimating weights. It is well beyond the scope of this book to describe such detailed procedures. However, we will carry out a crude weight buildup that is more detailed than the weight estimation made in Section 8.3. This will serve to illustrate the philosophy of pivot point 5 in Fig. 7.3, and it will also give us a better weight estimate with which to finish our first design iteration.

Raymer (Ref. 25) gives an approximate weight buildup for a general aviation airplane as follows:

$$\text{Wing weight} = 2.5 S_{\text{exposed wing planform}} \qquad \textbf{[8.83a]}$$

$$\text{Horizontal tail weight} = 2.0 S_{\text{exposed horiz tail planform}} \qquad \textbf{[8.83b]}$$

$$\text{Vertical tail weight} = 2.0 S_{\text{exposed vert tail planform}} \qquad \textbf{[8.83c]}$$

$$\text{Fuselage weight} = 1.4 S_{\text{wetted area}} \qquad \textbf{[8.83d]}$$

$$\text{Landing gear weight} = 0.057 W_0 \qquad \textbf{[8.83e]}$$

$$\text{Installed engine weight} = 1.4(\text{Engine weight}) \qquad \textbf{[8.83f]}$$

$$\text{All else empty} = 0.1 W_0 \qquad \textbf{[8.83g]}$$

Here all areas are in units of square feet, and all weights are in the units of pounds. Because Eqs. (8.83e) and (8.83g) involve the takeoff gross weight, which is determined in part by the other elements of Eq. (8.83), the use of this list of relations involves an iterative approach to converge on the empty weight. Let us apply Eqs. (8.83a) to (8.83g) to our airplane.

The *exposed* planform areas of the wing and tail are the areas seen in the configuration layout, Fig. 8.27, and do *not* include the effective additional areas that project into the fuselage. For example, our calculated planform area of the wing, $S = 176$ ft^2 obtained in Section 8.4.2, includes that part of the wing which is projected inside the fuselage. The wing area shown in Fig. 8.19 includes the region covered by the fuselage; the area shown in Fig. 8.19 (since it is only one-half of the wing) is $176/2 = 88$ ft^2. In contrast, the value of $S_{\text{exposed wing planform}}$ is less than 176 ft^2. From Figs. 8.10, 8.22, and 8.27, we obtain

$$S_{\text{exposed wing planform}} = 148 \text{ ft}^2$$

$$S_{\text{exposed horiz tail planform}} = 35.3 \text{ ft}^2$$

$$S_{\text{exposed vert tail planform}} = 14.4 \text{ ft}^2$$

To estimate the wetted surface area of the fuselage, let us approximate the fuselage shape by two cylinders and a cone, as shown in Fig. 8.28. The forward section, section A, is simulated by an elliptical cylinder, where the elliptical cross section has semimajor and semiminor axes of 4.28 and 2.93 ft, respectively. The center fuselage section, section B, is represented by a circular cylinder of diameter 4.28 ft. The rearward section, section C, is approximated by a right circular cone with a base diameter of 4.28 ft and an altitude of 9 ft. For the purposes of this book, the simulation shown in Fig. 8.28 is simply a crude way of estimating the wetted surface area of

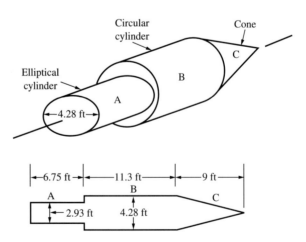

Figure 8.28   Model for the estimation of wetted surface area of the fuselage for our airplane design. Elliptical cylinder, circular cylinder, cone combination.

the fuselage; practicing professional design teams have more accurate methods for obtaining wetted surface area.

For section A, the surface area of the elliptical base is $\pi ab = \pi(4.28/2)(2.93/2)$ = 9.85 ft^2, where a and b are the semimajor and semiminor axes, respectively. The surface area of the side of the cylinder is the circumference times the length of the side. The circumference of the elliptical base is approximately $2\pi\sqrt{(a^2 + b^2)/2}$ = $2\pi\sqrt{[(2.14)^2 + (1.465)^2]/2} = 11.52$ ft. Hence, the surface area of the side of the elliptical cylinder is $(11.52)(6.75) = 77.8$ ft^2. The total wetted surface area of section A is therefore $9.85 + 77.8 = 87.63$ ft^2. For section B, the area of the base of the circular cylinder is $\pi d^2/4 = \pi(4.28)^2/4 = 14.39$ ft^2. The exposed wetted surface area of the base is that outside of the intersection with section A, namely, $14.39 - 9.85 = 4.54$ ft^2. The surface area of the side of the circular cylinder is $\pi(4.28)(11.3) = 151.9$ ft^2. Hence the total wetted surface area of section B is $4.54 + 151.9 = 156.5$ ft^2. The surface area of the cone designated as section C is given by $\pi r\sqrt{r^2 + h^2}$ where r is the radius of the base and h is the altitude. From Fig. 8.28, this surface area is $\pi(2.14)\sqrt{(2.14)^2 + (9)^2} = 62.2$ ft^2. Finally, the total wetted surface area of the geometric figure in Fig. 8.28 is

$$S_{\text{wetted area}} = \text{section A} + \text{section B} + \text{section C}$$

$$= 87.63 + 156.5 + 62.2 = 306.3 \text{ ft}^2$$

Since Fig. 8.28 represents a crude estimate for the wetted surface area of the fuselage, we will use for this wetted area the value of 306.3 ft^2 calculated above.

Returning to Eqs. (8.83a) to (8.83g) and inserting the above areas, we have

From Eq. (8.83a):

$$\text{Wing weight} = 2.5(148) = 370 \text{ lb}$$

From Eq. (8.83b):

$$\text{Horizontal tail weight} = 2.0(35.3) = 70.6 \text{ lb}$$

From Eq. (8.83c):

$$\text{Vertical tail weight} = 2.0(14.4) = 28.8 \text{ lb}$$

From Eq. (8.83d):

$$\text{Fuselage weight} = 1.4(306.3) = 428.8 \text{ lb}$$

From Eq. (8.83e):

$$\text{Landing gear weight} = 0.057(5,158) = 294 \text{ lb}$$

From Eq. (8.83f):

$$\text{Installed engine weight} = 1.4(547) = 765.8 \text{ lb}$$

From Eq. (8.83g):

$$\text{All else empty} = 0.1(5,158) = 515.8 \text{ lb}$$

$$\text{Total empty weight} \quad W_e = 2,474 \text{ lb}$$

In Eqs. (8.83e) and (8.83g), W_0 is our original estimate of 5,158 lb from Section 8.3. In Eq. (8.83f), the dry engine weight is 547 lb from Section 8.6.1. The gross takeoff weight is given by Eq. (8.1):

$$W_0 = W_{\text{crew}} + W_{\text{payload}} + W_{\text{fuel}} + W_{\text{empty}} \qquad \textbf{[8.1]}$$

From Section 8.3, we recall that $W_{\text{crew}} = 170$ lb, $W_{\text{payload}} = 970$ lb, and $W_{\text{fuel}} = 820$ lb. [Note that $W_f/W_0 = 0.159$ from Eq. (8.20). Hence, the value of $W_f = 820$ lb will change as W_0 changes in the iterative calculation we are now carrying out.] Thus, from Eq. (8.1), with our weight values obtained above, we have

$$W_0 = W_{\text{crew}} + W_{\text{payload}} + W_f + W_e$$

$$W_0 = 170 + 970 + 820 + 2,474 = 4,434 \text{ lb} \qquad \textbf{[8.84]}$$

With this new value of W_0, we return to Eqs. (8.83e and g) and recalculate W_e.

$$\text{Landing gear weight} = 0.057(4,434) = 252.7 \text{ lb}$$

$$\text{All else empty} = 0.1(4,434) = 443.4 \text{ lb}$$

This gives a new $W_e = 2,360$ lb. The new W_f is obtained from $W_f = 0.159(4,434) = 705$ lb. In turn, from Eq. (8.1) we obtain yet another value of W_0:

$$W_0 = 170 + 970 + 705 + 2,360 = 4,205 \text{ lb}$$

We repeat this process, recalculating W_e, W_f, and W_0, until convergence is obtained. The iterative process is summarized below.

Iteration	W_e (lb)	W_f (lb)	W_0 (lb)
1	2,474	820	4,434
2	2,360	705	4,205
3	2,324	668.6	4,132.6
4	2,313	657.1	4,110
5	2,309	653.5	4,103
6	2,308	652.4	4,100
7	2,308	651.9	4,100

The iteration converges to the following values:

$$W_e = 2,308 \text{ lb} \qquad W_f = 652 \text{ lb} \qquad W_0 = 4,100 \text{ lb}$$

We observe that the above weights are considerably different from the original values considered in our design calculations in the preceding sections. *We have just carried out the design philosophy associated with pivot point 5 in Fig. 7.3.* Based on the configuration layout, we have obtained a *better* weight estimate. Note that our new ratio of empty to gross weight is $W_e/W_0 = 0.56$. This is less than the value of 0.62 chosen in Section 8.3.1 based on the historical data shown in Fig. 8.2; the value of $W_e/W_0 = 0.56$ falls within the low side of the scatter of data points for airplanes with gross weights less than 10,000 lb in Fig. 8.1.

We now proceed to the next pivot point in Fig. 7.3, namely, a performance analysis using the better weight estimate obtained in the present section.

8.8 PERFORMANCE ANALYSIS

The estimate of $W_0 = 4,100$ lb obtained in Section 8.7 is lower than the initial estimate of $W_0 = 5,158$ lb used for our design calculations to this point. This is an encouraging trend, because the airplane shown in the configuration layout in Fig. 8.27 will have better performance with the lower W_0 than we have estimated so far. The function of pivot point 6 in Fig. 7.3 is to find out whether the design existing at pivot point 4 will meet or exceed the requirements. This is the subject of this section. Here we will carry out a performance analysis of the airplane shown in Fig. 8.27, using the improved weight estimates obtained from pivot point 5. We will use the performance analysis techniques discussed in Chapters 5 and 6.

The updated performance parameters are

$$\text{Wing loading} \qquad \frac{W}{S} = \frac{4,100}{176} = 23.3 \text{ lb/ft}^2$$

$$\text{Power loading} \qquad \frac{W}{P} = \frac{4,100}{360} = 11.39 \text{ lb/hp}$$

The aerodynamic coefficients have not been changed, by choice. In a more sophisticated design experience, at this stage in the design process better estimates for $C_{D,0}$, K, and $(C_L)_{max}$ would be made, using the configuration layout in Fig. 8.27. For simplicity, we choose not to do so here. Hence, we still assume

$$C_{D,0} = 0.017$$

$$K = 0.075$$

$$(C_L)_{max} = 2.34$$

$$\left(\frac{L}{D}\right)_{max} = 14$$

8.8.1 Power Required and Power Available Curves

Since cruise is set at 20,000 ft, the power required and power available are calculated for an altitude of 20,000 ft. Figure 8.29 gives the variation of drag with velocity, and Fig. 8.30 gives the variation of horsepower required and horsepower available

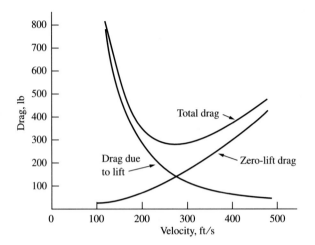

Figure 8.29   The variation of drag due to lift, zero-lift drag, and total drag with velocity at 20,000 ft. $W_0 = 4,100$ lb.

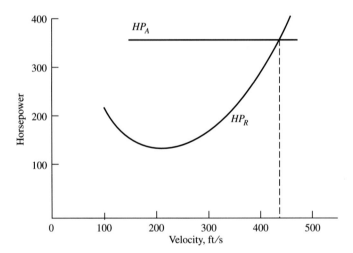

Figure 8.30   Horsepower required and horsepower available at 20,000 ft. $W_0 = 4,100$ lb.

at 20,000 ft. The graphical construction in Fig. 8.30 predicts $V_{max} = 437$ ft/s $= 298$ mi/h. This is considerably higher than the requirements of a maximum velocity of 250 mi/h, as given in Section 8.2. In fact, Fig. 8.30 assumes the weight to be the full $W_0 = 4,100$ lb, not the midcruise weight that is stipulated in the requirements; V_{max} at midcruise weight would be even higher. Clearly, our airplane design exceeds the V_{max} specification.

8.8.2 Rate of Climb

The variation of maximum rate of climb with altitude is shown in Fig. 8.31, where the weight at each altitude is assumed to be $W_0 = 4,100$ lb. At sea level, $(R/C)_{max} = 1,572$ ft/min. This far exceeds the required $(R/C)_{max} = 1,000$ ft/min. Once again, our airplane design exceeds specification.

At 18,000 ft, there is a kink in the rate-of-climb curve. This is due to the engine's being supercharged to sea-level density as high as 18,000 ft, and then above 18,000 ft the engine power decreases proportionately with ambient density. From Fig. 8.31, we obtain a graphical solution for the absolute and service ceilings as 33,600 and 32,400 ft, respectively. This far exceeds the requirement for a ceiling of 25,000 ft given in Section 8.2.

From the variation of $(R/C)_{max}$ with altitude shown in Fig. 8.31, the time to climb is calculated as described in Section 5.12. The results show that the time to climb to 20,000 ft is 14.02 min.

8.8.3 Range

Since we are assuming the same aerodynamic characteristics for the airplane in Fig. 8.27 as we have used during the earlier part of this chapter, the range also stays the same. For a range of 1,200 mi, $W_f/W_0 = 0.159$ as calculated in Section 8.3.2. However, because of the lighter gross weight, W_f is smaller. We have already calculated the new fuel weight in Section 8.7 to be 652 lb, down from our first estimate of 820 lb. Hence, our airplane design meets the specification for a range of 1,200 mi, and it does this with a smaller fuel load than had previously been calculated.

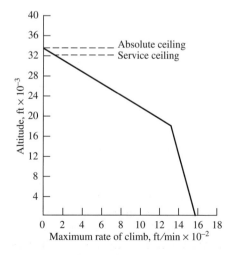

Figure 8.31 Maximum rate of climb as a function of altitude. $W_0 = 4,100$ lb.

8.8.4   Stalling Speed

The value of $(C_L)_{max} = 2.34$ obtained in Section 8.4.1 remains unchanged since we are assuming the same aerodynamic characteristics as utilized earlier. However, because W/S is now different, the stalling velocity will change from its earlier specified value. Specifically, from Eq. (5.67), we have

$$V_{stall} = \sqrt{\frac{2}{\rho_\infty} \frac{W}{S} \frac{1}{(C_L)_{max}}} = \sqrt{\frac{2(23.3)}{(0.002377)(2.34)}} = 91.5 \text{ ft/s} = \boxed{62.4 \text{ mi/h}}$$

Hence, the specification in Section 8.2 that the stalling speed be 70 mi/h or less is clearly satisfied.

8.8.5   Landing Distance

As in Section 8.4.2, we will again adopt an approach angle $\theta_a = 3°$. The average velocity during flare is $V_f = 1.23 V_{stall} = (1.23)(91.5) = 112.5$ ft/s. From Eq. (6.107), the flight path radius during flare is

$$R = \frac{V_f^2}{0.2g} = \frac{(112.5)^2}{(0.02)(32.2)} = 1,965 \text{ ft}$$

From Eq. (6.106), the flare height h_f is given by

$$h_f = R(1 - \cos \theta_a) = 1,965(1 - \cos 3°) = 2.69 \text{ ft}$$

From Eq. (6.108), the approach distance to clear a 50-ft obstacle is

$$s_a = \frac{50 - h_f}{\text{Tan } \theta_a} = \frac{50 - 2.69}{\text{Tan } 3°} = 902.7 \text{ ft}$$

The flare distance s_f is given by Eq. (6.109).

$$s_f = R \sin \theta_a = 1,965 \sin 3° = 102.8 \text{ ft}$$

The ground roll is approximated by Eq. (8.28).

$$s_g = jN \sqrt{\frac{2}{\rho_\infty} \frac{W}{S} \frac{1}{(C_L)_{max}}} + \frac{j^2(W/S)}{g\rho_\infty(C_L)_{max}\mu_r} \qquad \textbf{[8.28]}$$

where $j = 1.15$, $N = 3$ s, and $\mu_r = 0.4$. Using the updated value of $W/S = 23.3$ lb/ft^2, Eq. (8.28) yields

$$s_g = (1.15)(3)\sqrt{\frac{2(23.3)}{(0.002377)(2.34)}} + \frac{(1.15)^2(23.3)}{(32.2)(0.002377)(2.34)(0.4)} = 745.9 \text{ ft}$$

Hence,

$$\text{Total landing distance} = s_a + s_j + s_g = 902.7 + 102.8 + 745.9 = \boxed{1,751 \text{ ft}}$$

This is well within the specified landing distance of 2,200 ft given in Section 8.2.

8.8.6 Takeoff Distance

An estimate of the ground roll can be obtained from Eq. (6.95):

$$s_g = \frac{1.21(W/S)}{g\rho_\infty (C_L)_{max}(T/W)} \qquad \textbf{[6.95]}$$

In Eq. (6.95), an average value of T/W during takeoff is that value at $V_\infty = 0.7V_{LO}$, where $V_{LO} = 1.1V_{stall}$. Hence, T/W is evaluated at a velocity of

$$V_\infty = 0.7V_{LO} = 0.77V_{stall} = 0.77(91.5) = 70.4 \text{ ft/s}$$

Since the power available is, from Eq. (3.13),

$$P_A = \eta_{pr}P = T_A V_\infty$$

we have (recalling $HP = 360$, and 550 ft·lb/s is 1 hp)

$$T_A = \frac{\eta_{pr}P}{V_\infty} = \frac{(0.8)(360)(550)}{70.4} = 2,250 \text{ lb}$$

Hence,

$$\left(\frac{T}{W}\right)_{0.7V_{LO}} = \frac{2,250}{4,100} = 0.549$$

Returning to Eq. (6.95), and recalling that $(C_L)_{max} = 1.98$ with the flaps in the takeoff position, and $W/S = 23.3$ lb/ft^2, we have

$$s_g = \frac{1.21(W/S)}{g\rho_\infty(C_L)_{max}(T/W)} = \frac{1.21(23.3)}{(32.2)(0.002377)(1.98)(0.549)} = 338.9 \text{ ft}$$

To obtain the distance covered while airborne to clear an obstacle, we first calculate the flight path radius from Eq. (6.98).

$$R = \frac{6.96(V_{stall})^2}{g} = \frac{(6.96)(91.5)^2}{32.2} = 1,810 \text{ ft}$$

From Eq. (6.99), the included flight path angle is

$$\theta_{OB} = \text{Cos}^{-1}\left(1 - \frac{h_{OB}}{R}\right) \qquad \textbf{[6.99]}$$

where h_{OB} is the obstacle height, $h_{OB} = 50$ ft.

$$\theta_{OB} = \text{Cos}^{-1}\left(1 - \frac{50}{1,810}\right) = 13.5°$$

From Eq. (6.100), the airborne distance is

$$s_a = R\sin\theta_{OB} = 1,810\sin 13.5° = 422.5 \text{ ft}$$

The total takeoff distance is then

$$\text{Takeoff distance} = s_g + s_a = 338.9 + 422.5 = \boxed{761.4 \text{ ft}}$$

This is far less than the specified takeoff distance of 2,500 ft as stated in Section 8.2.

8.8.7   Interim Summary

We have just finished a performance analysis of the airplane shown in the configuration layout in Fig. 8.27. We have now completed pivot point 6 in Fig. 7.3. *Question:* Does our airplane design meet or exceed the requirements? *Answer:* Emphatically yes. In every respect, the design outperforms the specifications, in some cases by a considerable margin. An obvious reason for this excellent performance is the considerably reduced gross weight of 4,100 lb compared to the original estimate of 5,158 lb. Since the engine was originally sized to meet the specifications with the larger weight, the lighter airplane has a smaller power loading, namely 11.39 lb/hp compared to the earlier value of 14.3 lb/hp. Our lighter airplane is simply a "hot" airplane compared to the earlier stage of our design. For this reason, returning to Fig. 7.3, there is no need to iterate the design to obtain better performance.

However, moving to pivot point 7 in Fig. 7.3, we ask the question: Is it the *best* design? We do not know the answer to this question without carrying out the optimization study called for by pivot point 7. But it is virtually certain that we do *not* have the *best* design for the specifications given in Section 8.2. Indeed, our airplane appears to be greatly overdesigned for the given specifications. In particular, with the lighter gross weight of 4,100 lb, we can choose a less powerful, more light weight engine and still meet the specifications. In such a case, the airplane will be less expensive, and hence a "better" design. So pivot point 7 is absolutely critical. The performance parameter space needs to be examined (various choices of W/S, W/P, etc.) in order to find the *best* airplane that will meet the specifications. To carry out such an optimization here is beyond our scope. However, in terms of the design philosophy discussed in Chapter 7, you need to appreciate the importance of pivot point 7.

8.9   SUMMARY

The purpose of this chapter has been to illustrate the design philosophy discussed in Chapter 7, especially as highlighted in Fig. 7.3. We chose to design a propeller-driven airplane in this chapter. However, the general philosophy of design is the same, whatever type of airplane is considered—propeller-driven, jet-powered, subsonic, or supersonic. Since Chapters 5 and 6 used a turbofan-powered airplane as an example, this chapter provides a balance by dealing with a propeller-driven airplane.

We end this chapter with two short design case histories of perhaps the most important propeller-driven airplanes ever designed—the 1903 *Wright Flyer* and the Douglas DC-3 from the 1930s. One of the purposes of these case histories is to illustrate the role of the design philosophy as constructed in Chapter 7 in the design of these historic aircraft.

8.10   DESIGN CASE STUDY: THE *WRIGHT FLYER*

This section is an adjunct to Section 1.2.2, where the design features of the *Wright Flyer* are summarized and where the attributes of Wilbur and Orville Wright as the first true aeronautical engineers are highlighted. In this section, we reexamine the design of the *Wright Flyer* relative to the design philosophy discussed in Chapter 7, and we address the question of how closely the Wright brothers followed the intellectual pivot points listed in Fig. 7.3. Before you continue, please review Section 1.2.2 and Fig. 7.3.

There were no customer requirements specified for the *Wright Flyer*. There were only the requirements set by the Wright brothers themselves, namely, to design a powered flying machine that would lift a human being off the ground and fly through the air without loss of speed in a fully controlled fashion. For the Wright brothers, this was intellectual pivot point 1 in Fig. 7.3.

We have stated earlier that aircraft design is more often evolutionary than revolutionary. The design of the *Wright Flyer* was *both*. Let us explain. The Wrights did not operate in a vacuum. They inherited the bulk of aeronautical progress that occurred during the nineteenth century, including the work of Lilienthal, Langley, and Chanute (see Chapter 1). Jakab states in Ref. 1:

> An important beginning step of the Wrights' engineering approach to human flight was to become acquainted with the work of previous experimenters. By the time the brothers began their study of flight at the close of the nineteenth century, a growing community of aeronautical experimenters had emerged. As the field slowly organized, publication and dissemination of aeronautical research grew more widespread. Through contact with several key individuals and sources of information, the brothers were able to digest the work of generations of experimenters. Familiarization with these prior developments aided the Wrights in defining the basic obstacles to human flight and outlining their initial approach to the problem. Their literature search enabled them to take advantage of already established principles and to avoid dead-end paths pursued by others.

In many respects, the *Wright Flyer* evolved from this earlier work. Most likely, if the first successful airplane had not been designed and flown by the Wrights in 1903, someone else would have done it within the decade. Moreover, the Wrights began with three glider designs, the first two of which in 1900 and 1901 were based almost entirely on the existing bulk of aeronautical knowledge. These two glider designs were not successful, and the Wrights ultimately blamed the failure on errors in the existing data, particularly on a table of normal and axial force coefficients generated by Lilienthal. However, I have shown in Ref. 8 that the Wrights made three distinct errors in the interpretation of the Lilienthal tables which account for the failure of their 1900 and 1901 gliders. Nevertheless, the Wrights made the decision in the fall of 1902 to throw away the existing data and generate their own. To this end, they built a small wind tunnel and tested over a hundred different wing and airfoil shapes, finding out for themselves what constituted "good aerodynamics" for their purposes. On the basis of their wind tunnel results, they designed a new glider which in 1902 flew beautifully. The next step, in 1903, was the design of a powered machine—the *Wright Flyer*. Hence, the *Wright Flyer* indirectly inherited some of the aeronautical features developed in the nineteenth century and directly inherited the knowledge generated with the Wrights' 1902 glider. In these respects, the design of the *Wright Flyer* can be considered *evolutionary*.

However, the *Wright Flyer* was also *revolutionary* in the sense that it *worked*. It was the first flying machine to successfully fly, whereas all prior attempts by other inventors had met with failure. And the Wrights' design approach was unique for that time, which adds to the revolutionary nature of the *Wright Flyer*. This uniqueness is nicely caught by Jakab (Ref. 1):

> The Wrights' persistent attention to the overall goal of a completely successful flying machine during every phase of the work was also an important aspect of their inventive method. Each experimental glider and powered airplane they built, as well as every individual element of each aircraft, was seen and valued in terms of the ultimate aim of building a practical aircraft. The Wrights' approach was distinct among aeronautical experimenters in that they believed no specific component to be more important than any other. They recognized that every aspect of a workable flying machine must be designed to coordinate with every other. No matter how advanced the wing, without an adequate control system, an aircraft will not fly. No matter how effective the control system, without a sound structural design to carry the flight loads, an aircraft will not fly. And so on. Wilbur and Orville understood that an airplane is not a single device, but a series of discrete mechanical and structural entities, that, when working in proper unison, resulted in a machine capable of flight. Moreoever, realizing that the pilot is a part of this system, they devoted as much attention to learning to fly their aircraft as they did in designing and building them.

For all these reasons, the *Wright Flyer* was *revolutionary*.

For the *Wright Flyer*, the intellectual process embodied in pivot points 2 and 3 in Fig. 7.3 was a combination of experience with the 1902 glider and new, innovative thinking by the Wrights. First, consider the estimation of weight and wing surface area (hence wing loading W/S). The 1901 glider, with pilot, weighed 240 lb and had a wing area of 290 ft^2. Although the 1902 glider was redesigned with different aerodynamics, for their calculations the Wrights kept the weight essentially the same as that of the 1901 glider, namely 240 lb. Their wind tunnel tests had identified a wing with aspect ratio 6, curvature (camber) of $\frac{1}{20}$, and a parabolic airfoil shape as the most efficient aerodynamic shape. Moreover, the maximum lift-to-drag ratio for this wing was achieved with an angle of attack of 5°. This is perhaps one reason why the Wrights felt that a "proper" range of flight angle of attack was 4° to 8°, as stated by Wilbur in a paper delivered to the Society of Western Engineers in Chicago on June 24, 1903. (This was Wilbur's second paper to the Society, the first being delivered on September 18, 1901.) The Wrights calculated lift in pounds, using the formula

$$L = kSV^2C_L \qquad\qquad \textbf{[8.85]}$$

where k is Smeaton's coefficient, measured by the Wrights to be 0.0033 (see Ref. 8 for the role of Smeaton's coefficient in history), S is the wing area in square feet, V is velocity in miles per hour, and C_L is their measured value for the lift coefficient. At the lowest angle of attack deemed proper, namely, 4°, their measured lift coefficient was 0.433 (interpolated from their tables in Ref. 58). Thus, from Eq. (8.85), putting $L = W$, we have for the calculated wing loading

$$\frac{W}{S} = kV^2C_L \qquad\qquad \textbf{[8.86]}$$

The Wrights considered a wind of 25 mi/h to be an average for the region around Kill Devil Hills. Hence, from Eq. (8.86), we obtain

$$\frac{W}{S} = 0.0033(25)^2(0.433) = 0.89 \text{ lb/ft}^2$$

When it was finally constructed, the 1902 glider weighed about 260 lb including the pilot and had a total wing area of 305 ft^2. The resulting wing loading was 0.85 lb/ft^2, very close to (but slightly more conservative than) the calculated design value of 0.89 lb/ft^2. Finally, these conditions for the 1902 glider were translated to the design characteristics for the 1903 *Wright Flyer*. When the Wrights began to design their powered machine, they allotted no more than 180 lb for the weight of the engine, which they felt must also produce a minimum of 8 to 9 brake horsepower (bhp). (These are specifications stated by the Wrights in their letters that were mailed to a number of engine manufacturers in December 1902. Nobody could meet these specifications, so Orville along with Charlie Taylor, a mechanic at their bicycle shop, took on the design and fabrication of the engine themselves.) With the increased weight due to the engine and propellers, the new flying machine had to be larger than their 1902 glider, which increased the weight even more. They converged on a total estimated design gross weight of 625 lb. With this, they had executed pivot point 2 in Fig. 7.3. To obtain the wing loading (hence wing area), Eq. (8.86) was used. The actual design velocity used by the Wrights for the machine cannot be found in their correspondence (at least not by this author). However, it is most likely that they would have designed for 30 mi/h, which would allow them to make demonstrable forward progress over the ground in the face of the assumed average 25 mi/h headwinds at Kill Devil Hills. From Eq. (8.86) with $V = 30$ mi/h, and using the same lift coefficient of 0.433 which led to their successful 1902 glider design, the resulting wing loading is

$$\frac{W}{S} = (0.0033)(30)^2(0.433) = 1.29 \text{ lb/ft}^2$$

The Wrights built the *Wright Flyer* with a wing area of 510 ft^2, which gives a design wing loading of 1.23 lb/ft^2, very close to the above result. With this calculation, the Wrights were carrying out an important aspect of pivot point 3 in Fig. 7.3.

Another design calculation was for the thrust required, which is equal to the drag. The Wrights made a rather detailed drag breakdown for their machine, calculating what at that time was called "head resistance." The details are too lengthy to discuss here; they can be pieced together from the voluminous correspondence in Ref. 58. The net result was a calculation of 90 lb for thrust required. This meant that the engine horsepower had to translate into 90 lb of thrust from the propellers. In essence, the design thrust-to-weight ratio was $T/W = 0.144$. The Wrights' propeller design was a masterstroke of engineering brilliance, as described in Section 1.2.2 and discussed in Ref. 1. In the final result, the Wrights were elated when they measured 136 lb of thrust from their engine/propeller combination. And a good thing it was, because the actual fabricated weight of the *Wright Flyer* was slightly over 700 lb, considerably greater than the design figure of 625 lb. (This progressive increase in weight during the course of the design is a trend that has plagued most airplanes since the *Wright*

Flyer.) Once again, with this thinking, the Wrights were following pivot point 3 in Fig. 7.3. For the design of the *Wright Flyer*, the Wrights were aiming at a thrust-to-weight ratio of 0.144. What they achieved, due to the high efficiency of their propellers and in spite of the increase in weight, was an actual thrust-to-weight ratio of 0.19. (In regard to the efficiency of their propellers, the Wrights had calculated a value of 55% from their propeller theory; what was achieved was much higher. Later, in 1909, a Captain Eberhardt in Berlin made detailed measurements of the propeller efficiency of the Wrights' propellers used on their Type A Flyer of 1908, and found it to be 76%. This was by far the most efficient propeller for its day.)

The Wrights moved on to pivot point 4—a configuration layout. Their three-view sketch of the *Wright Flyer*, drawn in pencil on brown wrapping paper, with Wilbur's handwriting and notations, is shown in Fig. 8.32. The original sketch, mounted on cardboard, is now in the Franklin Institute in Philadelphia.

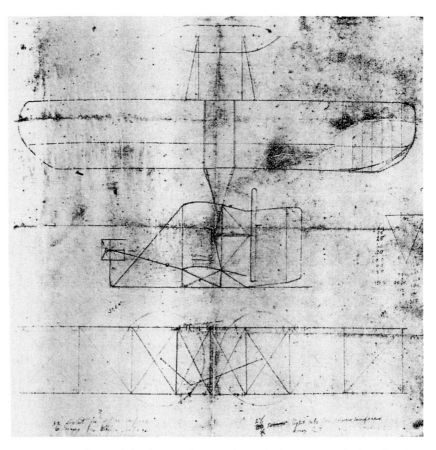

Figure 8.32   The Wright brothers' configuration layout for the 1903 *Wright Flyer*, drawn in pencil on brown wrapping paper. The notations were written by Wilbur Wright. The original sketch (with smudges), mounted on cardboard, is in the Franklin Institute, Philadelphia, Pennsylvania.

The Wrights did not go through pivot points 5, 6, and 7 in Fig. 7.3—they felt they did not have to. They had confidence that they had designed a machine that would do the job, that would fly. With the conceptual design finished, the Wrights essentially truncated the processes of preliminary and detailed design (as defined in Chapter 7) by carrying out the fabrication of the machine themselves, hand-crafting the individual parts to their satisfaction. The rest is history, made on the sand dunes of Kill Devil Hills on December 17, 1903, when, indeed, their machine did fly.

Wilbur and Orville Wright designed the first successful airplane. In so doing, without realizing it consciously, they followed the design philosophy discussed in Chapter 7. This design philosophy is basically innate. It was followed by the Wrights because it was simply the natural approach to take. However, the Wrights were consummate aeronautical engineers. What was natural for them was not always natural for others; the plethora of homespun airplane designs that followed during the next decade were not all products of the design philosophy we have set forth. However, by the end of World War I, aeronautical engineering had come into its own, and virutally all airplanes designed since then have embodied the design philosophy discussed in Chapter 7. The case history described in the next section is a perfect example.

8.11 DESIGN CASE STUDY: THE DOUGLAS DC-3

The genesis of many airplane designs is competition. So it was in 1932, when Boeing was putting the final touches to the prototype of its 247 airliner—a pioneering, low-wing monoplane, all metal, with twin engines wrapped in the new NACA low-drag cowling and with retractable landing gear. The Boeing 247 carried 10 passengers in a soundproof cabin at speeds near 200 mi/h. This airplane was expected to revolutionize commercial air travel. Because of this, the airlines were standing in line for orders. However, Boeing at that time was a member of the United Aircraft Group, which included Pratt & Whitney Engines and United Airlines. Hence, United Airlines was first in line, and was programmed to receive the first 70 new 247s to come off the production line. This put the other airlines in an untenable competitive position.

Because of this, on August 5, 1932, Donald W. Douglas, president of Douglas Aircraft Corporation, received a letter from Transcontinental and Western Air, Inc. (TWA). Dated August 2, the same letter had been sent to Glenn Martin Company in Baltimore and Curtiss-Wright Corporation in St. Louis as well as to Douglas in Santa Monica. A facsimile of the letter to Douglas, signed by Jack Frye, a vice president of TWA, is shown in Fig. 8.33. Frye was inquiring about Douglas's interest in designing a new commercial transport airplane; since TWA could not readily obtain the new Boeing 247, then in an aggressive fashion they went after their own state-of-the-art airplane. Attached to Frye's letter was a one-page list of general performance specifications for the new airplane; this list is reproduced in Fig. 8.34. [Recall that the U.S. Army's list of specifications that led to the purchase of the Wright *Military Flyer* (see Fig. 7.5) was also one page long; clearly, 25 years later airplane specifications could still be given in a short, concise, clear-cut manner.] The specifications given by TWA and shown in Fig. 8.34 illustrate for the case of the DC-3 the first step in

TRANSCONTINENTAL & WESTERN AIR INC.
10 RICHARDS ROAD
MUNICIPAL AIRPORT
KANSAS. CITY. MISSOURI

August 2nd,
19 32

Douglas Aircraft Corporation,
Clover Field,
Santa Monica, California.

Attention: <u>Mr. Donald Douglas</u>

Dear Mr. Douglas:

       Transcontinental & Western Air is interested
in purchasing ten or more trimotored transport planes.
I am attaching our general performance specifications,
covering this equipment and would appreciate your advising
whether your Company is interested in this manufacturing
job.

       If so, approximately how long would it take
to turn out the first plane for service tests?

Very truly yours,

Jack Frye
Vice President
In Charge of Operations

JF/GS
Encl.

N.B. Please consider this information confidential and
return specifications if you are not interested.

SAVE TIME — USE THE AIR MAIL

Figure 8.33   Facsimile of the letter from TWA to Donald Douglas. Douglas is later quoted as saying this letter was "the birth certificate of the modern airliner."

TRANSCONTINENTAL & WESTERN AIR, INC.

General Performance Specifications
Transport Plane

1. **Type:** All metal trimotored monoplane preferred but combination structure or biplane would be considered. Main internal structure must be metal.

2. **Power:** Three engines of 500 to 650 h.p. (Wasps with 10-1 supercharger; 6-1 compression O.K.).

3. **Weight:** Gross (maximum)   14,200 lbs.

4. **Weight** allowance for radio and wing mail bins   350 lbs.

5. **Weight** allowance must also be made for complete instruments, night flying equipment, fuel capacity for cruising range of 1080 miles at 150 m.p.h., crew of two, at least 12 passengers with comfortable seats and ample room, and the usual miscellaneous equipment carried on a passenger plane of this type. Payload should be at least 2,300 lbs. with full equipment and fuel for maximum range.

6. **Performance**

   Top speed sea level (minimum)   185 m.p.h.
   Cruising speed sea level - 79 % top speed 146 m.p.h. plus
   Landing speed not more than   65 m.p.h.
   Rate of climb sea level (minimum)   1200 ft. p.m.
   Service ceiling (minimum)   21000 ft.
   Service ceiling any two engines   10000 ft.

   This plane, fully loaded, must make satisfactory take-offs under good control at any TWA airport on any combination of two engines.

Kansas City, Missouri.
August 2nd, 1932

Figure 8.34 Facsimile of the specifications from TWA, attached to letter in Fig. 8.33.

the intellectual design process discussed in Section 7.3.1, given as pivot point 1 in Fig. 7.3.

The specifications called for an all-metal trimotor airplane that would have a cruising range of 1,080 mi at a cruise velocity of 150 mi/h. Of greatest importance, however, was the requirement listed at the bottom of the page, that is, that the new airplane at a full takeoff gross weight of 14,200 lb be able to safely takeoff from any TWA airport with one engine out. At that time, the highest airport in the TWA system was in Winslow, Arizona, at an elevation of 4,500 ft. Other aspects of the specifications called for a maximum velocity of at least 185 mi/h, landing speed of not more than 65 mi/h, a minimum rate of climb at sea level of 1,200 ft/min, and a minimum service ceiling of 21,000 ft, compromised downward to 10,000 ft with one engine out.

Donald Douglas and Jack Frye had met several times before, at various aviation functions in the Los Angeles, California, area. They held a strong mutual respect for each other. Since the formation of his company in 1921, most of Douglas's business dealt with designing and constructing military airplanes, especially a successful line of torpedo airplanes for the Navy. However, he had been recently thinking about venturing as well into the commercial airplane market (airline passenger service had skyrocketed since Charles Lindbergh's historic solo flight across the Atlantic Ocean in 1927). So Douglas paid serious attention to Frye's letter. He took it home with him that night, staying awake until 2 AM pondering the ramifications. The next day he met with his core engineering design group and went over the TWA specifications one by one. The discussion lasted well into the evening. It was Tuesday. Douglas suggested they think about it and meet again that Friday. The group had already made the decision to submit a proposal to TWA; Friday's discussion was to be about the basic nature of the airplane design itself.

The TWA specifications (Fig. 8.34) called for a "trimotored monoplane" as being preferred, but held out the possibility of the design's being a biplane. Trimotor monoplanes were not new; the Fokker F-10 and the Ford trimotor had been flying in airplane service for almost 5 years. However, this airplane configuration suffered a public setback on March 31, 1931, when a TWA Fokker trimotor crashed in a Kansas wheat field, killing among the passengers the famous Notre Dame football coach Knute Rockne. As for the biplane configuration, the early 1930s was a period of drag reduction via streamlining, and biplanes with their higher drag were on the way out.

So when the Friday meeting started out, it was no surprise that the chief engineer, James H. "Dutch" Kindelberger, stated emphatically:

> I think that we're damn fools if we don't shoot for a twin-engined job instead of a trimotor. People are skeptical about the trimotors after the Rockne thing. Why build anything that even looks like a Fokker or Ford? Both Pratt & Whitney and Wright-Aeronautical have some new engines on the test blocks that will be available by the time we're ready for them. Lots of horses . . . any two of them will pull more power than any trimotor flying right now.

Douglas agreed. As essential design decision was made without making a single calculation.

Arthur Raymond, Kindelberger's assistant, who had earned a master's degree in aeronautical engineering at Massachusetts Institute of Technology in 1921 (one of the few people with graduate degrees in aeronautical engineering at that time), was immediately thinking about the wing design. He suggested: "Why not use a modified version of Jack Northrop's taper wing? Its airfoil characteristics are good. The taper and slight sweepback will give us some latitude with the center of gravity." Raymond was referring to the innovative wing design by Jack Northrop, who had worked for Douglas between 1923 and 1927 and then left for Lockheed, finally forming his own company in 1931. (This is the same company that today builds the B-2 stealth bomber.) Northrop had developed a special cantilever wing which derived exceptional strength from a series of individual aluminum sections fastened together to form a multicellular structure. The wing is the heart of an airplane, and Raymond's thinking was immediately focused on it. He also wanted to place the wing low enough on the fuselage that the wing spars would not cut through the passenger cabin (as was the case with the Boeing 247). Such a structurally strong wing offered some other advantages. The engine mounts could be projected ahead of the wing leading edge, placing the engines and propellers far enough forward to obtain some aerodynamic advantage from the propeller slipstream blowing over the wing, without causing the wing to twist. Also, the decision was to design the airplane with a retractable landing gear. In that regard Douglas said: "The Boeing's got one. We'd better plan on it too. It should cut down on the drag by 20 percent." Kindelberger then suggested: "Just make the nacelles bigger. Then we can hide the wheels in the nacelles." The strong wing design could handle the weight of both the engines and the landing gear.

The early 1930s was a period when airplane designers were becoming appreciative of the advantages of streamlining in order to reduce aerodynamic drag. (See Ref. 8 for a detailed discussion of this history.) Retracting the landing gear was part of streamlining. Another aspect was the radial engines. Fred Stineman, another of Douglas's talented designers, added to the discussion: "If we wrap the engines themselves in the new NACA cowlings, taking advantage of the streamlining, it should give us a big gain in top speed." This referred to the research at NACA Langley Memorial Laboratory, beginning in 1928, that rapidly led to the NACA cowling, a shroud wrapped around the cylinders of air-cooled radial engines engineered to greatly reduce drag and increase the cooling of the engine. At this stage of the conversation, Ed Burton, another senior design engineer, voiced a concern: "The way we're talking, it sounds like we are designing a racing plane. What about this 65 mi/h landing speed Frye wants?" This problem was immediately addressed by yet another senior design engineer, Fred Herman, who expressed the opinion: "The way I see it, we're going to have to come up with some kind of an air brake, maybe a flap deal that will increase the wing area during the critical landing moment and slow the plane down . . . . Conversely, it will give us more lift on takeoff, help tote that big payload."

The deliberations extended into days. However, after a week of give-and-take discussions, they all agreed that the airplane design would

1. Be a low-wing monoplane.
2. Use a modified version of the Northrop wing.

3. Be a twin-engine airplane, not a trimotor.

4. Have retractable landing gear, retracted into the engine nacelles.

5. Have some type of flaps.

6. Use the NACA cowlings.

7. Locate the engine nacelles relative to the wing leading edge at the optimum position as established by some recent NACA research.

The design methodology and philosphy exemplied by these early discussions between Douglas and his senior design engineers followed a familar pattern. *No new, untried technology was being suggested.* All the design features itemized above were not new. However, the *combination* of all seven items into the *same* airplane was new. The Douglas engineers were looking at past airplanes and past developments and were building on these to scope out a new design. To a certain extent, they were building on the Northrop Alpha (Fig. 8.35). Although the Alpha was quite a different airplane (single-engine transport carrying six passengers inside the fuselage with an open cockpit for the pilot), it also embodied the Northrop multicellular cantilevered wing and an NACA cowling. Also, it was not lost on Douglas that TWA had been operating Northrop airplanes with great success and with low maintenance.

During this first critical work of their deliberations, the small team of Douglas designers had progressed through a semblance of the intellectual pivot points in

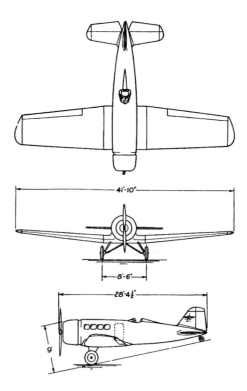

Figure 8.35   Northrop Alpha.

Fig. 7.3, simply in order to draw the overall design conclusions itemized above. However, they used more than a slide rule for calculations—they drew also on the collective intuitive feelings of the group, honed by experience. They practiced the *art* of airplane design to the extreme. At the end of that week, a proposal to TWA was prepared, and Arthur Raymond and Harry Wetzel (Douglas's vice president and general manager) took a long train ride across the country to deliver their proposal to the TWA executive office in New York. There, a three-week series of intense discussions took place; among the TWA representatives present at many of these meetings was Richard Robbins (president of TWA), Jack Frye, and Charles Lindbergh (the same Charles Lindbergh who had gained fame for his transatlantic solo flight in 1927 and who served as a technical consultant to TWA).

Although there were other competitors for the TWA contract, Raymond and Wetzel were successful in convincing TWA of the merits of a twin-engine ("bimotor") airplane over a trimotor. A major aspect of this consideration was the ability of the airplane to fly successfully on one engine, especially to takeoff at full gross weight from any airport along the TWA route and to be able to climb and maintain level flight over the highest mountains along the route. This was not a trivial consideration, and Raymond's calculations had a certain degree of uncertainty—the uncertainty that naturally is associated with the early aspects of the conceptual design process, as discussed in Chapter 7. Raymond called from New York to tell Donald Douglas about the critical nature of the one-engine-out performance's being a pivotal aspect of the discussions with TWA. When Douglas asked Raymond about his latest feelings as to whether the airplane design could meet this performance requirement, Raymond's reply was: "I did some slide-rule estimates. It comes out 90 percent yes and 10 percent no. The 10 percent is keeping me awake at nights. One thing is sure, it's never been done before with an aircraft in the weight class we're talking about." Douglas conferred with Kindelberger, who took the stand: "There's only one way to find out. Build the thing and try it." Douglas made the decision—Raymond should tell TWA that Douglas Company would be able to construct such an airplane.

On September 20, 1932, in Robbin's office, the contract was signed between TWA and Douglas to build the airplane. Douglas christened the project as the DC-1, the Douglas Commercial One. The contract called for the purchase by TWA of one service test airplane at the cost of $125,000, with the option (indeed, clear intent) of purchasing up to 60 additional airplanes, in lots of 10, 15, or 20 at $58,000 each. The contract was 42 typewritten pages long, 29 of which dealt specifically with the technical specifications. The first three pages of these technical specifications are reproduced in their original form in Fig. 8.36**a**, **b**, and **c**, so that you can obtain a better understanding of the detail to which the design had progressed by that time. Of particular interest is the detailed breakdown of the empty weight in Fig. 8.36**c**. The list of performance specifications had expanded as well by then, compared to the original one-page list shown in Fig. 8.34. Reproduced in Fig. 8.37 is just one page of the five from the contract dealing with performance. By comparing Figs. 8.36 and 8.37 with the original one-page document first sent out by TWA (Fig. 8.34), the effect of the conceptual design process on the details listed in the final contract is clearly seen. The contract even went to the detailed extent of specifying such items as this: "Air sickness container holders shall be located adjacent to each seat in such a position as to be easily reached with seat in any adjustment." (Actually, this was not

13

Schedule "A"

DOUGLAS BI-MOTORED TRANSPORT MATERIEL SPECIFICATIONS

I. Characteristics

1. General Type

This airplane shall be a low wing cantilever monoplane with retractable chassis, the general proportions being shown on Douglas Drawing No. 529289. It shall be powered with two Wright Cyclone Model 1820 F geared engines, each rated at 650 HP at sea level and supercharged to 600 HP at 1950 rpm at 8000 feet.

2. Construction

Fuselage, wings and control surfaces shall be of metal.

3. Requirements

The following characteristics and specifications shall be adhered to or bettered in the construction and performance of the airplane.

II. Materials and Workmanship

Materials and methods of construction approved by the Department of Commerce shall be used. In the absence of Department of Commerce material specifications those of the U. S. Army Air Corps shall be used. Seller will, at his own cost and expense, promptly remedy any structural weakness, defect of design, workmanship, or material that may evidence itself during the acceptance and/or service tests.

The Seller will provide an accurate and complete system covering the inspection of all materials, fabrication methods and finished parts. Records of all such inspection work shall be kept complete and shall be made available to Buyer's representative upon request.

Sufficient tests of materials shall be made by the Seller on all lots of stock in order to insure Buyer the use of approved aircraft material in the construction of said transport airplane.

(a)

Figure 8.36   Facsimile of the bimotor specifications from the contract signed between TWA and Douglas on September 20, 1932. *(continued)*

(continued)

14

III. General Requirements

1. Flying Characteristics

The airplane shall comply with Department of Commerce requirements with regard to general flying characteristics and shall be controllable to the satisfaction of the Buyer in all conditions of flight and taxiing, both when fully loaded and empty. Lateral, longitudinal, and directional stability shall comply with Department of Commerce requirements. The forces necessary to operate the controls shall be light and satisfactory to Buyer.

2. Load Factors

The airplane shall comply with Department of Commerce strength requirements and shall have an approved type certificate.

3. Interchangeability

All parts and assemblies subject to removal shall be interchangeable. The complete nacelle forward of the firewall, including the engine installed, oil tank and cowling, shall be interchangeable right and left.

IV. Detailed requirements

1. Weights

(a) Useful Load

The airplane shall be designed to carry the following useful load:

(1) Crew:
   Pilot and co-pilot @ 170 lbs. each...................... 340 lbs.

(2) Fuel and oil for either a normal range of 730 miles at 62.5% of power at 5,000 feet altitude with a payload of 3400 lbs. made up of:

Passengers, 12 @ 170 lbs.	2,040 lbs.
Baggage, 12 @ 30 lbs.	360 lbs.
Mail and cargo	1,000 lbs.
	3,400 lbs.

Or a payload of 2,000 lbs. with fuel and oil for a range of 1,000 miles at 62.5% of power at an altitude of 5,000 feet.

(b)

(concluded)

15

(b) Weight Empty

It is estimated that the weight empty of the airplane will be 8,475 lbs., subdivided as follows:

Wing	1,950 lbs.
Tail	220
Fuselage	1,125
Nacelles	220
Landing Gear	785
Surface controls	110
Instruments	95
Seats and Safety Belts	240
Floors and covering	125
Upholstering and Sound Insulation	225
Internal partitions and racks	75
Lavatory equipment and water	60
Heating and ventilating equipment	50
Fire Extinguishers and miscellaneous	60
Electrical Equipment	220
Radio	135
Flares	20
Engines	1,840
Propellers	330
Engine accessories	140
Starting system	80
Engine controls	60
Fuel system	250
Oil System	60
	8,475

(c) Gross Weight

Useful load (spec.)	6,125
Weight Empty (est.)	8,475
Gross Weight	14,600

(c)

39

Schedule "B."

DOUGLAS BI-MOTORED PERFORMANCE SPECIFICATIONS

Prior to delivery of the transport airplane, complete performance tests shall have been made at the plant of Seller at Santa Monica, California, in order to determine as hereinafter set forth that said transport airplane complies in all respects with the following:

(1) The airplane shall demonstrate its ability to meet the following performances when fully loaded to the licensed gross load. (Standard altitude is meant when altitude is referred to.)

    (a) High speed at zero feet   174 mph
    (b) High speed at 8000 ft.   183 mph
    (c) Cruising speed at zero feet, 75% power   155 mph
    (d) Cruising speed at 5000 ft., 75% power   159 mph
    (e) Cruising speed at 8000 ft., 75% power   162 mph
    (f) Cruising speed at zero ft., 62.5% power   144 mph
    (g) Cruising speed at 5000 ft., 62.5% power 148 mph
    (h) Cruising speed at 8000 ft., 62.5% power 150 mph
    (i) Landing speed at zero feet   64 mph
    (j) Rate of climb at zero feet   1030 ft/min
    (k) Rate of climb at 8000 ft.   920 ft./min
    (l) Cruising range, normal fuel, 5000 ft. at 62.5% power   730 miles
    (m) Cruising range, capacity fuel at 5000 ft. 62.5% power   1000 miles
    (n) Service ceiling   22800 ft.
    (o) Absolute ceiling   24800 ft.
    (p) Service ceiling on one engine   8000 ft.

The method of determining the foregoing speeds shall be as follows:

The airplane shall be flown over a speed course, which shall be at an altitude of 200 feet or less above sea level, appropriately laid out

Figure 8.37 Facsimile of the performance specifications from the contract signed between TWA and Douglas on September 20, 1932.

as trivial as it may seem today; the airplane was unpressurized, and hence it would be flying, as did all aircraft at that time, at low altitudes where there was plenty of air turbulence, especially in bad weather.)

The concern that the Douglas designers put into the aspect of one-engine-out flight is reflected in a detailed technical paper written by Donald Douglas, and presented by Douglas as the Twenty-Third Wilbur Wright Memorial Lecture of the Royal Aeronautical Society in London on May 30, 1935. The annual Wilbur Wright Lectures were (and still are) the most prestigious lectures of the Society. It was a testimonial to Douglas's high reputation that he had received the Society's invitation. The paper (Ref. 59) was entitled: "The Developments and Reliability of the Modern Multi-Engine Air Liner with Special Reference to Multi-Engine Airplanes after Engine Failure." Douglas began his paper with a statement that is as apropos today as it was then:

> Four essential features are generally required of any form of transportation: Speed, safety, comfort and economy.

However, today we would add *environmentally clean* to the list. Douglas went on:

> The airplane must compete with other forms of transportation and with other airplanes. The greater speed of aircraft travel justifies a certain increase in cost. The newer transport planes are comparable with, if not superior to, other means of transportation. Safety is of special importance and improvement in this direction demands the airplane designer's best efforts.

Douglas then concentrated on engine failure as it related to airplane safety. He wrote:

> Statistics show that the foremost cause of accident is still the forced landing. The multi-engine airplane, capable of flying with one or more engines not operating, is the direct answer to the dangers of an engine failure. It is quite apparent, however, that for an airplane that is not capable of flying with one engine dead the risk increases with the number of engines installed. Hence, from the standpoint of forced landings, it is not desirable that an airplane be multi-engine unless it can maintain altitude over any portion of the air line with at least one engine dead. Furthermore, the risk increases with the number of remaining engines needed to maintain the required altitude. In general, therefore, the greatest safety is obtained from—
>
> 1. The largest number of engines that can be cut out without the ceiling of the airplane falling below a required value;
> 2. The smallest number of engines on which the airplane can maintain this given altitude.
>
> For airplanes equipped with from one to four engines, it follows that the order of safety is according to the list following.

Douglas followed with a list of 10 options, starting with the category "four-engine airplane requiring 1 engine to maintain given altitude" as the most safe and "four-engine airplane requiring 4 engines to maintain given altitude" as obviously the least safe. Fourth down on the list was the two-engine airplane requiring one engine to maintain given altitude—this was the category of the DC-1 (and the DC-2 and DC-3 to follow). It is statistically safer than a three-engine airplane requiring two engines to maintain given altitude, which was fifth on Douglas's list.

Having made his point about the relative safety of a twin-engine airplane capable of flying on one engine, Douglas turned to the flight performance of such an aircraft. Of particular note was the stability and control of a twin-engine airplane with one engine out. Because of its relevance to airplane design, and because we have not studied the effects of engine-out performance on airplane design to this stage in this book, we pursue further some of Douglas's thoughts on this matter.

Consider a twin-engine airplane in straight and level flight. How can the airplane be controlled to maintain a straight and level flight path when an engine fails? Consider the airplane in Fig. 8.38, taken from Douglas's paper. Assume the right engine has

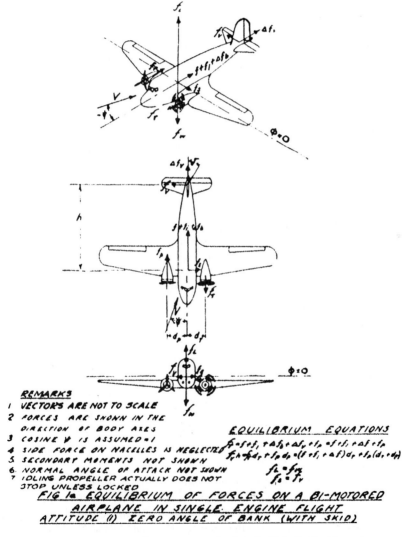

Figure 8.38 Engine-out performance—zero bank (with skid). Original figure by Donald Douglas.

failed, as indicated by the stationary propeller in Fig. 8.38. The thrust from the left engine is no longer balanced by an equal thrust from the right engine; instead, the thrust from the left engine creates a moment about a vertical axis through the center of gravity, which tends to yaw the airplane to the right. This yawing moment is counterbalanced by a horizontal force (f_v in Fig. 8.38) on the vertical tail, which acts through a moment arm to the center of gravity. In this way the moments about the vertical axis will be balanced, and the airplane will not rotate about the vertical axis. However, f_v is a right-side force, which if left unbalanced will cause the airplane to sideslip (translate) to the right. So f_v must be compensated by an equal side force toward the left, shown as f_s in Fig. 8.38. This is created by having the fuselage in a yawed position *to the left*, hence creating the left-side force f_s on the fuselage. In turn, this cocks the vertical tail in the wrong direction. Therefore, the *rudder* on the vertical tail must have enough control authority to produce the required f_v in the direction shown, even though the vertical tail is now at an unfavorable incidence angle. Note that in this attitude, the wings are still level, denoted by zero bank angle ϕ_0. So Fig. 8.38 illustrates one possible attitude of the airplane that produces a straight and level flight path after an engine failure, namely, a skid (fuselage yawed) with 0° angle of bank. The skid is in the direction of the operating engine.

Figure 8.39 taken from Douglas's paper illustrates another possible attitude of the airplane for a straight and level flight path after engine failure. Here, the side force on the vertical tail f_v necessary to counterbalance the yawing moment from the one operating engine is compensated by an equal and opposite side force obtained by banking the airplane (lowering the left wing) through the angle ϕ so that a component of the weight, $f_w \sin \phi$, is equal and opposite to f_v, thus *not* creating a sideslip. In this attitude, there is *no* yaw; all f_v is due to rudder deflection, not to any incidence angle for the vertical tail. Hence, Fig. 8.39 illustrates another possible attitude of the airplane that produces a straight and level flight path, namely, a bank with 0° angle of yaw. The bank is such that the lowered wing is on the same side as the operating engine.

Figure 8.40 is a third possible attitude, one which is necessary if there is insufficient rudder control authority to maintain either of the previous two. Here, the necessary f_v to counter the moment due to the one operating engine is produced mainly by the incidence angle of the vertical tail because the rudder is not powerful enough to do the job. This requires the fuselage to be yawed to the right. In turn, a side force f_s is produced on the yawed fuselage, pointing in the same direction as f_v. The sum $f_s + f_v$ must be balanced by an equal and opposite component of the weight obtained by banking the airplane to the left, such that $f_w \sin \phi = f_s + f_v$. This requires more bank angle than the case shown in Fig. 8.39. Hence, Fig. 8.40 shows yet another attitude that results in a straight and level flight path, namely, a combined yaw and angle of bank, with the yaw in the direction of the failed engine, and the lowered wing in the direction of the operating engine.

Note that in any of the three attitudes shown in Figs. 8.38 to 8.40, the drag is increased due to (1) the idling propeller, and (2) the increased aerodynamic drag on the vertical tail (the latter due to an increased induced drag on the tail).

The case of engine-out performance is not frequently discussed in basic design texts. However, it has been discussed here because one-engine-out performance was

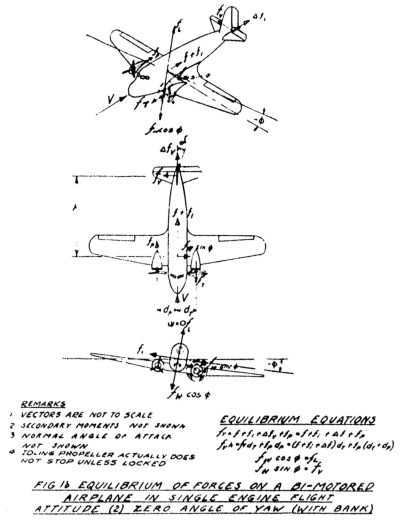

Figure 8.39 Engine-out performance—zero yaw (with bank). Original figure by Donald Douglas.

critical to the Douglas engineers as they embarked on building an airplane to satisfy the TWA specifications. From a design standpoint, this dictated the size of the vertical tail and the rudder. Indeed, even today the size of the vertical tail of multiengine airplanes, propeller- or jet-powered, is usually dictated by consideration of engine-out performance. Also, engine-out performance dictates in part the lateral location of the engines on the wing. The closer the engines to the fuselage, the smaller the moment about the vertical axis when an engine fails. Of course, for a propeller-driven airplane, the engines must be far enough away to allow sufficient propeller clearance with the fuselage.

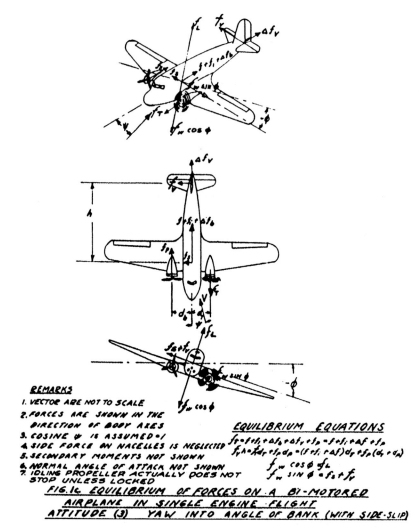

Figure 8.40   Engine-out performance—yaw into the angle of bank. Original figure by Donald Douglas.

Another hallmark governed the early design of the DC-1, namely, creature comfort. This was particularly emphasized by Art Raymond who, after the TWA contract negotiations were over in New York, chose to *fly* back to Santa Monica. Flying from coast to coast at that time was an endurance test, especially in the Ford trimotor that Raymond was on. Raymond suffered from the noise, vibration, cold temperature at altitude, small and primitive lavatory facilities, uncomfortable seats, and even mud splashed on his feet. Indeed, he complained later: "When the plane landed on the puddle-splotched runway, a spray of mud, sucked in by the cabin air vents, splattered everybody." After returning to the Douglas plant, Raymond stated: "We've got to build comfort, and put wings on it. Our big problem is far more than just building a satisfactory performing transport airplane." The team set about immediately to design an airplane which included soundproofing, cabin temperature control, improved plumbing, and no mud baths.

In 1932, the Guggenheim Aeronautical Laboratory at the California Institute of Technology (GALCIT) had a new, large subsonic wind tunnel. It was the right facility in the right place at the right time. Situated at the heart of the southern California aeronautical industry at the time when that industry was set for rapid growth in the 1930s, the California Institute of Technology (Cal Tech) wind tunnel performed tests on airplane models for a variety of companies that had no such testing facilities. Douglas was no exception. As conceptual design of the DC-1 progressed into the detailed design stage, wind tunnel tests on a scale model of the DC-1 were carried out in the Cal Tech wind tunnel. Over the course of 200 wind tunnel tests, the following important characteristics of the airplane were found:

1. The use of a split flap increased the maximum lift coefficient by 35% and increased the drag by 300%. Recall from Chapter 5 that both effects are favorable for landing; the increase in $(C_L)_{max}$ allowed a higher wing loading, and the corresponding decrease in L/D allowed a steeper landing approach.

2. The addition of a fillet between the wing and fuselage increased the maximum velocity by 17 mi/h.

3. During the design process, the weight of the airplane increased, and the center of gravity shifted rearward. For that case, the wind tunnel tests showed the airplane to be longitudinally unstable. The design solution was to add sweepback to the outer wing panels, hence shifting the aerodynamic center sufficiently rearward to achieve stability. The mildly swept-back wings of the DC-1 (also used on the DC-2 and DC-3 airplanes) gave these airplanes enhanced aesthetic beauty as well as a distinguishing configuration.

A photograph of the DC-1 model mounted upside-down in the Cal Tech wind tunnel is shown in Fig. 8.41. The upside-down orientation was necessary because the model was connected by wires to the wind tunnel balance above it, and in this position the downward-directed lift kept the wires taunt. Dr. W. Bailey Oswald, at

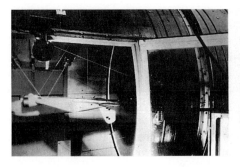

Figure 8.41   A model of the Douglas DC-1 mounted upside-down in the Cal Tech wind tunnel, late 1932.

that time a professor at Cal Tech who was hired by Douglas as a consultant on the DC-1 aerodynamics, said later on: "If the wind tunnel tests had not been made, it is very possible that the airplane would have been unstable, because all the previous engineering estimates and normal investigations had indicated that the original arrangement was satisfactory." [We note that this is the same Bailey Oswald who introduced the Oswald efficiency e_o defined in Eq. (8.43). Beginning in 1928, Arthur Raymond taught a class on the practical aspects of airplane design at Cal Tech, and Oswald attended the class in the first year. The two became trusted colleagues. Finally, reflecting on his first action upon returning to Santa Monica after his trip to the TWA offices in New York, Raymond wrote later (Ref. 60): "The first thing I did when I got back was to contact Ozzie (Oswald) and ask him to come to Santa Monica to help us, for that one-engine-out case still bothered me. I told him we only needed him for a little while, but he stayed until retirement in 1959, and ultimately had a large section working for him."]

On July 1, 1933, the prototype DC-1 was ready for its first flight. It took less than one year from the day the original TWA letter arrived in Douglas's office to the day that the DC-1 was ready to fly. The weather was bright and clear, with a gentle breeze blowing in from the ocean. At exactly 12:36 PM the DC-1, with test pilot Carl Cover at the controls, lifted off the runway at Clover Field in Santa Monica, California. The first flight almost ended in disaster. As Cover put the DC-1 in a climb about 30 s after takeoff, the left engine quit; a moment later, the right engine sputtered to a stop. However, as the airplane nosed over, the engines started again. Cover started to climb again, but once again the engines stopped. They started again when the nose dipped down. For the next 10 min, in a display of expert piloting, Cover was able to coax the DC-1 up to 1,500 ft, following a sawtooth flight path alternating between a climb, the engines cutting off, a noseover, the engines starting again, another climb until the engines again quit, etc. At 1,500 ft, the DC-1 was at a safe enough altitude to allow Cover to execute a gentle bank and to return safely to the runway.

Nobody knew what was wrong. The airplane and the engines appeared to be mechanically sound. Over the next 5 days the engines were taken apart and reassembled more than a dozen times. On the test block, the engines would run perfectly. Finally, the trouble was found in the carburetors that metered fuel to the engines—they had been installed backward, in such a fashion that when the airplane climbed, the gasoline could not flow uphill, and the fuel was automatically cut off. The carburetors were then rotated 180°, and the trouble disappeared. The rest of the DC-1 flight test program was carried out successfully. The airplane met all its flight specifications, including the one-engine-out performance at the highest altitudes encountered along the TWA routes. It was a wonderful example of successful, enlightened airplane design.

An interesting contrast can be made in regard to the time from design conception to the first flight. During World War I, some airplanes were designed by laying out chalk markings on the floor and rolling out the finished airplane 2 weeks later. Fifteen years later, the process was still relatively quick, that for the DC-1 being about 11 months. Compare this to the design time for today's modern civil and military airplanes, which sometimes takes close to a decade between design conception and first flight.

Only one DC-1 was built. The production version, which involved lengthening the fuselage by 2 ft and adding two more seats to make it a 14-passenger airplane, was labeled the DC-2. The first DC-2 was delivered to TWA on May 14, 1934. Altogether, Douglas manufactured 156 DC-2s in 20 different models, and the airplane was used by airlines around the world. It set new standards for comfort and speed in commercial air travel. But the airplane that really made such travel an *economic success* for the airlines was the next outgrowth of the DC-2, namely, the DC-3.

As in the case of the DC-1, the DC-3 was a result of an airline initiative, not a company initiative. Once again, the requirements for a new airplane were being set by the customer. This time the airline was American Airlines, and the principal force behind the idea was its tall, soft-spoken, but determined Texan Cyrus R. Smith. C. R. Smith had become president of American Airlines on May 13, 1934. American Airlines was operating sleeper service, using older Curtiss Condor biplanes outfitted with pullman-sized bunks. On one flight of this airplane during the summer of 1934, Smith, accompanied by his chief engineer, Bill Littlewood, almost subconsciously remarked, "Bill, what we need is a DC-2 sleeper plane." Littlewood said that he thought it could be done. Smith lost no time. He called Douglas to ask if the DC-2 could be made into a sleeper airplane. Douglas was not very receptive to the idea. Indeed, the company was barely able to keep up with its orders for the DC-2. Smith, however, would not take no for an answer. The long-distance call went on for 2 h, costing Smith over $300. Finally, after Smith virtually promised that American Airlines would buy 20 of the sleeper airplanes, Douglas reluctantly agreed to embark on a design study. Smith's problem was that he had just committed American Airlines to a multimillion-dollar order for a new airplane that was just in the imagination of a few men at that time, and the airline did not have that kind of money. However, Smith then traveled to Washington to visit his friend and fellow Texan Jesse Jones, who was the head of Reconstruction Finance Corporation, a New Deal agency set up by President Franklin Roosevelt to help U.S. business. Smith got his money—a $4,500,000 loan from the government. The new project, the Douglas Sleeper Transport, the DST, was on its way.

Design work on the DST, which was quickly to evolve into the DC-3, started in earnest in the fall of 1934. Once again, model tests from the Cal Tech wind tunnel were indispensable. The new design outwardly looked like a DC-2. But the fuselage had been widened and lengthened, the wingspan increased, and the shape of the rudder and vertical stabilizer were different. In the words of Arthur Raymond (Ref. 60): "From the DC-1 to the DC-2, the changes were minor; from the DC-2 to the DC-3, they amounted to a new airplane." The different plan view shapes of the DC-2 and DC-3 are shown in Fig. 8.42. The wind tunnel tests at Cal Tech were overseen by Professor A. L. Klein and Bailey Oswald. During the tests, a major stability problem was encountered. Klein stated: "The bigger plane with its change in the center of gravity had produced the stability of a drunk trying to walk a straight line." However, by slightly modifying the wing and changing the airfoil section, the airplane was made stable; indeed, the DST finally proved to be one of the most stable airplanes in existence at that time. The first flight of the DST was on December 17, 1935. After the efforts of over 400 engineers and drafters, the creation of 3,500 drawings, and some 300 wind tunnel tests, the airplane flew beautifully. American Airlines began service of the DST on June 25, 1936.

The distinguishing aspects of the DST compared to the DC-2 were that its payload was one-third greater and its gross weight was about 50% larger. These aspects did

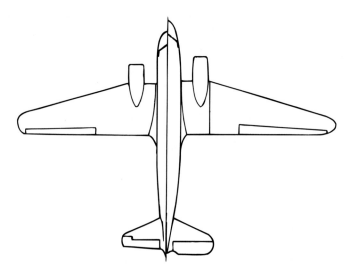

Figure 8.42   Comparison of the DC-2 (left) and DC-3 (right) planforms.

not go unappreciated by Douglas. If the bunks were taken out and replaced by seats, the airplane could carry 21 passengers in a relative state of luxury. This was yet another new airplane—the DC-3. In fact, by the time Douglas gave his 1935 annual report to his board of directors, the DC-3 was already moving down the production line in parallel with the DST.

Less than 100 airplanes in the sleeper configuration—DST—were produced. But when the DC-3 production line was finally shut down at the end of World War II, 10,926 had been built. The vast majority of these were for the military, 10,123, compared to 803 for the commercial airlines (see Ref. 61). The DC-3 was an amazing success, and today it is heralded by many aviation enthusiasts as the most famous airplane of its era. A three-view of the DC-3 is shown in Fig. 8.43.

The success of the DC-3 was due to the technology which was so artfully embodied in its design—the streamlined shape, NACA cowlings, retractable landing gear, split flaps, variable-pitch propellers, multicellular wing structure, etc. It was also due to the design objective of carrying more people in greater comfort with more safety at a faster speed than possible in other existing airplanes at that time. The flying public loved it; the DC-3 opened the doors for successful passenger-carrying airlines, greatly expanding the number of people flying and the number of routes flown during the late 1930s.

To be more specific, the DC-3 made money for the airlines. It did this through the combination of improved aerodynamic and engine efficiency, and the fact that its passenger capacity was higher than that of other existing transports (for example, 21 seats compared to 14 seats on the DC-2). The improved aerodynamic efficiency

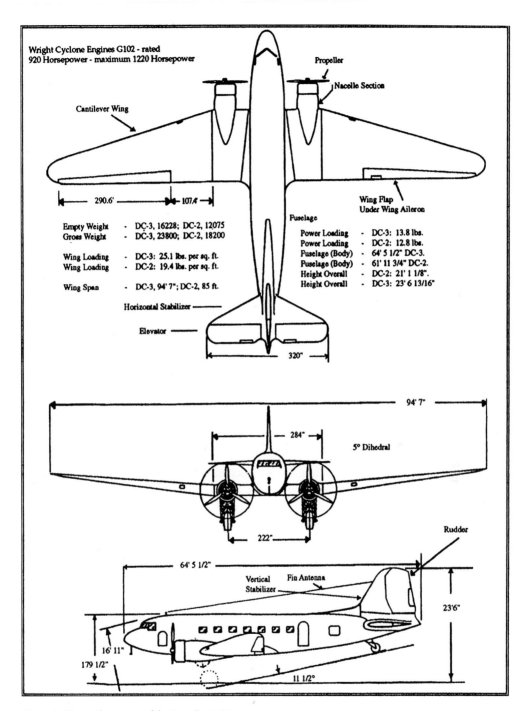

Figure 8.43 Three-view of the Douglas DC-3.

can be seen by comparing the values of maximum L/D for several contemporary airplanes.

Airplane	$(L/D)_{max}$
Ford 5-AT Trimotor	9.5
Northrop Alpha	11.3
Lockheed Vega	11.4
Boeing 247D	13.5
Douglas DC-3	14.7

Clearly, the DC-3 was the epitome of aerodynamic efficiency for its time. In terms of economics, a good metric is the direct operating cost (DOC) in cents per available seat-mile. The DOC for several airplanes is tabulated below, obtained from Ref. 62.

Airplane	DOC
Ford Trimotor	2.63
Lockheed Vega	2.51
Boeing 247	2.11
Douglas DC-3	1.27

The DC-3 was a major improvement in direct operating costs; it was a money maker for the airlines.

It is appropriate to end this section with some specifications and performance data for both the DC-2 and DC-3.

	DC-2	DC-3
Gross weight (lb)	17,880	24,000
Payload weight (lb)	2,180	3,890
Wingspan (ft)	85	95
Fuselage length (ft)	62	64.5
Airfoil section	NACA 2215 at root tapered to NACA 2209 at tip	NACA 2215 at root tapered to NACA 2206 at tip
Engines	Two Wright SGR-1820-F3, 1,420-hp total	Two Wright Cyclones, 1,700 hp total
Maximum speed (mi/h)	205	212
Cruising speed (mi/h)	180	188
Cruising range (mi)	1,200	1,260

The increase in fuselage length and wingspan for the DC-3 compared to the DC-2 is illustrated in Fig. 8.42. Finally, a partial cutaway of the DC-3 is shown in Fig. 8.44.

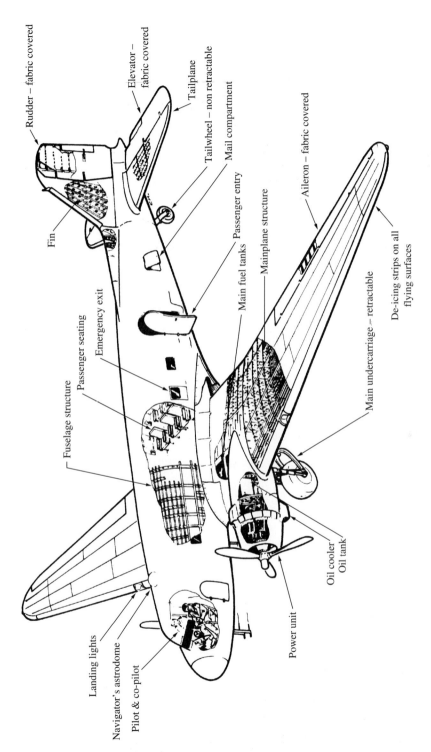

Figure 8.44 Cutaway drawing of the Douglas DC-3.

The design of the DC-1, DC-2, and DC-3 series is a classic case history from the era of the mature propeller-driven airplane, the period essentially between 1930 and 1945. Indeed, the DC-3 greatly helped to usher in that era. This case history shows how the design philosophy described in Fig. 7.3 was tailored by the Douglas intuitiveness, experience, and art of the engineers. Although airplane design at that time was much more organized than during World War I, there was still plenty of room for the inventiveness of an individual to play a strong role. Douglas DC-3s are still flying today (and will be into the twenty-first century), a testimonial to the design philosophy and methodology of the Douglas design team.

9

Design of Jet-Propelled Airplanes

The modern jet transport can be described as the largest integration of technology into a self-sufficient unit. All it needs to fly is a full fuel tank, a small crew, and a long runway. Its economic success depends on performance, low maintenance costs and high passenger appeal. It is unique in that all major sections are highly technical in content, from the wing tips to the nose and the tail. Designing the individual components and fitting them together into a cohesive whole is a long process that cannot be expressed in a formula. Airplane design is a combination of industrial art and technology. Usually the process of resolving the art precedes the application of formulae.

> William H. Cook, Retired Chief of the
> Technical Staff, Transport Division
> Boeing Airplane Company, 1991

The Skunk Works is a concentration of a few good people solving problems far in advance—and at a fraction of the cost—of other groups in the aircraft industry by applying the simplest, most straightforward methods possible to develop and produce new projects. All it is really is the application of common sense to some pretty tough problems.

> Clarence L. "Kelly" Johnson,
> Retired Director of the Lockheed
> Advanced Development Projects
> (The Skunk Works), 1985

9.1   INTRODUCTION

On August 27, 1939—five days before the beginning of World War II—a small airplane rolled sluggishly down the runway adjacent to Heinkel Aircraft Factory in Germany. Gaining speed, it finally left the ground and climbed to 2,000 ft. Heinkel's test pilot, Erich Warsitz, was at the controls. For 6 min Warsitz circled gracefully around the field, and then he came in for a landing. What was revolutionary about his flight is that the airplane had no propeller. The aircraft was the Heinkel 178—the first airplane to fly powered by a jet engine. Shown in Fig. 9.1, the Heinkel 178 achieved a maximum speed of 360 mi/h, not much different from the maximum velocity of some propeller-driven fighters at that time. However, it was an experimental airplane—the first jet airplane—and in the elated words of Ernst Heinkel, "He was flying! A new era had begun."

Indeed, that first flight of the He-178 on August 27, 1939, constituted the second revolution in flight in the twentieth century, the first being the flight of the *Wright Flyer* on December 17, 1903. When the small and relatively simple He-178 left the ground and circled the small onlooking crowd standing on the ramp of the airfield below, the jet age was born. Never mind that something was wrong with the landing gear, such that it would not retract. Erich Warsitz had to fly the entire 6 min flight with the landing gear down. But it did not matter, history had been made.

We are now deep into the jet age, as discusssed in Section 1.2.4, and nothing else more revolutionary appears to be on the horizon. Today, virtually all military aircraft and commercial transports are jet-powered. Most new executive aircraft are also jets, and the gas-turbine engine is even beginning to power a few small general aviation aircraft. Therefore, any consideration of airplane design today almost by

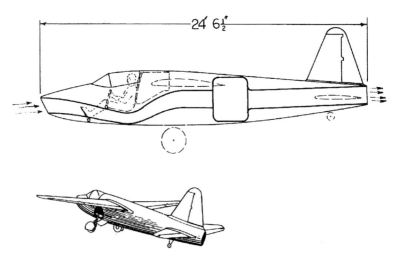

Figure 9.1   The He-178, the first jet-propelled airplane to successfully fly, on August 27, 1939. (*From Cook, Ref. 63, with permission.*)

default deals with a jet-propelled airplane. However, the design *philosophy* and *general methodology* for the design of jet airplanes are the same as described in Chapter 7, which is to say, generic in nature. We have explicitly illustrated this design philosophy in great detail in Chapter 8 for a propeller-driven airplane. For the design of a jet airplane, the intellectual approach is essentially the same, only some of the details are different. Therefore, there appears to be little to gain (except possibly a lot of repetition and boredom on the part of the reader) to illustrate the design of a jet airplane by following the same detailed path laid out in Chapter 8 for a propeller-driven airplane. Instead, in this chapter we will discuss the design of several pioneering jet airplanes; these discussions will essentially be case histories, but with a twist. The twist will be specific discussions, as appropriate, of some technical details of jet airplane design that are different from those covered in Chapter 8 for a propeller-driven airplane. In this way we aim to do justice to this chapter, giving you a good idea of how to design a jet-propelled airplane, but without repeating the type of detailed calculations illustrated in Chapter 8.

9.2 THE DESIGN OF SUBSONIC/TRANSONIC JET-PROPELLED AIRPLANES: A CASE STUDY OF THE BOEING 707 AND 727

The Boeing 707 is shown in Fig. 1.33; return to this figure and review the related short discussion of the 707 in Chapter 1. Examining Fig. 1.33, we see a sleek, swept-wing commercial jet transport that first entered airline service in 1958. As is usual with many airplane designs, the Boeing 707 was evolutionary; it was derived from Boeing's experience with the earlier designs for the B-47 and B-52 jet bombers. However, the B-47 itself was revolutionary—the first successful swept-wing jet bomber, with the engines housed in pods mounted underneath the wing. So the Boeing 707 was a derivative from an earlier airplane that was itself a revolutionary step. Furthermore, the 707 became the first successful civil jet airliner, and in that sense it was revolutionary because it dramatically changed airliner travel.

In this section we will explore the design philosophy of the Boeing 707, as well as that of the next Boeing jet transport, the trimotor 727. However, to appreciate this design philosophy, we should start at the beginning, with the revolutionary design of the B-47 jet bomber.

9.2.1 Design of the B-47—A Precursor to the 707

As discussed in Chapter 7 and shown in Fig. 7.3, the first pivot point in the design of a new airplane is a statement of the requirements. For the B-47, this first took the form of a study contract awarded to five aircraft companies in late 1943 by the Bombardment Branch of the U.S. Army Air Forces at Wright Field in Dayton, Ohio. The purpose of this contract was to have each company design a jet-propelled bomber, with the possibility that the Army would buy a prototype from each manufacturer.

Each of the aircraft was to be powered by the GE TG-180 axial-flow jet engine, then only in the design stage. The Army had such little experience with jet airplanes at the time that virtually the only requirement was that the airplanes in these studies be jet-powered. From these studies, in April 1944, the Army was able to establish a preliminary specification for a four-engine jet bomber. Convair, North American, and Boeing submitted design proposals which were very conventional airplanes with high-aspect-ratio straight wings and the jet engines mounted in nacelles that faired into the wings. Eventually, North American's design became the straight-wing B-45, the first U.S. jet bomber to go into service; the first flight of the B-45 was in March 1947, and 142 were manufactured.

The aerodynamicists at Boeing, however, were not satisfied with the performance of any of these straight-wing designs, including their own design. Wind tunnel data showed the critical Mach number for these designs to be lower than desired. As a result, they delayed submitting a detailed design to the Army Air Forces. Then, in May 1945, a technical intelligence team made up of U.S. scientists and engineers swept into a defeated Germany and discovered a mass of German test data on swept wings (see Section 1.2.4). One member of that team was George Schairer, a young Boeing aeronautical engineer who was participating in Boeing's jet bomber design. After studying the German data, Schairer quickly wrote to the design team, alerting them to the interesting design features of the swept wing and its potential for increasing the critical Mach number of the airplane. (For a more detailed discussion of the history of the swept wing and how its advantages were finally recognized by the U.S. aeronautical industry, see chapter 9 of Ref. 8.) The Boeing design team promptly dropped their straight-wing design and concentrated on the swept-wing configuration, not without some skepticism and opposition from other parts of the company. At that time, Boeing was fortunate to have its own large, high-speed wind tunnel, which went into service in 1944 after a 3-year period of development. The test section was 8 ft high, 12 ft wide, and 20 ft long. With an 18,000-hp synchronous motor, the tunnel was able to achieve Mach 0.975 with an empty test section. No other U.S. company had such a facility. The original decision in 1941 to build the wind tunnel was somewhat of a gamble on the part of Boeing executives—the gamble paid off royally after the war, because it was in this facility that Boeing was able to collect the necessary data for the design of a swept-wing jet bomber.

The design team worked through a number of different configurations. Figure 9.2 (from Ref. 63) shows the design evolution of what became the final configuration of the B-47. In addition to the swept wing, the location of the engines went through several design choices. Engine nacelles closely integrated with the wing (Fig. 9.2, top left) created an effective thickening of the wing, lowering the critical Mach number. When the engines were relocated to the top of the fuselage (Fig. 9.2, top and center part), the body had to be made wider and the jet exhaust would scrub the top of the airplane—both undesirable features. When the Boeing design team leaders Ed Wells and Bob Jewett took this configuration to Wright Field in Dayton in October 1945, the Army Air Forces Project Office resoundedly rejected it—and properly so. The Project Office wanted the engines to be mounted on the wing as was typical on past bombers. On the trip back to Seattle, Wells and Jewett conceived the idea of mounting

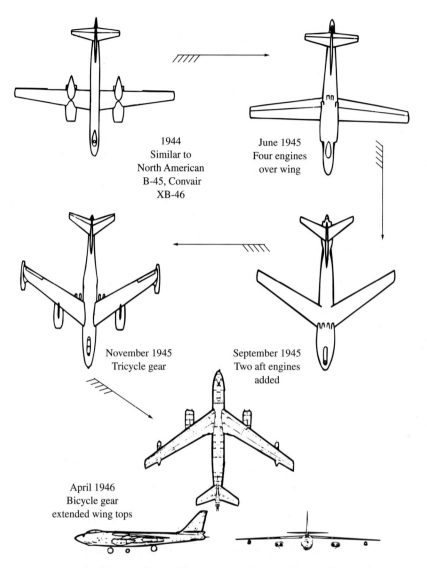

1944
Similar to
North American
B-45, Convair
XB-46

June 1945
Four engines
over wing

November 1945
Tricycle gear

September 1945
Two aft engines
added

April 1946
Bicycle gear
extended wing tops

Figure 9.2   The design evolution of the Boeing B-47. (*From Cook, Ref. 63, with permission.*)

the engines in pods suspended below the wing on struts. It was a radical idea for its time, but Boeing felt it had no choice, since Wright field had so firmly rejected the previous design. Once again, the Boeing high-speed wind tunnel was vital. The pod-on-strut configuration was tested and refined. The results showed that if the pods were located low enough under the wing that the jet exhaust did not impinge on the trailing-edge flaps when fully deployed, and if the location were forward enough such that the exit of the engine tailpipe was forward of, or just in line with, the wing leading edge, then there was virtually no unfavorable aerodynamic interference between the pods and the wing—in the words of Bill Cook (Ref. 63), "The wing was performing like the pods were absent." In essence, the underslung jet pods could be designed with very low drag characteristics.

With the design features of the swept wing and the podded engines underneath the wings, Boeing was way ahead of any of its competitors. But there were other unqiue design problems to solve. Not desiring to retract the landing gear into the relatively thin, 12%-thick wings, and avoiding fuselage side bulges if a conventional tricycle arrangement were retracted into the side of the fuselage, Boeing engineers chose a bicycle landing gear which retracted directly into the bottom of the fuselage. However, this meant that the airplane could not rotate on takeoff, so a large incidence angle of the wing relative to the fuselage had to be adopted—8° between the wing chord and the horizontal ground—to allow enough lift to be generated for takeoff. And then there were flexure problems. The B-47 wing had a very high aspect ratio of 9.43; no other swept-wing airplane since has had such a high aspect ratio. Also, the fuselage was long and thin. Both the wings and the fuselage flexed during gust loads in flight. The effects of these flexures were not major problems, but they needed to be taken into account in the stability and control aspects of the airplane. In addition, the high sweepback of the wing, 35°, gave the B-47 a substantial degree of lateral stability (a high *effective* dihedral) and an unacceptable Dutch-roll characteristic. The Boeing engineers designed a full-time stability augmentation system to solve the Dutch-roll problem; it consisted of a rate gyro that generated corrective rudder deflection. It was the first use of a full-time stability augmentation system on a production airplane, and the same technique is still used today.

The first flight of the XB-47 took place on December 17, 1947 (44 years to the day after the first flight of the *Wright Flyer*). The Air Force had agreed to purchase two prototypes, but amazingly enough did not show great enthusiasm for the new bomber of revolutionary design. This was mainly due to the poor performance of the earlier straight-wing jet bombers, such as the North American B-45, which had soured Wright Field on the idea of jet bombers in general. Even top Boeing management was cautious about the XB-47, and the flight tests which took place at Moses Lake airfield, about 120 mi from Seattle, were initially carried out without fanfare. The exception was the small flight test crew at Moses Lake, who immediately witnessed the tremendous performance characteristics of the airplane. Indeed, the early tests quickly proved that the drag of the XB-47 was 15% less than the predicted value—a cause for great celebration, since this meant the range of the airplane was greater than expected, something of real importance for a bomber. The low drag results finally got the attention of the Boeing management in Seattle, and after that, interest in the airplane suddenly picked up within the company. This was followed by an event that was essentially a happenstance. Although the Air Force test pilots flying the XB-47 were almost finished with their test program, the Air Force was still not showing great interest; the Project Office at Wright Field had turned its attention to turboprop bombers, thinking that turboprops were the only engines that would give the necessary long range for bombers. General K. B. Wolfe, head of bomber production at Wright

Field, made a brief visit to Moses Lake on his way back to Dayton from a meeting with Boeing in Seattle on the design of a new piston engine bomber labeled the B-54. General Wolfe took a 20-min flight in the XB-47. He was so impressed with the airplane's performance that immediately after landing he declared that the Air Force would by it "as is." In the end, the Air Force bought 2,000 B-47s. That 20-min flight by General Wolfe revolutionized strategic bombing.

The performance capability that caused this revolution is summarized in Fig. 9.3. Because of its aerodynamically clean, thin, high-aspect-ratio wing, the L/D of the B-47 was higher than that of either the B-17 or the B-29 from World War II, as shown in Fig. 9.3**a**. Moreover, because of the highly swept wing, the severe drag-divergence effect was not encountered until the Mach number was greater than 0.8. The high L/D of the B-47 was necessary to counter the poorer propulsive efficiency of the jet compared to that of the piston-engine airplanes, in Fig. 9.3**b**. Recall Eq. (5.152) for range:

$$R = \frac{V_\infty}{c_t} \frac{L}{D} \ln \frac{W_0}{W_1} \qquad \textbf{[5.152]}$$

Equation (5.152) is a generic equation that applies to a jet or a propeller-driven airplane, as long as c_t is the *thrust* specific fuel consumption for both types. Also, from Eq. (3.43) relating the specific fuel consumption in terms of thrust c_t to the specific fuel consumption in terms of power c, we have

$$\frac{V_\infty}{c_t} = \frac{\eta_{\text{pr}}}{c} \qquad \textbf{[9.1]}$$

So Eq. (5.152) can also be written as Eq. (5.153):

$$R = \frac{\eta_{\text{pr}}}{c} \frac{L}{D} \ln \frac{W_0}{W_1} \qquad \textbf{[5.153]}$$

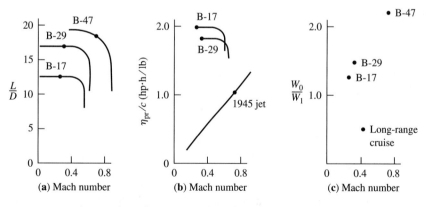

Figure 9.3 Three airplane characteristics that influence range: (**a**) lift-to-drag ratio; (**b**) propulsive efficiency; (**c**) ratio of gross weight to fuel-empty weight. Comparison between the B-17 and B-29 from World War II with the B-47 jet bomber.

From Eq. (9.1) the terms V_∞/c_t and η_{pr}/c are entirely equivalent; they are the same measure of propulsive efficiency. The dimensions are power multiplied by time per unit weight of fuel consumed, for example, hp·h/lb. The higher this value is, the more horsepower for a longer time is obtained from 1 lb of fuel. The units of η_{pr}/c on the ordinate of Fig. 9.3**b** are hp·h/lb. Note from this figure that the propulsive efficiency for the two famous Boeing propeller-driven bombers from World War II, the B-17 and the B-29, was on the order of 2 hp·h/lb at low Mach numbers, whereas the propulsive efficiency for a jet in 1945 was considerably smaller, on the order of 0.2 hp·h/lb, at the same low Mach numbers. However, note that the variations of η_{pr}/c and V_∞/c_t with V_∞ are totally different (Mach number rather than V_∞ is used as the abscissa in Fig. 9.3**b**, but it makes no difference in the variations shown). For the term η_{pr}/c, c is relatively constant with velocity (see Section 3.3.1), but the propeller efficiency dramatically drops at higher speeds due to compressibility effects (shock waves) at the propeller tips. For the cases of the B-17 and B-29 in Fig. 9.3**b**, η_{pr}/c is therefore relatively constant with M_∞ until these compressibility effects are encountered, beyond which η_{pr} (hence η_{pr}/c) plummets. On the other hand, consider the term V_∞/c_t for a jet. The value of c_t is approximately constant with velocity, although it may increase mildly with velocity as indicated in Section 3.4.1. However, the major velocity variation of V_∞/c_t is due to the presence of V_∞ in the numerator, and this is why a linear increase of propulsive efficiency with velocity is shown for a jet in Fig. 9.3**b**. At higher Mach numbers, this increase in the jet propulsive efficiency makes the jet airplane more viable, although the propulsive efficiency is still lower than that for the propeller-driven bombers at lower speeds. During the design of the B-47, the propulsive efficiency was estimated to be about 1.1 hp·h/lb, still only about one-half that achieved by the propeller-driven piston-engine airplanes. This deficiency in propulsive efficiency for the jet caused many people to feel at the time that a long-range strategic bomber would still have to be propeller-driven, with either piston engines or turboprops.

Faced with this reality, the Boeing design team faced a challenge; their response was to work hard on the other terms in the range equation. Given a relatively poor value of V_∞/c_t in Eq. (5.152), the only other possibilities to obtain a reasonable range are to increase both L/D and W_0/W_1 in Eq. (5.152). Hence this is the reason for the high-aspect-ratio wing of the B-47, shown in Fig. 9.3**a**. Also, the fuel capacity was made large, giving the B-47 a much larger W_0/W_1, as shown in Fig. 9.3**c**. With the larger L/D and W_0/W_1, the range of the B-47 was made even better than that of the B-17 and of the B-29; the data shown in Fig. 9.3 give the relative ranges of the B-47, B-29, and B-17 in the approximate ratio of 17 : 13 : 7. The design challenge had been met.

9.2.2   Design of the 707 Civil Jet Transport

The pioneering design technology that had been accrued during the B-47 project evolved into the first successful civil jet transport, the Boeing 707. These two airplanes are compared in three-view in Fig. 9.4. The design of the B-47 had been revolutionary;

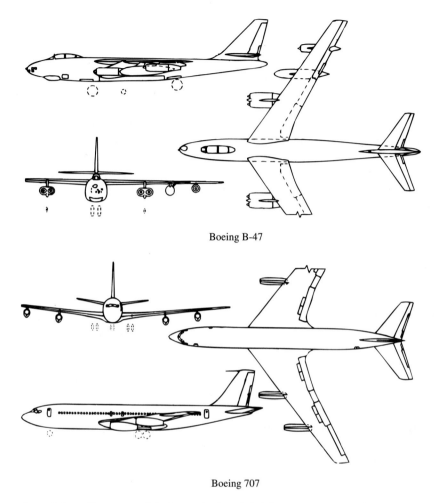

Boeing B-47

Boeing 707

Figure 9.4   Three-views of the Boeing B-47 and 707, for comparison.

the design of the 707 was evolutionary from the B-47. This is obvious from the comparison shown in Fig. 9.4. The dominant features of the 707—the 35° swept wing and the jet engines mounted in pods slung under the wing on struts—had been pioneered with the B-47. The major design changes reflected in the 707 were the low-wing configuration to allow a long body deck for carrying passengers or freight, and the use of a tricycle landing gear. The reason for the tricycle gear was purely and simply because airplane pilots were familiar with this type of landing gear; it allowed them to lift the nose at takeoff and depress the nose at landing.

At the time the Boeing 707 was designed, the only extant jet airliner was the British DeHavilland Comet; the Comet 4 is shown in three-view in Fig. 9.5. The Comet was a bold move on the part of the British; the first version, the Comet 1,

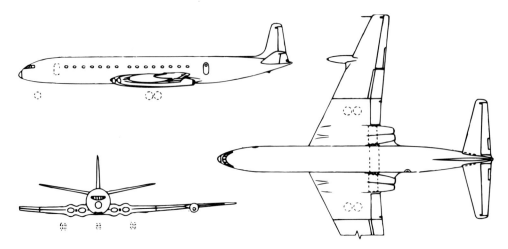

Figure 9.5   Three-view of the deHavilland Comet 4.

entered service with the British Overseas Airways Company (BOAC) on May 2, 1952. It was the first jet passenger service in the world. Passengers flocked to fly the Comet; the flights were smooth because it flew at high altitudes generally above the worst of the weather, and flight times between such distant cities as London and Johannesburg and London and Singapore were cut almost in half. However, in 1954 the Comets experienced two fatal accidents. On January 10 and April 8 of that year, two Comets virtually disintegrated at cruising altitude. The problem was structural failure near the corner of the nearly square windows on the fuselage, caused by repeated stress cycles during pressurization of the fuselage for each flight. Once a hole appeared in the fuselage, the pressurized vessel would explosively decompress, causing catastrophic failure of the airplane. The Comets were taken out of service. DeHavilland redesigned the Comet, and in 1958 the Comet 4 went into service. However, by then the British initiative in commercial jet transports had been lost forever. Only 74 Comet 4s were built.

The designs for the Comet and the 707 were quite different, as can be seen by comparing the three-views in Figs. 9.4 and 9.5. The Comet had a moderately swept wing of 20° but no sweep of the horizontal and vertical tails. The engines were buried in the root of the wing, which took up valuable internal wing volume that could have been used to store fuel. Also, with the engines buried in the wing, the wings had to be thick enough to accommodate the engines, hence reducing the critical Mach number of the wing. Hence, in no way was the 707 a derivative of the Comet—the design philosophies were quite different. So was the performance. The Comet cruised at Mach 0.74 at 35,000 ft, and the 707 cruised at Mach 0.87 at 30,000 ft. The Comet was a much smaller airplane; the gross weight of the Comet 1A was 115,000 lb with a wingspan of 115 ft, whereas the gross weight of the Boeing 707-320B was 336,000 lb with a wingspan of 146 ft. The Comet had a relatively short range of 1,750 mi, which required it to make intermediate stops for refueling. By comparison, the range of the Boeing 707-320B is 6,240 mi. The configuration of the Boeing 707 set a model for many subsequent jet transport designs; in contrast, the configuration of the Comet had virtually no impact on future design.

With the success of the B-47, and then later with the larger B-52, Boeing management knew it was in an advantageous position to produce the first jet airliner in the United States. However, the decision to go ahead with such a project did not come easily. The airlines were cautious, waiting to see how successful the Comet might be. Boeing felt that any initial orders for a new commercial jet transport would not be large enough to cover the development and tooling costs. This led to the idea for a military version to be used as a jet tanker for in-flight refueling, almost identical to the civil transport design. In this fashion, business from the Air Force would make up the start-up losses for the development of the airplane. But the Air Force dragged its heels on such an idea. Nevertheless, on April 22, 1952—the same year that the British Comet first went into airline service—Boeing management authorized the building of a prototype jet airliner. With that decision, the fortune and future destiny of Boeing Company were forever changed. A company that had produced mainly military airplanes for most of its existence was to become the world's leading manufacturer of civil jet transports in the last half of the twentieth century.

But in 1952 nobody knew this. The decision in 1952 was a bold one; the prototype jet transport was to be privately financed. The estimated cost of the prototype was $16 million. However, Boeing decided to use some of its independent research and development (IRD) funds, which came from prorated allotments taken from military contracts. These IRD funds were the government's way of providing some discretionary funding to companies to help them carry out their own research and advanced development. So the government would indirectly end up paying most of the cost of the prototype anyway; the direct cost to Boeing was estimated to be only $3 million. Boeing labeled the prototype with a company internal designation of 367-80; the airplane was quickly to be known as the "Dash-80."

The Dash-80 was powered by four Pratt & Whitney J-57 turbojet engines, which had proved to be very reliable on the B-52 bomber. The civil version of the J-57 was designated the JT3C; each engine produced 10,000 lb of thrust. Although the Dash-80 was an evolutionary derivative of the B-47, there were still some major design challenges. For one, the use of a tricycle landing gear in conjunction with a swept wing posed a problem: How would the main gear retract and be stowed in the swept wing? The structural design of the swept wing involved internal spars that were also swept, hence making it geometrically difficult for the main landing gear, which retracted in a line at right angles to the plane of symmetry, to be stowed in a convenient vacant space in the wing. The Boeing designers solved this problem by placing the main gear closer to the plane of symmetry and having the main gear retract into the bottom of the fuselage, as shown in Fig. 9.6.

Another challenge was flight control. A well-known aerodynamic characteristic of swept-back wings is that the backward sweep induces a spanwise component of flow over the wing which is toward the wing tip. Hence, the flow in the tip region tends to separate before that over other parts of the wing, with a consequent loss of control from ailerons placed near the tip. This problem had been noticed in both

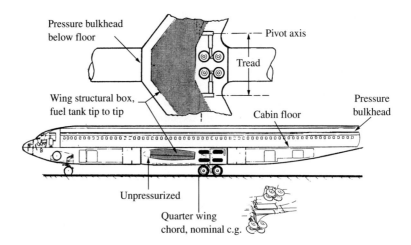

Figure 9.6   Landing gear retraction geometry for the Boeing 707. *(From Cook, Ref. 63, with permission.)*

the B-47 and the B-52, but was not dealt with in a totally acceptable way. However, what may have been acceptable in a military airplane was totally unacceptable for a civil transport. In order to provide proper and reliable lateral control, the Boeing aerodynamicists concentrated on that half of the wing closest to the root, where the spanwise induced flow was minimal and hence flow separation was not a problem. The control surfaces for the Dash-80 wing are shown in Fig. 9.7; shown here is Boeing's innovative solution for lateral control. Sandwiched between the inboard and outboard flaps was an inboard aileron positioned behind the jet exhaust from the inboard engine. Furthermore, two sets of spoilers, inboard and outboard, were positioned in front of the inboard and outboard flaps. Spoilers are essentially flat plates that deflect upward into the flow over the top surface of the wing, "spoiling" that flow and hence decreasing lift and increasing drag. The combination of the wing upper surface spoilers and the small aileron behind the inboard engine provided the necessary degree of lateral control at high speeds; the outboard aileron near the tip was locked in the neutral position except at low speeds with the flaps down, when the outboard aileron was reasonably effective. This lateral control arrangement proved to be quite successful on the Boeing 707.

From a philosophical point of view; the design of the Dash-80 followed in a general way the intellectual process described in Chapter 7, although the distinction between conceptual design and preliminary design was somewhat diffused. The Dash-80 evolved from the B-47, hence many of the fundamental configuration decisions that are usually made by numerous iterations in the conceptual design phase were not necessary; they were already in place for the Dash-80 (and hence the 707). Wind tunnel testing, which is usually brought in at the preliminary design phase to refine the configuration determined by conceptual design, played a strong role right from the start. The conceptual design of the B-47 itself was guided strongly by testing

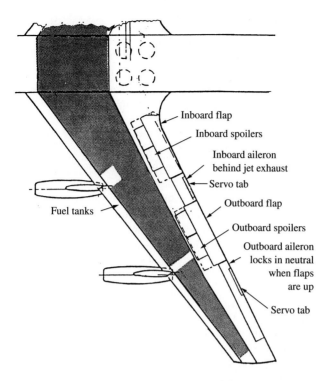

Inboard flap
Inboard spoilers
Inboard aileron behind jet exhaust
Servo tab
Outboard flap
Fuel tanks
Outboard spoilers
Outboard aileron locks in neutral when flaps are up
Servo tab

Figure 9.7 Trailing-edge flap, aileron, and spoiler locations for the Boeing 707. (*From Cook, Ref. 63, with permission.*)

different configurations in the tunnel. The swept-wing and podded engines were so new that even the basic conceptual design process for the B-47 needed data, and lots of them, from tunnel testing. Indeed, the Boeing high-speed tunnel was almost exclusively dedicated to the solution of problems with the B-47 during that airplane's design process. Nine months of intensive wind tunnel testing were necessary just during the conceptual design phase—many more tests followed in the preliminary and detailed design phases.

The first flight of the Dash-80 took place on July 15, 1954. Success followed success. The first production 707s were delivered to Pan American Airlines in September 1958. On October 26, the first jet service by a U.S. flag carrier was initiated when Pan Am flight 114 departed Idlewild Airport at 7:20 PM and landed at Paris's LeBourget about 9 h later, with an intermediate stop at Gander, Newfoundland, for refueling. (The early model 707 did not have quite the sufficient range, fully loaded, to make the trip from New York to Paris nonstop.) However, this was not the first transatlantic flight by a jet airliner. The British carrier BOAC beat Pan Am by a few weeks. On October 4, two redesigned deHavilland Comets, the Comet 4 (see Fig. 9.5), made simultaneous departures—one from Heathrow Airport in London, and the other from

Idlewild in New York—with full loads of passengers. Although these Comets crossed the Atlantic Ocean in both directions that day, the ultimate success belonged to the Boeing 707. Carrying 100 more passengers at 100 mi/h faster, the 707 outperformed the Comet 4, and it soon became the jet airliner of choice for airlines around the world. Also, the Air Force ordered a tanker version of the 707. Designated the KC-135, it became the natural airplane for air-to-air refueling of other jet airplanes. The first KC-135 was delivered to the Air Force in 1957.

The 707, in its earlier and later models, was a long-range airplane seating up to 189 passengers. It was a "hot" airplane in the sense that it had a high landing speed of 158 mi/h; it required a takeoff field length of 8,600 ft and a landing field length of 5,980 ft. Hence, only large airports could accept the 707; the vast number of regional airports with shorter runways were disenfranchised to jet transport aviation at the time. Recognizing this as a problem and an opportunity, in 1957 Boeing began the study of a smaller jet transport that could operate out of fields of 5,000 to 6,000 ft in length. Thus began the idea of the intermediate-range 727, the next Boeing success story.

9.2.3   Design of the Boeing 727 Jet Transport

As early as 1950, Boeing management had decided that the future of commercial transports lay in jets. With the subsequent success of the 707 jet airliner, Boeing embarked on a new jet transport design—the 727. This new airplane was part evolutionary and part revolutionary, as we shall see. The requirements for the 727 did not originate with the airlines. Rather, Boeing was astute enough to see a future need for a jet transport with the following characteristics:

1.   Short field capacity.
2.   Maximum passenger appeal.
3.   Low direct operating cost, which is enhanced by minimizing the time the airplane stays on the ground, maximizing the climb and descent rates, and having good reliability.
4.   Low community noise.
5.   All-weather operation.
6.   Operational flexibility and self-sufficiency.
7.   High profit potential.

The board of directors insisted that, before embarking on the project, the company have orders for such an airplane from at least two major airlines. So Boeing went to the airlines with a concept for a short- to medium-range airplane; ultimately United Airlines and Eastern Airlines showed some interest, albeit conflicting. United wanted a four-engine airplane because of its high-altitude operation at Denver; four engines would provide more safety in an engine-out situation. However, Eastern Airlines wanted a twin-engine airplane because it was more economical to operate. Boeing finally took the middle ground and chose to design a three-engine airplane; its studies had indicated that the increase in operating costs associated with three and four engines compared to twin engines was not linear; the increase associated with three engines was small compared to that for four engines. This nonlinear trend is shown in Fig. 9.8, based on the Boeing studies.

In terms of the design philosophy discussed in Chapter 7, the first step—setting the specifications—was taken by Boeing itself, based on the conviction that jet-powered transports were the wave of the future and that a short- to medium-range jet was going to be in demand as a complement to the long-range Boeing 707. Only after Boeing initiated discussions with the airlines did those companies have any input to the requirements that guided the design of the 727. The performance requirements for the initial version of the 727, labeled the 727-100, were set as follows:

1. Take off with full payload (100 passengers) from a 6,000-ft-long runway at sea level with an ambient air temperature of 90°F.

2. Cruise at 30,000 ft at Mach 0.80, with a range of 1,500 mi.

3. Land on a runway no longer than 4,900 ft.

4. Catering to the special requirements of United Airlines, carry a useful payload (75 passengers) from Denver (takeoff ambient temperature of 90°F) to Chicago.

5. Handle a 35-knot (kn) crosswind for takeoff. This was more severe than that for the larger 707, and reflected the need for uninterrupted service into smaller airports.

The second and third design pivot points, an estimate of the weight and design performance parameters, were dictated by the competing requirements for short-field takeoff and landing, and high-speed cruise. The former requirement calls for a small wing loading, and the latter for a large wing loading. However, recall that the stalling velocity (which dictates landing and takeoff speeds) is given by Eq. (5.67)

$$V_{stall} = \sqrt{\frac{2}{\rho} \frac{W}{S} \frac{1}{(C_L)_{max}}} \qquad \textbf{[5.67]}$$

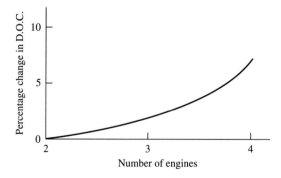

Figure 9.8 Results of a Boeing parametric study showing the percentage change in direct operating costs (D.O.C.) caused by using three and four engines, compared to two engines.

Hence, for a given stall velocity, the wing loading can be increased as long as $(C_L)_{max}$ is increased by the same factor. The Boeing designers took this tack. They designed for a large wing loading and then compensated by going to measures that could almost be called extreme to achieve a very high $(C_L)_{max}$. During the conceptual design, the maximum takeoff weight was taken as 142,000 lb, and the wing area was 1,650 ft², giving $W/S = 86.7$ lb/ft². (The weight of the 727 changed throughout the design and production process; the initial takeoff weight grew to 160,000 lb for the first production model.)

The extreme measures to achieve a high $(C_L)_{max}$ took the form of an advanced high-lift system, involving triple-slotted Fowler flaps at the wing trailing edge, and leading-edge flaps and slats. Boeing pioneered the triple-slotted Fowler flaps, driven by the design requirements for the 727. These flaps are shown in Fig. 9.9 in both their retracted (dashed lines) and fully extended positions. Figure 9.9**a** shows the inboard flaps, close to the fuselage, and Fig. 9.9**b** shows the outboard flaps. The position of these flaps along the span is illustrated in the wing planform views, also shown at the right in Fig. 9.9**a** and **b**. The triple-slotted Fowler flaps are essentially three separate flap surfaces which, when extended, greatly increase the effective wing area. The slots are gaps between each flap surface, allowing some of the higher-pressure air on the bottom surface to flow through the slots to the lower-pressure region on the top surface and delaying flow separation over the top of the flaps. The high-lift performance of various flap systems is shown in Fig. 9.10, which gives the variation of C_L versus wing angle of attack with flaps fully extended and deflected to 40°. Four configurations are compared; curves D and C are for single- and double-slotted flaps, respectively, and curves B and A are for triple-slotted flaps, with curve A for a slightly more extended geometry. Clearly the triple-slotted flaps (curves A and B) are superior to the single- and double-slotted flaps. Also, note that the data in Fig. 9.10 show the principal aspects of trailing-edge flaps in general, namely, to make the zero-lift angle of attack more negative (shift the lift curve to the left) and to increase

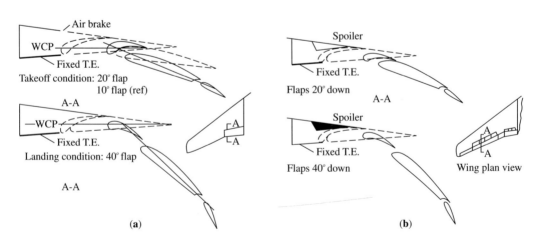

Figure 9.9   Flap movement and deflection for the Boeing 727: (**a**) inboard flap and (**b**) outboard flap.

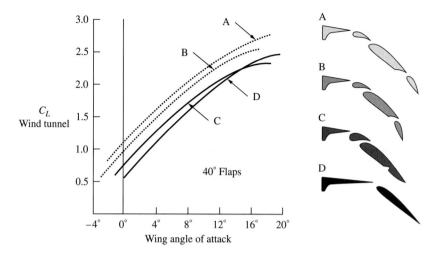

Figure 9.10 Boeing wind tunnel data for lift curves for four different trailing-edge flap configurations.

$(C_L)_{max}$. By choosing the triple-slotted flap, Boeing designers were reaching for the best high-lift performance from any trailing-edge device—better than any previously employed trailing-edge flap system. (The B-47 used a single-slotted flap; the 707, a double-slotted flap. The first production airplane to use a double-slotted flap was the Douglas A-26, produced at the end of World War II.)

However, more high-lift performance can be obtained by using leading-edge (LE) devices in conjunction with trailing-edge flaps. Boeing designers chose a combined leading-edge system of Kruger leading-edge flaps and leading-edge slats. The relative performance of these leading-edge mechanisms is shown in Fig. 9.11, compared to the alternative devices of a leading-edge droop and a simple leading-edge slot. Both the leading-edge slat and the Kruger leading-edge flap are superior to the other two, as seen in Fig. 9.11.

The generic wind tunnel results shown in Figs. 9.10 and 9.11 were obtained by Boeing during a series of intensive studies on high-lift devices aimed at finding the optimum high-lift combination for use on the 727. The final arrangement of these devices on the 727 wing is shown in Fig. 9.12, which gives a planform view showing the inboard and outboard trailing-edge flaps, the Kruger flaps on the inboard leading edge, and the slats on the outboard leading edge. Also shown in Fig. 9.12 is the location of the spoilers; the outboard spoilers are for lateral control at high speeds, and the inboard spoilers are for destroying the lift at touchdown upon landing. This arrangement of spoilers is a carryover from the 707 design. Recall that the spoilers can also be used during the landing approach to reduce L/D and hence steepen the glide slope of the approach path. The resulting variation of C_L with angle of attack for the 727 wing design is given in Fig. 9.13. This figure shows the relative role of trailing-edge and leading-edge devices for the 727 wing. The trailing-edge flaps

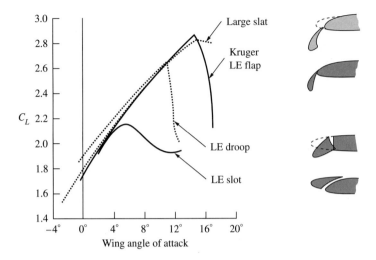

Figure 9.11   Boeing wind tunnel data for lift curves for four different leading-edge high-lift devices.

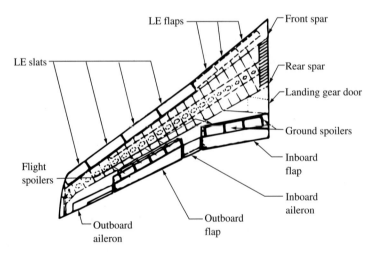

Figure 9.12   The wing configuration for the Boeing 727. (AIAA, with permission.)

serve to push the lift curve to the left (make the zero-lift angle of attack more negative) compared to the case with flaps up. The leading-edge devices serve to extend the lift curve to a higher $(C_L)_{max}$ than would be available with just the trailing-edge flaps. The design goal during conceptual design of the 727 was to achieve a $(C_L)_{max}$ of

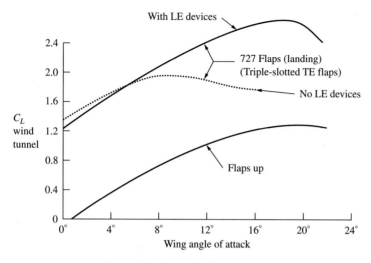

Figure 9.13 Wind tunnel data for wing lift curves for the Boeing 727, comparing the cases with and without the high-lift devices deployed.

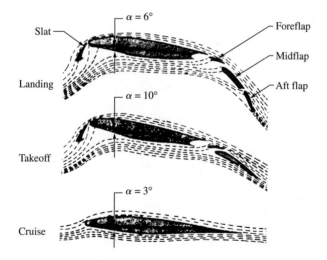

Figure 9.14 Streamline patterns over the Boeing 727 airfoil with and without high-lift devices deployed, comparing the cases for landing, takeoff, and cruise. (AIAA, with permission.)

2.9. The Boeing wind tunnel data given in Fig. 9.13 showed that the design goal was achievable with the combination of trailing-edge and leading-edge high-lift devices.

As a final note regarding the high-lift systems for the 727, Fig. 9.14 shows the deployed configurations for landing and takeoff, as well as the corresponding computed streamline patterns for each high-lift configuration and for the cruise configuration (no high-lift devices deployed). The angle of attack at cruise is 3°. The angle of attack for takeoff is 10°, with the trailing-edge flaps only partially deployed in order to obtain a reasonably high lift without too much of an increase in drag for takeoff. The angle of attack for landing is 6°, with the trailing-edge flaps fully deployed to obtain high lift and high drag.

Pivot point 4 in the design philosophy outlined in Fig. 7.3—the configuration layout—was in some respects an evolutionary process and in other respects a revolutionary process. The evolutionary process involved direct carryovers from the Boeing 707. For example, the 727 used the same cross section for the passenger compartment as the 707. This is shown in Fig. 9.15; the portion of the fuselage cross section above the floor of the passenger cabin is the same in the 707 and the 727. However, the portion of the fuselage cross section below the floor was a different shape for the two airplanes. Indeed, the 727 lower section shapes were different between that portion of the fuselage forward of the wing and that portion aft of the wing. Boeing designers felt that a short- to medium-range airplane required less baggage storage space than the longer-range 707, hence the fuselage cross section below the cabin floor was made smaller for the 727. Because the main wheels retracted into the fuselage in the same manner as on the 707, the fuselage bottom cross section for the 727 had to be expanded in the wing region to accommodate the wheels, hence the two different cross-section sizes forward and aft for the 727. Another evolutionary aspect was the amount of wing sweep. The 707 wing was swept at 35°, as were the earlier B-47 and B-52 designs. The Boeing aerodynamicists originally planned the same sweep angle for the 727; indeed, Boeing had amassed a large bulk of wind tunnel data for this sweep angle. However, United Airlines felt strongly that a smaller sweep angle was necessary to obtain the shorter field lengths required of the 727; they argued for a 30° sweep. Finally, Boeing designers compromised by splitting the difference; the 727 wing shown in Fig. 9.12 has a sweep angle of 32.5°.

The major revolutionary design challenges for the 727 were associated with the placement of the three engines and with the high-lift system. The latter has already been discussed. The use of three engines dictated that at least one engine be located in

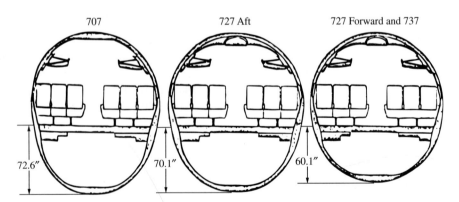

Figure 9.15   Comparison of the passenger cabin cross sections of the Boeing 707, 727, and 737. (AIAA, with permission.)

the symmetry plane of the aircraft. It made no sense to strut-mount this engine above or below the fuselage; instead, Boeing designers chose to bury the engine in the rear of the fuselage, with an inlet for the engine located at the root of the vertical tail. The air was ducted from the inlet to the entrance of the engine through a novel S-shaped duct sketched in Fig. 9.16. In regard to the placement of the other two engines, Boeing designers went through two major studies, one with the engines strut-mounted below the wing, in the time-honored style pioneered by Boeing, and the other with the engines mounted on the rear sides of the fuselage. The former configuration is shown in Fig. 9.17 and the latter in Fig. 9.18. The decision was not easy. Boeing set up two separate competitive design teams, one to study and optimize each configuration shown in Figs. 9.17 and 9.18. At the end of protracted and deep arguments, it was decided that there were advantages and disadvantages to both configurations, and that although there was no clearly decisive aspect, the fuselage-mounted engine configuration shown in Fig. 9.18 was finally chosen. A disadvantage of this configuration was that it was more cumbersome to load. However, wind tunnel tests indicated a slight drag reduction for the fuselage-mounted aft engine configuration. Also, the aft engine configuration appeared to be slightly cheaper to manufacture because the auxiliary systems were closer together. Besides, the engine noise in the front half of the passenger cabin was greatly reduced when the engines were aft-mounted. However, none of these considerations were compelling. Nevertheless, the final choice was the fuselage-mounted aft engine design shown in Fig. 9.18.

In a further departure from previous Boeing practice, the horizontal tail was mounted on top of the vertical tail—the T tail configuration shown in Fig. 9.18. With the engines mounted in the rear of the fuselage, especially the third engine buried in the back of the fuselage, the T tail was a good choice aerodynamically, albeit requiring a stronger, hence heavier, structure for the tail.

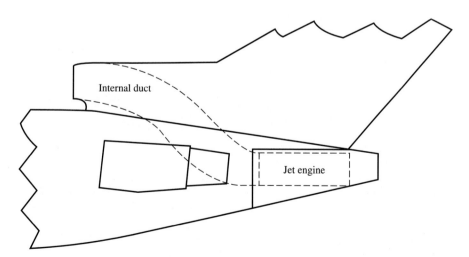

Figure 9.16   The S-shaped duct from the tail inlet to the engine is buried in the end of the fuselage for the Boeing 727.

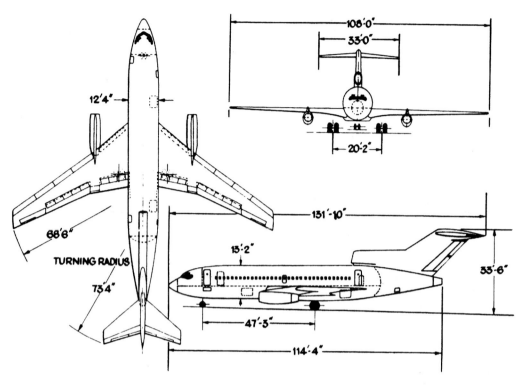

Figure 9.17   The alternate, wing-mounted engine arrangement studied during the Boeing 727 design process. (AIAA, with permission.)

After the configuration layout was determined, performance predictions were made, based on aerodynamic data from a substantial number of wind tunnel tests. For the most part, the predictions were conservative, because in all respects the actual flight performance of the airplane exceeded that predicted. At the source was the drag, which in flight tests proved to be from 4% to 7% less than predicted, and the engine specific fuel consumption, which was 2% less than predicted. Everything was going in the right direction for the 727 performance. The predicted variation of $(L/D)_{max}$ with M_∞ for the 727 is shown in Fig. 9.19, along with that for the Boeing 720, a smaller version of the 707. At $M_\infty = 0.8$, the predicted $(L/D)_{max}$ was 19, a very reasonable value. The range performance of the 727 is given in Fig. 9.20, where the dashed curves are the predicted values and the solid curves are the flight test data. The actual range is clearly better than the predicted values. In regard to the high-lift system, Fig. 9.21 gives the stall lift coefficient versus trailing-edge flap angle; again, the flight data were better than the predicted values of $(C_L)_{max}$. Note that the design value of $(C_L)_{max} = 2.9$ was exceeded in actual flight. Also, the fact that the 727 contained the most powerful high-lift system of any previous Boeing aircraft is clearly seen by the comparison shown at the right side of Fig. 9.21, where $(C_L)_{max}$ is given by the dots for other Boeing airplanes. The landing performance is given in Fig. 9.22; again, the flight data used for certification were better than those predicted

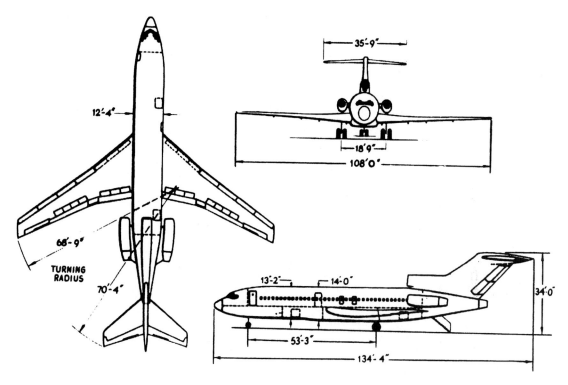

Figure 9.18 The fuselage-mounted engine arrangement studied during the Boeing 727 design process; the designer's choice. (AIAA, with permission.)

during the design study. Finally, the lower drag of the actual airplane compared to the wind tunnel data used for predictions can be seen in the drag polars for the 727, plotted in Fig. 9.23. A number of drag polars, each for a different flap deflection, are shown in Fig. 9.23. For any given flap deflection, at a fixed C_L, the value of C_D is smaller for the flight data (solid curves) than for the wind tunnel data (dashed curves). In some cases the wind tunnel data overpredict the drag by more than 25%.

In the design of the Boeing 727, Boeing engineers followed the general phases described in Section 7.2, namely, conceptual design, preliminary design, and detail design. The actual design schedule for these phases is shown in Fig. 9.24, as described in Ref. 64 by Fred Maxam, the Boeing chief project engineer on the 727. Note that after the general conceptual design was completed, wherein the overall configuration as shown in Fig. 9.18 was determined, the preliminary design phase took less than a year. In October 1960, the company made the formal go-ahead decision based on the results of the preliminary design phase, and a commitment was made to produce the 727. What followed was a protracted detailed design phase (called design development by Boeing), leading to the first flight on February 9, 1963, with FAA certification awarded on December 24, 1963. During the entire design process, Boeing carried out over 5,000 h of wind tunnel testing, 1,500 h of which occurred during conceptual and preliminary design in order to predict the performance of the airplane.

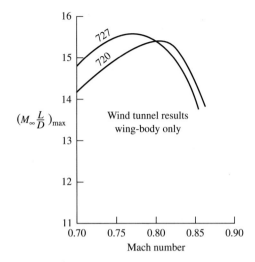

Figure 9.19 Variation of the maximum of the product of the free-stream Mach number and lift-to-drag ratio versus free-stream Mach number. Comparison between the Boeing 727 and the earlier 720.

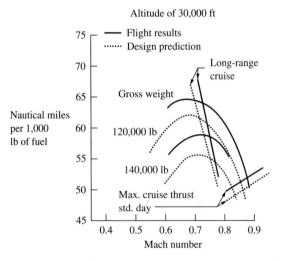

Figure 9.20 Range performance for the Boeing 727; comparison between actual flight results and the predictions made during the design process.

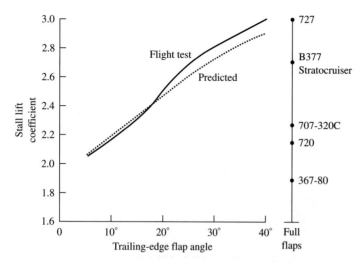

Figure 9.21 Variation of the stall lift coefficient $(C_L)_{max}$ with the tailing-edge flap deflection angle for the Boeing 727. Comparison between actual flight results and the design predictions. Also comparisons with other airplanes with full-flap deflection are made along the vertical line at the right.

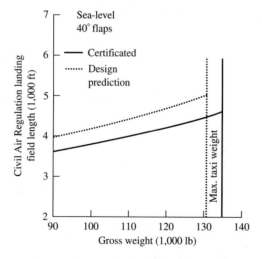

Figure 9.22 Landing performance for the Boeing 727. Comparison of the actual certificated flight results with the design prediction.

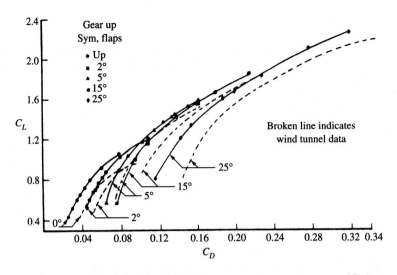

Figure 9.23 Drag polar for the Boeing 727. Comparison between actual flight results and wind tunnel data for different flap deflections.

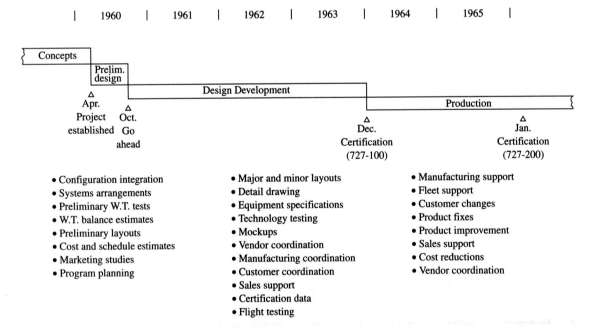

Figure 9.24 Phases of the design development of the Boeing 727.

9.2.4 Interim Summary

The three airplanes considered in this section are good examples of subsonic/transonic jet aircraft. The following is a summary of the design and performance characteristics of these airplanes.

	B-47	707	727
S (ft^2)	1,428	2,400	1,560
b (ft)	116	130	108
AR	9.43	7.04	7.5
W_0 (lb)	206,700	257,000	160,000
V_∞ (mi/h)	606	623	632
Range (mi)	4,100	4,650	2,690
Takeoff distance to clear 35-ft obstacle (ft)	—	10,550	7,800
Landing distance from 50-ft height (ft)	—	6,320	4,910
Service ceiling (ft)	40,500	31,500	36,500
R/C at sea level (ft/min)	—	1,400	2,900
V_{stall} (flaps down) (mi/h)	—	128	114
W/S (lb/ft^2)	145	107	103
T/W	0.21	0.21	0.26

One of the most important intellectual results obtained from the design of any airplane is usually a list of "lessons learned"—experience that can be applied to future designs. John Steiner, Boeing vice president, summarizes in Refs. 64 and 65 the lessons learned by Boeing during the course of its design of the early jet transports. In terms of the market for such airplanes, he itemizes:

1. The most important single element of success is to listen closely to what the customer perceives as her or his requirements and to have the will and ability to be responsive.

2. An airline airplane is only as good as its economics—its cost per seat-mile, its appeal to its passengers and the resultant load factor benefit, and its return on investment to its airline owner.

These lessons learned are not necessarily technically based, but they are important to the design of a successful jet transport, and hence are important to highlight in this book. On a more technical basis, Steiner continued with his lessons learned as follows:

3. Small elements of technical superiority are important, if, in aggregate, they add up to winning (versus losing) a sale—as they frequently will.

4. Technical superiority need not mean higher cost. In fact, it may mean just the opposite—lower cost.

5. As much effort is necessary in achieving lower-cost manufacturing as in achieving a superior technical design. A very close working relationship between engineering and manufacturing is essential.

Finally, Steiner addressed the aspects of goals.

6. A winning program will have technical and economic goals barely within reach and requiring engineering creativity to achieve.

7. A substantial continuous product improvement program providing for improved airplane reliability, flight characteristics, performance, operating economics, noise characteristics, and similar benefits is essential to achieving long-term program success.

8. Significant, or even major, derivatives must be continuously considered from program initiation, and a constant customer dialog maintained, not avoided.

In our discussion we have seen the importance of a good high-lift system for subsonic jet airplanes. A summary of the progress in high-lift systems is given in Fig. 9.25, which shows the geometry of different systems, along with $(C_L)_{\max}$ for the landing configuration and the value of L/D in the takeoff configuration. An interesting historical trend is seen here. Starting with the B-47 with its relatively simple single-slotted Fowler flap, the technical sophistication and mechanical complexity increased with each succeeding design, culminating in the triple-slotted trailing-edge flaps and leading-edge slats and flaps for the 727 and 747. However, the high-lift system for the more recent Boeing 767 steps back to a single-slotted trailing-edge flap along with leading-edge slats. Here, Boeing is stepping back to a simpler (hence

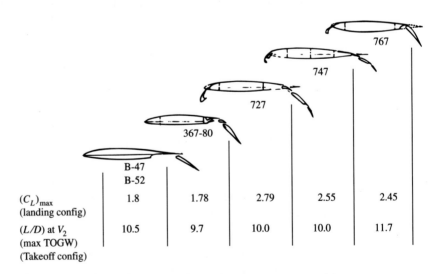

	B-47 B-52	367-80	727	747	767
$(C_L)_{\max}$ (landing config)	1.8	1.78	2.79	2.55	2.45
(L/D) at V_2 (max TOGW) (Takeoff config)	10.5	9.7	10.0	10.0	11.7

Figure 9.25   The evolution of wing high-lift systems. Comparison of the geometry, $(C_L)_{\max}$, and L/D at takeoff for five different airplanes from the Boeing B-47 to the 767.

less costly) mechanical high-lift system, following lesson learned number 5 listed earlier. In the process, Boeing is accepting a lower wing loading for the 767 to partly compensate for the lower $(C_L)_{max}$. Based on maximum takeoff weight, $W/S = 145$ lb/ft^2 for the 747 compared to $W/S = 131$ lb/ft^2 for the newer 767. This trend is continuing with Boeing's newest jet transport, the 777, which also uses a simpler single-slotted trailing-edge flap and has an even lower wing loading of 126 lb/ft^2. This trend is being driven by cost considerations.

Finally, the evolution of wing design for subsonic jet bombers and transports is shown in Fig. 9.26. Note the similarity in sweep angles—all in the 32° to 37° range. Also note the higher aspect ratios for the jet bombers compared to the civil transports shown in Fig. 9.26; the civil transports have aspect ratios on the order of 7.0 to 7.5, except for the more recent 767 which has a somewhat higher aspect ratio of 7.9. In the quest for improved aerodynamic efficiency, Boeing designers have gone to higher aspect ratios for their more recent designs. For example, the 777 has an aspect ratio of 8.68, getting closer to the high aspect ratio of the B-47, which started the entire line of Boeing jet airplanes in the first place. Somehow, this is a fitting end to this section, which started with a discussion of the B-47.

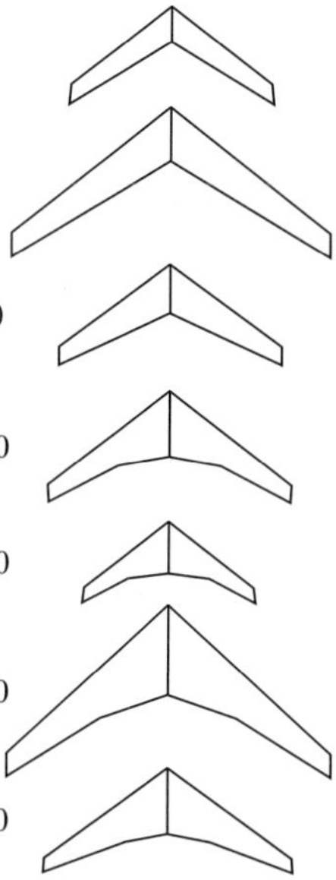

	Area (ft^2)	Aspect ratio	Sweep (c/4)
B-47	1,428	9.43	35°
B-52	4,000	8.55	35°
367-80	2,400	7.0	35°
707-320	2,892	7.35	35°
727-200	1,560	7.5	32°
747-200	5,550	7.0	37.5°
767-200	3,050	7.9	31.5°

Figure 9.26 The evolution of wing planform design, from the Boeing B-47 to the 767.

9.3   SUBSONIC JET AIRPLANE DESIGN: ADDITIONAL CONSIDERATIONS

The design philosophy set forth in Chapter 7 calls for an almost immediate first estimate of the gross weight of the airplane, as noted in pivot point 2 in Fig. 7.3. In Chapter 8 we illustrated how this estimation could be made for a propeller-driven airplane; the procedure is no different for a jet-propelled airplane. However, the database given in Fig. 8.1 used for the estimate of W_e/W_0 is exclusively for propeller-driven airplanes. A similar database for subsonic jet airplanes is given in Fig. 9.27. Unlike the data in Fig. 8.1, which allowed us to make a choice of $W_e/W_0 = 0.62$, independent of the value of W_0, the data in Fig. 9.27 show a decreasing trend for W_e/W_0 as W_0 becomes larger. For lighter jet airplanes with gross weights on the order of 10,000 to 20,000 lb, W_e/W_0 is on the order of 0.6, whereas for heavy jet transports and bombers, W_e/W_0 is more on the order of 0.45. Of course, there is some scatter in the data shown in Fig. 9.27. The dashed line drawn through this scatter in Fig. 9.27 can be used for a first estimate of W_e/W_0. Note that the dashed line is not horizontal, as was the case in Fig. 8.1. The estimation of W_0 for the jet airplane can be carried out using the same approach as discussed in Section 8.3. However, because W_e/W_0 is a function of W_0, the estimation of W_0 requires an iterative approach, as noted, but not executed, in Section 8.3.

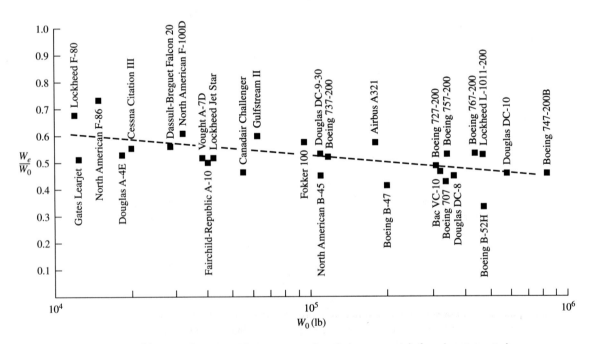

Figure 9.27   Variation of the ratio of empty weight to gross weight with the gross weight for subsonic jet airplanes.

Another consideration for jet airplane design is the integration of the airframe with the engines—the problem of *airframe-propulsion integration.* For conventional propeller-driven airplanes, airframe-propulsion integration is not a driving aspect of airplane design. The propeller wash from tractor-configuration propellers flowing downstream over the fuselage and/or wings is a consideration in the calculation of skin friction over these surfaces, but it is hardly a driver that determines the location, shape, and orientation of the propellers relative to the airframe; nor is the design of the airframe shape influenced by the propellers. To some extent, the location of engine nacelles on the wing—how far forward of the leading edge the front of the nacelle is located, and how centered the nacelle is in the vertical direction relative to the airfoil section—has an influence on the nacelle drag, as studied in the 1930s. However, for the design of propeller-driven airplanes, the engines and the airframe are usually treated as distinct entities; there is little reason to be concerned with airframe-propulsion integration in the true sense of that term.

For jet airplanes, airframe-propulsion integration becomes a more serious design consideration, mildly important for subsonic jets, important for supersonic jets, and essential for future hypersonic airplanes. Since this section deals with subsonic jets, we will limit our comments here to such airplanes.

For subsonic jet airplanes, the engine and airframe can still be treated as somewhat distinct entities. However, the following aspects should be considered in the conceptual design process.

If the jet engine is buried inside the fuselage, care should be taken to provide good-quality flow into the inlet. Good-quality flow means flow that has relatively uniform properties entering the inlet with as high a total pressure as possible. Boundary layer flow by this standard is low-quality flow; the velocity profiles are highly nonuniform, and the viscous shear stresses decrease the total pressure. Hence it is good practice not to place the inlet at a location where it will ingest a sizable boundary layer coming from another part of the airplane. Generally, two types of inlet configurations have been used for fuselage design. One is the simple nose inlet, such as used on the Republic F-84, shown in Fig. 9.28. Here the inlet is as far forward as it can be. It essentially ingests the free-stream flow, which is of high quality. However, this flow must pass through a relatively long duct through the fuselage to get to the engine mounted at the center rear of the airplane, with consequent frictional losses and hence losses in total pressure. To decrease these internal flow losses, the duct to the engine can be made shorter by putting the inlets farther back on the fuselage, one on each side of the airplane. Side-mounted inlets were used for the Lockheed F-80, the first U.S. fully operational jet fighter, shown in Fig. 9.29. Although such side-mounted inlets decrease the internal duct length to the engine, the boundary layer that builds up along the fuselage, if ingested by the inlet, promotes poor-quality flow into the engine duct. However, this can be mitigated by mounting the inlet lip slightly away from the fuselage surface so that the boundary layer passes between the fuselage and the inlet. A disadvantage of side inlets that provide flow to a single engine is that the flow path is split between the two inlets, and pressure instabilities may arise that cause the engine to stall. There are other considerations associated with the internal ducts, such as their weight and the volume they occupy inside the fuselage. So the

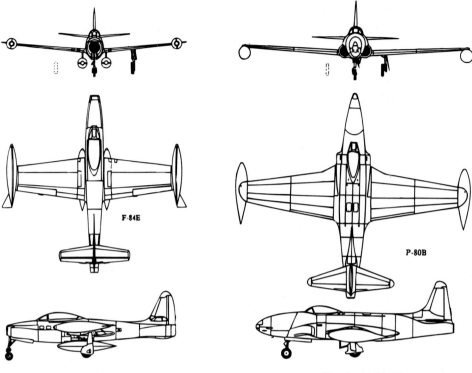

Figure 9.28   Republic F-84.   **Figure 9.29**   Lockheed P-80 (F-80).

choice between side inlets and nose inlets is not entirely clear-cut. As usual in the design process, compromises occur and decisions have to be made based on such compromises.

Recall that Boeing designed the 727 fuselage inlet at the rear of the fuselage, mounted on the top just ahead of the vertical stabilizer. The inlet was connected to the engine via an S-shaped duct, shown in Fig. 9.16. The proper design of an S-shaped duct is a challenge in aerodynamics, so as to avoid flow separation and consequent total pressure losses and nonuniform flow going into the engine.

For multiengine subsonic jet planes, the airframe-propulsion integration problem is usually treated in one of two ways. One is to bury the engines in the wing root region. This approach was followed particularly by the British, and the Comet airliner (Fig. 9.5) is a good example. An advantage of this approach is that the total wetted surface area of the airplane can be reduced compared to installations that require pods, struts, or any other type of separate inlet cowl. However, a disadvantage is that the wing must be thicker to accommodate the engines, hence causing a lower critical Mach number. Also, valuable space inside the wing is taken up that could otherwise be devoted to fuel tanks. The second installation is the pod configuration, already described in Section 9.2. We have already discussed how Boeing engineers learned to locate the pods relative to the wing such that there was virtually no aerodynamic interference between the two. The podded engines have very short inlet ducts, and the inlets are easily placed in regions of almost uniform flow.

9.4 SUPERSONIC AIRPLANE DESIGN

The physics of supersonic flow is completely different from that of subsonic flow—about as great as the difference between night and day. This causes many of the details of supersonic airplane design to be different from those for subsonic airplane design. However, the design *philosophy* as discussed in Chapter 7 is essentially the same, as we will see.

Almost all the supersonic airplanes designed to date are military airplanes; the single exception is the Concorde supersonic transport, designed during the 1960s and still in commercial service. Because of the strong military flavor on supersonic airplane design, in the following sections we highlight the design case histories of three military airplanes. The first is the General Dynamics (now Lockheed-Martin) F-16 lightweight fighter. This is followed by the Lockheed SR-71 Blackbird reconnaissance airplane. Finally, we examine the design of the most advanced fighter (at this date of writing), the Lockheed-Martin F-22.

9.4.1 Design of the F-16

The cost of military airplanes, from initial conceptual design to rolling off the production line, has increased almost exponentially during the last half of the twentieth century. (Some tongue-in-cheek extrapolations have shown that the price tag on a new airplane by the year 2020 would take the entire budget of the Defense Department of the United States.) Concerned about this trend, the Air Force initiated a project in the 1960s to design a comparatively low-cost, lightweight fighter. The requirements for the design (pivot point 1 in Fig. 7.3) were rather broadly stated, and are summarized by Buckner et al. (Ref. 66) as follows:

> The intent of the contract was to demonstrate the feasibility of a highly maneuverable, lightweight fighter aircraft through a prototype design, fabrication, and flight test program. The design objective was to maximize the usable maneuverability and agility of the aircraft in the air combat arena within the constraint that system cost, complexity, and utility are prudently considered and balanced. Emphasis was to be placed on small size and low weight/cost design techniques. The performance goal was to provide maximum maneuvering capability in the 0.8–1.6 Mach combat arena.

General Dynamics was one of the companies responding to this request for proposal from the Air Force; its design was labeled the YF-16. Ultimately, two prototypes were produced by General Dynamics, and in January 1975, the design was selected for full-scale development and production as the F-16. A three-view of the F-16 is shown in Fig. 9.30.

After the requirements for the airplane are established, the next step is the initial weight estimate (pivot point 2 in Fig. 7.3). As usual, historical data are very useful for this weight estimate. Such data for supersonic airplanes are shown in Fig. 9.31, which is the usual plot of W_e/W_0 versus W_0. As in the case of the subsonic jet

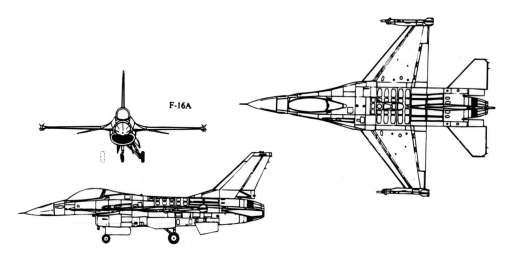

Figure 9.30   General Dynamics (now Lockheed-Martin) F-16.

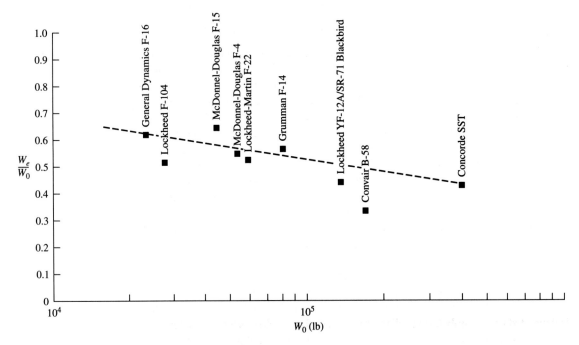

Figure 9.31   Variation of the ratio of empty weight to gross weight with the gross weight for supersonic jet airplanes.

airplane shown in Fig. 9.27, we see for supersonic jets a trend of decreasing W_e/W_0 as W_0 increases. The dashed line in Fig. 9.31 is faired through the data; of course, the usual scatter exists in the data. The YF-16 designers had these data available; most of the airplanes noted in Fig. 9.31 had been designed before the start of the YF-16, with the exception, of course, of the F-16 data point at the left. It is interesting to note that the resulting W_e/W_0 for the final F-16 design falls right on the dashed curve. Also, the General Dynamics designers accomplished a relatively lightweight design as specified; the F-16 is the most lightweight supersonic airplane shown in Fig. 9.31. The estimation of W_0 for the start of the conceptual design process follows that discussed in Section 8.3, recognizing that, because W_e/W_0 is a function of W_0, an iterative process is required to obtain W_0.

The particular emphasis on light weight and small size for the F-16 design was driven by the following considerations:

1. The smaller (size) and lighter (weight) for a fixed engine size (fixed thrust), the greater the maneuverability of the airplane. This can be seen in the discussion and equations in Section 6.2, where the turn radius is shown to decrease and the turn rate is shown to increase when the load factor is increased. For a steady, level turn, the maximum allowable load factor expressed by Eq. (6.18) increases with an increase in $(T/W)_{max}$, which is made larger by making the weight smaller. Also, in terms of energy considerations for accelerated performance, by reducing the size and weight, the specific excess power, given by Eq. (6.59), is increased.

2. The smaller the size, the smaller the total cost of producing the airplane.

3. The smaller the size, the smaller the radar cross section for detection.

The third item, dealing with reducing radar detectability, was becoming an increasingly important aspect for military airplane design in the 1960s. The aspect of stealth was beginning to take hold. We will see more of this in the ensuing sections.

Analytical studies supporting the conceptual design of the F-16 were carried out during the period beginning in late 1970, and into 1971. According to Buckner et al. (Ref. 66): "For the first time a realistic combat task was defined as part of the mission rules allowing combat performance to be a primary force equal to cruise range in the configuration design optimization process." In other words, the critical performance parameters representing pivot point 3 in Fig. 7.3 were dictated primarily by the accelerated performance criteria of maximum turn rate, minimum turn radius, and specific excess power rather than more conventional criteria such as takeoff length, range, etc. In particular, the wing shape and size were dictated by weight and accelerated performance. The wing shape and size are defined by wing loading W/S, aspect ratio AR, sweep angle Λ, taper ratio λ, and airfoil shape and thickness-to-chord ratio t/c.

Results from a parametric study using these wing characteristics are shown in Figs. 9.32 and 9.33. In both these figures, a baseline wing is defined with $W/S = 60$ lb/ft^2, AR $= 3$, $\Lambda = 35°$, $\lambda = 0.2$, and $t/c = 0.04$. In Fig. 9.32, the weight of the airplane at start of combat, normalized by that for the baseline wing, is plotted as a function of the five wing parameters. When these curves dip below unity, the weight at start of combat is less than the baseline value; and when the curves exceed unity,

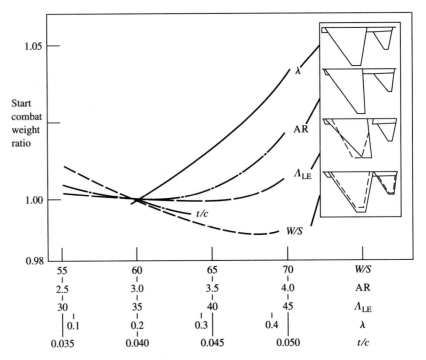

Figure 9.32   Resutls of a parametric study during the F-16 design process. Start combat weight ratio versus wing loading, aspect ratio, wing sweep angle, taper ratio, and thickness-to-chord ratio.

the weight at start of combat is greater than the baseline value. Everything else being equal, a start combat weight ratio less than unity represents an improvement over the baseline value. Figure 9.33 is in the same vein; here the time to accelerate through a given velocity increment and the turn rate, both normalized by the baseline, are plotted as a function of the five wing parameters. The turn rate is shown for $M_\infty = 0.8$ and 1.2. In Fig. 9.33, when the acceleration time ratio is less than unity and the turn rate ratio is greater than unity, the performance is better than the baseline.

For pivot point 4 in our design philosophy—the configuration layout—the wing shape and size for the F-16 were directly influenced by the previous parametric studies. Examining Figs. 9.32 and 9.33, the taper ratio λ should be as small as practical, limited by reasonable structural strength at the tip and early tip stall. The F-16 designers choose $\lambda = 0.227$. The baseline aspect ratio of 3 was chosen, since it minimized the start combat weight ratio and the acceleration time ratio. Increasing the wing sweep was favorable, especially for increasing the turn rate at supersonic speeds; clearly the reduction of supersonic wave drag by increasing the sweep angle enhances accelerated performance. However, from Fig. 9.32, if Λ is made larger than about 43°, the start combat weight ratio increases. For the F-16, a sweep angle of 40° was chosen. In regard to the wing thickness ratio, an increased t/c results in a more lightweight

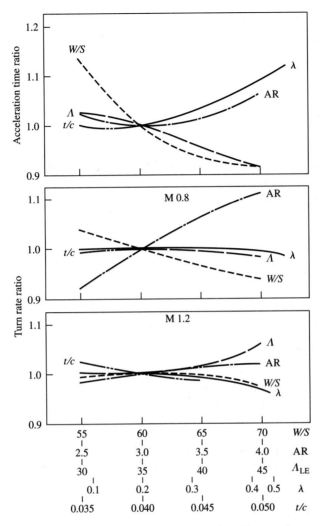

Figure 9.33 Results of a parametric study during the F-16 design process. Acceleration time ratio and turn rate ratio versus the same parameters as in Fig. 9.32.

airplane (the wing structural design is easier for thicker wings and results in a lighter wing). However, supersonic turn rate is improved with a smaller t/c. The balance between subsonic and supersonic maneuverability, consistent with flutter and aileron reversal considerations, resulted in a choice of $t/c = 0.04$. Finally, increasing the wing loading resulted in a lighter airplane with larger acceleration, but decreasing the wing loading increased the turn rate. For minimizing the airplane weight, an optimum of $W/S = 68$ lb/ft^2 is indicated in Fig. 9.32. However, by choosing a lower value of $W/S = 60$ lb/ft^2, only a 1% increase in combat weight was incurred while obtaining a beneficial 4% increase in subsonic turn rate. Such is the essence of the design compromise. The conceptual design of the F-16 was carried out for $W/S = 60 \, \text{lb/ft}^2$.

The airfoil section for the F-16 was chosen after a series of wind tunnel tests were carried out using a supercritical airfoil, a symmetric biconvex shape, and an NACA 64A204 airfoil. A wing using the NACA 64A204 airfoil with a leading-edge flap resulted in the best drag polar. Even though the NACA airfoil has a blunt nose, the 40° wing sweep ensures a subsonic leading edge. Although the supercritical wing provided a 5% gain in mission radius and a 13% increase in subsonic turn rate, at supersonic speeds it decreased the turn rate by 3% and decreased the supersonic acceleration by a drastic 69%. The NACA 64A204 airfoil was chosen for the F-16 design.

Continuing with the configuration layout, the General Dynamics designers examined two classes of configuration: the conventional wing-body arrangement and a *blended* wing-body. These two concepts are illustrated in Fig. 9.34. The blended wing-body configuration provided two important advantages. It was relatively natural to include forebody strakes in such a blended configuration, and the area ruling was more easily carried out. The forebody strakes are clearly seen in Fig. 9.30; they are essentially long, forward extensions of the wing leading edge in the region near the wing root. The strakes tend to promote symmetric vortex shedding from the forebody, improving directional stability and increasing the forebody lift. The normal cross-sectional area distribution of the F-16 is shown in Fig. 9.35. This shows the relative area contributions of different parts of the airplane and indicates a rather smooth total cross-sectional area distribution due to the blending of the wing and body. The area rule for minimizing transonic drag, as discussed in Section 2.8.3, calls for a smooth variation of the normal cross-sectional area of the airplane. The area rule also applies at supersonic speeds, but here the relevant cross-sectional area is not that perpendicular to the free-stream relative wind, but rather the area section cut by an oblique plane at the free-stream Mach angle. For the F-16, the distribution of this oblique area with distance along the fuselage is shown in Fig. 9.36 for both $M_\infty = 1.2$ and 1.6. Note the smoothness of these area distributions.

In regard to airframe-propulsion integration, the F-16 designers made a choice based on simplicity—a simple normal shock inlet. However, as seen in Fig. 9.30,

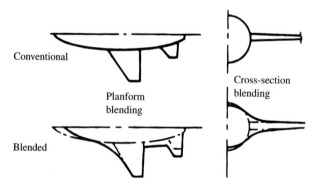

Figure 9.34   Schematic showing a conventional configuration with a blended wing-body configuration.

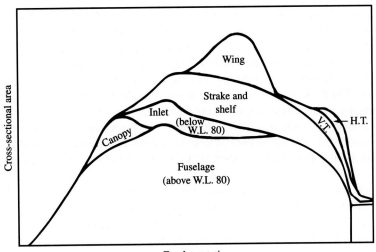

Figure 9.35 Transonic area ruling for the F-16. Variation of normal cross-sectional area as a function of location along the fuselage axis.

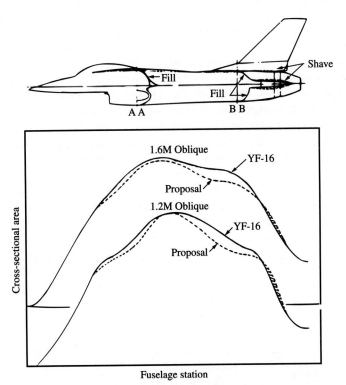

Figure 9.36 Supersonic area ruling for the F-16. Variation of *oblique* cross-sectional area as a function of location along the fuselage axis. Comparison between the actual area distribution and that proposed in an early design study.

the inlet is not at the nose of the airplane; rather, it is strategically placed underneath the fuselage more than one-quarter of the fuselage length downstream of the nose. Wind tunnel tests indicated that with the inlet underneath the fuselage, the fuselage provided a shielding effect for the inlet which was beneficial at the high angles of attack that would be encountered during dogfighting. The rearward placement of the inlet was to allow as short a duct to the engine as possible, thus saving both duct and fuselage weight (a savings of 1 lb was obtained per linear inch of duct length reduced). There was also a synergistic effect. With the duct in a more forward position, there was an increased directional destabilization which would have required a large (hence heavier) vertical tail. With the more rearward inlet location, there was a reduction of the destabilization effect, and the vertical tail was made smaller, hence saving additional weight. In the design of the F-16, the duct length was ultimately made the absolute minimum consistent with quality flow into the engine.

On April 14, 1972, the Air Force awarded contracts to General Dynamics and Northrop to build two prototypes each of a lightweight fighter; the YF-16 was the General Dynamics design, and the Northrop candidate was labeled the YF-17. Over the next 20 months, General Dynamics completed the preliminary and detail design phases, and the first of the two YF-16 prototypes was rolled out of the factory on December 13, 1973. On January 20, 1974, during one of the high-speed taxi tests, the all-moving horizontal tail was damaged, and the test pilot elected to takeoff; hence, the first flight of the F-16 was totally unscheduled. The official first flight took place on February 2, 1974. By February 5, it had flown beyond Mach 1. After a competitive fly-off program between the YF-16 and YF-17, on January 13, 1975, the Air Force announced that the winner was the F-16. The first production item, the F-16A, entered active service with the Air Force on January 6, 1979. Since then, more than 4,000 F-16s in various versions have been produced.

The airplane has gone through many design modifications since its early conceptual design discussed above; as expected, among these was a growth in weight. The maximum weight at which the prototype YF-16 was flown was 27,000 lb; the maximum takeoff weight of the F-16C is 42,300 lb with full external fuel tanks and ordnance. The performance capability of the F-16 is a maximum level speed at 40,000 ft of above Mach 2, a service ceiling of more than 50,000 ft, and a radius of action, depending on external stores, of between 230 and 850 mi.

As a final note in our discussion of the F-16, the airplane has been very long-lived. At the time of writing, it was almost 15 years ago that the YF-16 made its first flight. Today, the F-16 is still in production. This is a major example of the *longevity* of modern airplane designs, in contrast to the 1930–1940 period when large numbers of new airplane designs were surfacing every year, and the effective lifespan of a given airplane was closer to 5 years than 25 years. More about this phenomenon will be said in the Postface at the end of this book. For the F-16, even the manufacturer's name has changed, not once but twice, during its production history. On March 1, 1993, the Tactical Military Aircraft Division of General Dynamics at Fort Worth, Texas, which has designed and manufactured the F-16, was bought by Lockheed and became Lockheed Fort Worth Company. Two years later, Lockheed was bought by Martin-Marietta, becoming Lockheed-Martin Company. The F-16 started life as the General Dynamics F-16. Today, it lives on as the Lockheed-Martin F-16.

9.4.2 Design of the SR-71 Blackbird

The historical development of the airplane has always been dominated by the quest for speed and altitude. In this section, we will highlight the design case history of the F-12/SR-71 series of aircraft, an airplane that holds both the maximum speed and maximum altitude records for a production and in-service flying machine. Named the Blackbird, this airplane set a speed record of 2,070.1 mi/h at 80,258 ft on May 1, 1965, at a Mach number of 3.14. Although still classified, it is rumored that the Blackbird can exceed Mach 3.3. Because the Blackbird represents the epitome of supersonic airplane design today, we include it in our overall discussion of supersonic aircraft.

A three-view of the YF-12A and a side view of the SR-71 are shown in Fig. 9.37. The airplane was designed and built by the Lockheed "Skunk Works," an elite, small design group that has operated with great autonomy outside of the normal administrative organization of Lockheed Aircraft Company. The Skunk Works is legendary for a series of novel, innovative airplane designs since World War II. Operating in a shroud of secrecy, this design group has produced such pacesetting airplanes as the very high-altitude subsonic U-2 reconnaissance airplane in the 1950s and the super-secret F117 stealth fighter in the late 1970s. Led by Clarence "Kelly" Johnson

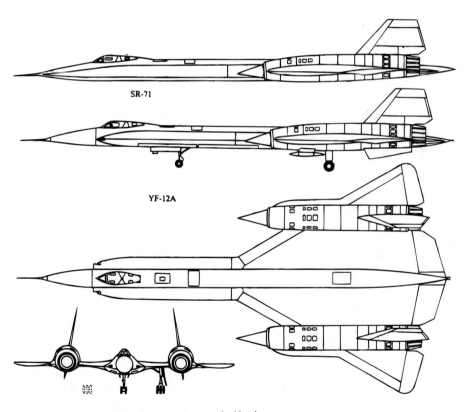

Figure 9.37 The Lockheed YF-12A/SR-71 Blackbird.

until 1975, and Ben Rich between 1975 and 1991, both lengendary men in their own right, the Skunk Works became perhaps the most elite and pioneering airplane design group in history, with the single exception of the Wright brothers. The interesting, indeed, riveting story of the Skunk Works can be found in the autobiographical books by Johnson and Smith (Ref. 67) and by Rich and Janos (Ref. 68).

One of the basic aspects of the design philosophy discussed in Chapter 7—that in the design of a new airplane much preliminary information can be obtained from the characteristics of previous airplanes—was dashed by the design of the Blackbird. There simply was no previous airplane that could serve as guidance for the Blackbird—the desired flight characteristics were so far advanced beyond those of any other aircraft. In Kelly Johnson's words (Ref. 19): "I believe I can truly say that everything on the aircraft from rivets and fluids, up through materials and power plants had to be invented from scratch." The design experience associated with the Blackbird was later elegantly stated by Rich and Janos (Ref. 68):

> The Blackbird, which dominated our work in the sixties, was the greatest high-performance airplane of the twentieth century. Everything about this airplane's creation was gigantic: the technical problems that had to be overcome, the political complexities surrounding its funding, even the ability of the Air Force's most skilled pilots to master this incredible wild horse of the stratosphere. Kelly Johnson rightly regarded the Blackbird as the crowning triumph of his years at the Skunk Works' helm. All of us who shared in its creation wear a badge of special pride. Nothing designed and built by any other aerospace operation in the world, before or since the Blackbird, can begin to rival its speed, height, effectiveness, and impact. Had we built Blackbird in the year 2010, the world would still have been awed by such an achievement. But the first model, designed and built for the CIA as the successor to the U-2, was being test-flown as early as 1962. Even today, that fact seems nothing less than miraculous.

The design concept for the Blackbird stemmed from an earlier study of a hydrogen-fueled high-speed spy plane by the Skunk Works in the late 1950s. The brainchild of Kelly Johnson, this airplane was to fly above 100,000 ft at greater than Mach 2. The problems associated with using hydrogen proved to be insurmountable—the airplane was essentially a flying fuel tank, and even so it could not achieve the desired range because the predicted L/D ratio was 16% less than required. Johnson personally canceled the design activity on the hydrogen-fueled airplane. However, in April 1958, this effort metamorphosed into a related, but distinctly different airplane—one that used conventional fuels and conventional engines, but could fly faster and higher than any Russian missile. Since the Skunk Works had designed the Lockheed F-104, the first fighter airplane capable of sustained flight at Mach 2, a Mach 3 airplane flying at 90,000 ft seemed like a logical extension. The first preliminary design was labeled the A-1 for internal Lockheed reference. Twelve designs later, the A-12 appeared to be satisfactory to Johnson, and on August 28, 1959, the CIA agreed to purchase five A-12 spy planes. The first flight of the A-12 took place on April 26, 1962. Johnson also pushed a version of the airplane as a high-speed interceptor for the Air Force, which was designated as the YF-12A. The first YF-12A flew on August 7, 1963. The existence of the YF-12A was publically announced by President Lyndon B. Johnson on February 29, 1964, and later that year, on July 24, the President revealed the reconnaissance version, designated the SR-71.

To the present time of writing, many aspects of the Blackbird are still classified. However, enough is known about the design of the airplane that we can cast it in light of the design philosophy discussed in Chapter 7.

To begin with, weight was always a major concern, as in all the airplane designs we have examined in this book. Aluminum had been the metal of choice for previous jet airplanes, but at the Mach 3 conditions for the Blackbird, the aerodynamic heating was so severe that the surface temperatures of the Blackbird exceeded that beyond which aluminum lost its strength. Stainless steel could withstand the heat, but it was heavy. This led to the pioneering use of titanium for the Blackbird; titanium was as strong as stainless steel, but was half its weight. Most importantly, titanium could withstand the surface temperatures to be encountered at sustained Mach 3 speeds. Although there were tremendous problems with the machining and availability of titanium, eventually 93% of the structural weight of the Blackbird was built of advanced titanium alloys. It is estimated that the takeoff gross weight of the YF-12A is over 140,000 lb, and its empty weight is about 60,000 lb. This data point is included in Fig. 9.31; although it falls slightly below the data, the value of $W_e/W_0 = 0.43$ for the Blackbird is still quite "conventional" for supersonic jet airplanes.

Since speed, altitude, and range were the primary performance goals for the Blackbird, high values of L/D and W/S were important. The wing area was chosen as 1,800 ft^2, which gives a maximum wing loading of 77.8 lb/ft^2. The variation of $(L/D)_{max}$ with M_∞ is shown in Fig. 9.38. Here we see an example of how dramatically the aerodynamic characteristics of an airplane change when going from

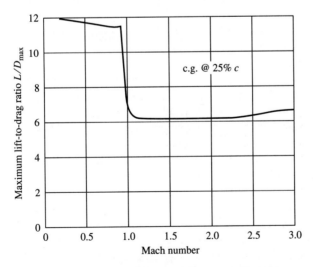

Figure 9.38 Variation of the trimmed maximum lift-to-drag ratio as a function of free-stream Mach number for the Blackbird.

subsonic to supersonic speeds; the value of $(L/D)_{max}$ is cut almost in half by the drag-divergence/wave drag effects at supersonic speeds. On the other hand, the resulting value of $(L/D)_{max} = 6.5$ at Mach 3 is quite reasonable for a supersonic vehicle.

Another aspect that dramatically changes when the airplane accelerates from subsonic to supersonic speeds is the aerodynamic center (the neutral point) of the airplane. Recall that the aerodynamic center of a flat plate theoretically is at the quarter-chord point for subsonic flow, but moves to the mid-chord point for supersonic flow. An airplane going through Mach 1 experiences a similar shift in the aerodynamic center. The variation of the neutral point for the YF-12A with M_∞ is shown in Fig. 9.39. This figure also illustrates one of the beneficial aspects of a major design feature of the Blackbird, namely, the use of chines along the fuselage. Returning to Fig. 9.37, the chines are essentially long strakes extending forward of the wing leading edge along the fuselage, but are much more integrated with the fuselage than the conventional strakes, as can be seen in the front view in Fig. 9.37. For the YF-12A, the chines stop at the canopy location, so as not to interfere with the nose radome. However, for the SR-71, the chines extend all the way to the nose, giving the fuselage a "cobralike" appearance. The chines have several important aerodynamic advantages. For one, they tend to decrease the travel of the neutral point as M_∞ is increased. This is clearly seen in Fig. 9.39, where there is an almost 35% rearward shift of the neutral point for the case with no chines, compared to the much more limited travel of the neutral point when chines are included. Why is this more limited travel of the neutral point an advantage? Recall that, for static longitudinal stability, the neutral point must be located behind the airplane's center of gravity. The normalized distance between the center of gravity and the neutral point is called the *static margin*. A *positive static margin* exists when the neutral point is behind the center of gravity, which as stated earlier is necessary for static longitudinal stability. The larger the positive

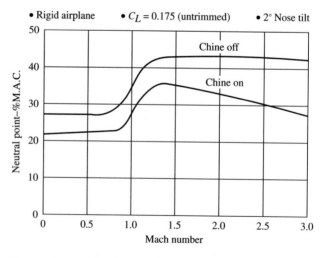

Figure 9.39   Shift in the neutral point for the Blackbird from subsonic to supersonic speed.

static margin, the more stable is the airplane. However, too much of a positive static margin is not good, because the airplane will be too stable for maneuvering and will require a large elevator deflection to trim the airplane because of the large distance between the neutral point and the center of gravity. This results in a trim drag which is unacceptably large. *For this reason, the sudden and marked rearward shift of the neutral point when an airplane goes from subsonic to supersonic speeds is one of the major problems in the design of supersonic airplanes.* If the center of gravity is located so as to achieve a proper static margin at subsonic speeds, then the static margin becomes too large at supersonic speeds. If the center of gravity is located so as to achieve a proper static margin at supersonic speeds, the subsonic static margin most likely would be negative (neutral point ahead of the center of gravity), hence making the airplane unstable in subsonic flight. With the use of chines, the designers of the Blackbird found a reasonable design solution to this problem.

Another aerodynamic advantage of chines at supersonic speeds is the favorable effect on directional stability (yaw stability). A cylindrical fuselage at a small yaw angle to the flow will experience crossflow separation, as shown at the top of Fig. 9.40, with a consequent large side force. In contrast, the blended chine-body cross section shown at the bottom of Fig. 9.40 shows an attached crossflow, with a much lower side force. In this way the chines are beneficial in designing for directional stability. This has a synergistic effect, because the vertical tail surface can be made smaller, with a consequent reduction in weight and skin-friction drag.

The design of a supersonic airplane is essentially the design of two different airplanes combined into one. The airplane must be optimized for its supersonic mission, whatever that may be. But it must also spend time flying at subsonic speeds, especially for takeoff and landing. So some attention must be paid to obtaining satisfactory (not necessarily optimum) low-speed characteristics. In essence, a supersonic

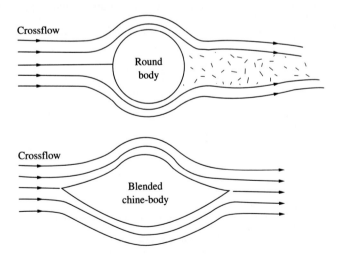

Figure 9.40 Schematic showing crossflow streamlines over a cylindrical body and a blended body with chines.

airplane is designed for double duty—reasonable flight characteristics at both sub-sonic and supersonic speeds. This is a compelling reason to choose a highly swept delta wing—high sweep to minimize supersonic wave drag, but a delta planform for satisfactory low-speed performance. The design choice for the Blackbird was a delta wing with 60° sweep.

The low-speed aerodynamic characteristics of a delta wing are discussed in Section 2.8.1. In particular, Eq. (2.25), repeated below, is an approximate expression for the variation of low-speed normal force coefficient with angle of attack for delta wings.

$$\frac{C_N}{(s/l)^2} = 2\pi \frac{\alpha}{s/l} + 4.9 \left(\frac{\alpha}{s/l}\right)^{1.7} \qquad \textbf{[2.25]}$$

In Eq. (2.25), s is the semispan, l is the length, and α is the angle of attack in radians. For the 60° swept wing of the Blackbird, $s/l = \sin(90° - 60°) = 0.5$. Let us use Eq. (2.25) to calculate the lift coefficient at an angle of attack of 10°. From Eq. (2.25), with $\alpha = 10° = 0.1745$ rad, and hence $\alpha/(s/l) = 0.349$,

$$C_N = (0.5)^2 \left[2\pi(0.349) + 4.9(0.349)^{1.7}\right]$$
$$= (0.5)^2(2.1932 + 0.8187) = 0.753$$

Hence, the lift coefficient is

$$C_L = C_N \cos\alpha = (0.753)(0.9848) = 0.742$$

For the designers of the Blackbird, the above calculation was optimistic because in the configuration layout the engines were placed on the wings (see Fig. 9.37), with a consequent decrease in the lifting power of that portion of the wings. On the other hand, the chined fuselage provided some additional lift at angle of attack. Measured values of C_L from wind tunnel tests of the Blackbird are shown in Fig. 9.41, along

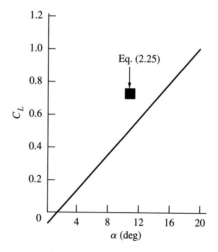

Figure 9.41   Wind tunnel results for the low-speed lift coefficient for the Blackbird. Comparison with one data point from Eq. (2.25).

with the calculated result from Eq. (2.25). Note that the lift curve does not go through zero, but rather shows a small positive zero-lift angle of about 1°. This is because the wings are placed at a small negative incidence angle relative to the fuselage due to the effective lifting characteristics of the chined forebody at supersonic speeds. (Here is an example of the configuration of the Blackbird being driven by supersonic cruise considerations.) The angle of attack given on the abscissa of Fig. 9.41 is based on the mean aerodynamic chord of the entire blended wing-body configuration. For this reason, the data point in Fig. 9.41 obtained from Eq. (2.25) for a wing angle of attack of 10° is plotted at an angle of attack of 11° in Fig. 9.41; this would seem to be a more valid comparison.

These results have a strong effect on landing and takeoff speeds for the Blackbird. As discussed in Section 2.8.1, $(C_L)_{max}$ for a delta wing is achieved at high angles of attack, on the order of 30° to 40°. This angle of attack was considered too high for the Blackbird; the pilot would have his or her ground view obstructed, plus the landing gear struts would have to be very long, which would add weight. The design angle of attack for approach was fixed at $\alpha = 8.35°$, which from Fig. 9.41 gives a value of $C_L = 0.37$. With this lift coefficient, the approach velocity for a weight of 100,000 lb is

$$V_\infty = \sqrt{\frac{2W}{\rho_\infty S C_L}} = \sqrt{\frac{2(100,000)}{(0.002377)(1,800)(0.37)}} = 355.4 \text{ ft/s}$$

The landing and takeoff performance of the Blackbird is given in Fig. 9.42. The approach speed as a function of weight is shown at the bottom of Fig. 9.42, given in knots. Noting that 1 kn = 1.689 ft/s, the calculated value of $V_\infty = 355.4$ ft/s is equivalent to 210 kn, which agrees with the value of the approach shown in Fig. 9.42 for a weight of 100,000 lb. Figure 9.42 also shows that the actual touchdown speed is less than the approach speed because of the favorable ground effect. In ground effect, C_L is increased at a fixed angle of attack. For the Blackbird, when the airplane is in ground effect, C_L increased to slightly above 0.5. Hence, maintaining the same angle of attack, the pilot can slow down at touchdown. However, looking in general at Fig. 9.42, we see that the landing and liftoff speeds and distances are quite large compared to those for conventional airplanes. This is the penalty accepted by the Blackbird designers for optimizing the airplane for Mach 3+ speeds. This serves as a graphic example of the compromises to be made in supersonic airplane design. Mach 3+ performance was paramount; landing and takeoff speeds and distance were secondary, especially since the airplanes were intended to only use the long runways of major military airbases.

Aerodynamic heating grows exponentially with Mach number. It is not a driver in the design of subsonic airplanes, but it becomes a factor for supersonic airplanes and a dominant aspect for hypersonic vehicles. At the Mach 3+ speeds of the Blackbird, it was important enough to dictate the material used for construction of the airplane, and it even determined the color of the external surface. To obtain a feeling for the magnitude of the heating problem for the Blackbird, consider the variation of stagnation temperature T_0 with M_∞, as given in the equation from Ref. 16

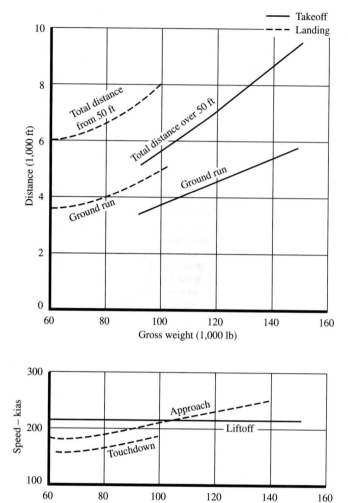

Figure 9.42   Takeoff and landing performance for the Blackbird.

$$\frac{T_0}{T_\infty} = 1 + \frac{\gamma - 1}{2} M_\infty^2$$

where T_∞ is the ambient static termperature and γ is the ratio of specific heats $\gamma = c_p/c_v$. For air below a temperature of $1500°R$, $\gamma = 1.4$. At $M_\infty = 3.3$, we have

$$\frac{T_0}{T_\infty} = 1 + 0.2(3.3)^2 = 3.18$$

At 80,000 ft, $T_\infty = 390°R$, hence $T_0 = (3.18)(390) = 1240°R = 779°F$. This is essentially the temperature encountered at the leading edges and inside the inlet of the Blackbird—a temperature hotter than the average soldering iron. The wing and fuselage encounter surface temperatures on the order of 450° to 500°F—hotter than the maximum available in a household oven. These surface temperatures dictated the use of titanium rather than aluminum for the airplane's skin and internal structure, as already mentioned. To handle the aerodynamic heating, two measures were taken. The fuel was used as a heat sink, to precool the hot compressor bleed air for the air conditioning for the cockpit, and then the hot fuel was fed directly to the engine. Also, radiative cooling of the surface was used. Recall that a surface at a temperature T radiates thermal energy which is given by

$$E_R = \epsilon \sigma T^4$$

where E_R is the rate of radiative energy emitted per unit area, σ is the Stefan-Boltzmann constant, and ϵ is the emissivity which varies from 0 to 1. The higher the emissivity, the more the surface is cooled by radiation. This is the reason why the Blackbird is painted a very dark blue-black color, to increase the emissivity and hence the radiative cooling. Even though the paint added close to an extra 100 lb to the airplane, it lowered the wing temperature by 35°F, allowing the use of a softer titanium alloy and hence improving the manufacturing processes for the airplane. Here is yet another design compromise—trading weight for an increase in manufacturing ease, something very important when titanium is being used.

The Blackbird has all-moving vertical tails, with no rudder surfaces. An investigation of conventional rudders early in the conceptual design stage showed that very large rudder deflections would be required to balance an engine-out condition. This was considered an inadaquate control authority. Moreover, at such large rudder deflections, the rudder hinge line exposed to the flow would encounter a large stagnation temperature, causing local aerodynamic heating problems. The solution to both these problems was to dispense with rudders and use all-moving vertical tails. Although all-moving horizontal tails (equipped also with elevators) had been used as early as 1947 (e.g., on the Bell X-1 and the North American F-86), the use of all-moving vertical tails (without rudders) for the Blackbird appears to be an innovative first. The vertical tails were also not vertical. Figure 9.43 shows the front view of the airplane with the orientations of the vertical tails, one with the tails exactly vertical and one with the vertical tails canted inward by a 15° angle. When there is a side force on the vertical tail, the center of pressure on the tail is above the longitudinal axis through the center of gravity, hence causing a rolling moment about that axis. This is shown in Fig. 9.43. By canting the tails inward, the side force acts through a smaller moment arm, hence reducing the rolling moment. The final design configuration of the Blackbird incorporated the canted vertical tails, as seen in Fig. 9.37.

The Blackbird is powered by two Pratt & Whitney J-58 bleed bypass turbojet engines, especially designed for use on this airplane. Each engine produces more than 30,000 lb of thrust at sea-level static conditions. The engine also uses a special low-vapor-pressure hydrocarbon fuel called JP-7. The combined inlet-engine combination is an interesting example of airframe-propulsion integration in the following sense. The inlet is an axisymmetric spike inlet, with a center cone that translates forward and backward. The location of the spike is automatically changed during flight to maintain

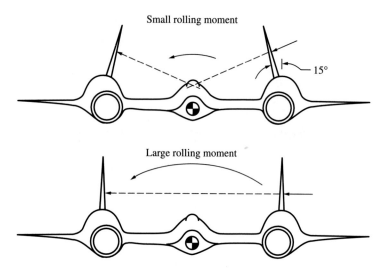

Figure 9.43 Effect of vertical stabilizer cant on rolling moment.

the optimum shock wave location on the edge of the inlet cowl, hence striving for minimum drag. The inlet-engine nacelle is also designed for effective bleeding of the boundary layer on both the spike and the outside of the inlet, in order to enhance the quality of the internal flow and stabilize the airflow pattern. The airflow patterns in the nacelle for both low speed (takeoff) and high speed (Mach 3+ cruise) are sketched in Fig. 9.44. In the low-speed case, the forward position of the spike allows the entering subsonic air to pass through a divergent passage, thus slowing the air inside the inlet. In the high-speed case, the more rearward position of the spike allows the entering supersonic air to pass through a convergent-divergent passage, thus slowing the air inside the inlet.

However, what is most interesting about this inlet-engine arrangement is the breakdown of where the thrust is coming from—a type of thrust budget that is shown in Fig. 9.45. This figure is somewhat analogous to the generic sketch shown in Fig. 3.10e, which shows the amount of thrust produced by each section of the jet engine. In Fig. 9.45, the inlet-engine combination is divided into four sections, as sketched at the top of the figure. The percentage of the thrust produced by each section is plotted versus speed, from low-speed subsonic to high-speed Mach 3+ cruise. Recall that the thrust of each section is due to the integration of the pressure distribution over that section. A negative percentage contributes drag, a positive percentage contributes thrust. Section 0–1 is the forward portion of the conical spike, and the pressure distribution there will always produce drag, as shown by the curve labeled 0–1 in Fig. 9.45. In contrast, section 1–2 includes the back end of the spike, and the pressure distribution there will always create a force in the forward direction, for example, thrust. Indeed, the percentage of the thrust generated in section 1–2 increases with Mach number, and this section produces almost 70% of the total thrust at high speed.

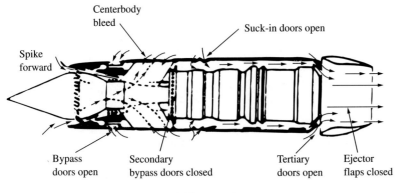

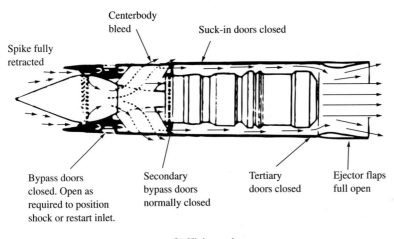

Figure 9.44 Nacelle airflow: (**a**) low speed at takeoff and (**b**) high speed at cruise. (AIAA, with permission.)

Section 2–3 is the engine itself—compressor, burner, turbine, and nozzle. Finally, section 3–4 is the ejector for both the primary flow through the engine core and the bleed air external to the core. It is intersting to note that, at Mach 3+, the engine core itself produces only about 17% of the total thrust. The rest of the thrust is produced by the aerodynamics of the nacelle, especially in section 1–2. Quoting Kelly Johnson (Ref. 69): "My good friends at Pratt & Whitney do not like me to say, that at high speeds, their engine is only a flow inducer, and that after all, it is the nacelle pushing the airplane." This phenomenon is not a unique characteristic of just the Blackbird; for any very high-Mach-number airplane of the future, such as scramjet-powered (supersonic combustion ramjet) hypersonic aircraft, the inlet will produce most of the thrust. This simply increases the importance of proper airframe-propulsion integration for such airplanes.

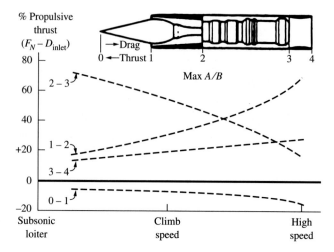

Figure 9.45   Contribution of various parts of the engine to the generation of thrust for the Blackbird.

Finally, we note that although stealth (low radar cross section) was not a driving aspect of the design of the Blackbird, it certainly was a consideration. The canted vertical tail surfaces tend to reflect incident radar beams away from the receiver, hence reducing the radar cross section. Also, when the blended chines were added (Fig. 9.40) to an otherwise cylindrical forebody, the radar cross section dropped by 90%; the chines turn the bottom of the fuselage into an almost flat surface which also reflects incident radar beams away from the receiver. Hence the Blackbird had a strong flavor of stealth considering the time at which it was designed.

In conclusion, the YF-12A/SR-71 Blackbird series of aircraft incorporated many unique design features never seen before on an operational aircraft. The design of this aircraft points the way for the design of future very high-Mach-number airplanes.

9.4.3   Design of the Lockheed F-22 Advanced Tactical Fighter

With this section, we end our discussion of the design of supersonic airplanes. We will highlight the design of the Lockheed-Boeing-General Dynamics (now Lockheed-Martin) F-22, which represents the most recent supersonic airplane design at the time of writing. Because of the newness of the F-22 and the high military classification still surrounding the airplane, less is known in the open literature about its design characteristics. However, enough information is available to piece together some aspects of its design philosophy. A four-view (including top and bottom views) of the F-22 is shown in Fig. 9.46.

Pivot point 1 in our design philosophy—establishing the requirements—was carried out by the Air Force in 1984 when the Advanced Tactical Fighter System Program Office at Wright Field in Dayton, Ohio, issued the following specifications

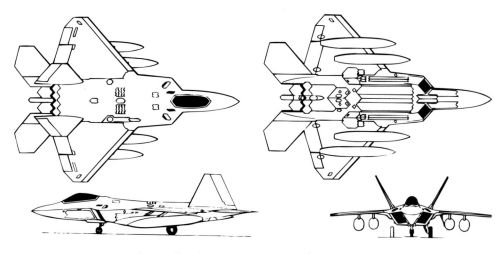

Figure 9.46 Four-view of the Lockheed-Martin F-22. (AIAA, with permission.)

for a new, advanced, tactical fighter:

Radius of action: 800 mi

Supersonic cruise: Mach 1.4 to 1.5

Gross takeoff weight: 50,000 lb

Takeoff length: 2,000 ft

Unit cost: No more than $40 million

The Air Force issued concept definition studies to seven manufacturers, with the idea of assessing on paper seven different designs. However, in May 1986, it was decided to make the final choice of the manufacturer on the basis of a prototype fly-off between the two top designs (much in the same vein as the fly-off that resulted in the choice to produce the F-16). These two top designs were from Lockheed, with General Dynamics and Boeing as partners, and from Northrop. Lockheed's airplane was designated the YF-22, and Northrop's entry was the YF-23. The Northrop YF-23 was the first to fly, getting into the air in September 1990. The YF-22 first flew in October 1990. After a lengthy series of flight tests for both airplanes, the Lockheed YF-22 was announced as the winner on April 23, 1991.

During the design of the YF-22, the target gross weight of 50,000 lb was missed; the gross weight grew to 58,000 lb, a normal trend in airplane design. The empty weight of the YF-22 was 31,000 lb, giving a value of $W_e/W_0 = 0.534$. This data point is shown in Fig. 9.31; it falls very close to the dashed curve faired through the data. As for other designs before, the designers of the F-22 could have used such historical data to make an initial weight estimate.

The design of the F-22 did not follow the trend of faster and higher; its function was not to better the YF-12/SR-71 discussed in the previous section. Rather, the comparatively low specified cruise Mach number of 1.4 to 1.5 was a recognition of a turnaround in supersonic fighter design, where speed was recognized as not as important as maneuverability and agility. Also, *stealth* capability was becoming of paramount importance; if the airplane is essentially invisible to radar, then how fast it can fly is not quite so important.

A major design feature which enhanced both maneuverability and stealth was the use of two-dimensional (in contrast to the standard axisymmetric) exhaust nozzles from the two jet engines; moreover, the two-dimensional nozzles could be tilted up or down to vector the thrust in the plane of symmetry of the aircraft. This feature is particularly useful for high-angle-of-attack maneuvers. A simple sketch comparing an axisymmetric nozzle with a two-dimensional nozzle is shown in Fig. 9.47. The F-22 is the first production airplane to use two-dimensional, thrust-vectoring exhaust nozzles. The thrust vectoring is made all the more powerful by the two Pratt & Whitney F119-PN-100 advanced-technology turbofan engines, capable of a combined thrust at sea level of 70,000 lb. This gives the F-22 a thrust-to-weight ratio greater than 1: $T/W_0 = 70,000/58,000 = 1.2$.

The designers of the F-22 chose a diamond planform wing with a taper ratio of 0.169 and a leading-edge sweep of 42° (see Fig. 9.46). The use of computational fluid dynamics (CFD) expedited the configuration design. (See Ref. 21 for an introduction to computational fluid dynamics and its use in design.) For example, the airfoil section for the F-22 was custom-designed using CFD; it is a biconvex shape with a thickness-to-chord ratio of 0.0592 at the wing root and 0.0429 at the wing tip. The wingspan is

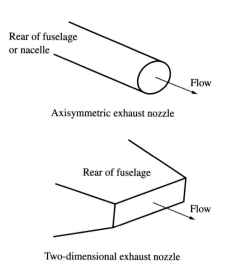

Rear of fuselage or nacelle

Flow

Axisymmetric exhaust nozzle

Rear of fuselage

Flow

Two-dimensional exhaust nozzle

Figure 9.47   Schematic of an axisymmetric exhaust nozzle and a two-dimensional exhaust nozzle.

44.5 ft, and the planform area is 840 ft^2, giving an aspect ratio of 2.36. The choice of a low aspect ratio is driven by the supersonic performance (the supersonic wave drag is reduced by reducing the aspect ratio). The wings have full-span leading-edge flaps. The vertical tails are canted outward by 28° and incorporate conventional rudders. The vertical tails are all-moving, slab "taileron" surfaces.

More than 19,000 h of wind tunnel time was invested in the design of the YF-22. These tests were instrumental in obtaining preflight predictions of the performance of the YF-22. Such preflight predictions were requested by the Air Force in early 1990 so that they could be compared with actual test flight data to be obtained later that year. Although the detailed comparisons are classified, the flight tests carried out in late 1990 provided the following results (Ref. 71).

1. Supersonic cruise was as predicted. Maximum level speed at 30,000 ft was $M_\infty = 1.58$; with afterburning, it was $M_\infty = 1.7$.

2. Up to $M_\infty = 0.9$, the subsonic drag was as predicted.

3. Supersonic drag for low angle of attack was as predicted. (Insufficient flight data were obtained for comparison at high angles of attack at supersonic speeds.)

4. The drag rise at transonic speeds was lower than predicted.

5. Specific excess power was as predicted.

6. Range at all test conditions was within 3% of predictions (which means that L/D values were well predicted).

7. Maximum speed was as predicted and was achieved in flight on December 28, 1990.

8. Maximum roll rates were smaller than predicted, and time to specific bank angles at subsonic and supersonic speeds was larger than predicted, but was judged to be satisfactory.

9. Flying qualities at high angles of attack (above 20°) were judged to be excellent with the use of thrust vectoring.

10. The $(C_L)_{max}$ was higher than predicted.

The detailed design and manufacturing processes that led to the first production F-22 were lengthy, taking 6 years. The high technology embodied in the design is partly responsible. Even the materials were of an advanced mix; the F-22 structure is 35% composite material, 33% titanium, 11% aluminum, 5% steel, and the other 16% miscellaneous materials. Finally, the rollout of the first production F-22 occurred in April 1997. The first flight of this production airplane was on September 7, 1997, lasting 58 min at altitudes of 15,000 to 20,000 ft, speeds up to 300 mi/h, and angles of attack during maneuvers of up to 14°.

The F-22 is considered to be the best fighter airplane anywhere in the world for the beginning of the twenty-first century. At the time of writing, at least 480 airplanes are anticipated to be manufactured.

9.5   SUMMARY

Supersonic airplane design is different from subsonic airplane design. Supersonic airplanes are two airplanes in one; they are optimized for their supersonic mission, but they must have satisfactory subsonic characteristics to allow safe and reasonable flight at low speeds. Because the physics of supersonic flow is so different from the physics of subsonic flow, every supersonic airplane design is a careful compromise between what is good for supersonic flight and what is good for subsonic flight.

Some of the characteristics and challenges that pertain to the design of supersonic airplanes are as follows:

1.   Shock waves, which occur at transonic and supersonic speeds, cause a large increase in drag—supersonic wave drag. Supersonic airplanes are designed with slender bodies and thin wings with sharp leading edges to reduce wave drag. These same features are not good for subsonic flow and can cause flow separation and hence large pressure drag at subsonic speeds.

2.   The center of pressure (hence the aerodynamic center) for an airplane shifts dramatically rearward when it accelerates from subsonic to supersonic flight. This creates a major design challenge for proper stability and control characteristics of the airplane.

3.   Attention must be paid to *where* the shock waves are on the airplane and where shocks may be impinging on the airplane's surface. Shock impingement can cause flow separation and local spots of intense aerodynamic heating.

4.   Airframe-propulsion integration becomes more of a design challenge, and a necessity, as the design Mach number of supersonic airplanes increases.

5.   Aerodynamic heating, usually negligible for subsonic airplane design, becomes one of the important design considerations for supersonic airplanes. Aerodynamic heating increases approximately as the cube of the velocity (see, e.g., chapter 8 of Ref. 3). Aerodynamic heating was enough of a problem in the design of the Mach 3+ YF-12A/SR-71 Blackbird that it dictated the material the airplane is made of (titanium) and its exterior color (dark blue-black to increase surface cooling by thermal radiation).

These are only some of the items that distinguish supersonic airplane design from that of subsonic airplanes. However, the overall design *philosophy* as described in Chapter 7 is essentially the same, whether the airplane is intended to be subsonic or supersonic. This chapter has been devoted exclusively to design case histories of three quite different supersonic airplanes in order to illustrate this point.

POSTFACE

At the end of *Skunk Works* (Ref. 68), Ben Rich, vice president of Lockheed and director of the Skunk Works from 1975 to 1991, has the following to say:

> In my forty years at Lockheed I worked on twenty-seven different airplanes. Today's young engineer will be lucky to build even *one*. The life cycle of a military airplane is far different from the development and manufacturing of anything else. Obsolescence is guaranteed because, outside of a secret, high-priority project environment like the Skunk Works, it usually takes eight to ten years to get an airplane from the drawing board into production and operational. Every combat airplane that flew in Operation Desert Storm in 1991 was at least ten to fifteen years old by the time it actually proved its worth on the battlefield, and we are now entering an era in which there may be a twenty- to thirty-year lapse between generations of military aircraft.

The purpose of this book is to present the fundamental aspects of airplane performance and to discuss and illustrate the philosophy of airplane design. However, in light of Rich's comments, what is the likelihood that you will ever have a chance to participate in the design of a new airplane? It is a fact that the number of new airplanes designed in a given year has decreased dramatically from the literally hundreds per year in the heyday of the 1930s to a very few per year today, and this is counting the design of a new variant of an existing aircraft, such as the design of the Boeing 747-400 as distinct from the earlier 747-200 version. The reasons for this situation are straightforward. First, modern military and civilian airplanes incorporate a level of sophisticated technology that was undreamed of 50 years ago, and it takes great effort, tremendous expense, and much time to design new, high-technology airplanes. Second, the cost of a new airplane today, even after the cost of development is subtracted, is considerably more than that of 50 years ago. So it is no surprise that the number of new airplane designs today is far smaller than that of 50 years ago. Compensating for this, and perhaps as a partial consequence, the lifespan of major airplanes today is on the order of 30 years, in contrast to just a few years for the average airplane from the 1930s. An extreme example is the Boeing B-52, designed and first built in the early 1950s; today, the B-52 is still in service as the primary strategic bomber for the Air Force, and the Air Force is projecting that it will continue in service well into the twenty-first century, at least until 2035—which would be a service life of 80 years! So again we ask: What is the likelihood that you will ever get a chance to participate in the design of a new airplane?

I believe the answer is, plenty of chances. First, even though the number of major high-technology designs for new, but rather conventional civilian aircraft is small, the design activity requires more people for longer periods, hence increasing your opportunity to participate in such designs. Second, and here is some real excitement, the vistas for new, unconventional airplane designs are expanding rapidly at the time of writing. For example, a whole new class of flight vehicles—micro air vehicles—is coming on the scene. These are ultra-miniature airplanes, with wing spans usually less than 15 cm, for purposes of detailed reconnaissance for military and law enforcement agencies. Imagine "mechanical birds" or "mechanical insects" that fly through hallways, poking around corners, buzzing by windows. These micro air vehicles pose dramatic new challenges in airplane design. The aerodynamics is totally different—very low Reynolds numbers. The miniature power plants, control and stability, and flight management (avionics) are all different.

Another class of flight vehicles, one that is both old and new but which has a spectacular new future, is uninhabited air vehicles (UAVs). Remotely piloted airplanes have been used since World War I at different times and places, but their serious use as battlefield reconnaissance vehicles began in 1982 when the Israelis employed them successfully in the Lebanon conflict. Until recently these flight vehicles were designated as RPVs, and they were, for the most part, overgrown model airplanes (e.g., with 6-ft wingspans). However, these have now become a subclass of a much larger array of pilotless vehicles under the designation of UAVs. In addition to short-range tactical reconnaissance, new UAVs are now being designed for very high-altitude, long-endurance strategic intelligence missions. These are full-size airplanes. For example, the Teledyne Ryan Global Hawk has a wingspan of 116 ft, and the Lockheed-Martin/Boeing Dark Star spans 65 ft, both designed for high-altitude, high-endurance flight. Another subclass of UAVs is uninhabited *combat* air vehicles (UCAVs), full-size pilotless aircraft designed for strike and fighter roles. There are several advantages of using pilotless aircraft for combat. By removing the pilot, space and weight are saved, which has a synergistic benefit that the airplane can be smaller, hence reducing drag. Also, the airplane can be made much more maneuverable; a 9-g maneuver is the maximum that a human pilot can endure without passing out, and even that for only a few seconds, whereas without the pilot the airplane can be designed for 25-g maneuvers and better. Another advantage of using pilotless airplanes in combat is that much more aggressive tactics can be employed that would otherwise not be used if a pilot's life hung in the balance.

Other new airplane designs will push the frontiers of flight in the twenty-first century. New supersonic airplanes for commercial use—a new generation of supersonic transport and supersonic executive general aviation airplanes—will very likely appear in the first decade of the new century. And the dream of hypersonic airplanes will be pursued, although most likely for military rather than for commercial purposes.

So a final word to you. Yes, aeronautical engineering has matured; the aeronautical engineers of the past century have done their job admirably and made great progress. However, there is much yet to do and to accomplish. The second century of flight will be full of interesting design challenges. Indeed, if you desire, you *will* have the opportunity to participate in the design of new airplanes, and many more than just one. I hope that you will find the experience of reading this book to have been rewarding when you press on to these new design challenges. If this book has helped to give you insight into airplane performance and the philosophy of airplane design, then I can rest easy. My task is done. Yours is just beginning.

John D. Anderson, Jr.

Standard Atmosphere, SI Units

Altitude		Temperature T, K	Pressure p, N/m^2	Density ρ, kg/m^3
h_G, m	h, m			
−5,000	−5,004	320.69	1.7761 + 5	1.9296 + 0
−4,900	−4,904	320.03	1.7587	1.9145
−4,800	−4,804	319.38	1.7400	1.8980
−4,700	−4,703	318.73	1.7215	1.8816
−4,600	−4,603	318.08	1.7031	1.8653
−4,500	−4,503	317.43	1.6848	1.8491
−4,400	−4,403	316.78	1.6667	1.8330
−4,300	−4,303	316.13	1.6488	1.8171
−4,200	−4,203	315.48	1.6311	1.8012
−4,100	−4,103	314.83	1.6134	1.7854
−4,000	−4,003	314.18	1.5960 + 5	1.7698 + 0
−3,900	−3,902	313.53	1.5787	1.7542
−3,800	−3,802	312.87	1.5615	1.7388
−3,700	−3,702	212.22	1.5445	1.7234
−3,600	−3,602	311.57	1.5277	1.7082
−3,500	−3,502	310.92	1.5110	1.6931
−3,400	−3,402	310.27	1.4945	1.6780
−3,300	−3,302	309.62	1.4781	1.6631
−3,200	−3,202	308.97	1.4618	1.6483
−3,100	−3,102	308.32	1.4457	1.6336
−3,000	−3,001	307.67	1.4297 + 5	1.6189 + 0
−2,900	−2,901	307.02	1.4139	1.6044
−2,800	−2,801	306.37	1.3982	1.5900
−2,700	−2,701	305.72	1.3827	1.5757
−2,600	−2,601	305.07	1.3673	1.5615

Altitude		Temperature T, K	Pressure p, N/m^2	Density ρ, kg/m^3
h_G, m	h, m			
−2,500	−2,501	304.42	1.3521	1.5473
−2,400	−2,401	303.77	1.3369	1.5333
−2,300	−2,301	303.12	1.3220	1.5194
−2,200	−2,201	302.46	1.3071	1.5056
−2,100	−2,101	301.81	1.2924	1.4918
−2,000	−2,001	301.16	1.2778 + 5	1.4782 + 0
−1,900	−1,901	300.51	1.2634	1.4646
−1,800	−1,801	299.86	1.2491	1.4512
−1,700	−1,701	299.21	1.2349	1.4379
−1,600	−1,600	298.56	1.2209	1.4246
−1,500	−1,500	297.91	1.2070	1.4114
−1,400	−1,400	297.26	1.1932	1.3984
−1,300	−1,300	296.61	1.1795	1.3854
−1,200	−1,200	295.96	1.1660	1.3725
−1,100	−1,100	295.31	1.1526	1.3597
−1,000	−1,000	294.66	1.1393 + 5	1.3470 + 0
−900	−900	294.01	1.1262	1.3344
−800	−800	293.36	1.1131	1.3219
−700	−700	292.71	1.1002	1.3095
−600	−600	292.06	1.0874	1.2972
−500	−500	291.41	1.0748	1.2849
−400	−400	290.76	1.0622	1.2728
−300	−300	290.11	1.0498	1.2607
−200	−200	289.46	1.0375	1.2487
−100	−100	288.81	1.0253	1.2368
0	0	288.16	1.01325 + 5	1.2250 + 0
100	100	287.51	1.0013	1.2133
200	200	286.86	9.8945 + 4	1.2071
300	300	286.21	9.7773	1.1901
400	400	285.56	9.6611	1.1787
500	500	284.91	9.5461	1.1673
600	600	284.26	9.4322	1.1560
700	700	283.61	9.3194	1.1448
800	800	282.96	9.2077	1.1337
900	900	282.31	9.0971	1.1226

Altitude		Temperature T, K	Pressure p, N/m^2	Density ρ, kg/m^3
h_G, m	h, m			
1,000	1,000	281.66	8.9876 + 4	1.1117 + 0
1,100	1,100	281.01	8.8792	1.1008
1,200	1,200	280.36	8.7718	1.0900
1,300	1,300	279.71	8.6655	1.0793
1,400	1,400	279.06	8.5602	1.0687
1,500	1,500	278.41	8.4560	1.0581
1,600	1,600	277.76	8.3527	1.0476
1,700	1,700	277.11	8.2506	1.0373
1,800	1,799	276.46	8.1494	1.0269
1,900	1,899	275.81	8.0493	1.0167
2,000	1,999	275.16	7.9501 + 4	1.0066 + 0
2,100	2,099	274.51	7.8520	9.9649 − 1
2,200	2,199	273.86	7.7548	9.8649
2,300	2,299	273.22	7.6586	9.7657
2,400	2,399	272.57	7.5634	9.6673
2,500	2,499	271.92	7.4692	9.5696
2,600	2,599	271.27	7.3759	9.4727
2,700	2,699	270.62	7.2835	9.3765
2,800	2,799	269.97	7.1921	9.2811
2,900	2,899	269.32	7.1016	9.1865
3,000	2,999	268.67	7.0121 + 4	9.0926 − 1
3,100	3,098	268.02	6.9235	8.9994
3,200	3,198	267.37	6.8357	8.9070
3,300	3,298	266.72	6.7489	8.8153
3,400	3,398	266.07	6.6630	8.7243
3,500	3,498	265.42	6.5780	8.6341
3,600	3,598	264.77	6.4939	8.5445
3,700	3,698	264.12	6.4106	8.4557
3,800	3,798	263.47	6.3282	8.3676
3,900	3,898	262.83	6.2467	8.2802
4,000	3,997	262.18	6.1660 + 4	8.1935 − 1
4,100	4,097	261.53	6.0862	8.1075
4,200	4,197	260.88	6.0072	8.0222
4,300	4,297	260.23	5.9290	7.9376
4,400	4,397	259.58	5.8517	7.8536

Altitude		Temperature T, K	Pressure p, N/m^2	Density ρ, kg/m^3
h_G, m	h, m			
4,500	4,497	258.93	5.7752	7.7704
4,600	4,597	258.28	5.6995	7.6878
4,700	4,697	257.63	5.6247	7.6059
4,800	4,796	256.98	5.5506	7.5247
4,900	4,896	256.33	5.4773	7.4442
5,000	4,996	255.69	5.4048 + 4	7.3643 − 1
5,100	5,096	255.04	5.3331	7.2851
5,200	5,196	254.39	5.2621	7.2065
5,400	5,395	253.09	5.1226	7.0513
5,500	5,495	252.44	5.0539	6.9747
5,600	5,595	251.79	4.9860	6.8987
5,700	5,695	251.14	4.9188	6.8234
5,800	5,795	250.49	4.8524	6.7486
5,900	5,895	249.85	4.7867	6.6746
6,000	5,994	249.20	4.7217 + 4	6.6011 − 1
6,100	6,094	248.55	4.6575	6.5283
6,200	6,194	247.90	4.5939	6.4561
6,300	6,294	247.25	4.5311	6.3845
6,400	6,394	246.60	4.4690	6.3135
6,500	6,493	245.95	4.4075	6.2431
6,600	6,593	245.30	4.3468	6.1733
6,700	6,693	244.66	4.2867	6.1041
6,800	6,793	244.01	4.2273	6.0356
6,900	6,893	243.36	4.1686	5.9676
7,000	6,992	242.71	4.1105 + 4	5.9002 − 1
7,100	7,092	242.06	4.0531	5.8334
7,200	7,192	241.41	3.9963	5.7671
7,300	7,292	240.76	3.9402	5.7015
7,400	7,391	240.12	3.8848	5.6364
7,500	7,491	239.47	3.8299	5.5719
7,600	7,591	238.82	3.7757	5.5080
7,700	7,691	238.17	3.7222	5.4446
7,800	7,790	237.52	3.6692	5.3818
7,900	7,890	236.87	3.6169	5.3195
8,000	7,990	236.23	3.5651 + 4	5.2578 − 1
8,100	8,090	235.58	3.5140	5.1967

Altitude		Temperature T, K	Pressure p, N/m^2	Density ρ, kg/m^3
h_G, m	h, m			
8,200	8,189	234.93	3.4635	5.1361
8,300	8,289	234.28	3.4135	5.0760
8,400	8,389	233.63	3.3642	5.0165
8,500	8,489	232.98	3.3154	4.9575
8,600	8,588	232.34	3.2672	4.8991
8,700	8,688	231.69	3.2196	4.8412
8,800	8,788	231.04	3.1725	4.7838
8,900	8,888	230.39	3.1260	4.7269
9,000	8,987	229.74	3.0800 + 4	4.6706 − 1
9,100	9,087	229.09	3.0346	4.6148
9,200	9,187	228.45	2.9898	4.5595
9,300	9,286	227.80	2.9455	4.5047
9,400	9,386	227.15	2.9017	4.4504
9,500	9,486	226.50	2.8584	4.3966
9,600	9,586	225.85	2.8157	4.3433
9,700	9,685	225.21	2.7735	4.2905
9,800	9,785	224.56	2.7318	4.2382
9,900	9,885	223.91	2.6906	4.1864
10,000	9,984	223.26	2.6500 + 4	4.1351 − 1
10,100	10,084	222.61	2.6098	4.0842
10,200	10,184	221.97	2.5701	4.0339
10,300	10,283	221.32	2.5309	3.9840
10,400	10,383	220.67	2.4922	3.9346
10,500	10,483	220.02	2.4540	3.8857
10,600	10,582	219.37	2.4163	3.8372
10,700	10,682	218.73	2.3790	3.7892
10,800	10,782	218.08	2.3422	3.7417
10,900	10,881	217.43	2.3059	3.6946
11,000	10,981	216.78	2.2700 + 4	3.6480 − 1
11,100	11,081	216.66	2.2346	3.5932
11,200	11,180	216.66	2.1997	3.5371
11,300	11,280	216.66	2.1654	3.4820
11,400	11,380	216.66	2.1317	3.4277
11,500	11,479	216.66	2.0985	3.3743
11,600	11,579	216.66	2.0657	3.3217
11,700	11,679	216.66	2.0335	3.2699

Altitude		Temperature T, K	Pressure p, N/m^2	Density ρ, kg/m^3
h_G, m	h, m			
11,800	11,778	216.66	2.0018	3.2189
11,900	11,878	216.66	1.9706	3.1687
12,000	11,977	216.66	1.9399 + 4	3.1194 − 1
12,100	12,077	216.66	1.9097	3.0707
12,200	12,177	216.66	1.8799	3.0229
12,300	12,276	216.66	1.8506	2.9758
12,400	12,376	216.66	1.8218	2.9294
12,500	12,475	216.66	1.7934	2.8837
12,600	12,575	216.66	1.7654	2.8388
12,700	12,675	216.66	1.7379	2.7945
12,800	12,774	216.66	1.7108	2.7510
12,900	12,874	216.66	1.6842	2.7081
13,000	12,973	216.66	1.6579 + 4	2.6659 − 1
13,100	13,073	216.66	1.6321	2.6244
13,200	13,173	216.66	1.6067	2.5835
13,300	13,272	216.66	1.5816	2.5433
13,400	13,372	216.66	1.5570	2.5036
13,500	13,471	216.66	1.5327	2.4646
13,600	13,571	216.66	1.5089	2.4262
13,700	13,671	216.66	1.4854	2.3884
13,800	13,770	216.66	1.4622	2.3512
13,900	13,870	216.66	1.4394	2.3146
14,000	13,969	216.66	1.4170 + 4	2.2785 − 1
14,100	14,069	216.66	1.3950	2.2430
14,200	14,168	216.66	1.3732	2.2081
14,300	14,268	216.66	1.3518	2.1737
14,400	14,367	216.66	1.3308	2.1399
14,500	14,467	216.66	1.3101	2.1065
14,600	14,567	216.66	1.2896	2.0737
14,700	14,666	216.66	1.2696	2.0414
14,800	14,766	216.66	1.2498	2.0096
14,900	14,865	216.66	1.2303	1.9783
15,000	14,965	216.66	1.2112 + 4	1.9475 − 1
15,100	15,064	216.66	1.1923	1.9172
15,200	15,164	216.66	1.1737	1.8874
15,300	15,263	216.66	1.1555	1.8580

Altitude		Temperature T, K	Pressure p, N/m^2	Density ρ, kg/m^3
h_G, m	h, m			
15,400	15,363	216.66	1.1375	1.8290
15,500	15,462	216.66	1.1198	1.8006
15,600	15,562	216.66	1.1023	1.7725
15,700	15,661	216.66	1.0852	1.7449
15,800	15,761	216.66	1.0683	1.7178
15,900	15,860	216.66	1.0516	1.6910
16,000	15,960	216.66	1.0353 + 4	1.6647 − 1
16,100	16,059	216.66	1.0192	1.6388
16,200	16,159	216.66	1.0033	1.6133
16,300	16,258	216.66	9.8767 + 3	1.5882
16,400	16,358	216.66	9.7230	1.5634
16,500	16,457	216.66	9.5717	1.5391
16,600	16,557	216.66	9.4227	1.5151
16,700	16,656	216.66	9.2760	1.4916
16,800	16,756	216.66	9.1317	1.4683
16,900	16,855	216.66	8.9895	1.4455
17,000	16,955	216.66	8.8496 + 3	1.4230 − 1
17,100	17,054	216.66	8.7119	1.4009
17,200	17,154	216.66	8.5763	1.3791
17,300	17,253	216.66	8.4429	1.3576
17,400	17,353	216.66	8.3115	1.3365
17,500	17,452	216.66	8.1822	1.3157
17,600	17,551	216.66	8.0549	1.2952
17,700	17,651	216.66	7.9295	1.2751
17,800	17,750	216.66	7.8062	1.2552
17,900	17,850	216.66	7.6847	1.2357
18,000	17,949	216.66	7.5652 + 3	1.2165 − 1
18,100	18,049	216.66	7.4475	1.1975
18,200	18,148	216.66	7.3316	1.1789
18,300	18,247	216.66	7.2175	1.1606
18,400	18,347	216.66	7.1053	1.1425
18,500	18,446	216.66	6.9947	1.1247
18,600	18,546	216.66	6.8859	1.1072
18,700	18,645	216.66	6.7788	1.0900
18,800	18,745	216.66	6.6734	1.0731
18,900	18,844	216.66	6.5696	1.0564

Altitude		Temperature T, K	Pressure p, N/m^2	Density ρ, kg/m^3
h_G, m	h, m			
19,000	18,943	216.66	6.4674 + 3	1.0399 − 1
19,100	19,043	216.66	6.3668	1.0238
19,200	19,142	216.66	6.2678	1.0079
19,300	19,242	216.66	6.1703	9.9218 − 2
19,400	19,341	216.66	6.0744	9.7675
19,500	19,440	216.66	5.9799	9.6156
19,600	19,540	216.66	5.8869	9.4661
19,700	19,639	216.66	5.7954	9.3189
19,800	19,739	216.66	5.7053	9.1740
19,900	19,838	216.66	5.6166	9.0313
20,000	19,937	216.66	5.5293 + 3	8.8909 − 2
20,200	20,136	216.66	5.3587	8.6166
20,400	20,335	216.66	5.1933	8.3508
20,600	20,533	216.66	5.0331	8.0931
20,800	20,732	216.66	4.8779	7.8435
21,000	20,931	216.66	4.7274	7.6015
21,200	21,130	216.66	4.5816	7.3671
21,400	21,328	216.66	4.4403	7.1399
21,600	21,527	216.66	4.3034	6.9197
21,800	21,725	216.66	4.1706	6.7063
22,000	21,924	216.66	4.0420 + 3	6.4995 − 2
22,200	22,123	216.66	3.9174	6.2991
22,400	22,321	216.66	3.7966	6.1049
22,600	22,520	216.66	3.6796	5.9167
22,800	22,719	216.66	3.5661	5.7343
23,000	22,917	216.66	3.4562	5.5575
23,200	23,116	216.66	3.3497	5.3862
23,400	23,314	216.66	3.2464	5.2202
23,600	23,513	216.66	3.1464	5.0593
23,800	23,711	216.66	3.0494	4.9034
24,000	23,910	216.66	2.9554 + 3	4.7522 − 2
24,200	24,108	216.66	2.8644	4.6058
24,400	24,307	216.66	2.7761	4.4639
24,600	24,505	216.66	2.6906	4.3263
24,800	24,704	216.66	2.6077	4.1931
25,000	24,902	216.66	2.5273	4.0639

Altitude		Temperature T, K	Pressure p, N/m^2	Density ρ, kg/m^3
h_G, m	h, m			
25,200	25,100	216.96	2.4495	3.9333
25,400	25,299	217.56	2.3742	3.8020
25,600	25,497	218.15	2.3015	3.6755
25,800	25,696	218.75	2.2312	3.5535
26,000	25,894	219.34	2.1632 + 3	3.4359 − 2
26,200	26,092	219.94	2.0975	3.3225
26,400	26,291	220.53	2.0339	3.2131
26,600	26,489	221.13	1.9725	3.1076
26,800	26,687	221.72	1.9130	3.0059
27,000	26,886	222.32	1.8555	2.9077
27,200	27,084	222.91	1.7999	2.8130
27,400	27,282	223.51	1.7461	2.7217
27,600	27,481	224.10	1.6940	2.6335
27,800	27,679	224.70	1.6437	2.5484
28,000	27,877	225.29	1.5949 + 3	2.4663 − 2
28,200	28,075	225.89	1.5477	2.3871
28,400	28,274	226.48	1.5021	2.3106
28,600	28,472	227.08	1.4579	2.2367
28,800	28,670	227.67	1.4151	2.1654
29,000	28,868	228.26	1.3737	2.0966
29,200	29,066	228.86	1.3336	2.0301
29,400	29,265	229.45	1.2948	1.9659
29,600	29,463	230.05	1.2572	1.9039
29,800	29,661	230.64	1.2208	1.8440
30,000	29,859	231.24	1.1855 + 3	1.7861 − 2
30,200	30,057	231.83	1.1514	1.7302
30,400	30,255	232.43	1.1183	1.6762
30,600	30,453	233.02	1.0862	1.6240
30,800	30,651	233.61	1.0552	1.5735
31,000	30,850	234.21	1.0251	1.5278
31,200	31,048	234.80	9.9592 + 2	1.4777
31,400	31,246	235.40	9.6766	1.4321
31,600	31,444	235.99	9.4028	1.3881
31,800	31,642	236.59	9.1374	1.3455
32,000	31,840	237.18	8.8802 + 2	1.3044 − 2
32,200	32,038	237.77	8.6308	1.2646

Altitude		Temperature T, K	Pressure p, N/m^2	Density ρ, kg/m^3
h_G, m	h, m			
32,400	32,236	238.78	8.3890	1.2261
32,600	32,434	238.96	8.1546	1.1889
32,800	32,632	239.55	7.9273	1.1529
33,000	32,830	240.15	7.7069	1.1180
33,200	33,028	240.74	7.4932	1.0844
33,400	33,225	214.34	7.2859	1.0518
33,600	33,423	241.93	7.0849	1.0202
33,800	33,621	242.52	6.8898	9.8972 − 3
34,000	33,819	243.12	6.7007 + 2	9.6020 − 3
34,200	34,017	243.71	6.5171	9.3162
34,400	34,215	244.30	6.3391	9.0396
34,600	34,413	244.90	6.1663	8.7720
34,800	34,611	245.49	5.9986	8.5128
35,000	34,808	246.09	5.8359	8.2620
35,200	35,006	246.68	5.6780	8.0191
35,400	35,204	247.27	5.5248	7.7839
35,600	35.402	247.87	5.3760	7.5562
35,800	35,600	248.46	5.2316	7.3357
36,000	35,797	249.05	5.0914 + 2	7.1221 − 3
36,200	35,995	249.65	4.9553	6.9152
36,400	36,193	250.24	4.8232	6.7149
36,600	36,390	250.83	4.6949	6.5208
36,800	36,588	251.42	4.5703	6.3328
37,000	36,786	252.02	4.4493	6.1506
37,200	36,984	252.61	4.3318	5.9741
37,400	37,181	253.20	4.2176	5.8030
37,600	37,379	253.80	4.1067	5.6373
37,800	37,577	254.39	3.9990	5.4767
38,000	37,774	254.98	3.8944 + 2	5.3210 − 3
38,200	37,972	255.58	3.7928	5.170i
38,400	38,169	256.17	3.6940	5.0238
38,600	38,367	256.76	3.5980	4.8820
38,800	38,565	257.35	3.5048	4.7445
39,000	38,762	257.95	3.4141	4.6112
39,200	38,960	258.54	3.3261	4.4819
39,400	39,157	259.13	3.2405	4.3566

Altitude		Temperature T, K	Pressure p, N/m^2	Density ρ, kg/m^3
h_G, m	h, m			
39,600	39,355	259.72	3.1572	4.2350
39,800	39,552	260.32	3.0764	4.1171
40,000	39,750	260.91	2.9977 + 2	4.0028 − 3
40,200	39,947	261.50	2.9213	3.8919
40,400	40,145	262.09	2.8470	3.7843
40,600	40,342	262.69	2.7747	3.6799
40,800	40,540	263.28	2.7044	3.5786
41,000	40,737	263.87	2.6361	3.4804
41,200	40,935	264.46	2.5696	3.3850
41,400	41,132	265.06	2.5050	3.2925
41,600	41,300	265.65	2.4421	3.2027
41,800	41,527	266.24	2.3810	3.1156
42,000	41,724	266.83	2.3215 + 2	3.0310 − 3
42,400	41,922	267.43	2.2636	2.9489
42,400	42,119	268.02	2.2073	2.8692
42,600	42,316	268.61	2.1525	2.7918
42,800	42,514	269.20	2.0992	2.7167
43,000	42,711	269.79	2.0474	2.6438
43,200	42,908	270.39	1.9969	2.5730
43,400	43,106	270.98	1.9478	2.5042
43,600	43,303	271.57	1.9000	2.4374
43,800	43,500	272.16	1.8535	2.3726
44,000	43,698	272.75	1.8082 + 2	2.3096 − 3
44,200	43,895	273.34	1.7641	2.2484
44,400	44,092	273.94	1.7212	2.1889
44,600	44,289	274.53	1.6794	2.1312
44,800	44,486	275.12	1.6387	2.0751
45,000	44,684	275.71	1.5991	2.0206
45,200	44,881	276.30	1.5606	1.9677
45,400	45,078	276.89	1.5230	1.9162
45,600	45,275	277.49	1.4865	1.8662
45,800	45,472	278.08	1.4508	1.8177
46,000	45,670	278.67	1.4162 + 2	1.7704 − 3
46,200	45,867	279.26	1.3824	1.7246
46,400	46,064	279.85	1.3495	1.6799
46,600	46,261	280.44	1.3174	1.6366

Altitude		Temperature T, K	Pressure p, N/m^2	Density ρ, kg/m^3
h_G, m	h, m			
46,800	46,458	281.03	1.2862	1.5944
47,000	46,655	281.63	1.2558	1.5535
47,200	46,852	282.22	1.2261	1.5136
47,400	47,049	282.66	1.1973	1.4757
47,600	47,246	282.66	1.1691	1.4409
47,800	47,443	282.66	1.1416	1.4070
48,000	47,640	282.66	$1.1147 + 2$	$1.3739 - 3$
48,200	47,837	282.66	1.0885	1.3416
48,400	48,034	282.66	1.0629	1.3100
48,600	48,231	282.66	1.0379	1.2792
48,800	48,428	282.66	1.0135	1.2491
49,000	48,625	282.66	$9.8961 + 1$	1.2197
49,200	48,822	282.66	9.6633	1.1910
49,400	49,019	282.66	9.4360	1.1630
49,600	49,216	282.66	9.2141	1.1357
49,800	49,413	282.66	8.9974	1.1089
50,000	49,610	282.66	$8.7858 + 1$	$1.0829 - 3$
50,500	50,102	282.66	8.2783	1.0203
51,000	50,594	282.66	7.8003	$9.6140 - 4$
51,500	51,086	282.66	7.3499	9.0589
52,000	51,578	282.66	6.9256	8.5360
52,500	52,070	282.66	6.5259	8.0433
53,000	52,562	282.66	6.1493	7.5791
53,500	53,053	282.42	5.7944	7.1478
54,000	53,545	280.21	5.4586	6.7867
54,500	54,037	277.99	5.1398	6.4412
55,000	54,528	275.78	$4.8373 + 1$	$6.1108 - 4$
55,500	55,020	273.57	4.5505	5.7949
56,000	55,511	271.36	4.2786	5.4931
56,500	56,002	269.15	4.0210	5.2047
57,000	56,493	266.94	3.7770	4.9293
57,500	56,985	264.73	3.5459	4.6664
58,000	57,476	262.52	3.3273	4.4156
58,500	57,967	260.31	3.1205	4.1763
59,000	58,457	258.10	2.9250	3.9482
59,500	58,948	255.89	2.7403	3.7307

B

Standard Atmosphere, English Engineering Units

Altitude		Temperature T, °R	Pressure p, lb/ft^2	Density ρ, slugs/ft^3
h_G, ft	h, ft			
−16,500	−16,513	577.58	3.6588 + 3	3.6905 − 3
−16,000	−16,012	575.79	3.6641	3.7074
−15,500	−15,512	574.00	3.6048	3.6587
−15,000	−15,011	572.22	3.5462	3.6105
−14,500	−14,510	570.43	3.4884	3.5628
−14,000	−14,009	568.65	3.4314	3.5155
−13,500	−13,509	566.86	3.3752	3.4688
−13,000	−13,008	565.08	3.3197	3.4225
−12,500	−12,507	563.29	3.2649	3.3768
−12,000	−12,007	561.51	3.2109	3.3314
−11,500	−11,506	559.72	3.1576 + 3	3.2866 − 3
−11,000	−11,006	557.94	3.1050	3.2422
−10,500	−10,505	556.15	3.0532	3.1983
−10,000	−10,005	554.37	3.0020	3.1548
−9,500	−9,504	552.58	2.9516	3.1118
−9,000	−9,004	550.80	2.9018	3.0693
−8,500	−8,503	549.01	2.8527	3.0272
−8,000	−8,003	547.23	2.8043	2.9855
−7,500	−7,503	545.44	2.7566	2.9443
−7,000	−7,002	543.66	2.7095	2.9035
−6,500	−6,502	541.88	2.6631 + 3	2.8632 − 3
−6,000	−6,002	540.09	2.6174	2.8233
−5,500	−5,501	538.31	2.5722	2.7838

Altitude		Temperature T, °R	Pressure p, lb/ft^2	Density ρ, slugs/ft^3
h_G, ft	h, ft			
−5,000	−5,001	536.52	2.5277	2.7448
−4,500	−4,501	534.74	2.4839	2.7061
−4,000	−4,001	532.96	2.4406	2.6679
−3,500	−3,501	531.17	2.3980	2.6301
−3,000	−3,000	529.39	2.3560	2.5927
−2,500	−2,500	527.60	2.3146	2.5558
−2,000	−2,000	525.82	2.2737	2.5192
−1,500	−1,500	524.04	2.2335 + 3	2.4830 − 3
−1,000	−1,000	522.25	2.1938	2.4473
−500	−500	520.47	2.1547	2.4119
0	0	518.69	2.1162	2.3769
500	500	516.90	2.0783	2.3423
1,000	1,000	515.12	2.0409	2.3081
1,500	1,500	513.34	2.0040	2.2743
2,000	2,000	511.56	1.9677	2.2409
2,500	2,500	509.77	1.9319	2.2079
3,000	3,000	507.99	1.8967	2.1752
3,500	3,499	506.21	1.8619 + 3	2.1429 − 3
4,000	3,999	504.43	1.8277	2.1110
4,500	4,499	502.64	1.7941	2.0794
5,000	4,999	500.86	1.7609	2,0482
5,500	5,499	499.08	1.7282	2.0174
6,000	5,998	497.30	1.6960	1.9869
6,500	6,498	495.52	1.6643	1.9567
7,000	6,998	493.73	1.6331	1.9270
7,500	7,497	491.95	1.6023	1.8975
8,000	7,997	490.17	1.5721	1.8685
8,500	8,497	488.39	1.5423 + 3	1.8397 − 3
9,000	8,996	486.61	1.5129	1.8113
9,500	9,496	484.82	1.4840	1.7833
10,000	9,995	483.04	1.4556	1.7556
10,500	10,495	481.26	1.4276	1.7282
11,000	10,994	479.48	1.4000	1.7011
11,500	11,494	477.70	1.3729	1.6744
12,000	11,993	475.92	1.3462	1.6480
12,500	12,493	474.14	1.3200	1.6219
13,000	12,992	472.36	1.2941	1.5961

Altitude		Temperature T, °R	Pressure p, lb/ft^2	Density ρ, slugs/ft^3
h_G, ft	h, ft			
13,500	13,491	470.58	1.2687 + 3	1.5707 − 3
14,000	13,991	468.80	1.2436	1.5455
14,500	14,490	467.01	1.2190	1.5207
15,000	14,989	465.23	1.1948	1.4962
15,500	15,488	463.45	1.1709	1.4719
16,000	15,988	461.67	1.1475	1.4480
16,500	16,487	459.89	1.1244	1.4244
17,000	16,986	458.11	1.1017	1.4011
17,500	17,485	456.33	1.0794	1.3781
18,000	17,984	454.55	1.0575	1.3553
18,500	18,484	452.77	1.0359 + 3	1.3329 − 3
19,000	18,983	450.99	1.0147	1.3107
19,500	19,482	449.21	9.9379 + 2	1.2889
20,000	19,981	447.43	9.7327	1.2673
20,500	20,480	445.65	9.5309	1.2459
21,000	20,979	443.87	9.3326	1.2249
21,500	21,478	442.09	9.1376	1.2041
22,000	21,977	440.32	8.9459	1.1836
22,500	22,476	438.54	8.7576	1.1634
23,000	22,975	436.76	8.5724	1.1435
23,500	23,474	434.98	8.3905 + 2	1.1238 − 3
24,000	23,972	433.20	8.2116	1.1043
24,500	24,471	431.42	8.0359	1.0852
25,000	24,970	429.64	7.8633	1.0663
25,500	25,469	427.86	7.6937	1.0476
26,000	25,968	426.08	7.5271	1.0292
26,500	26,466	424.30	7.3634	1.0110
27,000	26,965	422.53	7.2026	9.9311 − 4
27,500	27,464	420.75	7.0447	9.7544
28,000	27,962	418.97	688.96	9.5801
28,500	28,461	417.19	6.7373 + 2	9.4082 − 4
29,000	28,960	415.41	6.5877	9.2387
29,500	29,458	413.63	6.4408	9.0716
30,000	29,957	411.86	6.2966	8.9068
30,500	30,455	410.08	6.1551	8.7443
31,000	30,954	408.30	6.0161	8.5841
31,500	31,452	406.52	5.8797	8.4261
32,000	31,951	404.75	5.7458	8.2704

Altitude		Temperature T, °R	Pressure p, lb/ft^2	Density ρ, slugs/ft^3
h_G, ft	h, ft			
32,500	32,449	402.97	5.6144	8.1169
33,000	32,948	401.19	5.4854	7.9656
33,500	33,446	399.41	5.3589 + 2	7.8165 − 4
34,000	33,945	397.64	5.2347	7.6696
34,500	34,443	395.86	5.1129	7.5247
35,000	34,941	394.08	4.9934	7.3820
35,500	35,440	392.30	4.8762	7.2413
36,000	35,938	390.53	4.7612	7.1028
36,500	36,436	389.99	4.6486	6.9443
37,000	36,934	389.99	4.5386	6.7800
37,500	37,433	389.99	4.4312	6.6196
38,000	37,931	389.99	4.3263	6.4629
38,500	38,429	389.99	4.2240 + 2	6.3100 − 4
39,000	38,927	389.99	4.1241	6.1608
39,500	39,425	389.99	4.0265	6.0150
40,000	39,923	389.99	3.9312	5.8727
40,500	40,422	389.99	3.8382	5.7338
41,000	40,920	389.99	3.7475	5.5982
41,500	41,418	389.99	3.6588	5.4658
42,000	41,916	389.99	3.5723	5.3365
42,500	42,414	389.99	3.4878	5.2103
43,000	42,912	389.99	3.4053	5.0871
43,500	43,409	389.99	3.3248 + 2	4.9668 − 4
44,000	43,907	389.99	3.2462	4.8493
44,500	44,405	389.99	3.1694	4.7346
45,000	44,903	389.99	3.0945	4.6227
45,500	45,401	389.99	3.0213	4.5134
46,000	45,899	389.99	2.9499	4.4067
46,500	46,397	389.99	2.8801	4.3025
47,000	46,894	389.99	2.8120	4.2008
47,500	47,392	389.99	2.7456	4.1015
48,000	47,890	389.99	2.6807	4.0045
48,500	48,387	389.99	2.2173 + 2	3.9099 − 4
49,000	48,885	389.99	2.5554	3.8175
49,500	49,383	389.99	2.4950	3.7272
50,000	49,880	389.99	2.4361	3.6391
50,500	50,378	389.99	2.3785	3.5531

Altitude		Temperature T, °R	Pressure p, lb/ft^2	Density ρ, slugs/ft^3
h_G, ft	h, ft			
51,000	50,876	389.99	2.3223	3.4692
51,500	51,373	389.99	2.2674	3.3872
52,000	51,871	389.99	2.2138	3.3072
52,500	52,368	389.99	2.1615	3.2290
53,000	52,866	389.99	2.1105	3.1527
53,500	53,363	289.99	2.0606 + 2	3.0782 − 4
54,000	53,861	389.99	2.0119	3.0055
54,500	54,358	389.99	1.9644	2.9345
55,000	54,855	389.99	1.9180	2.8652
55,500	55,353	389.99	1.8727	2.7975
56,000	55,850	389.99	1.8284	2.7314
56,500	56,347	389.99	1.7853	2.6669
57,000	56,845	389.99	1.7431	2.6039
57,500	57,342	389.99	1.7019	2.5424
58,000	57,839	389.99	1.6617	2.4824
58,500	58,336	389.99	1.6225 + 2	2.4238 − 4
59,000	58,834	389.99	1.5842	2.3665
59,500	59,331	389.99	1.5468	2.3107
60,000	59,828	389.99	1.5103	2.2561
60,500	60,325	389.99	1.4746	2.2028
61,000	60,822	389.99	1.4398	2.1508
61,500	61,319	389.99	1.4058	2.1001
62,000	61,816	389.99	1.3726	2.0505
62,500	62,313	389.99	1.3402	2.0021
63,000	62,810	389.99	1.3086	1.9548
63,500	63,307	389.99	1.2777 + 2	1.9087 − 4
64,000	63,804	389.99	1.2475	1.8636
64,500	64,301	389.99	1.2181	1.8196
65,000	64,798	389.99	1.1893	1.7767
65,500	65,295	389.99	1.1613	1.7348
66,000	65,792	389.99	1.1339	1.6938
66,500	66,289	389.99	1.1071	1.6539
67,000	66,785	389.99	1.0810	1.6148
67,500	67,282	389.99	1.0555	1.5767
68,000	67,779	389.99	1.0306	1.5395
68,500	68,276	389.99	1.0063 + 2	1.5032 − 4
69,000	68,772	389.99	9.8253 + 1	1.4678

Altitude		Temperature T, °R	Pressure p, lb/ft^2	Density ρ, slugs/ft^3
h_G, ft	h, ft			
69,500	69,269	389.99	9.5935	1.4331
70,000	69,766	389.99	9.3672	1.3993
70,500	70,262	389.99	9.1462	1.3663
71,000	70,759	389.99	8.9305	1.3341
71,500	74,256	389.99	8.7199	1.3026
72,000	71,752	389.99	8.5142	1.2719
72,500	72,249	389.99	8.3134	1.2419
73,000	72,745	389.99	8.1174	1.2126
73,500	73,242	389.99	7.9259 + 1	1.1840 − 4
74,000	73,738	389.99	7.7390	1.1561
74,500	74,235	389.99	7.5566	1.1288
75,000	74,731	389.99	7.3784	1.1022
75,500	75,228	389.99	7.2044	1.0762
76,000	75,724	389.99	7.0346	1.0509
76,500	76,220	389.99	6.8687	1.0261
77,000	76,717	389.99	6.7068	1.0019
77,500	77,213	389.99	6.5487	9.7829 − 5
78,000	77,709	389.99	6.3944	9.5523
78,500	78,206	389.99	6.2437 + 1	9.3271 − 5
79,000	78,702	389.99	6.0965	9.1073
79,500	79,198	389.99	5.9528	8.8927
80,000	79,694	389.99	5.8125	8.6831
80,500	80,190	389.99	5.6755	8.4785
81,000	80,687	389.99	5.5418	8.2787
81,500	81,183	389.99	5.4112	8.0836
82,000	81,679	389.99	5.5837	7.8931
82,500	82,175	390.24	5.1592	7.7022
83,000	82,671	391.06	5.0979	7.5053
83,500	83,167	391.87	4.9196 + 1	7.3139 − 5
84,000	83,663	392.69	4.8044	7.1277
84,500	84,159	393.51	4.6921	6.9467
85,000	84,655	394.32	4.5827	6.7706
85,500	85,151	395.14	4.4760	6.5994
86,000	85,647	395.96	4.3721	6.4328
86,500	86,143	396.77	4.2707	6.2708
87,000	86,639	397.59	4.1719	6.1132
87,500	87,134	398.40	4.0757	5.9598
88,000	87,630	399.22	3.9818	5.8106

Altitude		Temperature T, °R	Pressure p, lb/ft^2	Density ρ, slugs/ft^3
h_G, ft	h, ft			
88,500	88,126	400.04	3.8902 + 1	5.6655 − 5
89,000	88,622	400.85	3.8010	5.5243
89,500	89,118	401.67	3.7140	5.3868
90,000	89,613	402.48	3.6292	5.2531
90,500	90,109	403.30	3.5464	5.1230
91,000	90,605	404.12	3.4657	4.9963
91,500	91,100	404.93	3.3870	4.8730
92,000	91,596	405.75	3.3103	4.7530
92,500	92,092	406.56	3.2354	4.6362
93,000	92,587	407.38	3.1624	4.5525
93,500	93,083	408.19	3.0912 + 1	4.4118 − 5
94,000	93,578	409.01	3.0217	4.3041
94,500	94,074	409.83	2.9539	4.1992
95,000	94,569	410.64	2.8878	4.0970
95,500	95,065	411.46	2.8233	3.9976
96,000	95,560	412.27	2.7604	3.9007
96,500	96,056	413.09	2.6989	3.8064
97,000	96,551	413.90	2.6390	3.7145
97,500	97,046	414.72	2.5805	3.6251
98,000	97,542	415.53	2.5234	3.5379
98,500	98,037	416.35	2.4677 + 1	3.4530 − 5
99,000	98,532	417.16	2.4134	3.3704
99,500	99,028	417.98	2.3603	3.2898
100,000	99,523	418.79	2.3085	3.2114
100,500	100,018	419.61	2.2580	3.1350
101,000	100,513	420.42	2.2086	3.0605
101,500	101,008	421.24	2.1604	2.9879
102,000	101,504	422.05	2.1134	2.9172
102,500	101,999	422.87	2.0675	2.8484
103,000	102,494	423.68	2.0226	2.7812
103,500	102,989	424.50	1.9789 + 1	2.7158 − 5
104,000	103,484	425.31	1.9361	2.6520
104,500	103,979	426.13	1.8944	2.5899
105,000	104,474	426.94	1.8536	2.5293
106,000	105,464	428.57	1.7749	2.4128
107,000	106,454	430.20	1.6999	2.3050
108,000	107,444	431.83	1.6282	2.1967

Altitude		Temperature T, °R	Pressure p, lb/ft^2	Density ρ, slugs/ft^3
h_G, ft	h, ft			
109,000	108,433	433.46	1.5599	2.0966
110,000	109,423	435.09	1.4947	2.0014
111,000	110,412	136.72	1.4324	1.9109
112,000	111,402	438.35	1.3730 + 1	1.8247 − 5
113,000	112,391	439.97	1.3162	1.7428
114,000	113,380	441.60	1.2620	1.6649
115,000	114,369	443.23	1.2102	1.5907
116,000	115,358	444.86	1.1607	1.5201
117,000	116,347	446.49	1.1134	1.4528
118,000	117,336	448.11	1.0682	1.3888
119,000	118,325	449.74	1.0250	1.3278
120,000	119,313	451.37	9.8372 + 0	1.2697
121,000	120,302	453.00	9.4422	1.2143
122,000	121,290	454.62	9.0645 + 0	1.1616 − 5
123,000	122,279	456.25	8.7032	1.1113
124,000	123,267	457.88	8.3575	1.0634
125,000	124,255	459.50	8.0267	1.0177
126,000	125,243	461.13	7.7102	9.7410 − 6
127,000	126,231	462.75	7.4072	9.3253
128,000	127,219	464.38	7.1172	8.9288
129,000	128,207	466.01	6.8395	8.5505
130,000	129,195	467.63	6.5735	8.1894
131,000	130,182	469.26	6.3188	7.8449
132,000	131,170	470.88	6.0748 + 0	7.5159 − 6
133,000	132,157	472.51	5.8411	7.2019
134,000	133,145	474.13	5.6171	6.9020
135,000	134,132	475.76	5.4025	6.6156
136,000	135,199	477.38	5.1967	6.3420
137,000	136,106	479.01	4.9995	6.0806
138,000	137,093	480.63	4.8104	5.8309
139,000	138,080	482.26	4.6291	5.5922
140,000	139,066	483.88	4.4552	5.3640
141,000	140,053	485.50	4.2884	5.1460
142,000	141,040	487.13	4.1284 + 0	4.9374 − 6
143,000	142,026	488.75	3.9749	4.7380
144,000	143,013	490.38	3.8276	4.5473
145,000	143,999	492.00	3.6862	4.3649

Altitude		Temperature T, °R	Pressure p, lb/ft^2	Density ρ, slugs/ft^3
h_G, ft	h, ft			
146,000	144,985	493.62	3.5505	4.1904
147,000	145,971	495.24	3.4202	4.0234
148,000	146,957	496.87	3.2951	3.8636
149,000	147,943	498.49	3.1750	3.7106
150,000	148,929	500.11	3.0597	3.5642
151,000	149,915	501.74	2.9489	3.4241
152,000	150,900	503.36	2.8424 + 0	3.2898 − 6
153,000	151,886	504.98	2.7402	3.1613
154,000	152,871	506.60	2.6419	3.0382
155,000	153,856	508.22	2.5475	2.9202
156,000	154,842	508.79	2.4566	2.8130
157,000	155,827	508.79	2.3691	2.7127
158,000	156,812	508.79	2.2846	2.6160
159,000	157,797	508.79	2.2032	2.5228
160,000	158,782	508.79	2.1247	2.4329
161,000	159,797	508.79	2.0490	2.3462

ANSWERS TO SELECTED PROBLEMS

2.2 $L = 529.2$ N, $D = 5.788$ N

2.4 $C_{m_{a.c.}} = -0.0415$

2.6 $C_L = 0.261$

2.8 (a) $c_l = 0.0605$;   (b) $C_L = 0.0536$;   (c) $C_L = 0.061$

2.10 $L/D \rightarrow \infty$

2.12 $C_{D,0} = 0.0105$

3.2 $P_A = 513$ hp

3.4 0.223 h

3.6 $P_{es} = 5,163$ hp

5.2 (a) $V_{max} = 467.3$ ft/s;   (b) $V_{max} = 461.1$ ft/s

5.4 $\left(\dfrac{L}{D}\right)_{max} = \dfrac{1}{(M_\infty^2 - 1)^{1/4}\sqrt{c_{d,f}}}$

$\alpha = \dfrac{(M_\alpha^2 - 1)^{1/4}\sqrt{c_{d,f}}}{2}$

5.6 $V_{stall} = 103.4$ ft/s

5.9 $(R/C)_{max} = 33.65$ ft/s; $V_{(R/C)_{max}} = 284.2$ ft/s; $\theta_{max} = 8.05°$; $V_{\theta_{max}} = 193$ ft/s

5.11 $\theta_{min} = 4.03°$; $d_{max} = 26.9$ mi; $V_{(L/D)_{max}} = 225.7$ ft/s at 10,000 ft;
   $V_{(L/D)_{max}} = 194$ ft/s at sea level

5.13 30,422 ft

5.15 $R_{max} = 820$ mi; $V_{(C_L^{1/2}/C_D)_{max}} = 297.1$ ft/s

5.17 1,112 mi

5.19 $V_{max} = 857.8$ ft/s at sea level, $V_{max} = 911.1$ ft/s at 30,000 ft

5.22 $(R/C)_{max} = 85.23$ ft/s at sea level, $(R/C)_{max} = 26.8$ ft/s at 30,000 ft

6.1 $R_{min} = 538$ ft; $\omega_{max} = 24.52$ deg/s

6.3 286.7 ft/s

6.5 47,839 ft

6.7 Total takeoff distance = 2,033 ft

REFERENCES

1. Peter L. Jakab, *Visions of a Flying Machine; The Wright Brothers and the Process of Invention*, Smithsonian Press, Washington, 1990.

2. John D. Anderson, Jr., "Faster and Higher; The Quest for Speed and Altitude," in *Milestones of Aviation*, National Air and Space Museum, Smithsonian Institution, edited by John T. Greenwood, produced by Hugh Lauter Levin Associates, Inc., New York, distributed by Macmillan Publishing Company, New York, 1989, pp. 80–146.

3. John D. Anderson, Jr., *Introduction to Flight*, 3rd ed., McGraw-Hill, New York, 1989.

4. John D. Anderson, Jr., "Sir George Cayley," in *Great Lives from History: British and Commonwealth Series*, Salem Press, 1987.

5. C. H. Gibbs-Smith, *Sir George Cayley's Aeronautics, 1796–1855*, Her Majesty's Stationery Office, London, 1962.

6. J. Lawrence Pritchard, *Sir George Cayley: The Inventor of the Airplane*, Max Parrish and Company, London, 1961.

7. Charles H. Gibbs-Smith, *Aviation: An Historical Survey from Its Origins to the End of World War II*, Her Majesty's Stationery Office, London, 1970.

8. John D. Anderson, Jr., *A History of Aerodynamics and Its Impact on Flying Machines*, Cambridge University Press, New York, 1997.

9. John D. Anderson, Jr., "The Wright Brothers: The First True Aeronautical Engineers," in *The Wright Flyer: An Engineering Perspective*, edited by Howard S. Wolko, National Air and Space Museum, Washington, 1987, pp. 1–17.

10. Tom D. Crouch, *A Dream of Wings*, Norton, New York, 1981.

11. Tom D. Crouch, *The Bishop's Boys*, Norton, New York, 1989.

12. F. E. C. Culick and H. R. Jex, "Aerodynamics, Stability and Control of the 1903 Wright Flyer," in *The Wright Flyer: An Engineering Perspective*, edited by Howard S. Wolko, National Air and Space Museum, Washington, 1987, pp. 19–43.

13. Laurence K. Loftin, *Quest for Performance: The Evolution of Modern Aircraft*, NASA SP-468, Washington, 1985.

14. R. Miller, and D. Sawers, *The Technical Development of Modern Aviation*, Praeger Publishers, New York, 1970.

15. James R. Hansen, "Bigger: the Quest for Size," in *Milestones of Aviation*, National Air and Space Museum, Smithsonian Institution, edited by John T. Greenwood, produced by Hugh Lauter Levin Associates, Inc., New York, distributed by Macmillan, New York, 1989, pp. 150–221.

16. John D. Anderson, Jr., *Fundamentals of Aerodynamics*, 2nd ed., McGraw-Hill, New York, 1991.

17. S. F. Hoerner, *Fluid Dynamic Drag*, Hoerner Fluid Dynamics, Brick Town, NJ, 1965.

18. S. F. Hoerner, and H. V. Borst, *Fluid Dynamic Lift*, Hoerner Fluid Dynamics, Brick Town, NJ, 1975.

19. Ira H. Abbott, and Albert E. Von Doenhoff, *Theory of Wings Sections*, McGraw-Hill, New York, 1949; also, Dover edition, New York, 1959.

20. David A. Lednicer and Ian Gilchrist, Jr., "A Retrospective: Computational Aerodynamic Analysis Methods Applied to the P-51 Mustang," AIAA Paper 91-3288-CP, in the *Proceedings of the AIAA 9th Applied Aerodynamics Conference*, September 23–25, 1991, pp. 688–700.

21. John D. Anderson, Jr., *Computational Fluid Dynamics: The Basics with Applications*, McGraw-Hill, New York, 1995.

22. Joseph Katz and Allen Plotkin, *Low-Speed Aerodynamics*, McGraw-Hill, New York, 1991.

23. H. B. Helmbold, "Der unverwundene Ellipsenflugel als tragendi Flache," J 1942 DL I, 1, 1942 (German Wartime Report).

24. Dietrich Kuchemann, *The Aerodynamic Design of Aircraft*, Pergamon Press, Oxford, 1978.

25. Daniel P. Raymer, *Aircraft Design: A Conceptual Approach*, 2nd ed., AIAA Education Series, American Institute of Aeronautics and Astronautics, Washington, 1992.

26. John J. Bertin and Michael L. Smith, *Aerodynamics for Engineers*, 2nd ed., Prentice-Hall, Englewood Cliffs, NJ, 1989.

27. Charles E. Jobe, "Prediction and Verification of Aerodynamic Drag, Part I: Prediction," in *Thrust and Drag: Its Prediction and Verification*, edited by Eugene Covert, Progress in Astronautics and Aeronautics Series, vol. 98, American Institute of Aeronautics and Astronautics, Washington, 1985.

28. L. T. Goodmanson and L. B. Gratzer, "Recent Advances in Aerodynamics for Transport Aircraft," *Aeronautics and Astronautics*, vol. 11, no. 12, December 1973, pp. 30–45.

29. R. T. Whitcomb and J. R. Clark, "An Airfoil Shape for Efficient Flight at Supercritical Mach Numbers," NASA TMX-1109, July 1965.

30. John D. Anderson, *Hypersonic and High Temperature Gas Dynamics*, McGraw-Hill, New York, 1989.

31. Otto Lilienthal, *Der Vogelplug als Grundlage der Fliegekunst*, R. Gaertners Verlagsbuchhandlung, Berlin, 1889.

32. Gustave Eiffel, *The Resistance of the Air and Aviation: Experiments Conducted at the Champ-de-Mars Laboratory*, Dunot and Pinat, Paris, 1910; English translation by Jerome C. Hunsaker, comprising the 2nd ed. (revised and enlarged), Houghton Mifflin, Boston, 1913.

33. P. G. Hill and C. R. Peterson, *Mechanics and Thermodynamics of Propulsion*, Addision-Wesley, Reading, MA, 1965.

34. Jack D. Mattingly, *Aircraft Propulsion*, McGraw-Hill, New York, 1995.

35. Egbert Torenbeck, *Synthesis of Subsonic Airplane Design*, Delft University Press, Delft, Holland, 1982.

36. John W. R. Taylor, ed., *Jane's All the World's Aircraft*, 80th ed., Jane's Information Group Limited, Coulsdon, England, 1989–1990.

37. E. P. Hartman and David Biermann, "The Aerodynamic Characteristics of Full-Scale Propellers Having 2, 3 and 4 Blades of Clark Y and R.A.F. 6 Airfoil Sections," NACA TR 640, November 1937.

38. C. C. Carter, *Simple Aerodynamics and the Airplane*, 5th ed., Ronald Press, New York, 1940.

39. John D. Anderson, Jr., *Computational Fluid Dynamics: The Basics with Applications*, McGraw-Hill, New York, 1995.

40. Walter J. Hesse and Nicholas V. S. Mumford, Jr., *Jet Propulsion for Aerospace Applications*, 2nd ed., Pitman Publishing Corp., New York, 1964.

41. W. Austyn Mair and David L. Birdsall, *Aircraft Performance*, Cambridge University Press, Cambridge, England, 1992.

42. John D. Anderson, Jr., *Modern Compressible Flow: With Historical Perspective*, 2nd ed., McGraw-Hill, New York, 1990.

43. J. D. Mattingly, W. H. Heiser, and G. H. Daley, *Aircraft Engine Design*, American Institute of Aeronautics and Astronautics, Washington, 1987.

44. H. I. H. Saravanamuttoo, "Modern Turboprop Engines," *Progress in Aerospace Sciences*, vol. 24, 1987, pp. 225–248.

45. Nguyen X. Vinh, *Flight Mechanics of High-Performance Aircraft*, Cambridge University Press, Cambridge, England, 1993.

46. W. Sears, "On Projectiles of Minimum Wave Drag," *Quarterly of Applied Mathematics*, vol. 4, no. 4, January 1947, pp. 361–366.

47. A. E. Winkelmann and J. B. Barlow, "Flowfield Model for a Rectangular Planform Wing beyond Stall," *AIAA Journal*, vol. 18, no. 8, August 1980, pp. 1006–1008.

48. D. J. Peake and M. Tobak, "Three-Dimensional Flows about Simple Components at Angle of Attack," *High Angle-of-Attack Aerodynamics*, AGARD/VKI Lecture Series No. 121, von Karman Institute, Brussels, Belgium, March 1982.

49. Francis, J. Hale, *Introduction to Aircraft Performance, Selection and Design*, Wiley, New York, 1984.

50. Barnes W. McCormick, *Aerodynamics, Aeronautics, and Flight Mechanics*, Wiley, New York, 1979.

51. B. H. Carson, "Fuel Efficiency of Small Aircraft," AIAA Paper No. 80-1847, American Institute of Aeronautics and Astronautics, Washington, 1980.

52. Leland M. Nicoli, *Fundamentals of Aircraft Design*, published by the author and distributed by METS, Inc., San Jose, CA, 1975.

53. Jan Roskam, *Airplane Design*, Roskam Aviation and Engineering Corp., Ottawa, KS, 1985.

54. Darrol, Stinton, *The Design of the Airplane*, Van Nostrand Reinhold, New York, 1983.

55. Gerald Corning, *Supersonic and Subsonic Airplane Design*, self-published, College Park, MD, 1970.

56. N. Currey, *Aircraft Landing Gear Design: Principles and Practices*, American Institute of Aeronautics and Astronautics, Washington, 1988.

57. H. Conway, *Landing Gear Design*, Chapman and Hall, London, 1958.

58. Marvin W. McFarland, ed., *The Papers of Wilbur and Orville Wright*, vol. 1, McGraw-Hill, New York, 1953.

59. Donald W. Douglas, "The Developments and Reliability of the Modern Multi-Engine Air Liner with Special Reference to Multi-Engine Airplanes after Engine Failure," *The Aeronautical Journal*, vol. 39, November 1935, pp. 1010–1042; also reprinted in *Journal of the Aeronautical Sciences*, vol. 2, no. 4, July 1935, pp. 128–52.

60. Arthur E. Raymond, "Recollections of Douglas 1925–1960," *Journal of the American Aviation Historical Society*, Summer 1987, pp. 110–25.

61. D. J. Ingells, *The Plane that Changed the World*, Aero Publishers, Fallbrook, CA, 1966.

62. Ronald Miller and David Sawers, *The Technical Development of Modern Aviation*, Praeger Publishers, New York, 1970.

63. William H. Cook, *The Road to the 707*, TYC Publishing Company, Bellevue, WA, 1991.

64. J. E. Steiner, G. M. Bowes, F. G. Maxam, S. Wallick, and M. C. Gregoire, *Case Study in Aircraft Design: The Boeing 727*, AIAA Professional Study Series, American Institute of Aeronautics and Astronautics, Washington, September 1978.

65. J. E. Steiner, "Jet Aviation Development: A Company Perspective," in *The Jet Age*, edited by Walter J. Boyne and Donald S. Lopez, Smithsonian Press, Washington, 1979, pp. 141–83.

66. J. K. Buckner, D. B. Benepe, and P. W. Hill, "Aerodynamic Design Evolution of the YF-16," AIAA Paper No. 74-935, American Institute of Aeronautics and Astronautics, Washington, 1974.

67. Clarence L. Johnson and Maggie Smith, *Kelly*, Smithsonian Institution Press, Washington, 1985.

68. Ben R. Rich, and Leo Janos, *Skunk Works*, Little, Brown, Boston, 1994.

69. Clarence L. Johnson, "Some Development Aspects of the YF-12A Interceptor Aircraft," AIAA Paper No. 69-757, American Institute of Aeronautics and Astronautics, Washington, 1969.

70. Ben R. Rich, "The F-12 Series Aircraft Aerodynamic and Thermodynamic Design in Retrospect," AIAA Paper No. 73-820, American Institute of Aeronautics and Astronautics, Washington, 1973.

71. Richard Abrams and Jay Miller, *Lockheed (General Dynamics/Boeing) F-22*, Aerofax, Inc., Arlington, TX, 1992.

INDEX

X

Y